Derived SI units with speci

Quantity

Activity (of a radionuclide)
Celsius temperature
Conductance
Electric potential, potential
  difference, electromotive
  force
Electric resistance
Energy, work, quantity of
  heat
Force
Frequency
Power, radiant flux
Pressure, stress
Quantity of electricity, elec-
  tric charge

se units

$kg^{-1} s^3 A^2$
$g s^{-3} A^{-1}$

$g s^{-3} A^{-2}$
$g s^{-2}$

s^{-2}

$g s^{-3}$
$kg s^{-2}$

Prefixes and abbreviations f

No.		
10^{-15}		
10^{-12}		
10^{-9}		
10^{-6}	micro	μ
10^{-3}	milli	m
10^{-2}	centi	c
10^{-1}	deci	d
10	deca	da
10^2	hecto	h
10^3	kilo	k
10^6	mega	M
10^9	giga	G
10^{12}	tera	T
10^{15}	peta	P

Principles and Applications
of Soil Microbiology

♦

Edited by

David M. Sylvia
Jeffry J. Fuhrmann
Peter G. Hartel
David A. Zuberer

Prentice Hall
Upper Saddle River
New Jersey 07458

Library of Congress Cataloging-in-Publication Data

Principles and applications of soil microbiology / edited by David M.
   Sylvia . . . [et al].
        p. cm.
     Includes bibliographical references and index.
     ISBN 0-13-459991-8
     1. Soil microbiology. I. Sylvia, D. M. (David M.)
   QR111.S674 1998
   579′.1757—DC21   98-12659

                                                              CIP
                                                              Rev.

Acquisitions Editor: *Charles Stewart*
Production Editor: *Lori Harvey*
Production Liaison: *Eileen M. O'Sullivan*
Director of Manufacturing and Production: *Bruce Johnson*
Managing Editor: *Mary Carnis*
Manufacturing Manager: *Ed O'Dougherty*
Production Manager: *Marc Bove*
Assistant Editor: *Kate Linsner*
Marketing Manager: *Melissa Bruner*
Editorial Assistant: *Kimberly Yehle*
Cover Design: *Miguel Ortiz*
Cover Artist: *Kim Luoma*
Formatting/page make-up: *Carlisle Publishers Services*
Printer/Binder: *RR Donnelley & Sons Company*

© 1998 by Prentice-Hall, Inc.
Simon & Schuster / A Viacom Company
Upper Saddle River, New Jersey 07458

Printed in the United States of America

10 9 8 7 6 5 4 3 2 1

ISBN 0-13-459991-8

Prentice-Hall International (UK) Limited, *London*
Prentice-Hall of Australia Pty. Limited, *Sydney*
Prentice-Hall Canada Inc., *Toronto*
Prentice-Hall Hispanoamericana, S.A., *Mexico*
Prentice-Hall of India Private Limited, *New Delhi*
Prentice-Hall of Japan, Inc., *Tokyo*
Simon & Schuster Asia Pte. Ltd., *Singapore*
Editora Prentice-Hall do Brasil, Ltda., *Rio de Janeiro*

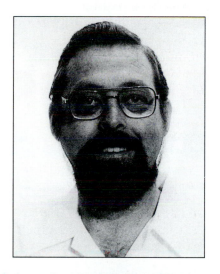

This book is dedicated to the memory of David H. Hubbell (1937—1995). Dr. Hubbell was a valued colleague and a rare philosophical voice among soil microbiologists.

Brief Contents

♦

Contents

♦

List of Methods

♦

Preface

♦

This book grew out of discussions among scientists in Southern Regional Research Project S–226 (and later S–262) who wanted a comprehensive textbook for teaching soil microbiology. The field is now so broad that it is almost impossible for any one individual to stay current in all aspects of the science. Thus adopting King Solomon's adage that "as iron sharpens iron, so one man sharpens another," we took the nontraditional approach of having multiple authors write this textbook. Many authors are members of the Regional Project, while others were invited to write chapters related to their specific area of research expertise. In this way we hope to better capture the rapid advances in both fundamental knowledge and potential applications of soil microbiology. We also believe a multiauthored approach will serve to bridge students from a traditional, introductory, single-authored text to a typical multiauthored scientific treatise. Students need to learn the concept of consulting the works of individual scientists to keep abreast of current developments. Of course, this approach has a potential drawback in that the book may read too much like a multi-authored text. For this reason, we have taken care in the editing process to ensure a uniform style and consistent usage of terminology.

Principles and Applications of Soil Microbiology is targeted at advanced undergraduate and beginning graduate students who require a comprehensive treatment of the field of soil microbiology. Professionals in agricultural, environmental, and industrial fields should also find this book a valuable reference.

This book is divided into three major sections: habitat and organisms, microbially mediated transformations, and applied and environmental topics. Instructors may find that this book contains more information than they cover in a typical, one-semester or one-quarter course in soil microbiology. Some chapters (Chapter 2, The Soil Habitat; Chapter 10, Microbial Metabolism) are provided as background information for students who do not have prior courses in soil science or biochemistry. Some chapters on the important groups of microorganisms found in soil may include more detail than is required of beginning students, but they will serve as important references on the taxonomy, physiology, and ecology of microorganisms for advanced students and professionals. Our short-term goal was to provide chapters on topics where active research shows promise of practical application. Our long-term goal is to provide timely updates of this book so students can keep abreast of the most recent advances in soil microbiology.

Although many methods are referred to in individual chapters in connection with discussions of specific principles and applications, and several methods are presented in more detail (see List of Methods, page xiv), the reader should be aware that this is not a book of methods. This is the reason why we avoided a specific chapter on methods. For a detailed treatise on methods in soil microbiology, the reader should consult other works (e.g., *Methods of Soil Analysis, Part 2. Microbiological and Biochemical Properties,* published by the Soil Science Society of America).

Our collective experience in teaching soil microbiology is that many students stumble over the many new terms that are introduced during a typical course in soil microbiology. To help students overcome this obstacle, we have included a comprehensive glossary of soil microbiology terms at the end of the book. The reader should note that bolded terms in the text are defined in the glossary.

List of Contributors

♦

Steve L. Albrecht
Columbia Plateau Conservation Research Center
USDA, PWA
Pendleton, Ore. 97801

David B. Alexander
Department of Biology
University of Portland
Portland, Ore. 97203

J. Scott Angle
Department of Agronomy
University of Maryland
College Park, Md. 20742

Peter J. Bottomley
Department of Microbiology
Oregon State University
Corvallis, Ore. 97331

Jeffry J. Fuhrmann
Department of Plant and Soil Sciences
University of Delaware
Newark, Del. 19717

Joel V. Gagliardi
Department of Agronomy
University of Maryland
College Park, Md. 20742

James J. Germida
Department of Soil Science
University of Saskatchewan
Saskatoon, Canada S7N 0W0

James H. Graham
Citrus Research Station
University of Florida
Lake Alfred, Fla. 33850

Peter H. Graham
Department of Soil Science
University of Minnesota
St. Paul, Minn. 55108

Peter G. Hartel
Department of Crop and Soil Sciences
University of Georgia
Athens, Ga. 30602

William J. Hickey
Department of Soil Science
University of Wisconsin
Madison, Wis. 53706

Elisabeth A. Holland
Atmospheric Chemistry Division
National Center for Atmospheric Chemistry
Boulder, Colo. 30307

Elaine R. Ingham
Department of Botany and Plant Pathology
Oregon State University
Corvallis, Ore. 97331

Karen L. Josephson
Department of Soil and Water Science
University of Arizona
Tucson, Ariz. 85721

Ann C. Kennedy
USDA-ARS
Land Management & Water Conservation Res.
Washington State University
Pullman, Wash. 99164

Kim Luoma
8759 Enclid Ave., N.E.
Salem, Ore. 97305

David J. Mitchell
Plant Pathology Department
University of Florida
Gainesville, Fla. 32611

Joseph B. Morton
Division of Plant and Soil Sciences
West Virginia University
Morgantown, W.V. 26506

Michael D. Mullen
Department of Plant and Soil Science
University of Tennessee
Knoxville, Tenn. 37901

David D. Myrold
Department of Crop and Soil Science
Oregon State University
Corvallis, Ore. 97331

Ian L. Pepper
Department of Soil and Water Science
University of Arizona
Tucson, Ariz. 85721

Joshua Schimel
Department of Biology
University of California
Santa Barbara, Calif. 93106

David M. Sylvia
Soil and Water Science Department
University of Florida
Gainesville, Fla. 32611

Horace D. Skipper
Department of Agronomy and Soils
Clemson University
Clemson, S.C. 29634

George H. Wagner
Soil and Atmospheric Sciences
University of Missouri
Columbia, Mo. 65211

Duane C. Wolf
Department of Agronomy
University of Arkansas
Fayetteville, Ark. 72701

Arthur G. Wollum
Department of Soil Science
North Carolina State University
Raleigh, N.C. 27695

Larry Zibilske
Department of Applied Ecology and
    Environmental Sciences
University of Maine
Orono, Maine 04469

David A. Zuberer
Department of Soil and Crop Sciences
Texas A & M University
College Station, Texas 77843

Acknowledgments

◆

We thank Kim Luoma for her many excellent drawings. We thank the many internal and external reviewers of the individual chapters and many soil microbiology students (particularly those in the editors' classes) for providing constructive criticism of the various drafts of the chapters. The external reviewers were Abid Al-Agely, David Bezdicek, John Doran, Rhae Drijber, Samuel Farrah, Melvin Finstein, Pauline Grierson, Paul Hendrix, Frank Hons, Joseph Kloepper, David Kuykendall, Bill Lindermann, Murray Milford, Murray Miller, Li Ou, David Quinn, Mark Radosevich, Charles Rice, Amy Rossman, Eduardo Schroder, David Smith, Tom Staley, and Kristi Westover. We thank Mona Freer and Karen Bankston for their copyediting. Finally, we thank other members of the S-226 and S-262 Regional Research Projects for their stimulating discussions. This work was partially funded under the Regional Research Projects S-226 and S-262.

About the Editors

♦

David M. Sylvia is a professor in the Department of Soil and Water Science at the University of Florida. He teaches a senior undergraduate/beginning graduate course in soil microbiology. He is currently leading a task force on technology-assisted graduate instruction. His research interest is the microbial ecology of the rhizosphere with an emphasis on mycorrhizal fungi. He has served as an associate editor for the Soil Science Society of America Journal and is on the editorial board of Biology and Fertility of Soils. He resides in Gainesville, Florida, with his wife, Jeanne, and four children, Peter, Anna, Emily, and Stephen.

Jeffry J. Fuhrmann is an associate professor in the Department of Plant and Soil Sciences at the University of Delaware. He teaches graduate courses in soil microbiology, soil microbial ecology, and co-teaches a graduate course on research techniques and an undergraduate course in environmental soil microbiology. His major research interests are rhizosphere ecology, N_2 fixation, and microbial diversity in soils. He currently serves as an associate editor for the Soil Science Society of America Journal. He resides near Newark, Delaware, with his wife, Beth, and two children, Marlee and Dustin.

Peter G. Hartel is an associate professor in the Department of Crop and Soil Sciences at the University of Georgia. He teaches courses in soil microbiology, research methods, and agricultural ethics. He is also the coordinator for the University of Georgia's nationally recognized certificate program in environmental ethics. His primary research interest is in plant-microbe interactions in the rhizosphere. He resides in Athens, Georgia, with his wife, Mary Jean, and two children, Benjamin and Ruthanne.

David A. Zuberer is a professor in the Department of Soil and Crop Sciences at Texas A&M University. He teaches a senior undergraduate course in soil microbiology and lectures in several environmental science courses. His major research interests are rhizosphere microbiology, N_2 fixation, microbial ecology, and microbiological aspects of the reclamation of disturbed lands. He has served as a member of the editorial board for Applied and Environmental Microbiology and as section editor (Microbial Ecology) for the Canadian Journal of Microbiology. He resides in College Station, Texas, with his wife Karen. He has one daughter, Sandra, living in Irving, Texas.

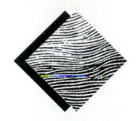

Chapter Quotations

♦

Part 1

◆

Habitat and Organisms

Chapter 1

♦

Introduction and Historical Perspective

Arthur G. Wollum, II

♦

Society has its roots in the soil.
Charles Kellogg

Soil microbiology is a branch of soil science concerned with microorganisms found in the soil and their relationship to soil management, agricultural production, and environmental quality. Hence, the soil microbiologist studies the numbers and kinds of microorganisms found in soil and the effect of these and introduced microorganisms on soil-ecological processes (e.g., nutrient cycling). The applications of these studies have important consequences for crop production, environmental quality, and the restoration of compromised environments.

The Soil Habitat

The soil is a complex habitat for microbial growth. It differs markedly from the environment microorganisms encounter in traditional microbiological culture media in two crucial ways:

- In its natural state, the soil is a heterogeneous medium of solid, liquid, and gaseous phases, varying in its properties, both across the landscape and in depth.

- In soil, **competition** exists among a wide variety of organisms for nutrients, space, and moisture. Competition occurs among bacteria, actinomycetes, and fungi, as well as with other living forms in soil, including animals and plant roots.

If we are to understand soil microorganisms, then developing a knowledge of the habitat in which they grow is of utmost importance.

The Nature of Cellular Organisms

The basis of living matter is the cell. Each cell is a unique entity made up of a complex mixture of chemical materials and subcellular components. The cell is bounded by the **cytoplasmic membrane,** separating the interior of the cell, known as the **cytoplasm,** from the external environment.

Box 1–1

Characteristics of Living Cells. Madigan et al. (1997) recognize five key characteristics that separate living cells from nonliving chemical systems:

- *Self-feeding or nutrition:* The capacity to take up and use chemicals from the environment and transform these chemicals into usable products, including energy to grow or survive.

- *Self-replicating or growth:* The capacity to self-direct synthesis, growing by division, forming two cells from one.

- *Differentation:* The capacity to undergo change in form or function, often in response to environmental changes or normal growth processes.

- *Chemical signaling:* The capacity to interact with other cells through chemical signals.

- *Evolution:* The capacity to change genetically, which may affect the overall fitness of the cell to survive in a particular environment.

Two fundamental types of living cells are recognized: **prokaryote** (from *pro,* meaning "before," and *karyon,* meaning "nucleus") and **eukaryote** (from *eu,* meaning "true"). Major structural differences exist between the two types of cells. The nucleus of the eukaryote is in the cytoplasm, bounded by a nuclear membrane and containing several DNA molecules. The eukaryote undergoes division by the well-known process of mitosis. The prokaryote has no nucleus; a nuclear region is recognized, but it is not bounded by a membrane and consists of a single, circular DNA molecule (chromosome). Cell division in the prokaryote is usually by binary division (i.e., nonmitotic). Additional differences between prokaryotic and eukaryotic cells are presented in Table 1–1. **Bacteria** (including cyanobacteria and actinomycetes) and **Archaea** are prokaryotes, while all other organisms are eukaryotes.

Classification of Organisms

The study and use of microorganisms is based on our ability to recognize and establish the identity of individuals. Most classification schemes are organized to show relationships among organisms. This orderly arrangement allows us to communicate descriptive information about the organism to others. These data can also be entered into various microbial databases, allowing retrieval of information about related organisms.

Table 1–1 A structural comparison of prokaryotic and eukaryotic cells.

Organelle	Prokaryotes Bacteria & Archaea	Eukaryotes Fungi	Algae	Protozoa
Cytoplasmic membrane	+	+	+	+
Nuclear division	+	+	+	+
Nuclear membrane	–	+	+	+
Ribosomes	70S	80S	80S	80S*
Endoplasmic reticulum	–	+	+	+
Golgi complexes	–	+	+	+
Mitochondria	–	+	+	+
Cytosketeton	–	+	+	+
Chloroplasts	–	–	+	–
Vacuole	–	+	+	+
Cell wall	+	+	+	+

*S = Svedberg unit

Microbiologists use the Linnean system of **binomial nomenclature** to name the microorganisms with which they work. An organism's name is made up of genus and species. In higher organisms, species are defined as groups of interbreeding or potentially interbreeding natural populations; however, many microorganisms do not reproduce sexually so this definition is not very useful. Microbiologists define **species** as a group of *similar* individuals that are sufficiently *different* from other individuals to be considered a recognized taxonomic group. A collection of species that share a major property (or properties), making them a distinct grouping, permits the group to be considered a **genus** (plural, genera). Hence, the Latin binomial name, *Thiobacillus thiooxidans* (abbreviated *T. thiooxidans* after it is used the first time in the text) is representative of a group of individuals (species: *thiooxidans*) that have the capacity to oxidize sulfur and share some common characteristics with other organisms in the genus *Thiobacillus*. Often microorganisms are named for an outstanding feature they possess (e.g., *T. thiooxidans;* a rod-shaped bacterium capable of oxidizing reduced sulfur for the generation of energy). In other cases organisms are named to commemorate the contributions of an outstanding scientist in the field (e.g., *Nitrobacter winogradskyi,* named in honor of the Russian soil microbiologist Sergei Winogradsky).

Historically, microorganisms were classified on the basis of taxonomic features, which were relatively easy to measure. These characters include structure, morphology, staining reactions, and physiological parameters (e.g., ability to use a particular carbohydrate). However, these features are **phenotypic** (based on physical characteristics) rather than **phylogenetic** (based on genetic relationships) and may obscure important relationships among related groups of organisms.

The technology of molecular sequencing has introduced a totally new way of determining relationships among organisms. Phylogenetic "trees," showing relationships among organisms, are constructed directly from comparisons of informational macromolecules, such as ribosomal RNA (rRNA) genes, occurring in living cells. The traditional classification scheme recognized five kingdoms of organisms: bacteria, fungi, protista (including algae and protozoa), animals, and plants. However, molecular phylogeny shows that there are three **domains** of living organisms: **Bacteria, Archaea**

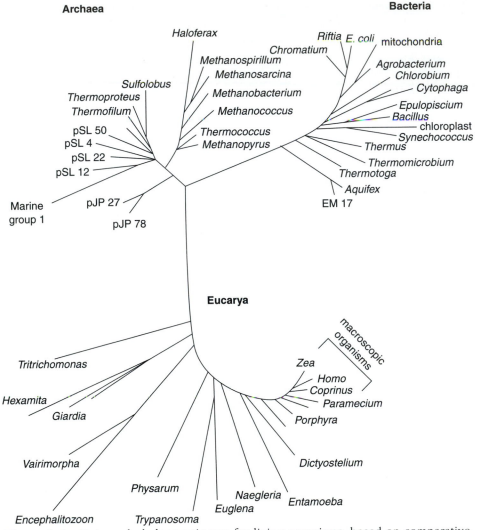

Figure 1–1   Universal phylogenetic tree for living organisms, based on comparative sequencing of 16s or 18s ribosomal RNA. Note that microorganisms comprise most of the biological diversity found on earth. *From Pace (1996). Used with permission.*

and **Eucarya** (Fig. 1–1; Woese et al., 1990). Although the placement of organisms into three domains was defined by differences within the rRNA gene, subsequent studies reveal that organisms in these domains also differ in cell wall properties, lipid composition, and protein synthesis. Below the three domains, eight or more kingdoms have been recognized. Our new understanding of the universal tree of life contradicts several long-held beliefs (Pace, 1996). For example:

- The deep divergence between Archaea and Bacteria shatters the notion of evolutionary unity among prokaryotes.

- It appears that the Eucarya line is as old as the prokaryotic lines.

- Most genetic diversity is microbial. Indeed, from the standpoint of rRNA variation, the multicellular life forms are relatively minor twigs at the tip of the eucaryal branch.

In these classification schemes there is no place for **viruses.** Viruses are not cells because they lack a cytoplasmic membrane with internal cytoplasm. Only when viruses are associated with another organism (e.g., bacterium, plant, animal) are they able to fulfill the basic life processes as stated in Box 1–1.

Organisms in the Soil

Organisms in the soil are both numerous and diverse. Many soil organisms are small and cannot be seen without the aid of magnification (Table 1–2). The smallest organisms—bacteria, actinomycetes, fungi, and algae—are referred to collectively as the **microflora.** Soil animals range in size from microscopic (**microfauna**) to earthworms and small mammals (**macrofauna**). With the exception of some soil animals and fungi, most soil organisms are single cells. Chapters 3 to 7 of this book describe the microorganisms present in soil.

Bacteria are the most abundant microorganisms in soil, attaining populations in excess of one hundred million (10^8) individuals per gram (g^{-1}) of soil and representing perhaps as many as 10^4 to 10^6 different species. The actinomycetes and fungi are the next most numerous microorganisms in soil, numbering 10^6 to 10^7 and 10^4 to 10^6 g^{-1} soil, respectively. Numbers of soil animals vary widely in the soil, ranging from just a few to as many as 10^6 g^{-1} soil. It is important to note, however, that we must consider more than the number of individuals in a gram of soil if we are to understand microbial function in soil. Microorganisms have a wide range of sizes and morphologies; thus, numbers alone may not provide a very good indication of the importance of a microbial group in the soil. For example, even though bacterial numbers are usually several orders of magnitude greater than fungi, the fungi generally have a greater total biomass in the soil (Table 1–2).

The relative position and size of soil microorganisms within the soil habitat is illustrated in Figure 1–2. The physical, chemical, and even biological properties of the soil habitat and their interactions with the resident community of soil

Table 1–2 **Microbial groups with representative size, numbers,[†] and biomass[†] found in soil.**

Microbial group	Example	Size (μm)	Numbers (no. g^{-1} of soil)	Biomass (kg wet mass ha^{-1} of soil)
Viruses	Tobacco mosaic	0.02×0.3	$10^{10} - 10^{11}$	
Bacteria	*Pseudomonas*	0.5×1.5	$10^8 - 10^9$	$300 - 3,000$
Actinomycetes	*Streptomyces*	$0.5 - 2.0$[‡]	$10^7 - 10^8$	$300 - 3,000$
Fungi	*Mucor*	8.0[‡]	$10^5 - 10^6$	$500 - 5,000$
Algae	*Chlorella*	5×13	$10^3 - 10^6$	$10 - 1,500$
Protozoa	*Euglena*	15×50	$10^3 - 10^5$	$5 - 200$
Nematodes	*Pratylenchus*	$1,000$[§]	$10^1 - 10^2$	$1 - 100$
Earthworms	*Lumbricus*	$100,000$[§]		$10 - 1,000$

[†]Data from Metting (1993).
[‡]diameter of hyphae
[§]length

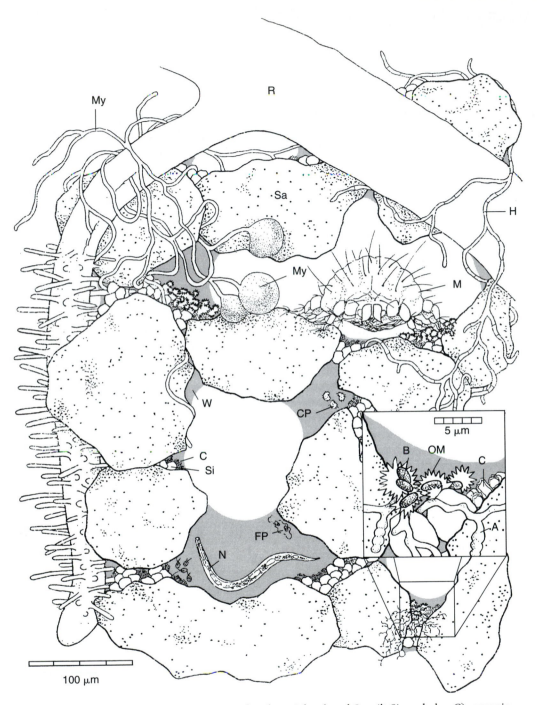

Figure 1–2 A soil habitat containing mineral soil particles (sand-Sa, silt-Si, and clay-C), organic matter (OM), water (W), plant root with root hairs (R), and soil organisms (bacteria-B, actinomycetes-A, mycorrhizal spores and hyphae-My, hyphae of a saprophytic fungus-H, a nematode-N, ciliate protozoa-CP, flagellate protozoa-FP, and a mite-M. This soil can be a habitat of enormous complexity and diversity even over small distances. For example, the actual size of the soil in this drawing is < 1 mm in both directions yet may contain habitats that are acid to basic, wet to dry, aerobic to anaerobic, reduced to oxidized, and nutrient-poor to nutrient-rich. Realizing this complexity and diversity is the key to understanding soil microbiology. *Original drawing by Kim Luoma.*

Scientific Notation. For convenience, soil microbiologists express the numbers of microorganisms in soil in an exponential manner, a convention known as scientific notation. Two million organisms per gram of soil is written, $2 \times 10^6 \text{ g}^{-1}$. The same number can also be expressed as a logarithmic number, $\log_{10} 6.30 \text{ g}^{-1}$.

microorganisms have a significant impact on growth and activity of microorganisms. As our understanding of these complex relationships develops, we should be better able to manage the soil and its microorganisms for the maintenance and improvement of soil without damaging the soil as a resource.

Microorganisms have an enormous diversity of functions in the soil. For example microorganisms decompose organic compounds releasing inorganic elements, oxidize reduced forms of elements (e.g., elemental sulfur → sulfate) (Chapters 11, 12, 15, and 16), and reduce oxidized forms of elements (e.g., nitrate → dinitrogen) (Chapters 11, 12, 15, and 16). Also the reduction of dinitrogen to a biologically usable form, ammonia (Chapters 13 and 14), or degradation of organic wastes and pollutants to carbon dioxide and water (Chapters 20, 21, and 22) are important functions of soil microorganisms. Interactions among different organisms, including plant roots, can lead to benefits for some participating organisms, while others have a detrimental effect on growth or development (Chapters 17, 18, and 19). Other microorganisms may alter levels of global gases (Chapter 23). In the second and third sections of this book we explore all these functions.

The Historical Context of Soil Microbiology

The development of soil microbiology is inextricably linked with the development of microbiology as a science and cannot be understood in isolation. Here we highlight significant accomplishments before 1950 that led to the development of both microbiology and soil microbiology (Table 1–3). After 1950 the list of those making significant contributions becomes too long to mention individual accomplishments; thus the contributions are considered more generally.

Pre-Nineteenth Century

The first historical mention of the presence of microscopic organisms in soil is attributed to a Roman writer in about 60 B.C. (Waksman, 1927). Writing about marshes, Columella noted they gave up "noxious and poisonous steams," breeding "animals armed with poisonous stings," from which "hidden diseases are often contracted, the causes of which even physicians cannot properly understand." Even prior to that time, there are reports in the Old Testament of people who practiced strict isolation and cleanliness codes, particularly to those afflicted with leprosy, suggesting they understood that disease had some relationship to an unseen cause. Likewise, there is good evidence that the Romans recognized that leguminous plants enriched for soil productivity and practiced crop rotations with leguminous plants. However, no one really understood the involvement of microscopic organisms in these phenomena.

Although Robert Hooke reported on the fruiting structures of molds in 1664, we most often think of the Dutch microscope builder, Anton van Leeuwenhoek—with his newly constructed microscope—as the first individual to describe microorganisms

Table 1–3 **Some of the outstanding individuals contributing to the development of soil microbiology.**

Name	Country	Area of contribution
Leeuwenhoek	Netherlands	Inventor of the microscope
Pasteur	France	Repudiation of spontaneous generation, biological nature of nitrification
Tyndall	England	Repudiation of spontaneous generation, understanding of sterilization processes
Cohn	Germany	Understanding of sterilization processes and taxonomy of *Bacillus* (particularly the endospore)
Koch	Germany	Koch's Postulates, gelatin plates for studying soil microorganisms
Winogradsky	Russia	Isolation and taxonomy of chemoautotrophic bacteria, especially nitrifiers and sulfur oxidizers
Beijerinck	Netherlands	Isolation of legume root nodulating bacteria, Director of The Delft School of Microbiology
Gram	Denmark	Differential staining procedures
Lipman	USA	Concept of soil as a complex, living entity
Russell	England	Development of importance of soil-plant-microorganism interactions in agriculture and the environment
Starkey	USA	Microbiology of the sulfur bacteria
Waksman	USA (Russia)	Discovery of antibiotics from actinomycetes

(Madigan et al., 1997). In his paper to the Royal Society of London in 1684, van Leeuwenhoek reported the presence of "wee animalcules" in pond water. These observations were confirmed by others, but little was known about these organisms and their relation to the environment in which they grew.

The Nineteenth Century

During the nineteenth century, two important questions were answered that would lay the scientific foundations for both microbiology and soil microbiology:

- Does spontaneous generation occur?

- What is the origin of contagious disease?

The eminent French scientist, Louis Pasteur, in a simple and elegant experiment (Box 1–3), demonstrated that it was microorganisms present in the air that were capable of initiating growth in an exposed sterile culture medium. Microbes simply did not arise spontaneously in a suitable medium.

Pasteur's simple experiment effectively settled the controversy surrounding the prominently held theory of **spontaneous generation,** vigorously debated at the time. Similar experiments were also conducted by Spallanzani, the Italian scientist, and the Englishman John Tyndall. Tyndall's original flasks can still be viewed at the Royal Institute near Piccadilly Square in London. The rejection of the theory of spontaneous generation provided the foundations for **aseptic (sterile) technique.** While conducting his experiments, Tyndall also noted difficulties in trying to sterilize

Box 1–3

Pasteur's Experiment Disproving Spontaneous Generation. (a) Pasteur introduced a nonsterile broth into each of two flasks, drew out the flask to a swan-necked shape to provide a dust trap, and sterilized the contents of each by heating. The broth in one flask (b) was not allowed to contact the dust that settled in the trap, and no growth occurred even after a long incubation. After the broth had cooled in the second flask (c) it was brought into contact with dust collecting at the low point near the mouth of the swan neck. After a short period of incubation, the broth became turbid, indicating that growth had occurred in this flask.

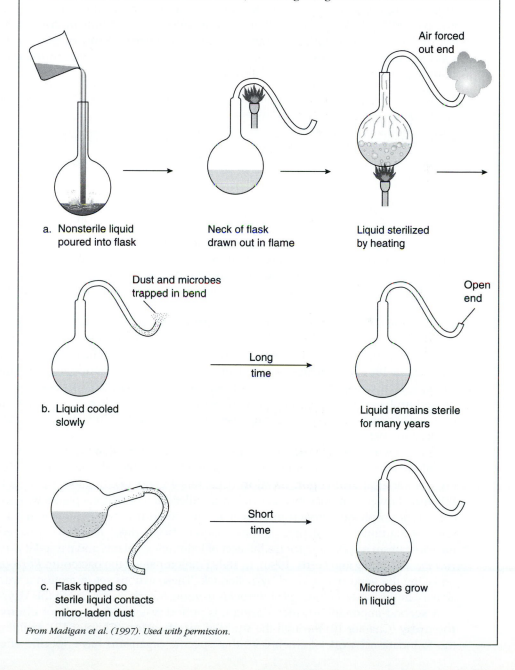

Air forced
out end

a. Nonsterile liquid
poured into flask

Neck of flask
drawn out in flame

Liquid sterilized
by heating

Dust and microbes
trapped in bend

Long
time

b. Liquid cooled
slowly

Open
end

Liquid remains sterile
for many years

Short
time

c. Flask tipped so
sterile liquid contacts
micro-laden dust

Microbes grow
in liquid

preparations from hay infusions. Further investigations by Tyndall, and also by Ferdinand Cohn of Germany, led to the discovery of organisms that were producing **endospores** in these difficult-to-sterilize preparations.

During this same period, there was a widely-held belief that disease was caused by something called "contagion." After microorganisms were discovered, they were accepted as the responsible agents, but rigorous proof was lacking. Although Ignaz Semmelweis and Joseph Lister (after whom the product Listerine was named) provided evidence that microorganisms caused human disease, it was not until the seminal work of Robert Koch that the **germ theory of disease** had a solid scientific basis. Koch reasoned that to prove that a microorganism was the causative agent of a disease, the following should apply:

- The organism should be consistently present in the subject exhibiting the disease symptoms, but not in the healthy subject.

- The organism should be grown in pure culture away from the subject.

- When the organism is used to inoculate a healthy susceptible subject, the disease symptoms should appear.

- The organism should be reisolated from the exposed subject and recultured in the laboratory to confirm its similarity with the original organism.

This series of steps became known as **Koch's Postulates,** which are important for several reasons. First, by following these steps, it is possible to demonstrate that a specific organism is responsible for a specific disease or for some microbiological process (e.g., sulfur oxidation). Second, they suggest the importance of the laboratory culture of organisms, and third, they recognized that specific organisms have specific functions. Acceptance of Koch's Postulates brought the science of microbiology to the point that others could now make specific contributions in more applied fields, such as soil microbiology. Another of Koch's significant developments, reported in 1881, was the use of a gelatin culture medium to study microorganisms *in vitro*. Five years later the Danish physician Christian Gram described a staining procedure based on differential properties of cell walls, leading to important taxonomic decisions. Now soil microorganisms could be studied in a systematic fashion (Waksman, 1927).

In the few remaining years until the turn of the century, significant developments in soil microbiology came in rapid succession. One notable advance was the discovery of biological **dinitrogen fixation** (Chapters 13 and 14). In 1886, Hellriegel and Wilfarth showed that microbial activity in nodules on leguminous plants was able to convert atmospheric nitrogen to a form a plant could use. Subsequently, the Dutch scientist Martinus W. Beijerinck (pronunciation: "buy-a-rink," Fig. 1–3a) isolated the microorganism responsible for nodulation of leguminous plants and named it *Bacillus radicicola* (Chung and Ferris, 1996). In 1889 Frank renamed the bacterium *Rhizobium,* replacing the genus name used by Beijerinck. These actions set in motion a wave of activity in the study of biological nitrogen fixation, which continues to this day.

A second important advance during this period was the discovery of **chemoautotrophy** (Chapter 10) through the study of **nitrification** (Chapter 12). Building on the foundational work of Koch (i.e., that specific organisms have specific effects),

a.

b.

c.

d.

Figure 1–3 Several important individuals in the development of soil microbiology. (a) Martinus Beijerinck, (b) Sergei Winogradsky, (c) Jacob Lipman, and (d) Selman Waksman. *Figures a, b, and d are courtesy of the Waksman Institute at Rutgers University. Figure c is from the American Society of Microbiology Archives.*

Sergei Winogradsky (Fig. 1–3b) began studying the nitrifying bacteria. He was able to demonstrate that nitrification was really a two-step process, the first being the oxidation of ammonium (NH_4^+) to nitrite (NO_2^-) and the second being the oxidation of NO_2^- to nitrate (NO_3^-). These two steps were mediated by two distinctly different bacteria. Winogradsky isolated representatives of each group and named them, *Nitrosomonas* for the NH_4^+ oxidizers and *Nitrobacter* for the NO_2^- oxidizers. Given the meager understanding of bacterial metabolism at that time, this was a truly remarkable discovery. For these and many other ground-breaking accomplishments, Winogradsky is considered by many to be the Father of Soil Microbiology.

By 1895 developments in microbiology had reached such a point that the Delft School of Microbiology in the Netherlands was established. First headed by Beijerinck— considered by some to be the Father of Microbial Ecology—and then by A. J. Kluyver and C. B. van Niel, the Delft School had a lasting influence on the study of soil microbiology through discoveries in microbial biochemistry, biodiversity, and biotechnology. These individuals posed the following basic questions about microbial physiology that drew the study of microbiology and the environment closer together:

• How does the intact organism interact with its abiotic and biotic environment?

• How can the fundamental principles of microbiology be brought to bear on applied problems?

• What is the place of microorganisms in the natural world?

These questions are as valid now as they were in the nineteenth century. Certainly the ubiquity and diversity of physiological function of microorganisms in the environment is underscored by *Beijerinck's Rule:* "Everything is everywhere and the milieu selects."

The Twentieth Century

If Winogradsky is considered the Father of Soil Microbiology, then Jacob G. Lipman (Fig. 1–3c) is the founder of American soil microbiology. He considered soil as a complex and living entity which needed to be understood and studied from the standpoint of soil fertility and crop production. This revolutionary concept stands as a milestone in soil microbiology (Clark, 1977) and has been carried across the United States and throughout the world.

Two seminal works on soil microbiology were published in the early 1900s: *Handbuch der landwirtschaftlichen Bakteriologie* (Handbook of Agricultural Bacteriology), a comprehensive treatise on soil bacteriology by F. Löhnis, and *Bacteria in Relation to Country Life* by Lipman. Although these works emphasized the role of bacteria in soil fertility, they also directed attention to other organisms residing in soil, including fungi, actinomycetes, algae, protozoa, nematodes, and insect larvae. Sir John Russell, Director of the Rothamsted Experiment Station in Great Britain, was a particularly strong advocate of the importance of protozoa in soil fertility, suggesting that when soil protozoa were absent or few in number, soil fertility would be low, and conversely, when protozoa were numerous, soil fertility would be high.

Two prominent individuals who shared Lipman's revolutionary concept of soil were Selman A. Waksman (Fig. 1–3d) and Robert L. Starkey. Besides promoting the "Lipman Philosophy," Waksman's book, *Principles of Soil Microbiology,* and a later work by

Waksman and Starkey, *The Soil and the Microbe,* were standard soil microbiology texts for much of the period between 1925 and 1950. However, in many circles Waksman is remembered less for his contributions to soil microbiology than for his discovery of the antibiotic streptomycin, for which he was awarded the 1952 Nobel Prize in physiology and medicine. It should be noted that the native environment of the streptomycin-producing organism (an actinomycete) was the soil! Thus, the quote attributed to Waksman, "From the earth will come our salvation," was prophetic.

Perhaps the greatest contribution of these noteworthy individuals was not so much their scientific papers, published books, inventions, or even patents, but the students they trained, who themselves went on to productive and noteworthy careers. The contributions of these individuals, now too numerous to mention, have enhanced crop production and fostered sound use of the environment worldwide.

Current and Future Directions

No look to the future can be made without first looking to see where we have been. If we neglect this first step, then we are relegated to repeat the mistakes of the past. Given the urgency of some of the issues we face and the shrinking resources we have at our disposal, it is prudent to step back and look at where we have been, before deciding on the next meaningful step. In the previous section we discussed the historical roots of soil microbiology. Here we summarize important topics in soil microbiology and suggest future research priorities.

Nutrient Transformations

From the moment Winogradsky first characterized the nitrifying bacteria, soil microbiologists have studied organisms involved in nutrient transformations and tried to understand the factors affecting the various processes. The poor nutrient-use efficiencies of common crops and the expense of fertilizers for standard cropping systems drove soil microbiologists to characterize the magnitude of nutrient losses, the conditions under which poor nutrient-use efficiencies might occur and, to a lesser extent, fertilizer management to reduce losses.

Emphasis on nutrient transformations has now shifted to new concerns. Environmental issues, such as how much nitrous oxide (N_2O), a "greenhouse" gas, is produced during denitrification and nitrification (Chapters 12 and 23), became important topics studied by soil microbiologists. Other environmental issues include the microbial transformations not only of nitrogen, but of sulfur (Chapter 15), phosphorus (Chapter 16), and more recently some of the metallic cations (Chapter 16) such as copper, mercury, iron, and aluminum. These issues grow more urgent as soils are increasingly used to recycle or dispose of a variety of waste products (Chapter 22).

Organic Matter

For most of the twentieth century, studies were dominated by attempts to characterize the chemical composition of the **soil organic matter** (referred to by many as **humus**). Such research is still underway and has been complemented by newer studies designed to provide a clearer understanding of soil organic matter as a group of functional pools (i.e., which pools turn over most rapidly and what is in them). These ideas have brought a fresh approach to this complex topic. This research was driven initially by the importance of soil organic matter in agriculture, but more recently a strong interest in global-scale transformations of carbon with respect to climate change has contributed to the importance of soil organic matter studies. In years to come, these ideas should help us to understand the potential for nutrient cycling from within the soil organic fraction and its relationship to sustainable agricultural practices.

As we move closer to a better understanding of soil organic matter, perhaps we may gain new insights for further studies. A challenge of the next century and beyond will be to manage our organic resources while maintaining clean air and water. The emphasis must be on salvaging resources such as plant nutrients through recycling. For example, municipalities have captured methane from landfills and water treatment facilities for supplemental energy generation. Another topic deserving attention is the disposal of organic waste by converting it to high-grade, single-cell protein.

Perhaps other opportunities exist at the frontiers of space. If space travel is ever to become a reality, waste products must be recycled in closed systems. For the soil microbiologist, who already understands the processes of decomposition of organic substances, the application of that knowledge to a closed spaceship would not require newly developed technologies (Alexander et al., 1989). The knowledge transfer to this problem is only a matter of scale and control in a closed environment.

Biological Dinitrogen Fixation

Soil microbiologists have received great acclaim for their work in the area of biological dinitrogen (N_2) fixation (Chapters 13 and 14). In fact, as early as the 1960s, some scientists stated categorically that we knew all we needed to know about biological N_2 fixation! Funding for research became more difficult to obtain, and some individuals were diverted into other fields of soil microbiology. However, the fuel crisis of the early 1970s in the United States changed all this. Suddenly biological N_2 fixation was "rediscovered" and new opportunities abounded for enterprising scientists.

For a long time this research topic was dominated by selection of superior strains of rhizobia (i.e., the N_2-fixing bacteria) and refinement of inoculation practices. Many questions were posed during this period, and some still await satisfactory answers:

- Why is it so difficult to displace the indigenous and decidedly inferior strains of rhizobia with superior stains?

- What is the composition of the indigenous population of rhizobia?

- Do manageable plant-rhizobia combinations exist?

- What are the biochemical mechanisms of N_2 fixation?

- What is the exact taxonomic composition of the family Rhizobiaceae?

In the last question, molecular genetic procedures have revealed new species of rhizobia, particularly as scientists have examined rhizobia from tropical regions. Undoubtedly, there are exciting times ahead as the full story of the rhizobia continues to unfold and the field of biological N_2 fixation remains a meeting ground, uniting scientists from a variety of disciplines.

Mycorrhizae

In 1885 A.B. Frank first applied the term **mycorrhizae** to the **symbiotic** associations between tree roots and fungi. Foresters found that certain fungi grew between the cortical cells of the feeder roots and covered the root surface; these fungal-plant associations were termed **ectomycorrhizae.** Not until the 1930s was the role of mycorrhizae in plant nutrition appreciated. Scientists confirmed that mycorrhizal plants grew better and had greater nutrient contents, especially phosphorus, than nonmycorrhizal plants. Through a series of carefully designed experiments, researchers proved that phosphorus from the soil was taken up by the soil hyphae of the mycorrhizal fungus and transferred to the host plant. Conversely, the **mutualistic** nature of the association was confirmed by the introduction of radio-labeled carbon dioxide ($^{14}CO_2$) to the top of the plant and the detection of radioactive carbon (^{14}C) in the fungal portion of the mycorrhiza.

A broader significance of these findings was realized when we discovered that most plants had some sort of mycorrhizal association. Although the ectomycorrhizae were present on only a small proportion of all plants (i.e., certain trees and shrubs), **endomycorrhizae** were found on a wide range of plants, including many of agronomic importance. Endomycorrhizae also assist in nutrient absorption and are especially important for plants growing under environmental stress (e.g., drought conditions).

There is now a growing awareness of the important contribution of mycorrhizae to soil structure because of the large amounts of carbonaceous materials that flow to the soil through the network of fungal hyphae. Reports of interactions with plant pathogens also suggest the potential to use mycorrhizal fungi as biological control agents. Currently soil microbiologists interact with scientists from numerous disciplines (e.g., agronomy, ecology, genetics, forestry, molecular biology, and plant physiology) to gain understanding of the role of mycorrhizae in agricultural and natural landscapes.

Biodiversity of Soil Populations

The early phases of soil microbiology were often dominated by attempts to characterize soil populations using **selective media.** While these attempts produced valuable information, little more could be said other than that there was some number of **colony-forming units** (cfu) g^{-1} of soil. As we approach the end of the 1990s, we are becoming increasingly aware that our knowledge of the soil microbial community is far from complete. During the past 10 years we learned that many microorganisms existing in the soil are viable (alive) but nonculturable (Chapter 9). Some suggest that these organisms may exceed 99% of the total soil population. This means that in our studies of microbial communities, we perhaps have observed less than 1% of the soil population. Increasingly, questions are being raised about the effects of soil management on the composition of the microbial community, particularly in relation to soil quality. Thus, it remains important to develop procedures to characterize soil microbial populations. Until such time, we will be unable to determine whether species are becoming extinct or are undergoing evolutionary change.

Some of the more recently developed molecular tools are proving useful in characterizing soil populations. For example, techniques that rely on amplifying DNA, such as various **polymerase chain reaction** (PCR) procedures, will contribute to our understanding of the diversity of the soil microbial community (Chapters 8 and 9).

Biological Control

Compared to the chemical control of pests, the advantages of **biological control** (or biocontrol) are numerous. Biological control is a natural mechanism that leaves no **xenobiotic** residues after treatment (Chapter 19). Early attempts at biological control using soil microorganisms followed Waksman's discovery of antibiotic-producing microbes in the soil. Unfortunately most, if not all, of these attempts failed outright.

Since that time, soil microbiologists learned that it is not sufficient just to increase the number of biocontrol agents in the soil to achieve success, but that a basic understanding of the ecology of the biocontrol agent is needed. Amelioration of the factor limiting the growth or survival of the control agent is necessary for biocontrol to work. While successful biocontrol experiments are few in number in comparison to failures, biocontrol research continues to be an area where soil microbiologists make significant contributions. As we learn more about biocontrol agents and their pest-suppression mechanisms, ecology, and genetics, this area of soil microbiology research should continue to increase in importance.

Biotechnology

Soil microbiologists are interested in discovering microorganisms with superior traits, including better N_2 fixers, superior biocontrol agents, and agents for bioremediation processes. In the past, we relied on natural variation and selection from natural populations to obtain organisms with better environmental fitness; however, with the various molecular techniques now available, it is becoming possible to make specific modifications in the organism's genome to obtain specific phenotypic properties. This capacity can affect processes carried out by soil microorganisms—from methane production to denitrification in groundwater.

As these potential opportunities move to reality, questions on the fate of genetically engineered or modified microorganisms deliberately released into the environment assume primary importance. How do these genetically modified organisms grow, compete, move, and survive in the soil environment? Can the traditional models of growth describing carrying capacity of organisms help us understand what might happen upon release of genetically modified organisms? Questions and opportunities abound and the likelihood that soil microbiology will remain an important field of investigation is very strong.

Summary

Many microorganisms have their origin in the soil or are closely associated with the soil environment. Throughout history these microorganisms have had a substantial impact on humankind. In some instances the impacts have been beneficial while others have been detrimental (Doyle and Lee, 1986). On the beneficial side, microorganisms have a major role in nutrient cycling and thereby contribute to the sustainability of life. On the detrimental side, microorganisms contribute to some environmental problems, such as global warming and groundwater contamination with nitrate.

As we approach the twenty-first century, soil microbiologists are faced with at least two choices. Will our emphasis be on those areas that have brought us to our present position, or will we be willing to take advantage of opportunities that will permit us to step aggressively into the next century? Traditionally, soil microbiologists have been opportunists in their work for the betterment of society. In light of the environmental problems associated with unesthetic substrates such as sewage, animal manures, food processing wastes, and industrial wastes, the efforts of soil microbiologists will be needed to make these substrates less objectionable. Solutions to these problems are likely to come from existing technologies and those yet to be developed. Along with the time-tested techniques of traditional soil microbiology, new analytical procedures and genetic manipulations will contribute to the solution of environmental problems, whether these are associated with agricultural or other environmental issues. In the future, we should be able to advance our understanding of the microbial ecology of soil, enhance opportunities for bioremediation and understand the diversity of the soil population. The future of soil microbiology seems unlimited. Perhaps the solution of a food-production or environmental issue somewhere in the future will propel a soil microbiologist (perhaps yourself) into the ranks of the Nobel Prize winners, just as Selman Waksman's work was honored.

Cited References

Alexander, D.B., D.A. Zuberer, and D.H. Hubbell. 1989. Microbiological considerations for lunar-derived soils. pp. 245–255. *In* D.W. Ming and D.L. Henninger (eds.), Lunar base agriculture: Soils for plant growth. American Society of Agronomy, Madison, Wis.

Chung, K.-T., and D.H. Ferris. 1996. Martinus Willem Beijerinck (1851–1931). Pioneer of general microbiology. Am. Soc. Microbiol. News 62:539–543.

Clark, F.E. 1977. Soil microbiology—It's a small world. Soil Sci. Soc. Am. J. 41:238–241.

Doyle, R.J., and N.C. Lee. 1986. Microbes, warfare, religion and human institutions. Can. J. Microbiol. 32:193–200.

Lipman, J.G. 1908. Bacteria in relation to country life. Macmillian, New York.

Löhnis, F. 1910. Handbuch der landwirtschaftlichen Bakteriologie. Borntrager, Berlin.

Madigan, M.T., J.M. Martinko, and J. Parker. 1997. Brock biology of microorganisms. 8th ed. Prentice Hall, Upper Saddle River, N.J.

Metting, F.B. 1993. Structure and physiological ecology of soil microbial communities. pp. 3–25. *In* F.B. Metting (ed.), Soil microbial ecology: Applications in agricultural and environmental management. Marcel Dekker, Inc., New York.

Pace, N.R. 1996. New perspective on the natural microbial world: Molecular microbial ecology. Am. Soc. Microbiol. News 62:463–470.

Waksman, S.A. 1927. Principles of soil microbiology. Williams and Wilkins, Baltimore, Md.

Waksman, S.A., and R.L. Starkey. 1931. The soil and the microbe: An introduction to the study of the microscopic population of the soil and its role in soil processes and plant growth. John Wiley and Sons, Inc., New York.

Woese, C.R. 1990. Towards a natural system of organisms. Proposal for the domains Archaea, Bacteria, and Eucarya. Proc. Natl. Acad. Sci. USA 87:4576–4579.

General References

Alexander, M. 1986. Highlights of research in Division S-3—Soil Microbiology and Biochemistry since 1961. Soil Sci. Soc. Am. J. 50:839–840.

Allison, F.E. 1961. Twenty-five years of soil microbiology and a look to the future. Soil Sci. Soc. Am. Proc. 25:432–439.

Lowdermilk, W.C. 1975. Conquest of the land through seven thousand years. USDA Soil Conservation Service, Agricultural Information Bulletin No. 99.

Lynch, J.M., and J.E. Hobbie. 1988. Microorganisms in action: Concepts and applications in microbial ecology. 2nd ed. Blackwell Scientific Publications, Boston.

Worldwide Web Sites of Interest to Soil Microbiology

American Society of Microbiology: http://www.asmusa.org/

Digital Learning Center for Microbial Ecology: http://commtechlab.msu.edu/CTLProjects/dlc-me/

International Culture Collection of VA Mycorrhizal Fungi (INVAM): http://invam.caf.wvu.edu/

Microbial Underground: http://www.ch.ic.ac.uk/medbact/microbio.html

Soil Science Society of America: http://www.soils.org/sssa.html

The Tree of Life: http://phylogeny.arizona.edu/tree/phylogeny.html

The Worldwide Web Virtual Library (for Microbiology): http://golgi.harvard.edu/ biopages /micro.html

World Data Center for Microbiology: http://www.wdcm.riken.go.jp/

Study Questions

1. In what ways is soil a unique environment for microorganisms?

2. What individual(s) and event(s) have had the greatest influence on the development of soil microbiology?

3. Why is Sergei Winogradsky often considered the Father of Soil Microbiology?

4. What is the contribution of Jacob G. Lipman to soil microbiology?

5. What are some of the challenges for soil microbiology in the twenty-first century?

Chapter 2

♦

The Soil Habitat

Peter G. Hartel

♦

*Whatever our accomplishments, our sophistication, our artistic pretension, we owe
our very existence to a six-inch layer of topsoil—and the fact that it rains.*

Anonymous, *The Cockle Bur*

This chapter covers soil description, soil physical and chemical characteristics, and soil abiotic factors. Together, these elements help define the habitat for soil microorganisms. Although it may seem obvious that one needs to understand "soil" in order to understand soil microbiology, this is not an easy concept. Soil is dynamic. Because soil forms from the interaction of climate (especially temperature and rainfall) and living organisms (especially native vegetation) as influenced by topography (e.g., elevation) and type of parent material (i.e., the original composition of the minerals and organic matter) over time, soil is the most complex and variable of all microbial habitats. Therefore, soil does not conform easily to our conclusions and rules. The key to understanding the role of "soil" in soil microbiology is always to think of soil as the sum of many interrelated parts. By thinking constantly of these interdependencies, the science of soil microbiology will be much easier to understand.

Soil Description

Definition and Types of Soil

Soil is defined as a mantle of weathered rock which, in addition to organic matter, contains minerals and nutrients capable of supporting plant growth. Soil scientists may describe a particular soil this way: "The Ap horizon of a Tifton loamy sand

Figure 2–1 A soil profile with unusually distinct horizons. This forest soil, from the lower coastal plain of Georgia (U.S.), has a thin O horizon (2 to 3 cm of pine needles in various stages of decomposition), an A horizon (10 to 12 cm; light black), an E horizon (the bleached layer of variable thickness), and a B horizon (black layer). Soil corer, 122 cm (48 in.). *Photograph courtesy of L. West, Univ. of Georgia.*

(pH 5.6; 18.0 g organic matter kg^{-1}, 54 g of clay kg^{-1}, and 825 g of sand kg^{-1}) from Tifton, Georgia, was collected and passed through a 2 mm sieve." This single sentence describes several important characteristics of soil: horizons, series name, and texture.

There are two broad types of soil: mineral and organic. The definition of an organic soil varies according to the amount of clay and water saturation, but generally an **organic soil** contains at least 20% organic carbon; a **mineral soil** does not. Only 0.9% of the world's soils are organic (Miller and Donahue, 1995); therefore, the vast majority of soils in the world are mineral soils. Edwards muck (pH 7.6, 572 g organic matter kg^{-1}) contains 57.2 g of organic matter 100 g^{-1} of soil or 57.2% organic matter; it is an organic soil. Because Tifton loamy sand has 18.0 g of organic matter kg^{-1} (1.8 g of organic matter 100 g^{-1} of soil or 1.8%), it is not an organic soil but a mineral soil.

Horizons

A soil is composed of layers, each with distinct characteristics, called **horizons.** A slice of soil is called a **profile** (Fig. 2–1). Each horizon is identified with the letters (beginning from top to bottom) O, A, E, B, or C. Not every soil has all horizons. The topmost layer, the *O horizon,* is formed from plant and animal (organic) litter; it is an organic horizon. Because it is easily disrupted by human activity, the O horizon often does not exist in many soils. A forest is a good place to find a soil with an O horizon. The next horizon, the *A horizon,* is the first mineral horizon. The A horizon is distinguished from the O horizon by having less organic matter. Because minerals and nutrients leach down from this horizon, the A horizon is referred to as the zone of **eluviation.** In some soils, an *E horizon* underlies the A horizon. In this case, both horizons are eluviated, but the A horizon is darkened by organic matter whereas the E horizon is more lightly colored. Beneath the A or E horizon is the *B horizon,* called the zone of **illuviation,** because here minerals and nutrients accumulate. At the bottom of the soil layer, below the B horizon, is the *C horizon,* or unconsolidated parent material.

Horizons can have subdivisions that differentiate one horizon from another or show a transition zone from one horizon to another. In the case of Tifton loamy sand, the soil has an Ap horizon, where "p" stands for "plowed." Typically, the Ap horizon is the depth of a plow—a "furrow slice"—or about 20 cm (2.54 cm = 1 inch; 20 cm ≅ 8 inches).

Box 2–1

Furrow Slice. In old notation, the approximate weight of a "furrow slice" was 2,000,000 lb of soil acre $^{-1}$; in new notation, this is approximately 2,200,000 kg of soil ha $^{-1}$ (1 hectare = 2.47 acres and 1 kg = 2.2 lb; hence, multiply lb of soil acre^{-1} by 1.12 to convert to kg of soil ha^{-1}). The weight of a "furrow slice" is useful for determining application rates of soil amendments (e.g., fertilizer).

Soil Names

Tifton loamy sand is a **soil phase** name. The soil phase name includes the **series** name and the texture of the A horizon. The series name represents the lowest level of soil classification and is usually taken from a locale where the series was first described (i.e., the name of a town, county, or some local feature). In this case, Tifton is Tifton, Georgia. Names and complete classification of soils in the United States may be found in Natural Resources Conservation Service (formerly the Soil Conservation Service) publications (e.g., Soil Conservation Service, 1959) or State Experiment Station bulletins (e.g., Perkins et al., 1986). These publications are available in libraries and are arranged by county or counties. In addition to classifying soils (information that is often required by scientific journals), these publications provide important information on the chemical and physical characteristics of each soil and locate the soils on a map. The vast majority of soils in the United States have already been mapped.

Soil Physical Characteristics

Soil Texture

The **texture** of a soil is determined by the size distribution of the individual inorganic grains in soil. The grains are separated into three particle-size fractions: sand, silt, and clay. For the U.S. Department of Agriculture system of classification:

- Sand is soil particles with diameters from 0.05 to 2.0 mm.

- Silt is soil particles with diameters from 0.002 to 0.05 mm.

- Clay is soil particles with diameters < 0.002 mm.

In the case of Tifton loamy sand, a 2 mm sieve eliminated all the larger-size groups (i.e., stones and gravel).

Based on the particle-size distribution, all mineral soils can be placed into one of 12 major textural classes. The textural class of a soil is determined by means of a soil textural triangle (Fig. 2–2). The sample from the Ap horizon of Tifton soil contained 54 g of clay kg^{-1} (5.4%) and 825 g of sand kg^{-1} (82.5%) and is therefore a loamy sand (sand + silt + clay totals 100% of the < 2 mm inorganic particles, so it is understood that the remaining percentage, 12.1%, is silt). The texture of a soil does not change quickly with time and is considered a basic soil property. Soils may also be described as coarse or fine; a coarse-textured soil has more sand, whereas a fine-textured soil has more clay. A soil whose properties are equally influenced by sand, silt, and clay is called a *loam* or *loamy soil*. The properties of texture are important in the aeration and drainage of soil. It is important to note that these classes of soil texture do not apply to organic soils: an organic soil is classified as a muck, peaty muck, mucky peat, or peat depending on the state of decomposition of its organic matter. Mucks have well-decomposed organic matter; peats do not.

Soil Density

A soil has both a particle density and a bulk density. Density is the weight per unit volume. The **particle density** is determined by the weight of the solid soil particles divided by the volume of the solid soil particles. If 1 cm^3 of solid soil particles weighs 2.65 g, its particle density is 2.65 g cm^{-3}. To get solid soil particles, one can imagine compressing a soil sample so as to eliminate the pore spaces (Fig. 2–3a). Because quartz, feldspar, and colloidal silicates make up the major portion of mineral soils, the particle density for mineral soils is relatively narrow and typically ranges from 2.60 to 2.75 g cm^{-3}. The **bulk density** is determined by dividing the weight of the soil by the total volume of the soil, including the weight of the solid soil particles and the pore spaces (Fig. 2–3b). If 1 cm^3 of soil weighs 1.38 g, its bulk density is 1.38 g cm^{-3}. The bulk density of mineral soils typically ranges from 1.00

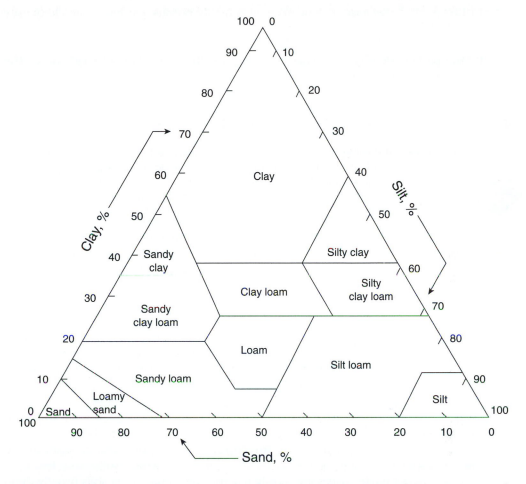

Figure 2–2 Soil textural triangle. To use the triangle, determine any two of the percentages of sand, silt, and clay in a soil. Follow the arrow in the direction of the tic marks at the appropriate percentage. The texture of the soil is identified at the intersection of the two lines. *Adapted from Soil Survey Staff (1975).*

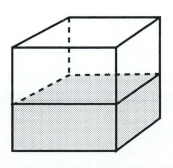

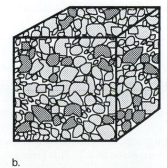

a.   b.

Figure 2–3 Particle density (a) versus bulk density (b). Particle density considers only the space occupied by soil solids; bulk density considers the total soil space.

to 1.80 g cm^{-3}. Because organic material is highly porous and has a particle density of only 1.20 to 1.50 g cm $^{-3}$, the incorporation of organic matter into the soil decreases both the particle and the bulk densities of a soil. The bulk density divided by the particle density, multiplied by 100, gives the percentage of solid space. The remaining space is pore space.

Box 2–2

Calculating Soil Pore Space. If a soil has a bulk density of 1.38 g cm $^{-3}$ and a particle density of 2.65 g cm $^{-3}$, the percentage of solid space would be (1.38/2.65)(100)= 52%; the pore space would be 48%.

A typical mineral soil contains 50% solid material (45% minerals and 5% organic matter) and 50% pore space. The pore space will be occupied by air or water, and these two are inversely related: as the volume of soil water increases, the volume of soil air decreases, and vice versa.

Soil Pores

Soil pores play a major role in water and air movement. Also, soil microorganisms reside in pores. Coarse-textured (sandy) soils have higher bulk densities and less total pore space (35% to 50%) than fine-textured (clay) soils that have lower bulk densities and more pore space (40% to 60%). The size of the pores, however, is just as important as the total quantity of pore space. Two classes of pore sizes are recognized: macropores and micropores. The minimum diameter of a **macropore** has been a source of debate, but is generally accepted to be between 30 and 100 μm. Pores smaller than this are **micropores.** Macropores characteristically allow the rapid movement of soil gases and soil water. Sandy soils have less total pore space, but those spaces are mostly macropores; thus, sandy soils usually drain rapidly. In contrast, clayey soils have more total pore space, but these spaces are mostly micropores. Soils high in clay usually drain slowly because the micropores restrict the water flow. This is why a sandy soil has a relatively low water-holding capacity and a clayey soil has a relatively high water-holding capacity.

Box 2–3

Soil Nanopores. One interesting recent development has been the concept of soil **nanopores** (nano = 10^{-9}; Pignatello and Xing, 1996). There are numerous examples of the long-term persistence of intrinsically biodegradable compounds in soil even when environmental conditions are not limiting for microbial growth. One possible mechanism for this reduced bioavailability is the slow (weeks to months) diffusion of organic chemicals into soil nanopores—soil pores so small that any organic compounds within them are beyond microbial or even enzymatic attack. This emerging concept may have important ramifications for biodegradation and bioremediation.

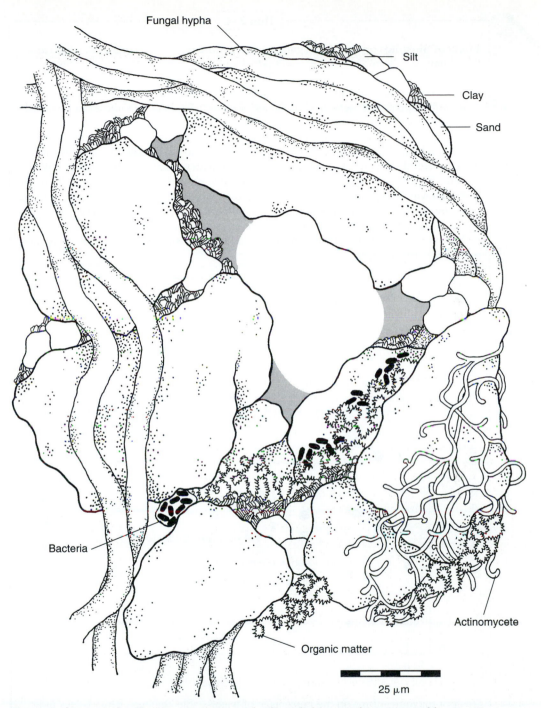

Fungal hypha

Silt

Clay

Sand

Bacteria

Actinomycete

Organic matter

25 μm

Figure 2–4 A typical soil aggregate. Here sand, silt, and clay particles, cemented by organic matter, precipitated inorganic materials, and microorganisms, bind the soil particles together to form an aggregate. Note how the water forms a meniscus surrounding the air space (*center*). Bacteria (rods in organic matter, rods in a polysaccharide "plug," and actinomycete) and fungus (hyphae only), as well as the sand, silt, and clay particles are all to scale. Also note how an aggregate can offer a diverse set of microsites for microbial habitation over very small distances. *Original drawing by Kim Luoma.*

| Box 2–4 |

Effect of Biosolids on Soil Aggregation. Several researchers (Metzger et al., 1987) tested the ability of sewage sludge (i.e., biosolids) to promote water-stable aggregates in soil. (Water-stable aggregates are aggregates that do not fall apart when raindrops hit them. This is measured by determining the percentage of soil aggregates remaining on a sieve after repeatedly dipping the sieve in a container of water.) To determine the extent to which microbes were responsible for this aggregation, the sludge was amended with various antimicrobial agents; the controls included one treatment without any antimicrobial agents (sludge only) and one treatment of nonamended soil (no sludge added). The researchers then measured the percentage of water-stable aggregates in each treatment over time.

After 30 days, the largest increase in water-stable aggregates was measured in a) soil amended with sludge and a bacteriocide, to a lesser extent in b) soil amended with sludge only, and c) soil amended with sludge and a fungicide. No increase in water-stable aggregates was observed in nonamended soil or in soil amended with sludge and formaldehyde. These results suggest that following the addition of sewage sludge, fungi were most responsible for soil aggregation.

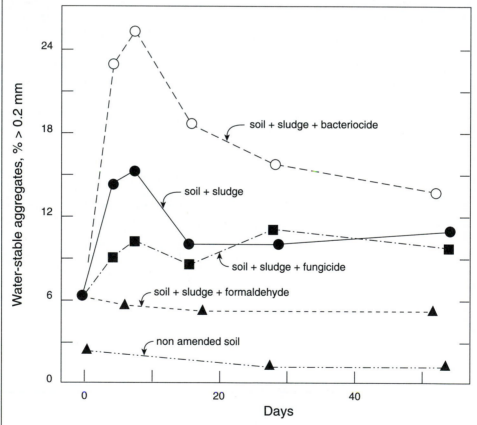

Effect of microorganisms on the percentage of water-stable aggregates in a Gilat sandy clay loam mixed with sewage sludge. The sewage sludge promotes microbial growth. The soil-sludge mixture with fungicide kills fungi (and allows the preferential growth of bacteria); the soil-sludge mixture with bacteriocide kills bacteria (and allows the preferential growth of fungi); and the soil-sludge mixture with formaldehyde kills all soil microorganisms. *Adapted from Metzger et al. (1987). Used with permission.*

Soil Structure

When groups of soil particles cohere more strongly to each other than to other adjoining particles, these groups form a **soil aggregate.** These aggregates can range in size from 0.5 to 5 mm in diameter and can even form clusters of aggregates (see Chapter 1, Fig. 1–2). Depending on their shape, these aggregates define **soil structure.** For example, spheroid aggregates have more pore space and more rapid permeability than aggregates that are block-like or prism-like. Soil scientists are interested in soil structure and soil aggregation because these attributes influence soil productivity.

Integrating Soil Aggregation and Soil Microorganisms

Although abiotic factors like the parent material, climate, tillage practices, and adsorbed cations (e.g., Na^+ ions tend to disperse soil particles, and Ca^{2+} ions tend to flocculate soil particles) are important in the formation of soil aggregates, biotic factors also play a major role. Plant roots disrupt the soil and promote granulation. Organic matter not only lightens the soil, but also binds it. Most important for this discussion, soil microorganisms can promote soil aggregation though the production of extracellular polysaccharides and hyphae. In this manner, soil microorganisms can physically bind soil particles (Fig. 2–4). Thus, soil microorganisms are fundamental to soil structure and as an aspect to soil formation.

Soil Chemical Characteristics

Soil pH

Soil pH is important because microorganisms and plants respond markedly to chemicals in their environment. Acid soils are most prevalent where rainfall is sufficient to leach bases from the soil; where rainfall is insufficient, the soils are usually alkaline. Not surprisingly, the majority of alkaline soils are found in arid and semi-arid regions.

| Box 2–5 |

Understanding Soil pH. pH is defined as the negative log of the hydrogen ion (H^+) concentration in solution. This is the same as pH = $\log_{10} 1/[H^+]$, where the brackets indicate concentration in moles per liter. Because the $[H^+]$ in pure water is 1×10^{-7} at 25°C, pure water has a pH of $\log 1/[1 \times 10^{-7}] = 7$; this is considered neutral. The pH scale is based on the dissociation of water (H^+ and OH) from 1.0 $M H^+$ (pH 1; acid) to 1.0 $M OH^-$ (pH 14; basic; 0.00000000000001 $M H^+$ remaining). As an example, the pH of 0.001 M HCl is $\log 1/[1 \times 10^{-3}] = 3$. It is important to remember that pH is a log scale; if two solutions differ by 1 pH unit, then one solution has 10 times more H^+ ions than the other.

Most soil microorganisms and plants prefer a near-neutral pH range of 6 to 7 because the availability of most soil nutrients is best in this pH range. For example, actinomycetes prefer neutral conditions and do not tolerate acid conditions well. Nevertheless, microorganisms can be found in soils from pH 1 to 13. Most fungi are acid tolerant and commonly are found in acid soils. Microorganisms also have the ability to alter soil pH. Under anaerobic conditions, some microorganisms produce organic acids; under aerobic conditions, some microorganisms can oxidize ammonia and sulfur with the concomitant production of H^+.

Soil Anion and Cation Exchange Capacity

Soils can possess both positive and negative charge. The ability of positively charged materials in soil to hold negative ions (e.g., orthophosphate, $H_2PO_4^-$) is the **anion exchange capacity** (AEC) of the soil, and the corresponding ability of soil to hold positive ions (e.g., K^+, Ca^{2+}) is the **cation exchange capacity** (CEC). Because the ability of a soil to hold cations often exceeds its ability to hold anions, soil scientists typically report only the CEC of a soil. This does not mean that the AEC of soils is unimportant; the AEC is particularly important in the subsoil of highly weathered soils. The CEC is important because it can alter soil physical properties, and it affects soil pH and fertility (most plants obtain K^+, Ca^{2+}, and Mg^{2+} from exchangeable sites).

The exchange capacity of a soil is determined by the type and amount of clay and organic matter in the soil. In this section, the CEC of clay is considered; the CEC of organic matter follows. An understanding of clay mineral structure is necessary to understand the source of the exchange capacity in clay minerals. Most clays are composed of crystalline sheets of silica and alumina; hence, they are aluminosilicate clays. One sheet of silica and one sheet of alumina give a 1:1 clay like kaolinite (Fig. 2–5). One alumina sheet between two silica sheets gives a 2:1 clay like smectite. These sheets give clays their characteristic layered effect (Fig. 2–6) and contribute to their large surface areas.

Clays have two sources of charges. One source is **isomorphous substitution,** which is the substitution in the crystalline sheet of one atom by a similarly sized atom of lower valence. In the case of a sheet of silicon tetrahedra, Si^{4+} (radius of 0.041 nm) is replaced by Al^{3+} (radius of 0.051 nm) or Fe^{3+} (radius of 0.064 nm). In

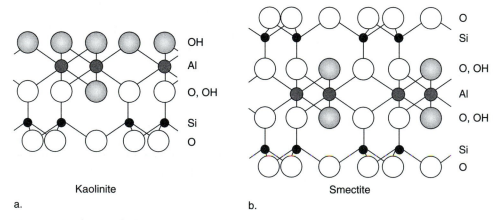

Figure 2–5 Sheets of silica and alumina layered together to form (a) kaolinite, a 1:1 clay, and (b) smectite, a 2:1 clay. The O atoms are shown as large, open circles; the OH^- atoms are shown as large, shaded, bold circles. *Adapted from Brady and Weil (1996). Used with permission.*

Figure 2–6 A scanning electron micrograph of kaolinite, a 1:1 clay. Bar, 100 μm. *Micrograph courtesy of N. White, Texas A&M Univ.*

each tetrahedron where this occurs, the tetrahedron will have an overall net -1 charge because the total negative charges of the O are no longer satisfied. This same type of substitution can occur in a sheet of aluminum octahedra where Mg^{2+} (radius of 0.066 nm), Fe^{2+} (radius of 0.070 nm), and Zn^{2+} (radius of 0.074 nm) can replace the Al^{3+} atom with the same overall net -1 charge. These charges are considered permanent charges and are *not* affected by soil pH.

A second source of charge on clays is broken edges. These are the actual edges of the silica and alumina sheets where ionizable H^+ atoms, as part of the hydroxyl ions, are held tightly by the O atoms under acid conditions (Fig. 2–7a). Here the charge of the broken edges is neutral. However, when the soil pH is > 6, the H^+ atoms are held more loosely and can be exchanged with such cations as Ca^{2+} and Mg^{2+} (Fig. 2–7b). This difference is the charge attributed to broken edges. This charge is pH-dependent, and this pH dependency is what distinguishes it from charge due to isomorphous substitution. Most of the charge in 2:1 clays is due to isomorphous substitution, and most of the charge in 1:1 clays is due to broken edges.

The AEC is also located at broken edges. Where OH^- is broken off or Al^{3+} or Si^{4+} is exposed, some anions (e.g., $H_2PO_4^-$) have the right size and geometry to be adsorbed; other anions (e.g., nitrate, NO_3^-) do not fit well and are not adsorbed.

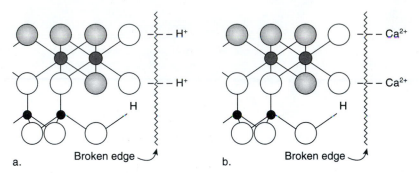

Figure 2–7 Broken edge of kaolinite under (a) acid and (b) alkaline conditions. *Adapted from Brady and Weil (1996). Used with permission.*

Table 2–1 **Relative size and surface areas of soil particles. For the sake of simplicity, the particles are assumed to be spherical and the largest size possible.**

Particle	Diameter (mm)	No. of particles (no. g^{-1})	Surface area ($cm^2 g^{-1}$)
Very coarse sand	2.00 to 1.00	90	11
Coarse sand	1.00 to 0.50	720	23
Medium sand	0.50 to 0.25	5,700	45
Fine sand	0.25 to 0.10	46,000	91
Very fine sand	0.10 to 0.05	722,000	227
Silt	0.05 to 0.002	5,776,000	454
Clay	<0.002	90,260,853,000	8,000,000

Adapted from Foth (1990). Used with permission.

There is an important relationship between the exchange capacity of a soil and soil texture. The adsorption of water, nutrients, and gases and the attraction of particles are all surface phenomena—and the greater the surface area of the soil particles, the greater the adsorption. Because the particle surface area per unit mass increases logarithmically as the particle diameter decreases, clay has 50 to 100 times more surface area than the equivalent amount of silt or fine sand (Table 2–1). For this reason, clay dominates the adsorption of water, nutrients, and gases and the attraction of particles in a soil.

The CEC of organic matter far exceeds that of clay. Unlike clay, the structure of organic matter is poorly understood. It is composed primarily of carbon, hydrogen, and oxygen and is chemically heterogeneous. The charge of organic matter is similar to that of broken edges, except the source of the charge is primarily carboxyl (—COOH) groups because these dissociate at the pH of most soils. Phenolic hydroxyls and other groups also play a less significant role. Like broken edges, the H^+ ions are strongly held under acid conditions and are not easily replaced by other cations. As the pH increases, the H^+ ions of the carboxylic acid are gradually replaced by other cations. For this reason, the CEC of organic matter is pH-dependent.

The CEC and AEC of a soil are expressed as centimoles (a centimole is 0.01 M) of positive or negative charge per kg of soil [i.e., cmol (+) kg^{-1} of soil or cmol ($-$) kg^{-1} of soil], respectively.

| Box 2–6 |

Units for Expressing the Ion Exchange Capacity of Soils. In old notation, the AEC and CEC were measured in milliequivalents (meq) 100^{-1} g of soil. The old and new terms are directly equivalent: 1 meq 100^{-1} g of soil = 1 cmol (+) kg^{-1}.

The number of centimoles of an anion or cation a soil can hold depends on the valency of the anion or cation. A monovalent cation (e.g., K$^+$) can satisfy one -1 charge; a divalent cation (e.g., Ca^{2+}) can satisfy two -1 charges. Therefore, a soil can hold only half as many Ca^{2+} cations as it can hold K$^+$ cations.

| Box 2–7 |

Calculating Ion Retention in a Soil. Assume a soil is at a constant pH, so that its CEC is fixed. The atomic weights of K$^+$ and Ca^{2+} are 39 (1 mole = 39 g) and 40 (1 mole = 40 g), respectively. If a silt loam has a CEC of 10 cmol (+) kg^{-1} of soil, it can hold (39 g × 10 cmol) or 3.9 g of K$^+$ kg^{-1} of soil. Similarly, the same silt loam can hold (40 g × 10 cmol) or 4.0 g of Ca^{2+}kg^{-1} of soil. But Ca cations are divalent, and each cation can satisfy two negative charges, so the soil can hold only 2.0 g of Ca^{2+} kg^{-1} of soil.

The CEC and AEC of some "soils" are shown in Table 2–2, and the relation of the CEC of organic matter and smectite to soil pH is shown in Fig. 2–8. Because soils are mixtures, the CEC and AEC of typical soils are much lower than pure clays or organic matter.

Table 2–2 **Cation and anion exchange capacity for selected "soils."**

"Soil"	CEC (cmol (+) kg^{-1})	AEC (cmol ($-$) kg^{-1})
Pure organic matter	240	1
Pure smectite	118	1
Pure kaolinite	7	4
Typical sand	5	ND
Typical loam	15	ND
Typical clay	30	ND
Tifton sandy loam	3	ND

Adapted from Brady and Weil (1996) and Miller and Donahue (1995). Used with permission.
ND, not determined.

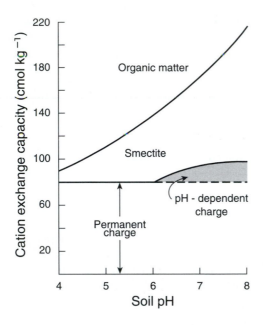

Figure 2–8 Charge of smectite and organic matter with varying pH. *Adapted from Brady and Weil (1996). Used with permission.*

Integrating Soil Microorganisms with AEC and CEC

The anion and cation exchange sites of a soil are important not only because they can attract or repel anions or cations, but also because they can attract and repel charged organic molecules. Because the surfaces of microorganisms are composed of organic molecules, which are positively or negatively charged (e.g., —COOH → —COO⁻; NH_2 → NH_3^+) depending on the soil pH, soils have the capacity to attract and repel microorganisms. In contrast to physical binding, this means soil microorganisms can chemically adsorb or bind to soil particles. How this is done is not fully understood. At typical soil pH values (pH 5 to 8), soil microorganisms are negatively charged. Because clays and organic matter are negatively charged at these pH values as well, some soil microbiologists have suggested that divalent cations "bridge" soil microorganisms and clays (hence the term *divalent cation bridging*). Evidence for this is unconvincing (Stotzky, 1985). More likely a variety of mechanisms is responsible for microorganisms chemically adsorbing or binding to clay and organic matter. These mechanisms include:

- attraction by ion exchange (where pH affects the charge of various groups),

- weak attractive forces like van der Waals forces, where fluctuating dipoles give rise to an instantaneous attraction between nonpolar molecules,

- coordination bonding (sharing electrons), and

- hydrogen-bonding (bonding of H^+ to an electronegative atom like O) as a result of protonation or water bridging.

In protonation, protons from the surface of the clay are transferred to the soil microorganism to make the soil microorganism neutral or positively charged. In water bridging, soil microorganisms form hydrogen bonds with water molecules that form part of the hydration shell of an exchangeable cation.

Soil Abiotic Factors

Soil Water

Soil water is essential for soil microorganisms. Without some water, there is no microbial activity. Soil water also affects gas exchange and a variety of soil chemical reactions (e.g., as a reactant in hydrolysis). Water in soil flows from an area of higher energy to an area of lower energy (this is an expression of the Second Law of Thermodynamics), and this spontaneous flow of water is measured in terms of a water potential. Therefore, a **water potential** is the measure of the potential energy (per unit mass or volume) of water at a point in a system relative to the potential energy of pure, free water.

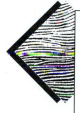

| Box 2–8 |

Kinetic versus Potential Energy. Energy is the ability to do work. Kinetic energy is the energy a body possesses because of its motion and mass. Potential energy is the energy a body possesses because of its position or arrangement with respect to other bodies. Therefore, potential energy is not a constant property but a relative measure with reference to an arbitrarily chosen zero level. The potential energy of an apple in a tree and the ground depends on the relative height of the apple to the ground. When the apple falls from a tree, the apple's kinetic energy increases and its potential energy decreases.

Because pure, free water is usually assigned a water potential of zero, and the water potential in soil is usually lower in potential energy than pure, free water, the water potential in soil is usually a negative number. The water potential of a soil is assigned the Greek letter Ψ (psi) and is the sum of various forces.

| Box 2–9 |

Water Potential versus Water Content. In addition to the energy term, Ψ, soil water can be measured in terms of its volumetric or gravimetric water content. In these cases, the Greek letter θ_v (theta) and θ_w are used, respectively.

Although there are many forces comprising the total water potential, the three major forces in soil are:

- osmotic potential (Ψ_π),

- matric potential (Ψ_m), and

- gravitational potential (Ψ_g).

The **osmotic potential** (Ψ_π) is primarily the attraction of solute ions for water molecules and is always a negative number. Because the osmotic potential arises from the dissolution of solutes (e.g., various salts), the potential is significant in saline soils or in soils amended with organic wastes or fertilizer. The **matric potential** (Ψ_m) is the sum of adsorption of water to the surfaces of soil particles and capillary forces arising from water being trapped in very fine pores. Like the osmotic potential, the matric potential is always a negative number. The matric potential is most significant in unsaturated soils. Water will move from a more saturated soil (high free energy; high potential) to a less saturated soil (low free energy; low potential).

Box 2–10

Capillary Water. To observe capillary water, place a fine glass tube in water. Because of the surface tension of water and the attraction of water molecules to the sides of the tube, the water rises in the tube according to the tube diameter. The smaller the tube, the greater the rise of water.

The **gravitational potential** (Ψ_g) is the force of gravity pulling water towards the earth's center and may have a positive or negative potential depending on the reference level of the water. If the reference level is the lower edge of the soil profile (the usual case), then the gravitational potential will be positive.

Box 2–11

Units for Expressing Water Potential. In practice, all energy potentials in soil, whether gravitational, matric, osmotic, or other potentials, are united in equivalents of pressure expressed as kiloPascals (kPa) or MegaPascals (MPa; a Pascal is a Newton m^{-2}; a Newton is a measure of force required to accelerate 1 kilogram 1 meter per second per second). Previously, soil water was measured in bars or atmospheres (1 bar = 0.987 atmospheres). To convert bars to kPa or MPa, multiply bars by 100 or 0.1, respectively (e.g., −1 bar = −100 kPa or −0.1 MPa). To convert atmospheres to kPa or MPa, multiply by 101 or 0.101, respectively.

Soil texture affects soil water relations (Fig. 2–9). Over the entire range of water pressure, a soil high in clay, with its greater percentage of micropores, retains a larger amount of water than a soil high in silt, which in turn retains a larger amount of water than a soil high in sand. These curves are called moisture characteristic curves, or **moisture release curves.** The moisture release curves of most agricultural soils lie between the curves for clay and sandy soils. If organic matter were added, the curve would shift upward. Recent soil microbiology literature reports both the soil water content and the water potential.

Although the water potential in soil is measured most accurately in terms of potential (or pressure), it may be measured less accurately in terms of its physical and biological characteristics. These terms have persisted because they are useful for relating soil water to plant growth. When a soil is saturated with water and the water is allowed to drain freely, the water drains only from the soil macropores. This is "gravitational water" and is of little use to plants because it reduces soil aeration (see

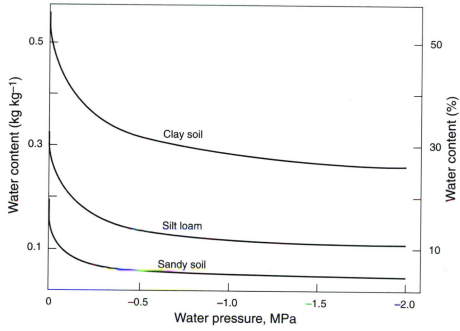

Figure 2–9 Water potential–water content relationship for a sandy soil, a silt loam, and a clay soil. *Adapted from Papendick and Campbell (1981). Used with permission.*

next section). When gravitational water and water that moves readily at high matric potential have drained (macropores now empty), the soil is at *field capacity* (also called water-holding capacity). In terms of water potential, field capacity is approximately -0.03 MPa (-33 kPa; -0.33 bar), except for sandy soils, where field capacity is approximately -0.01 MPa (-10 kPa; -0.10 bar). Most soil micropores are still full of water, which is available for plant growth. When a plant uses all of this water (micropores now empty), almost all water remaining in the soil is *hygroscopic water,* water bound too tightly to the soil solids for plants to use. At this point, plants permanently wilt and do not recover, even when water is added. This is the *permanent wilting point* and is approximately -1.5 MPa (-1500 kPa; -15 bar). While field capacity and permanent wilting point remain in use because they are still good terms for plant scientists to describe the upper and lower limits of plant available water, these terms should not be confused with water potential (or water pressure). They do not have the word "potential" (or pressure) associated with them.

| Box 2–12 |

Water Activity. The water requirement for microorganisms frequently was measured in water activity (a_w), which is the ratio of the vapor pressure of the solution over the vapor pressure of pure water at the same temperature. This term was used primarily in food microbiology. If a_w is multiplied by 100, it is the same as relative humidity (an a_w of 0.99 = 99% relative humidity). The use of a_w has now been replaced by water potential.

Table 2–3 **Microbial tolerance to matric–controlled (Ψ_m) water stress.**

Water potential (MPa)	Water activity (a_w)	Water film thickness	Microbial activity limited (example of genus)
−0.03	0.999	4.0 μm	movement of protozoa, zoospores, and bacteria
−0.1	0.999	1.5 μm	
−0.5	0.996	0.5 μm	
−1.5	0.990	3.0 nm	nitrification; sulfur oxidation
−4.0	0.97	<3.0 nm	bacterial growth (*Bacillus*)
−10.0	0.93	<1.5 nm	fungal growth (*Fusarium*)
−40.0	0.75	<0.9 nm	fungal growth (*Penicillium*)

Adapted from Harris (1981).

The water potential in soil also has a profound effect on soil microorganisms and soil microbial processes. Some microorganisms and processes are tolerant of moisture stress; others are not (Table 2–3). Also, as soil water becomes limited, microbial movement becomes limited. Some of these water-related effects are covered in more detail in later chapters.

Soil Aeration

Soil aeration is a measure of the oxygenation of the soil. Ideally, a well-aerated soil would have sufficient oxygen for the respiration of plant roots and the function of most aerobic microorganisms (i.e., **aerobes**). Under these conditions, roots and aerobic microorganisms oxidize organic compounds to CO_2. High CO_2 levels may indicate that the soil is poorly aerated. A poorly drained soil is not necessarily detrimental to all soil microorganisms; **facultative anaerobes** can grow both in the presence or absence of oxygen, whereas **obligate anaerobes** grow only in the absence of oxygen.

Soil aeration is highly dependent on soil moisture, soil texture, and soil porosity. The earth's atmosphere is the major source of oxygen; thus, oxygen can get into the soil only by **mass flow** or **diffusion.** Because mass flow is based on total air pressure differences, mass flow of oxygen into the soil is relatively unimportant beneath the top few centimeters of soil. Therefore, the major mechanism for replenishment of oxygen in the soil is diffusion. Soil texture affects this diffusion. If the soil has a high percentage of clay, then it will have a high percentage of micropores. The small diameter of the micropores will slow diffusion. Soil water will also affect diffusion of oxygen because the diffusion coefficient of oxygen in air is 0.189 $cm^2\ sec^{-1}$, but in water is only 0.000025 $cm^2\ sec^{-1}$ (e.g., Papendick and Campbell, 1981). In other words, oxygen diffuses through water 10,000 times more slowly than through air. This 10,000-fold difference means soil pores filled with water will reduce considerably the diffusion of oxygen into the soil. Because the moisture release curves are higher for clay soils than sandy soils, it is no surprise that clay soils, with their higher percentage of micropores, are often poorly aerated. Under these conditions, the CO_2 produced by soil animals, plant roots, and microorganisms accumulates, and it is possible to have clayey soils with CO_2 concentrations hundreds of times higher than the atmosphere.

Soil aeration is commonly measured in terms of a redox (*reduction—oxidation*) potential. The **redox potential** is the measure of the tendency of a compound to accept or donate electrons. As electrons are transferred, a potential difference is created, and this difference is measured in millivolts (mv); the potential itself is abbreviated

E_h. As a substance loses electrons (e.g., $Fe^{2+} \rightarrow Fe^{3+}$), it becomes more positive (more oxidized); as it gains electrons (e.g., $Fe^{3+} \rightarrow Fe^{2+}$), it becomes more negative (more reduced). The more oxidized (aerobic) a soil becomes, the more positive the millivolt reading. Waterlogged soils, especially those with an available carbon source, generally have a low redox potential, and these environments are conducive to anaerobic processes like methane production and sulfate reduction. In general, aerobic soils are "oxidizing" environments and anaerobic soils are "reducing" environments. For a more detailed description of redox reactions, refer to Chapter 10.

Box 2–13

Soil Aeration, Redox Potential, and Soil Color. There is an interesting relationship between soil aeration and soil color (a soil physical characteristic). When soil is well-aerated, elements like iron and manganese are oxidized and soil colors of red, yellow, and reddish-brown predominate. When soil is poorly aerated, these elements are reduced, and soil colors of blue and gray predominate. When soils have mixed zones of good and poor aeration, soils will have a mottled appearance.

Thus, the soil atmosphere differs from the overlying air in two important ways:

- it contains less oxygen and

- it contains a much higher concentration (10- to 100-fold) of carbon dioxide.

These differences arise from the respiration of roots, soil animals, and soil microbes as well as the physical constraints on diffusion.

Soil Temperature

Soil temperature greatly influences the rates of biological, physical, and chemical processes in the soil. It is well known that, within a limited range, the rates of chemical reactions and biological processes double for every 10°C increase in temperature. This is often stated as the "Q_{10} for biological systems" (i.e., $Q_{10} = 2$). In addition, soil temperature and soil moisture are inextricably linked. Water has a high specific heat; that is, it requires a considerable amount of energy to raise 1 cm^3 of water by 1°C. When water is added to soil, it is easy to understand how the high specific heat of water and the natural high density of soil combine to moderate rapid changes in soil temperature. This effect increases with soil depth. As soil depth increases, the temperature of the soil *below* the soil surface lags behind the temperature of the soil surface and the temperature fluctuation is reduced (Fig. 2–10). This is one important way that soil microbiology differs from aquatic microbiology: the mass of soil moderates the rapid fluctuation of environmental parameters more commonly found in aquatic systems (Alexander, 1977).

Integrating Soil Physical Characteristics and Soil Abiotic Factors with Microorganisms

Tillage is a mechanical stirring of the soil surface to provide a suitable environment for seed germination and root growth. In conventional tillage, almost 100%

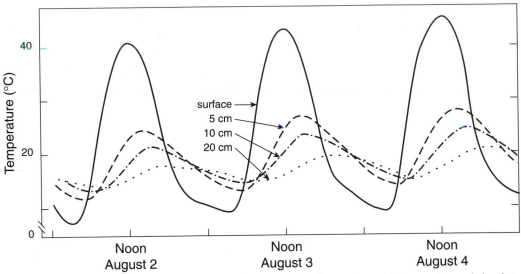

Figure 2–10 Increasing lag and moderation of temperature fluctuation with increasing soil depth in a bare soil from Rothamsted, U. K. *Adapted from Wild (1988). Used with permission.*

of the soil surface is overturned, usually with a moldboard plow (Fig. 2–11a and 2–11b). This helps to control weeds and other pests. In contrast, in reduced tillage, only a fraction of the soil surface is overturned, usually with a chisel plow (Fig. 2–11c). Although it is harder to control weeds and other pests with reduced tillage, the practice does leave large amounts of crop residues on the soil surface, which serve as a mulch to conserve moisture and reduce soil erosion. Also, by not turning over as much soil, reduced tillage saves time, fuel, labor, and equipment.

In a study of the effects of reduced tillage versus conventional tillage on soil microorganisms, soil microbiologists looked at several soil characteristics (Table 2–4, Part A; Linn and Doran, 1984). Because crop residues are substrates for the formation of organic matter, the total organic C and N are significantly higher in reduced till soils than in conventional till soils for the 0 to 7.5 cm depth. In addition, the crop residues, with their low particle and bulk densities, were not incorporated into the soil; consequently the bulk densities of the reduced till soils were significantly higher (i.e., the soils were more compact) than the conventional till soils at both soil depths. When combined with the mulching effect of organic matter, the higher bulk densities in the reduced till soils trap more of the soil moisture (and hence more water-filled pore space) than conventional till soils at both depths.

The increases in C and N (more nutrients) for the 0 to 7.5 cm depth of reduced till soil increase the bacterial populations of total aerobes, facultative anaerobes, and total anaerobes (Table 2–4, Part B). How were populations of anaerobes able to grow so near to the soil surface? Possibly, the higher amount of water-filled pore spaces prevented O_2 diffusion into the soil, and portions of the soil became more anaerobic (i.e., there were anaerobic microsites).

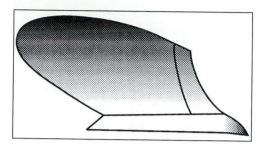

a.

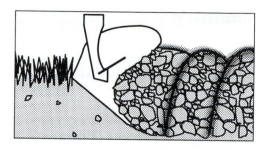

b.

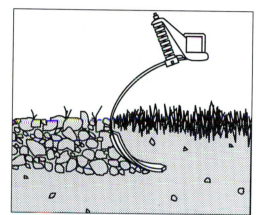

c.

Figure 2–11 A moldboard plow viewed from the side (a) and rear (b), and a chisel plow (c).

Table 2–4 **Physical and chemical (A) and microbiological (B) differences between reduced tillage and conventional tillage in soils at six locations**

Soil characteristic	Ratio of reduced tillage: conventional tillage at two soil depths	
	(0 to 7.5 cm)	(7.5 to 15.0 cm)
A. Physical and chemical properties		
Total organic C	1.41*	0.99
Total N	1.29	1.01
Bulk density	1.04	1.05
Water-filled pore space	1.28	1.11
B. Microbial group		
Total aerobes	1.35	0.66
Bacteria	1.41	0.68
Fungi	1.35	0.55
Facultative anaerobes	1.31	0.96
Total anaerobes	1.27	1.05

*A ratio of more than 1.00 means the value for reduced tillage was higher than conventional tillage.
Adapted from Linn and Doran (1984). Used with permission.

Summary

This chapter covered soil description and the principal soil characteristics and factors that define the soil habitat, the most complex of all microbial habitats. Soil description includes differences between mineral and organic soils, horizons, series name, and textural class; soil physical properties include bulk density, particle density, and pore spaces; soil chemical properties include pH and anion and cation exchange capacity; and soil abiotic factors include soil water, soil aeration, and soil temperature. Some of these characteristics and factors are integrated with soil microorganisms to illustrate how soil conditions can influence soil microorganisms. However, it is equally important to understand that soil microorganisms can, in turn, influence some of these conditions and characteristics positively or negatively. Knowing this interdependency helps explain the enormous heterogeneity in soil—how, within the space of a few millimeters, some microorganisms will grow and others will die and how some microbial processes will begin and others will stop. Knowing this interdependency is the key to understanding "soil" and, indeed, understanding soil microbiology.

Cited References

Alexander, M. 1977. Introduction to soil microbiology, 2nd ed. John Wiley & Sons, Inc., New York.

Brady, N.C., and R.R. Weil. 1996. The nature and properties of soils, 11th ed. Prentice Hall, Upper Saddle River, N.J.

Foth, H.D. 1990. Fundamentals of soil science, 8th ed. John Wiley & Sons, Inc., New York.

Harris, R.F. 1981. Effect of water potential on microbial growth and activity. pp. 23–95. *In* Parr, J.F., W.R. Gardner, and L.F. Elliott (eds.), Water potential relations in soil microbiology. SSSA Spec. Publ. 9. Soil Science Society of America, Inc., Madison, Wis.

Linn, D.M., and J.W. Doran. 1984. Aerobic and anaerobic microbial populations in no-till and plowed soil. Soil Sci. Soc. Am. J. 48:794–799.

Metzger, L., D. Levanon, and U. Mingelgrin. 1987. The effect of sewage sludge on soil structural stability: Microbiological aspects. Soil Sci. Soc. Am. J. 51:346–351.

Miller, R.W., and R.L. Donahue. 1995. Soils in our environment, 7th ed. Prentice Hall, Englewood Cliffs, N.J.

Papendick, R.I., and G.S. Campbell. 1981. Theory and measurement of water potential. pp. 1–22. *In* Parr, J.F., W.R. Gardner, and L.F. Elliott (eds.), Water potential relations in soil microbiology. SSSA Spec. Publ. 9. Soil Science Society of America, Inc., Madison, Wis.

Perkins, H.F., J.E. Hook, and N.W. Barbour. 1986. Soil characteristics of selected areas of the Coastal Plain Experiment Station and ABAC Research Farms. Georgia Agric. Exp. Stn. Res. Bull. 346.

Pignatello, J.J., and B. Xing. 1996. Mechanisms of slow sorption of organic chemicals to natural particles. Environ. Sci. Technol. 30:1–11.

Soil Conservation Service. 1959. Soil survey of Tift County, Georgia. U.S. Gov. Print. Office, Washington, D.C.

Soil Survey Staff. 1975. Soil taxonomy: A basic system of soil classification for making and interpreting soil surveys. USDA-SCS Agric. Handb. 436, U.S. Government Printing Office, Washington, D.C.

Stotzky, G. 1985. Mechanisms of adhesion to clays, with reference to soil systems. pp. 195–253. *In* D.C. Savage and M. Fletcher (eds.), Bacterial adhesion. Mechanisms and physiological significance. Plenum Press, New York.

Wild, A. 1988. Russell's soil conditions and plant growth. Longman Scientific & Technical, Essex, England.

Study Questions

1. A sample of the Eel soil series from upstate New York contains 158 g of clay kg^{-1} and 279 g of sand kg^{-1}. Using the soil triangle in Figure 2–2, determine the textural class of this soil.

2. The CEC of the Ap horizon of Tifton loamy sand is 2.6 cmol kg^{-1} of soil; the CEC of the Ap horizon of a Eel soil series is 10.6 cmol kg^{-1} of soil. What is CEC and how would the CEC differences between Eel and Tifton soils be important to soil microorganisms?

3. Discuss the source of electrical charge in (A) an aluminosilicate clay, (B) organic matter, and (C) a typical bacterium.

4. A 10 cm layer of sand is on top of a 10 cm layer of clay. Water is poured on the top of the sand. (A) Diagram how the water moves through the sand into the clay. (B) Which layer has the greater water potential and why?

5. A soil microbiologist is walking from one laboratory to another carrying a flask containing a sterile, nonselective broth for growing bacteria. He accidentally trips and falls. He's all right but the medium has spilled on a patch of bare soil. (A) In the portion of soil that has been wetted, how would the ability of soil microorganisms to physically and chemically bind the soil be affected? (B) What other soil properties and characteristics would be affected that would, in turn, affect soil microorganisms?

Chapter 3

♦

Bacteria and Archaea

David B. Alexander

♦

The true biologist deals with life, with teeming, boisterous life, and learns something from it, learns that the first rule of life is living.

John Steinbeck

The **domains** Bacteria and Archaea comprise a remarkably diverse group of single-celled, prokaryotic organisms which inhabit the soils of every terrestrial ecosystem—from the warm, moist, densely vegetated soils of tropical rain forests to the deserts, grasslands, and forests of temperate regions and the frigid tundra of the high latitudes. They are the smallest of all the cellular organisms that inhabit soils. Only the viruses, which exist not as cells but as inanimate particles outside of a susceptible host, are smaller. Despite their small size, the Bacteria and Archaea exhibit a greater variety of metabolic capabilities than any other groups of organisms and play crucial roles in soil formation, organic matter decomposition, remediation of contaminated soils, biological transformations of mineral nutrients in soils, mutualistic interactions with plants, and plant diseases.

Classification

The Bacteria and Archaea are difficult to classify because of their small size, limited range of morphological characteristics, complex array of physiological characteristics, and unique types of genetic recombination which make it difficult to determine genetic relationships. Although there is currently no universally accepted system for classifying these diverse microorganisms, each species does have one officially recognized name.

The most widely accepted system for classifying these single-celled, prokaryotic microorganisms is described in *Bergey's Manual of Determinative Bacteriology* (Holt, 1994). *Bergey's Manual* places them in the Kingdom *Prokaryotae* and describes four Divisions based upon the type of wall that the cells possess:

- Division I, *Gracilicutes* (prokaryotes with a **Gram-negative** cell wall),

- Division II, *Firmicutes* (prokaryotes with a **Gram-positive** cell wall),

- Division III, *Tenericutes* (prokaryotes that have no cell wall, commonly called **mycoplasmas**), and

- Division IV, *Mendosicutes* (prokaryotes with walls that do not contain the bacterial polymer **peptidoglycan;** previously called **archaebacteria**).

As noted in Chapter 1, current **phylogenetic** classification schemes place Divisions I-III in the domain Bacteria and Division IV in the Archaea. These two domains of life differ radically in many fundamental properties (Table 3–1), yet the generic term bacteria (note lower case) is commonly applied to both groups of prokaryotes. The Gram-negative and Gram-positive bacteria are the most abundant cellular organisms found in soil. The Archaea include microorganisms that grow in harsh environments (extreme halophiles and thermophiles) and strictly anaerobic methanogens, which can reduce carbon dioxide to methane gas.

The primary taxonomic unit in bacterial classification is the **species.** *Bergey's Manual* describes a bacterial species as "a collection of strains that share many features in common and differ considerably from other strains." In bacteriology, a **strain** is a culture of cells descended from a single pure isolate. This definition of a bacterial species is less explicit than the definitions of species of higher organisms because bacteria do not reproduce sexually.

Relationships among the higher taxa are often uncertain, and most microbiologists prefer to assign bacteria to descriptive groups that do not reflect a formal taxonomic ranking. The organization of *Bergey's Manual* reflects this type of

Table 3–1 **Differences between the domains Bacteria and Archaea.**

Character	Bacteria	Archaea
Lipids in membrane	Ester-linked, straight-chain fatty acids	Ether-linked, branched chain aliphatics
Cell walls	Peptidoglycan-muramic acids	Variety, no muramic acid
Transfer RNA	Thymine present	Thymine absent
Ribosome, response to:*		
Chloramphenicol	Sensitive	Insensitive
Kanamycin	Sensitive	Insensitive
Anisomycin	Insensitive	Sensitive
DNA-dependent, RNA polymerase		
Number of enzymes	One	Several
Structure	Simple subunit	Complex subunit
Rifamicin sensitivity*	Sensitive	Insensitive

*It is interesting to note that eukaryotic cells have antibiotic sensitivities similar to the Archaea.

Table 3–2 **Characteristics commonly used to classify bacteria.**

Characteristic	Description
Cellular morphology	Size, shape, and arrangement of cells; staining reactions; presence or absence of specific cellular structures.
Chemical characteristics	Chemical nature of cellular constituents.
Cultural characteristics	Nutritional and environmental requirements for growth; appearance of cultures in liquid or solid media.
Metabolism	Chemical reactions carried out by cells to satisfy nutritional and energy requirements.
Antigenic characteristics	Distinctive chemical components of cells that react specifically with antibodies produced by an animal.
Genetic characteristics	Base composition and nucleotide sequence of chromosomal or plasmid DNA (such as rRNA genes).
Pathogenicity	Ability to produce disease in plant or animal hosts.
Ecological characteristics	Normal habitat and distribution of the organism in nature; interactions with other organisms.

Adapted from Pelczar, Chan, & Krieg (1986).

descriptive grouping. The prokaryotes are grouped according to descriptive characteristics such as the "Gram-Negative Aerobic Rods and Cocci" and the "Endospore-forming Gram-Positive Rods and Cocci." The meanings of the terms used in these descriptions will be explained later in this chapter. Table 3–2 lists the characteristics that are commonly used to classify bacteria. Examples of some of the major groups of bacteria found in soils are listed in Table 3–3.

It is difficult to determine phylogenetic relationships among bacteria based upon physiological and morphological characteristics alone. Organisms that are only distantly related may exhibit similar phenotypes because they inhabit similar environments. The development in the past 30 years of new techniques for analyzing and comparing nucleic acids and proteins extracted from cells has greatly improved our ability to determine genetic relationships among bacteria. Information regarding genetic relationships remains fragmentary, but methods such as nucleic acid hybridization, ribosomal RNA oligonucleotide cataloging, and protein sequence analysis have greatly improved our knowledge in this area and promise to reveal new relationships in the years to come (Woese, 1994).

Structure and Function of Bacterial Cells

Cell Shapes

Most soil bacteria live as single cells, though some species form long, slender, branching filaments. The common shapes of bacterial cells found in soils are (Figs. 3–1 and 3–2):

- rod-shaped cells called rods or **bacilli** (singular, **bacillus**),

- spherical cells called **cocci** (singular, **coccus**),

- twisted or spiral-shaped rods called **spirilla** (singular, **spirillum**), and

- slender, branching filaments, collectively referred to as **actinomycetes.**

Table 3–3 **Major groups of Bacteria that include species which commonly inhabit soils. Descriptive names and taxonomic groups from** *Bergey's Manual of Determinative Bacteriology.*

Gram-Negative Aerobic Rods and Cocci	**Anoxygenic Phototrophic Bacteria**
Family I. Pseudomonadaceae	Purple Bacteria
Family II. Azotobacteriaceae	Family I: Chromatiaceae
Family III. Rhizobiaceae	Family II: Ectothiorhodospiraceae
Family IV. Methylococcaceae	Purple Nonsulfur Bacteria
Family V. Halobacteriaceae	Green Bacteria
Family VI. Acetobacteraceae	**Aerobic Chemolithotrophic Bacteria and**
Several other genera not grouped into families	**Associated Organisms**
Aerobic/Microaerophilic, Motile,	A: Nitrifying Bacteria
Helical/Vibrioid Gram-Negative	Family: Nitrobacteraceae
Bacteria	Includes *Nitrosomonas* and *Nitrobacter*
7 genera, not grouped into families	B: Colorless Sulfur Bacteria
Includes *Azospirillum* and *Bdellovibrio*	Includes *Thiobacillus* and *Thermothrix*
Facultatively Anaerobic Gram-Negative Rods	C: Obligate Hydrogen Oxidizers
Family I. Enterobacteriaceae	D: Iron and Manganese Oxidizing and/or
Several other genera not grouped into families	Depositing Bacteria
Dissimilatory Sulfate- or Sulfur-Reducing	E: Magnetotactic Bacteria
Bacteria	**Nocardioform Actinomycetes**
7 genera, not grouped into families	Form hyphae that segment into rod-shaped or spherical spores
Includes *Desulfovibrio, Desulfomonas,* and	**Actinomycetes with Multi-Locular**
Desulfococcus	**Sporangia**
Endosymbionts	Includes *Frankia,* an N_2-fixing actinomycete
A: Endosymbionts of Protozoa	**Actinoplanetes**
B: Endosymbionts of Insects	Actinomycetes that produce motile spores
C: Endosymbionts of Fungi and	**Streptomycetes and Related Genera**
Invertebrates other than Arthropods	Most common actinomycetes in soils
Endospore-forming Gram-Positive Rods	Produce several antibiotics and other use-
and Cocci	ful metabolites
6 genera, not grouped into families	**Thermonospora and Related Genera**
Includes *Bacillus* and *Clostridium*	Lignocellulose decomposers, important in
Irregular, Nonsporing, Gram-Positive Rods	primary degradation of organic matter
21 genera, not grouped into families	**Thermoactinomycetes**
Includes *Corynebacterium* and *Arthrobacter*	Thermophilic actinomycetes
Oxygenic Photosynthetic Bacteria	
Group I: Cyanobacteria	
Group II: Order Prochlorales	

Some bacteria that live in aquatic environments or in animal hosts exhibit additional shapes. Among these are the curved rods (vibrios), corkscrew-shaped cells (spirochetes), and long, slender filaments of rod-shaped cells enclosed within a sheath (sheathed bacteria).

Cell Grouping

In some species of bacteria, particularly among the cocci, individual cells often do not separate after undergoing cell division. Instead, the cells remain attached to one another and exhibit characteristic types of cell grouping. Cocci,

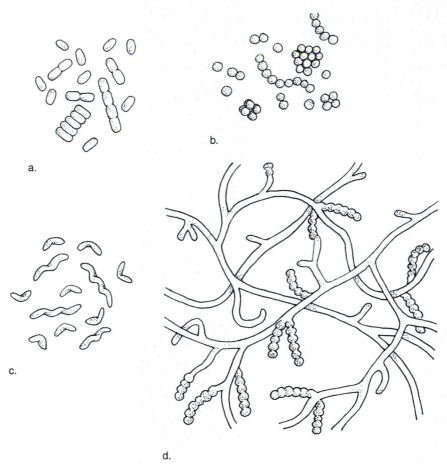

Figure 3–1 Common shapes and grouping of soil bacteria: (a) rods, occurring as single cells, in chains (streptobacilli), and palisade arrangement; (b) cocci, occurring as single cells and in chains (streptococci), irregular clusters (staphylococci), and tetrads; (c) spirilla; and (d) *Streptomyces,* an actinomycete, with chains of spores. *Original drawing by Kim Luoma.*

which consistently divide longitudinally, form chains of cells called **streptococci.** Those that divide randomly form irregular clusters of cells called **staphylococci.** A few species of cocci form planar packets of four cells, known as tetrads, or cuboidal packets of eight or more cells.

Some of the rod-shaped bacteria also exhibit characteristic types of grouping. Rods that remain attached end-to-end following cell division form chains of cells called **streptobacilli.** In some species the cells may align side-by-side rather than end-to-end, forming an arrangement called a *palisade.*

Several bacterial genera are named on the basis of cell shape and grouping, and it is important to recognize when a term is descriptive and when it is used as a formal name. When a term such as bacillus or staphylococcus denotes a genus, rather than the shape or grouping of bacterial cells, the name is capitalized and either italicized or underlined, as in the genus names *Bacillus* and *Staphylococcus.* Microbiologists also commonly use a nonitalicized, lower case form of a genus name

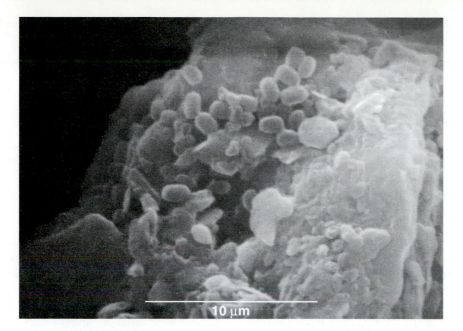

a.

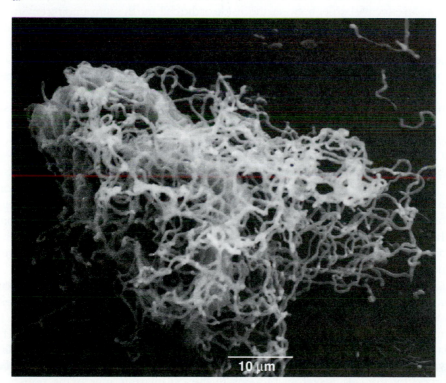

b.

Figure 3–2 Scanning electron micrographs of soil bacteria: (a) a colony of short, rod-shaped cells on the surface of a soil aggregate, (b) an actinomycete mycelium surrounding a soil particle. Compare the dimensions of the actinomycete hyphae with the rod-shaped cells in the lower right of the field. *From Dr. E. Florance, Lewis & Clark College. Used with permission.*

to refer to groups of bacteria in a less formal manner. As a result, bacteria of the genus *Rhizobium* are often referred to as rhizobia, and members of the genus *Pseudomonas* as pseudomonads.

Cell Structure

The bacteria differ from the other major groups of soil microbes in their distinctive cell structure and mechanisms of genetic recombination (bacterial recombination is discussed below). Bacteria are **prokaryotic.** They are much smaller than the **eukaryotic** soil microbes—the fungi, algae, protozoa, and microscopic animals—and lack the complex array of membrane-bound organelles found in eukaryotes. Most importantly, they lack a nucleus enclosed within a nuclear envelope.

Cytoplasmic membrane. The structure of a "typical" bacterial cell is shown in Fig. 3–3. As in all living cells, the cytoplasm of a bacterial cell is enclosed within a **cytoplasmic membrane** (also called **cell membrane**). The cytoplasmic membrane is a fluid structure consisting of two layers of phospholipid molecules oriented so that the lipids face one another in the interior of the membrane (Fig. 3–4). As a result, the interior of the membrane is **hydrophobic** (water-repellent) while the inner and outer surfaces are **hydrophilic** (water-soluble). This arrangement makes the cytoplasmic membrane an effective barrier between the inside of the cell and the surrounding environment.

The cytoplasmic membrane is the site of many essential functions in bacterial cells. These functions are carried out by protein molecules that are embedded within the membrane or closely associated with the membrane's inner or outer surface. One essential function of the cytoplasmic membrane is to transport nutrients into the cell, and waste products and certain types of enzymes out of the cell. Because bacteria are too small and rigid to engulf large particles of food, they must absorb dissolved nutrients from the soil solution. These nutrients enter the cell through channels formed by transport proteins that span the membrane. Only water, certain gases (O_2, N_2, and CO_2), and a few small lipid-soluble molecules can pass directly through the phospholipid bilayer by simple diffusion.

If the concentration of a nutrient in the soil solution is higher than the concentration inside a bacterial cell, then a bacterium does not have to expend energy to absorb the nutrient. Nutrients can enter the cell passively through a protein channel. The concentration of dissolved nutrients in soils is rarely high enough for passive absorption to occur. Most nutrients must be absorbed actively by processes that require energy (Ames, 1986).

The organic nutrients available to bacteria in soil are often large, complex polymers contained in the residues of dead plant or animal tissues. These substances cannot be absorbed through small protein channels in the cytoplasmic membrane in the same manner as simple water-soluble nutrients. To feed on these substances, bacteria must excrete hydrolytic enzymes into the soil solution. These **extracellular enzymes** break down the complex polymers to smaller molecules which can be absorbed by the cell. Secretory proteins in the cytoplasmic membrane transport the extracellular enzymes out of the cell (Pugsley, 1993). The manner in which the large organic polymers in soils are decomposed is described in Chapter 11.

Another essential function of the cytoplasmic membrane is to generate the energy that a bacterial cell needs to survive and grow. Bacteria that respire aerobically or anaerobically synthesize ATP by **oxidative phosphorylation** (Chapter 10), and the

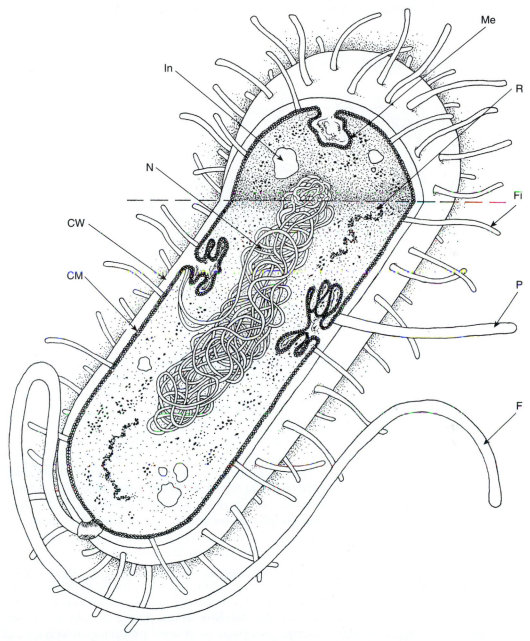

Figure 3–3 Generalized structure of a bacterial cell: CM = cytoplasmic membrane, CW = cell wall, N = nucleoid, In = inclusion, Me = mesosome, R = ribosomes occurring in the form of a polysome, Fi = fimbria, P = pilus, and F = flagellum. *Original drawing by Kim Luoma.*

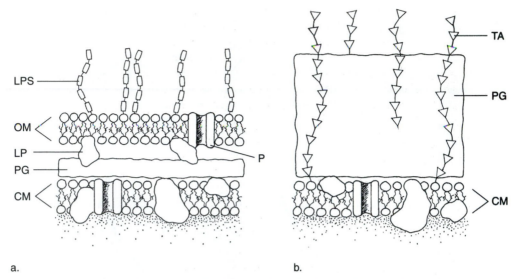

a.   b.

Figure 3–4 Structure of Gram-negative (a) and Gram-positive (b) bacterial cytoplasmic membranes and walls. CM = cytoplasmic membrane, PG = peptidoglycan, LP = lipoprotein, OM = outer membrane, LPS = lipopolysaccharide, P = porin, and TA = teichoic acid.

electron carriers that participate in these reactions are located in the cytoplasmic membrane. **Photosynthetic** bacteria capture the energy of sunlight and use it to synthesize ATP by a process known as **photophosphorylation.** The enzymes that catalyze these reactions are also located in the cytoplasmic membrane, or in small vesicles which extend from the cytoplasmic membrane.

Cell wall. A rigid **cell wall** surrounds the cytoplasmic membrane of all soil bacteria. Microbiologists divide the domain Bacteria into two large groups based upon the type of cell wall that they possess (Fig. 3–4, Box 3–1). In Gram-positive bacteria, the cell wall is a single thick layer composed largely of a rigid polymer called **peptidoglycan.** As the name implies, this polymer is part protein (peptide) and part polysaccharide (glycan). The polysaccharide portion of the molecule consists of long strands of alternating amino sugars, **N-acetylglucosamine** and **N-acetylmuramic acid.** The polysaccharide strands are cross-linked by short peptide chains, forming a rigid framework in much the same way as wooden beams form the framework of a house. This arrangement not only gives the cell wall its rigidity, but also makes the wall somewhat porous. Water and small soluble molecules and ions can pass directly through the cell wall to the cytoplasmic membrane. As a result, the cell wall is nonselectively permeable. Gram-positive cell walls also contain **teichoic acids,** polysaccharides composed primarily of repeating units of ribitol phosphate or glycerol phosphate, which extend outward from the wall into the surrounding medium. The function of teichoic acids is uncertain, but they may serve as recognition and binding sites in Gram-positive bacteria and provide a negatively charged cell surface for binding cationic nutrients.

The cell wall of Gram-negative bacteria has a more complex structure. A relatively thin layer of peptidoglycan surrounds the cytoplasmic membrane. The peptidoglycan layer is, in turn, enclosed within a second membrane known as the outer membrane. Numerous lipoprotein molecules anchor the outer membrane to the peptidoglycan layer.

The outer membrane differs from the cytoplasmic membrane in two important ways. The outer leaflet of the outer membrane (the outermost layer of the two layers of lipids that make up the membrane) is composed primarily of **lipopolysaccharides** rather than phospholipids. The polysaccharide portion of these molecules extends outward from the surface of the cell, giving these bacteria a distinctive "sugar-coating." These polysaccharides play an important role in the attachment of bacteria to soil particles and in their recognition by specific host plants. The outer membrane also contains numerous protein channels, called **porins,** which allow water and small dissolved molecules and ions to pass through the cell wall to the cytoplasmic membrane. Just as in the Gram-positive bacteria, then, the cell wall of Gram-negative bacteria is nonselectively permeable.

A gel-like **periplasm** fills the space between the cytoplasmic membrane and outer membrane of a Gram-negative bacterium. The gelatinous consistency of the periplasm results from the high concentration of proteins in this region of the cell. Periplasmic proteins perform several key functions in Gram-negative bacteria, such as catalyzing the initial steps in the hydrolysis of food molecules, transporting substances from the outer membrane to the cytoplasmic membrane, and conveying signals involved in chemotaxis (discussed later).

The primary function of the cell wall is to confer osmotic protection to soil bacteria. The concentration of dissolved substances in the cytoplasm of a bacterial cell (soluble nutrients, enzymes, lipids, nucleic acids, and so forth) is much higher than the concentration of dissolved substances in the soil solution. As a result, water diffuses into the cells and would cause the cells to burst if they were not enclosed by a rigid wall.

Box 3–1

The Gram Stain. The two common types of bacterial cell walls can be distinguished with a staining technique originally developed in 1884 by the Danish physician Hans Christian Gram. Gram experimented with various combinations of dyes and fixatives in an effort to make bacterial cells more easily visible in infected tissues. In the process, he discovered that different types of bacteria stain different colors when treated in a certain manner. The staining technique that he devised, now called the **Gram stain,** has great practical value for identifying and classifying bacteria.

To prepare a Gram stain, a smear of bacterial cells is first stained with crystal violet, a dye which imparts a deep purple color to the cells. After applying an iodine solution, which forms a complex with the crystal violet in the cytoplasm of the cell, the smear is carefully rinsed with a solution of acetone and ethanol. This solution removes the crystal violet-iodine complex from the cytoplasm of Gram-negative bacteria, but not from the cytoplasm of Gram-positive bacteria because of differences in the molecular structure of the peptidoglycan layer of the cell wall. The smear is then counterstained with a red dye called safranin. The counterstain makes Gram-negative bacteria appear pink, while Gram-positive bacteria appear dark purple.

Bacterial chromosome. The **chromosome,** or **nucleoid,** of a bacterial cell consists of a single, circular double-stranded DNA molecule which is suspended directly in the cytoplasm (Robinow and Kellenberger, 1994). Unlike the nucleus of a eukaryotic cell, the nucleoid in a bacterial cell is not enclosed within a nuclear envelope. The bacterial chromosome differs from eukaryotic chromosomes in several respects. The DNA of a bacterial chromosome is circular, whereas the DNA of a eukaryotic chromosome is linear. The DNA in a eukaryotic chromosome wraps around clusters of positively charged proteins (*histones*) to form threadlike structures called *chromatin* fibers. The DNA in a bacterial cell does not associate with histones in this manner, though studies have shown that the bacterial chromosome does contain histone-like proteins (Pettijohn, 1988). These proteins enable the large, circular molecule to twist and fold extensively to form a compact structure which occupies a distinct region within the cytoplasm. The bacterial chromosome is often attached to the cytoplasmic membrane at one or more sites, including the site at which DNA replication begins (known as the *origin*).

The chromosome contains the essential genetic information of a bacterial cell. Most of the genes in the nucleoid encode proteins that perform specific functions for the organism. Many of the proteins form structural components of the cell, while others catalyze chemical reactions. Some genes, such as those that encode **constitutive enzymes** (Chapter 10), are translated into proteins continuously, whereas others encode proteins that are produced only when they are needed by the organism. **Inducible enzymes** (Chapter 10), for instance, are produced only when the protein they encode is needed.

Most, if not all, bacteria possess one or more small DNA molecules called **plasmids** in addition to the DNA of the chromosome. Plasmids are circular molecules, like the DNA of the nucleoid, but they are much smaller and contain much less information than the nuclear DNA. Often they contain only a few genes that confer specific phenotypic characteristics on the bacteria carrying them, such as resistance to an antibiotic or toxic metal or the ability to break down a specific substrate (Couturier et al., 1988). Plasmids are dispersed throughout the cytoplasm of a bacterial cell and can replicate even when the cell itself is not reproducing. As a result, bacterial cells may contain multiple copies of one or more plasmids. Plasmids can be passed from generation to generation as a population of cells grows, and some can be transferred from cell to cell, even between different species of bacteria, during conjugation (discussed later in this chapter).

Cytoplasmic structures. The cytoplasm of a bacterial cell lacks the complex array of membrane-bound organelles which fills the cytoplasm of a eukaryotic microbe. Nonetheless, it is a region of intense biochemical activity. Much of this activity is mediated by the thousands of **ribosomes** that fill the cytoplasm. Ribosomes are small particles composed of granular proteins and ribosomal RNA (rRNA) which catalyze protein synthesis. They often form structures known as **polysomes** (strings of ribosomes attached by strands of mRNA) when a bacterium is actively synthesizing proteins. Bacterial ribosomes are smaller than the ribosomes in the cytoplasm of eukaryotic microbes, and they do not attach to the membranes of an endoplasmic reticulum (remember, prokaryotes do not have an endoplasmic reticulum). The difference in size is expressed in Svedberg (S) units, based upon the rate at which the particles sediment when centrifuged. Bacteria contain 70S ribosomes, whereas eukaryotic microbes contain larger 80S ribosomes in the cytoplasm and 70S ribosomes in mitochondria and chloroplasts.

Many bacteria store nutrients in the cytoplasm when an excess supply of the nutrient is available in the soil. The nutrients are stored as droplets or granules called *inclusions* which can later be hydrolyzed when the nutrient supply becomes depleted. Several species of bacteria store excess carbon in the form of poly-β-hydroxybutyrate, a lipid-like substance, or glycogen, a glucose polymer. Some store excess phosphorus as polyphosphate granules, also called metachromatic granules because they appear red under a light microscope when stained with a blue dye such as methylene blue. Some sulfur-oxidizing bacteria produce elemental sulfur granules, which they can use as a source of energy when external supplies of reduced sulfur become depleted.

A few species of bacteria, most notably members of the genera *Bacillus* and *Clostridium,* produce a unique structure called an **endospore** in response to nutrient depletion or other environmental stresses. Unlike the reproductive spores produced by the fungi and actinomycetes, bacterial endospores do not serve to increase the number of individuals in a population. Endospores are *survival* structures, not *reproductive* structures. Each bacterial cell in a population of spore formers is transformed into a single endospore during a period of stress. When favorable conditions for growth return, each endospore germinates to form a single cell, so there is no increase in the number of individuals in the population. Endospores are specially constructed to enable a bacterial population to withstand harsh environmental conditions, including extremely high temperatures, desiccation (drying), radiation, and exposure to toxic chemicals (Box 3–2).

Box 3–2

Twenty-Five-Million-Year-Old Bacteria? The bacterial endospore is probably nature's consummate survival structure. Recently, scientists succeeded in reviving endospores of what they believe to be *Bacillus sphaericus* which were contained in the stomach contents of an amber-encased, extinct Dominican bee (*Proplebeia dominicana*). The amber and its contents range from 25 to 40 million years old. In order to avoid contamination from modern-day bacteria in the laboratory, the amber was rigorously surface-sterilized before the bacteria were aseptically removed from the stomach contents of the bee. This discovery (and others like it) are exciting for microbiologists and biologists in general for two reasons. First, they extend our knowledge about the remarkable persistence of microbes on earth and, second, they confront us with the possibility of studying the evolution of bacterial species, through modern nucleic acid methods, over vast stretches of time. If this research is confirmed, then it will truly expand our thinking about the durability of life on earth.

Surface structures. Bacterial cells possess a variety of surface structures which play important roles in their activities in soils. An organism's survival in soil often depends on its ability to adhere to the surface of soil particles or to the cells of other organisms with which it forms a mutualistic or antagonistic association. Many soil bacteria produce structures known as **pili** (singular, **pilus**) or **fimbriae** (singular, **fimbria**) which enable them to adhere to surfaces. Both structures are slender protein filaments which extend through the cell wall into the surrounding medium. Fimbriae are shorter and more numerous than pili and seem to function primarily in adhesion. Pili are longer and fewer in number in most species (typically only one or a few per cell). A special type of pilus, known as a **sex pilus,** enables compatible cells to adhere and draw together during conjugation (see mechanisms of gene transfer).

Fimbriae and pili, along with certain components of the cell wall, also function in recognition between bacteria and other soil microbes or between bacteria and plants. Protozoa that feed on bacteria initiate **phagocytosis** by recognizing and binding specific macromolecules (polysaccharides or proteins) on the surface of bacterial cells (Ofek et al., 1992). Viruses that infect bacteria recognize susceptible hosts and initiate infection by attaching to specific macromolecules on the surface of bacterial cells (Madigan et al., 1997). Bacteria that colonize plant roots recognize and bind to specific macromolecules on the surface of root cells (Clarke et al., 1992). Fimbriae and pili serve as recognition sites for these types of interactions, as do the lipopolysaccharides in Gram-negative cell walls and the teichoic acids in Gram-positive cell walls. These interactions are often specific because the chemical composition of these structures varies among different species of bacteria. A variation of just one or a few sugars in a polysaccharide, for instance, can make the difference between recognition and nonrecognition.

Bacteria must contend with wide fluctuations in the moisture content of soils, and they must avoid predation by larger soil microbes and infection by lytic viruses in order to survive. Many species excrete a layer of polysaccharides and glycoproteins, or **capsule,** onto the cell surface to satisfy these requirements. The structure and consistency of the capsule varies from a thin slime layer to a thick gel, depending on the species of bacteria and the type and availability of organic nutrients in the soil. The capsule enables the organism to avoid desiccation and masks surface structures that phagocytic microbes would otherwise use to recognize and engulf the bacteria and that viruses would use to initiate an infection. Thick, sticky capsules may also play a role in the attachment of bacterial cells to surfaces and in the formation of biofilms.

To persist in the soil environment for long periods of time, bacteria must be able to adapt to changing environmental conditions. They can sense and respond to changes in the environment through the activities of receptor proteins in the cytoplasmic membrane and regulator proteins in the cytoplasm. Membrane receptors bind signal molecules at the cell surface and transmit a signal to the interior of the cell by activating regulator proteins in the cytoplasm (Stock et al., 1989). The signal molecule may be a nutrient that the organism can use, a toxic substance which has begun to accumulate outside the cell, an ion which indicates a change in pH or reduction potential, or any of a number of other signals. The bacteria may respond by moving toward an increasing concentration of the nutrient or away from a toxic substance (if they are motile), by initiating sporulation (if they are endospore formers), or by altering their metabolic activities.

The ability to "swim" in the soil solution from one location to another, an activity known as **motility,** is an important characteristic of many soil bacteria. Depending on soil moisture conditions, motility enables unattached bacteria to move from a site where conditions are less favorable for growth to a site where conditions are more favorable, though the distances traveled through the soil are not great. Most motile bacteria swim by rotating a long, slender, helically shaped protein filament called a **flagellum** (plural, **flagella**). Some species of bacteria produce a single flagellum, whereas others produce two or more in various characteristic arrangements. Flagella are anchored to the cytoplasmic membrane and extend outward through the cell wall into the surrounding medium (Shapiro, 1995). Motor proteins in the cytoplasmic membrane generate the energy needed to rotate the flagella by hydrolyzing ATP.

Movement of bacteria toward a chemical attractant (such as an essential nutrient or a more favorable oxygen concentration), or away from a chemical repellent (such as a toxic substance or an unfavorable pH), is referred to as **chemotaxis** (from *chemo,* meaning "chemical," and *taxis,* meaning "movement").

Mechanisms of Gene Modification and Transfer

Gene Modification

The primary source of gene modification is **mutation,** which can be defined as a heritable change in the base sequence of DNA. These random changes are the underlying source of genetic variation in microorganisms. Many mutations are harmful to the organism, although beneficial changes occur occasionally. Mutations occur spontaneously, or they may be induced in the laboratory. Spontaneous mutations result from errors in DNA replication or repair and occur about once in 10^6 replications. In other words, there is one chance in a million that a mutation will arise at some location in a given gene during one cell cycle (Madigan et al., 1997). The rate of mutation can be increased by exposing microorganisms to external agents such as chemicals (e.g., 5-bromouracil or nitrosoguanidine) or radiation (UV, X-rays, or gamma rays). Induced mutations have been used to produce useful phenotypes for soil microbiology research (Box 3–3).

| Box 3–3 |

Dual Antibiotic Resistance. To generate mutants which can be traced in natural environments, a soil microbiologist exposes a bacterial culture to UV radiation. The cells are then grown in the presence of two antibiotics, such as rifampicin and nalidixic acid, to select mutants that have acquired resistance to both compounds. The resistant strains are inoculated into soil and selectively recovered by adding dilute soil suspensions to a medium that contains the two antibiotics. The probability that the natural population will have resistance to both antibiotics is approximately 1 in 10^{12}. In this manner, a soil microbiologist can obtain information about the growth and survival of selected bacteria in soil. Other methods for marking and tracing bacteria are discussed in Chapter 9.

Another important mechanism of gene modification is **transposition,** a process in which copies of a DNA sequence move from one location to another on the same chromosome or between a chromosome and a plasmid. The ability to move in this manner is a property of genetic elements known as **transposable elements.** Insertion sequences (IS elements), the simplest of all transposable elements, are approximately 1,000 nucleotides in length and carry the minimal information required to duplicate themselves and integrate into DNA molecules. **Transposons** are larger transposable elements that carry genes other than those directly involved in the transposition process. Transposons insert randomly into the chromosome at sites that may be within or in close proximity to a gene. As a result, they inactivate or modify the characteristics of this gene.

Gene Transfer

Genetic material can be transferred from one bacterium to another by several mechanisms. Recent studies have shown that gene transfer occurs in soils (Veal et al., 1992), though the frequency of transfer is not known. It is also not known which transfer mechanisms contribute most to the overall rate of gene transfer. The frequency of gene transfer in natural populations is of great interest to microbiologists who study microbial diversity and to those who study the fate of microorganisms (or genes) introduced into soil for the purposes of bioremediation or biological control. In all prokaryotic mechanisms of gene transfer, the genetic material is transferred in one direction—from a donor cell to a recipient cell.

Transformation occurs when extracellular DNA (e.g., DNA released from lysed cells) is taken up by a bacterium and integrated into its chromosome by displacing a homologous segment of chromosomal DNA. Cells that can take up DNA from the environment are said to be **competent.** These cells synthesize specific proteins that enable them to bind, transport, and integrate DNA fragments during a specific stage of growth. Only a few bacterial species are naturally competent, including *Bacillus subtilis,* a common soil bacterium, and *Streptococcus pneumoniae,* a bacterial pathogen. Competence can be induced in other species under laboratory conditions. It is not known what proportion of soilborne species are transformable or whether soil conditions might induce or prolong the transformable state. Certainly, factors that affect the persistence of extracellular DNA play a pivotal role in the frequency with which transformation occurs.

Transduction is a process in which genes are transferred between bacterial cells by viruses known as **bacteriophages,** or simply **phages** (Chapter 7). Soils harbor many viruses that can attack soilborne bacteria (Veal et al., 1992). Some phages infect only one species, whereas others are capable of infecting several species. Viruses that do not immediately lyse the bacterial host, known as **temperate phages,** often integrate into the chromosome of the cell they have infected. If the bacterial DNA subsequently becomes damaged, the repair process can cause the integrated virus to replicate, resulting in lysis of the host cell. During viral replication, bacterial genes (or segments of genes) may be mistakenly incorporated into new virus particles and transferred to the next bacterial cell that the virus infects. Not enough is known about the natural frequency of transduction, or about the prevalence, survival, and host range of soil viruses to assess the role of transduction in gene transfer among soil bacteria.

Conjugation requires direct physical contact between bacterial cells. The genetic material transferred may be a plasmid, or it may be a portion of a chromosome that is mobilized by a plasmid. Many plasmids, termed **conjugative plasmids,** carry the genes that control their own transfer by conjugation. The complex process of conjugation involves cell wall fusion between cells and the transfer of DNA from donor to recipient. In the donor cell, one strand of the DNA is nicked and transferred through a conjugation tube into the recipient. DNA complementary to the transferred strand is synthesized in the recipient cell, while a strand complementary to the "stay-at-home" strand is synthesized in the donor cell. Some plasmids (**episomes**) can integrate stably into the chromosome of the host cell and at a later time initiate the transfer of chromosomal genes from the donor cell to another recipient. Usually only a part of the donor chromosome is transferred during conjugation, and the transferred genes must become integrated into the recipient chromosome to be stably inherited. Although conjugation and transfer of plasmids has been measured in soil (Veal et al., 1992), it is unclear to what extent this process drives gene flow in native soil populations.

Nutrition and Metabolism of Soil Bacteria

Nutritional Requirements

Bacteria have many of the same nutritional requirements as higher organisms. A **nutrient** is any substance that a bacterium must obtain from its surroundings to survive and grow. Some nutrients are assimilated into cellular constituents, while others are transformed to obtain energy. Substances required in large quantities are called **macronutrients.** These elements often occur as structural components of the biological molecules that are abundant in bacterial cells: carbohydrates, proteins, lipids, and nucleic acids. Carbon is required in the greatest amount, followed by nitrogen, phosphorus, and sulfur. Water is generally not regarded as a "nutrient" in the same sense as these elements, yet it is essential for all living cells and it serves as the primary source of hydrogen and oxygen. These elements account for about 95% of the dry mass of bacterial cells. Potassium, sodium, calcium, and magnesium are also required in substantial quantities.

Micronutrients are required in lesser amounts (a few ng g^{-1} dry mass). These elements often serve as structural components or activators of specific enzymes in bacterial cells. Iron is the micronutrient that is usually required in the greatest amount. Others include cobalt, zinc, molybdenum, copper, and manganese.

Some bacteria can synthesize all necessary cellular constituents and obtain energy for growth using a single carbon source and a few mineral nutrients. Others have more complex nutritional requirements. Many soil bacteria require specific organic compounds that they are unable to synthesize from simple starting materials. Organic nutrients of this type are called **growth factors** and are usually classified into one of the following categories: amino acids (precursors for protein synthesis), purines and pyrimidines (precursors for nucleic acid synthesis), and vitamins (precursors for the synthesis of certain types of enzymes).

Sources of Carbon and Energy

Most soil bacteria are **chemoheterotrophs.** They obtain energy by oxidizing the organic matter in soils and use the products of energy metabolism or other organic compounds as sources of carbon for growth (Table 3–4, see Chapter 10 for a more complete discussion of microbial metabolism). Most of these bacteria are **saprophytes** which feed on nonliving plant and animal residues or on humic substances in soils. A few species are **symbionts** or **pathogens** which invade the tissues of other living organisms. **Symbiotic** bacteria feed on organic compounds in the tissues of a host organism and usually benefit the host. **Pathogenic** bacteria harm the hosts whose tissues they invade, thereby producing disease.

Several genera of soil bacteria are **photoautotrophs.** They capture light energy and use it to synthesize carbohydrates (photosynthesis) with carbon dioxide as their source of carbon. The **cyanobacteria** resemble eukaryotic algae (and plants) in the manner in which they use light energy and water to fix

Table 3–4 **Major nutritional groups of soil bacteria.**

Nutritional classification	Sources of carbon and energy	Representative groups
Chemoheterotrophs	Carbon source: Organic	Saprophytic bacteria
	Energy source: Organic	Most symbiotic bacteria
Photoautotrophs	Carbon source: Carbon dioxide	Cyanobacteria
	Energy source: Light energy	Green bacteria
		Purple bacteria
Chemoautotrophs	Carbon source: Carbon dioxide	Nitrifying bacteria
	Energy source: Inorganic	Sulfur-oxidizing bacteria
		Hydrogen-oxidizing bacteria

carbon dioxide (forming oxygen in the process). The green bacteria produce green photosynthetic pigments and use reduced sulfur compounds (sulfide or thiosulfate) or hydrogen gas as electron donors. The purple bacteria produce purple photosynthetic pigments and are divided into two groups: (i) purple sulfur bacteria which use reduced sulfur compounds, hydrogen, or organic compounds as electron donors, and (ii) purple nonsulfur bacteria which generally do not use reduced sulfur as an electron donor and oxidize hydrogen or organic compounds when growing photoautotrophically. The green and purple bacteria grow photoautotrophically only under anaerobic conditions. A few species within each group can grow aerobically in the dark by oxidizing organic compounds for energy. Many of the purple bacteria, but none of the green bacteria, can also grow photoheterotrophically, using light energy to assimilate organic compounds as their primary source of carbon.

Several important groups of soil bacteria are **chemoautotrophs,** also known as **lithotrophs.** They obtain energy by oxidizing reduced inorganic compounds in soil and use carbon dioxide as their source of carbon to synthesize carbohydrates.

Two groups of chemoautotrophic bacteria, collectively known as the **nitrifying bacteria,** obtain energy by oxidizing inorganic nitrogen in soil (Table 3–5). The **ammonia-oxidizing bacteria** (*Nitrosomonas* and related genera) obtain energy by oxidizing ammonia (NH_3) to nitrite (NO_2^-). They are usually accompanied in soils by **nitrite-oxidizing bacteria** (*Nitrobacter* and related genera) which obtain energy by oxidizing nitrite to nitrate (NO_3^-). The interaction between these populations of soil bacteria is **synergistic** (Chapter 8). The nitrite-oxidizing bacteria benefit from the production of nitrite by the ammonia-oxidizing bacteria, though they could survive without the ammonia-oxidizers as long as an alternate source of nitrite was available. The ammonia-oxidizing bacteria benefit from the removal of nitrite ions from the soil solution by the nitrite-oxidizers. Both groups of nitrifying bacteria are primarily obligate aerobes, and both synthesize carbohydrates using carbon dioxide as a source of carbon. The nitrifying bacteria play a crucial role in the cycling of nitrogen in soil ecosystems and are discussed in more detail in Chapter 12.

The **sulfur-oxidizing bacteria** obtain energy by oxidizing sulfides, elemental sulfur, or thiosulfate. The oxidation of these substances results in the production of substantial amounts of sulfuric acid, significantly lowering the pH of the microenvironment in which sulfur oxidation occurs. Most of the sulfur-oxidizing bacteria are obligate aerobes, though at least one species (*Thiobacillus denitrificans*) is a facultative anaerobe that can use nitrate as a terminal electron acceptor. Most are also au-

Table 3–5 **Energy-yielding reactions of chemoautotrophic soil bacteria.**

Nitrifying bacteria

Ammonia-oxidizers $2NH_3 + 3O_2 \rightarrow 2NO_2^- + 2H^+ + 2H_2O$

Nitrite-oxidizers $2NO_2^- + O_2 \rightarrow 2NO_3^-$

Sulfur-oxidizing bacteria

Thiobacillus thiooxidans $2S^\circ + 3O_2 + 2H_2O \rightarrow 2SO_4^{2-} + 4H^+$

Thiobacillus ferrooxidans $2FeS_2 + 7O_2 + 2H_2O \rightarrow 2Fe^{2+} + 4SO_4^{2-} + 4H^+$

$4Fe^{2+} + 4H^+ + O_2 \rightarrow 4Fe^{3+} + 2H_2O$

Thiobacillus denitrificans $5S^\circ + 6KNO_3 + 2H_2O \rightarrow K_2SO_4 + 4KHSO_4 + 3N_2$

Hydrogen-oxidizing bacteria $2H_2 + O_2 \rightarrow 2H_2O$

$2H_2 + NAD^+ \rightarrow NADH + H^+$

totrophs, though some can grow heterotrophically, using the energy derived from the oxidation of inorganic sulfur compounds to assimilate organic carbon. Several sulfur-oxidizing bacteria are **obligate acidophiles** that not only tolerate the acid they produce but are actually unable to grow at pH greater than 4. The sulfur-oxidizing bacteria play an important role in the cycling of sulfur in soil ecosystems and are discussed further in Chapter 15.

The **hydrogen-oxidizing bacteria** are **facultative lithotrophs** which normally grow as chemoheterotrophs when organic substrates are available in the soil. In the absence of an oxidizable organic substrate, they can oxidize hydrogen for energy and synthesize carbohydrates with carbon dioxide as their source of carbon.

Oxygen Requirements

Oxygen concentrations in soils can vary widely from one microsite to another. Bacteria in a **macropore** filled with air may find plenty of oxygen to respire aerobically, while just a few millimeters away bacteria in a **micropore** filled with water may encounter strict anaerobic conditions. Oxygen concentrations can also vary widely with time, as soils undergo periods of saturation with water which may lead to temporary hypoxic or anoxic conditions and periods of dryness. It is not surprising, then, that soil bacteria exhibit a full range of adaptations to different oxygen concentrations.

Many soil bacteria are **obligate aerobes.** Obligate aerobes obtain energy exclusively by **aerobic respiration** and can only grow in microsites where oxygen is available to use as a terminal electron acceptor. **Obligate anaerobes** cannot survive in these sites because they lack the enzymes that are needed to rid themselves of the toxic products (hydrogen peroxide and superoxide) that are initially formed when oxygen serves as an electron acceptor. Obligate anaerobes grow exclusively in anaerobic microsites, obtaining energy either by **fermentation** or by **anaerobic respiration.**

Soil microbiologists recognize several groups of obligate anaerobes that generate energy by anaerobic respiration (Table 3–6). The **sulfate-reducing bacteria** use sulfate (SO_4^{2-}) as a terminal electron acceptor and produce hydrogen sulfide (H_2S) as a gaseous waste product. Along with the sulfur oxidizers, these bacteria play an important role in the sulfur cycle. The **methanogenic bacteria** use carbonate (CO_3^{2-}) as a terminal electron acceptor and produce methane (CH_4) as a gaseous waste product. The **acetogenic bacteria** also use carbonate as a terminal electron acceptor, but produce acetic acid (CH_3COOH) as a waste product rather than methane.

Table 3–6 Terminal electron acceptors utilized by bacteria that respire anaerobically.

Obligate anaerobes

Sulfate-reducing bacteria $SO_4^{2-} + 8e^- + 10H^+ \rightarrow H_2S + 4H_2O$

Methanogenic bacteria $CO_3^{2-} + 8e^- + 10H^+ \rightarrow CH_4 + 3H_2O$

Acetogenic bacteria $2CO_3^{2-} + 8e^- + 12H^+ \rightarrow CH_3COOH + 4H_2O$

Facultative anaerobes

Denitrifying bacteria $2NO_3^- + 10e^- + 12H^+ \rightarrow N_2 + 6H_2O$

Fumarate-respiring bacteria $^-OOC\text{-}CH=CH\text{-}COO^- + 2e^- + 2H^+ \rightarrow {}^-OOC\text{-}CH_2CH_2\text{-}COO^-$

Many soil bacteria can grow either in the presence or absence of oxygen. **Facultative anaerobes** respire aerobically when oxygen is available, but can alter their metabolism to grow anaerobically in the absence of oxygen. Some facultative anaerobes shift to fermentative metabolism under anaerobic conditions, whereas others shift to anaerobic respiration using nitrate or other inorganic compounds as the terminal electron acceptor. In either case aerobic respiration is the preferred mode of metabolism because it provides the bacteria with more energy than either fermentation or anaerobic respiration.

Aerotolerant anaerobes grow under both aerobic and anaerobic conditions, but do not shift from one mode of metabolism to another as conditions change. They lack the electron transport proteins that function in aerobic and anaerobic respiration and obtain energy exclusively by fermentation. Because they do not form toxic products by using oxygen as an electron acceptor, aerotolerant anaerobes are not poisoned by oxygen in the same manner as obligate anaerobes.

Ecology of Soil Bacteria

Every organism has certain requirements for growth and survival and lives in an environment that provides the necessary physical conditions and nutrients. The place that an organism inhabits is called its **habitat.** The functional role of an organism within its habitat is described as its **niche.** A complete description of an organism's niche requires a detailed understanding of the influence the organism has on its environment and the manner in which the environment affects the organism. Bacteria inhabit a wide variety of habitats and occupy many essential niches in soil ecosystems because of their extraordinary physiological diversity.

Trophic Levels

Each time the energy and nutrients in the tissues of an organism are consumed by another organism, they enter a higher **trophic level.** The first trophic level in any ecosystem consists of the *primary* **producers,** organisms that add biomass to the ecosystem by synthesizing organic molecules from carbon dioxide and simple inorganic nutrients. Photoautotrophic and chemoautotrophic bacteria function as producers in many soil ecosystems, though the amount of biomass that they add to soils is small compared to the input from plants.

Because they are too small to prey or graze upon other organisms, the primary role of soil bacteria is to function as **decomposers.** By feeding on nonliving organic matter in soils, saprophytic bacteria convert the complex organic molecules in plant and animal residues to carbon dioxide, water, ammonia, and other simple inorganic nutrients, thereby returning these nutrients to the soil in a form that plants and soil microbes can use. The return of inorganic nutrients to the soil in this manner is called **mineralization.** Of course, not all of the nutrients in decomposing organic matter are released for other organisms to use. Bacteria use many of the nutrients themselves to support their own growth and survival. The assimilation of inorganic nutrients into microbial biomass, rendering the nutrients unavailable to plants, is termed **immobilization.**

Nutrient Cycling and Soil Formation

Through their various metabolic activities, bacteria play critical roles in the cycling of several key inorganic nutrients in soils. Dinitrogen-fixing bacteria add nitrogen to soil ecosystems in a biologically usable form by converting dinitrogen to ammonia. Nitrifying bacteria convert ammonia to nitrate, an inorganic ion which is a source of nitrogen for plants but which is also susceptible to leaching and denitrification. Denitrification occurs when facultative anaerobes use nitrate as the final electron acceptor in anaerobic respiration, producing nitrogen, N_2O, and other nitrogenous gases that diffuse out of the soil. Sulfate-reducing bacteria have a similar role in the sulfur cycle. These obligate anaerobes cause sulfur to become unavailable to plants by converting SO_4^{2-} to H_2S as they respire anaerobically with SO_4^{2-} as their final electron acceptor. The H_2S produced by these bacteria usually precipitates as insoluble metal sulfides. Sulfur-oxidizing bacteria regenerate SO_4^{2-} by oxidizing reduced sulfur compounds in soils. Bacteria also regulate the availability of phosphorus in soils by solubilizing and immobilizing inorganic phosphates. The roles of bacteria in organic matter decomposition and in the cycling of nitrogen, sulfur, and phosphorus in soil ecosystems are discussed in greater detail in Chapters 11 through 16.

Bacteria rapidly metabolize the sugars, starches, and simple proteins in soil organic matter, but decompose other substances, such as the lignins, waxes, and oils in plant residues, much more slowly. These **recalcitrant** compounds, along with the products of microbial metabolism that resist further decomposition, become part of the stable organic fraction of soils known as **humus.** Humus improves soil structure, and its ability to retain nutrient ions and water greatly enhances the soil's capacity to support plant growth. Nutrients are slowly released from this fraction of soil organic matter as specific groups of bacteria, notably the actinomycetes, gradually break down its complex constituents.

Bacteria also contribute to soil formation and soil structure by producing acids that facilitate the weathering of soil minerals. Bacteria produce a variety of organic and inorganic acids as by-products of their metabolism that dissolve parent material, releasing soluble materials into the soil solution and leading to the formation of fine soil particles. Remember also that bacteria and other organisms excrete polysaccharides that help to cement soil particles into stable aggregates.

Habitats

Bacteria exhibit an exceptional variety of adaptations that enable them to survive and grow in habitats that no other organisms can tolerate. We have already seen that bacteria, through their various modes of energy metabolism, can grow in aerobic or anaerobic environments. Bacteria also tolerate temperature extremes far beyond those tolerated by other organisms. Most soil bacteria are **mesophiles** that grow optimally at temperatures in the range of 15 to 35°C, but some species, known as **thermophiles,** survive and often thrive at temperatures in excess of 40 to 50°C (and as high as 100°C). Most eukaryotes cannot survive temperatures above 45°C. Thermophilic bacteria typically produce heat-stable enzymes and structural proteins that do not denature at elevated temperatures. One such enzyme is Taq polymerase (produced by the bacterium *Thermus aquaticus*), a key ingredient in the **polymerase chain reaction (PCR)** used for DNA amplification (Chapter 9). Other species of bacteria have adapted to grow optimally at temperatures below 15°C. These **psychrophilic** bacteria often synthesize large quantities of unsaturated fatty acids that maintain the fluidity of the cytoplasmic membrane at low temperatures.

Several species of bacteria thrive in acidic soils that predominate in regions with high precipitation. Some, such as the sulfur-oxidizing bacteria discussed in the preceding section, tolerate extremely low pH that severely inhibit normal enzymatic activity. These **acidophiles** maintain a neutral cytoplasmic pH by actively transporting H^+ ions out of the cell. **Alkalophilic** bacteria use a similar strategy (ion pumps) to grow at pH as high as 10.5 in arid and semiarid regions. Actinomycetes frequently predominate in alkaline soils because of their tolerance of high pH.

Poor drainage and rapid surface evaporation in arid regions often lead to the formation of saline soils with stressful osmotic potential that inhibits the growth of many microbes. Many species of bacteria have adapted to saline or arid environments by developing membranes and enzymes that function in solutions of high ionic strength or by accumulating solutes (usually amino acids) in the cytoplasm to compensate for the low water potential of the soil solution. Bacteria that tolerate high salt concentrations are called **halophiles,** while those that tolerate dry habitats are called **xerophiles.**

Interactions with Other Organisms

Bacteria exhibit a broad spectrum of interactions with one another and with other soil microorganisms, ranging from neutral relationships at the center of the spectrum to symbiotic interactions at one extreme or antagonistic interactions at the other (see Chapter 8 for a more in-depth discussion of microbial interactions). Probably the most common type of interaction among soil microbes is **competition.** Bacteria, fungi, protozoa, and microscopic animals compete for water, food, shelter, and other vital resources that are present in short supply. Beneficial interactions involving soil bacteria are also common. Decomposition of the complex constituents of soil organic matter, for instance, typically requires the cooperative activity of a diverse community of microorganisms. One population may break down a complex substrate to an intermediate product that a second population can use as its source of carbon or energy for growth. Bacteria may have a negative impact on one another and on other soil microbes by producing **antibiotics** or other substances that harm the organisms that live nearby. Streptomycetes, a group of filamentous bacteria that commonly inhabit soils, are known to produce a number of antibiotics, though the quantities produced and the impact of these substances in soil ecosystems is not well understood.

Bacteria also have important interactions with plants. Bacteria colonize plant roots in greater numbers than any other group of soil microbes and profoundly influence plant growth and productivity in natural and agricultural ecosystems (Chapter 17). We have already seen that bacteria may increase the availability of inorganic nutrients in soils by mineralizing soil organic matter and solubilizing soil minerals, and that they may compete with plants for those same nutrients through the process of immobilization. Many bacterial species influence plant growth directly by producing **hormones** or **toxins** that stimulate or impede root function and morphology. A few species, notably those of the genera *Rhizobium, Bradyrhizobium,* and *Frankia,* form intricate mutualistic symbioses with selected plant species (Chapter 14), while other bacteria produce disease in the plants that they infect. Most of the bacteria that colonize plant roots are harmless saprophytes that feed on the organic nutrients in root exudates and protect the plant from infection by competing with pathogens for nutrients, water, and places to attach to the root, or by excreting substances that directly inhibit the pathogen. Several species of bacteria are currently being investigated as potential biological control agents because of their ability to inhibit the growth of plant pathogens (Chapter 19).

The Actinomycetes

The actinomycetes constitute a specialized group of soil bacteria that occurs in soils throughout the world. Coined in the late 1800s, the term **actinomycete** *(actinis,* meaning "ray" and *myces,* meaning "fungus") is a misnomer. Because actinomycetes form an aerial mycelium—albeit much smaller than that of fungi, and may produce abundant asexual spores called **conidia** which give the colonies a "powdery" or "chalky" appearance—they resemble miniature fungi. However, their resemblance to fungi is strictly morphological. The actinomycetes are classified with the bacteria because they:

- are prokaryotic,

- contain peptidoglycan in their cell walls,

- are sensitive to lysozyme which degrades the polysaccharide backbone of the peptidoglycan,

- are sensitive to antibacterial but not antifungal antibiotics, and

- possess flagella typical of bacterial flagella in the few species that show motility.

Within this group are organisms that produce colonies ranging from those typical of bacteria, such as species of *Mycobacterium* and *Corynebacterium,* to those that produce the tough, leathery mycelium characteristic of *Streptomyces* species.

The actinomycetes play important roles in soils and other environments such as compost piles. They are important agents in the degradation of organic materials in soil and contribute to the formation of stable humus. Recent research indicates they, along with certain fungi, may play an important role in the degradation of lignin. They are also causative agents of animal, plant, and human diseases.

The actinomycetes are often described as slow-growing organisms. For this reason, culture plates are often incubated for one to two weeks to allow differentiation of the actinomycete colonies. Although they do not compete well with the faster-growing bacteria and fungi for readily-available carbon substrates, they are thought to be important in mineralizing carbon and nitrogen formed during the early stages of decomposition, such as fungal cell walls. Many actinomycetes are good chitin degraders and simply because actinomycetes are "late colonizers" should not detract from the vital roles they play in the degradation processes and in the formation of humus.

Perhaps one of the most outstanding characteristics of the actinomycetes is their ability to produce antibiotics. Streptomycin, neomycin, erythromycin, and tetracycline are but a few of the medically important antibiotics derived from species of *Streptomyces*. This capacity of the actinomycetes and other microbes to produce antibiotics is the basis of a billion-dollar pharmaceutical industry and has saved countless human lives in the last half century. However, despite their outstanding medical significance, the significance of antibiotic production in ecological interactions among soil microbes is still largely unknown.

Actinomycetes tend to respond to environmental influences like bacteria. However, there are some exceptions. Actinomycetes tend to become more abundant in soils subjected to prolonged drying. This shift toward numerical dominance is generally attributed to the ability of actinomycete to produce conidia which withstand desiccation well. Thus, while more sensitive organisms succumb to low-moisture stress, actinomycetes persist. The conidia also confer a slight tolerance to increased soil temperature. Though most of the actinomycetes are mesophiles, thermophilic actinomycetes are important in the high-temperature transformations of organic substrates during composting. As a group, the actinomycetes tend to be sensitive to low pH. In general, as the soil pH decreases to 6.0 their numbers decline and below pH 5.0 they are almost absent. This sensitivity to soil acidity can have consequences for the microbial balance within the soil ecosystem.

Quantifying Bacterial Populations

Several methods can estimate the number of bacteria or other microbes in a soil community. Each method has certain advantages and limitations. The method chosen depends upon many factors, including the type of microbe being studied and the microbe's nutritional requirements and culture characteristics. Molecular methods are now also becoming available for quantifying bacteria in soil (Chapter 9).

The Dilution Plate Count Method

This method is often used to enumerate bacterial populations in soils because many soil bacteria have relatively simple nutritional requirements and grow readily on solid culture media. The underlying principle of the plate count method is that each bacterial cell extracted from a soil sample produces a visible colony of cells when inoculated onto a solid culture medium and incubated under the appropriate conditions. Therefore, this method only counts viable (living) cells. By counting the number of colonies that appear on the medium following incubation, the number of bacteria in the soil sample can be estimated. Bacteria often populate soils in very large numbers, ranging from 10^6 to $> 10^8$ cells g^{-1} of soil. Because it is not possible to isolate and accurately count such large numbers of organisms in a culture dish, the bacterial cells extracted from a soil sample must first be diluted to more manageable numbers before inoculating onto the culture medium.

The bacteria in a soil sample can be rapidly diluted several million-fold with a simple technique known as **serial dilution** (Fig. 3–5). A liquid suspension of the cells in the soil sample is prepared by mixing 10 g of soil (dry mass equivalent) in 95 ml of a sterile diluent (water or a dilute buffer solution). This ratio of soil to diluent dilutes the bacterial population by approximately one-tenth. The original cell sus-

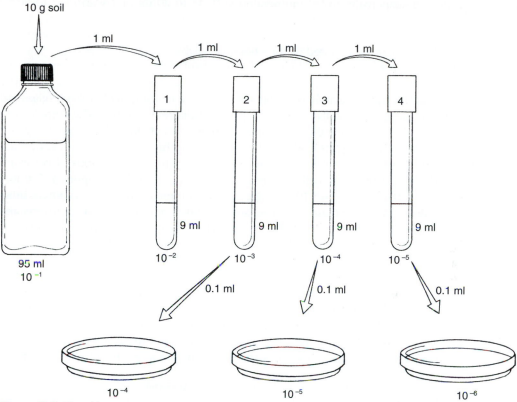

Figure 3–5 The dilution spread plate method for quantifying bacterial populations in soils. *Original drawing by Kim Luoma.*

pension is therefore designated a 10^{-1} dilution. The mixture is agitated vigorously to suspend the bacterial cells in the liquid, then allowed to settle briefly so that sand will sediment to the bottom of the dilution bottle.

Subsequent dilutions are prepared by transferring a defined volume of the cell suspension to tubes containing defined volumes of sterile diluent. Transfers are made sequentially from tube to tube to obtain increasingly dilute cell suspensions. In the example illustrated in Figure 3–5, a 10-fold dilution of the original cell suspension is prepared by transferring 1 ml of the cell suspension in the dilution bottle to 9 ml of sterile diluent in tube 1. The dilution ratio of this transfer is calculated as follows:

$$\text{Dilution ratio } = \text{ Volume transferred / total volume after the transfer}$$
$$= \quad 1 \text{ ml / } (9 \text{ ml} + 1 \text{ ml})$$
$$= \quad 1/10 \text{ or } 10^{-1}$$

The resulting cell suspension is diluted another 10-fold by transferring 1 ml of the liquid in tube 1 to 9 ml of sterile diluent in tube 2. The cell suspension in tube 2 is then diluted 10-fold in tube 3, and the suspension in tube 3 is diluted 10-fold in tube 4.

At this point, either of two methods may be used to inoculate a solid culture medium with the dilute cell suspensions. **Pour plates** are prepared by mixing 1 ml of each cell suspension to be enumerated with 15 to 20 ml of a warm (45 to 50°C) liquid culture medium containing melted agar and pouring the mixtures into sterile Petri plates. After the culture medium solidifies, the plates can be incubated and bacterial colonies can be counted as they become visible.

Spread plates are prepared by inoculating 0.1 ml of each cell suspension onto the surface of a sterile culture medium previously poured and solidified in a Petri plate. The suspensions are spread evenly over the surface of the medium with a sterile glass rod to disperse the bacterial cells so that they produce discrete colonies. The colonies can be counted after incubating the plates under conditions that allow the bacteria to grow.

Inoculating an agar plate with a 0.1-ml aliquot in this manner constitutes an additional 10^{-1} dilution of the cells in the suspension. The total dilutions of plates inoculated with 0.1 ml of the cell suspensions in tubes 2, 3, and 4 in Figure 3–5 would therefore be 10^{-4}, 10^{-5}, and 10^{-6}, respectively. Soil microbiologists routinely inoculate replicate plates with aliquots from several different dilute cell suspensions and use the average number of colonies on the plates to calculate the number of bacteria in the original soil sample. Plates that yield fewer than 30 or more than 300 colonies are not counted because calculations based on colony counts outside this range are not considered accurate.

The number of cells per gram of soil is calculated by multiplying the average number of colonies by the *reciprocal* of the total dilution of the plates. For example, if the plates inoculated with 0.1-ml aliquots from tube 3 using the spread plate technique yielded an average of 125 colonies per plate, then the number of bacteria in the soil sample would be 125×10^4 or, in scientific notation, 1.25×10^6 cells g^{-1}.

Usefulness and Limitations of the Dilution Plate Method

Soil microbiologists routinely use the dilution plate method to enumerate bacterial populations because the method is simple and reproducible and does not require expensive equipment or supplies. It also enables the microbiologist to count only **viable** cells (cells that are capable of growing and producing a visible colony) and yields isolated colonies which can be subcultured for further study. Direct microscopic counts often fail to distinguish between viable cells and dead cells, while biochemical detection methods do not lead to the isolation of pure cultures. Estimates of numbers do not necessarily provide information on microbial activity in soil. For example, high numbers of *Bacillus* species may have been derived from dormant endospores in soil.

Compared with other counting methods, however, plate counts often underestimate bacterial populations in soils (Amann et al., 1995). Several factors contribute to the relatively low numbers of bacteria detected with the plate count method. Only those species that can grow using the nutrients contained in the culture medium will produce a visible colony which can be counted, and no culture medium provides all of the essential nutrients that every species of soil bacteria might require for growth. In addition, incubation conditions are unavoidably restrictive. Obligate anaerobes cannot grow on plates that are incubated aerobically, and obligate aerobes cannot grow on plates that are incubated anaerobically. Incubation temperature and the pH of the culture medium may also restrict the growth of certain species and favor the growth of others. For these reasons and others, a plate count detects only a small proportion (typically $< 10\%$) of the total microbial community in a soil. Nonetheless, it is a useful method for determining the relative numbers of certain types of bacteria in different soils under different environmental conditions. For more information regarding the dilution plate count method, consult Zuberer (1994) in *Methods of Soil Analysis*.

Soil microbiologists often want to count or isolate specific groups of bacteria within a soil community, rather than studying the community as a whole. They do so with culture media that deliberately exclude or promote the growth of certain types of microbes. **Selective media** contain ingredients which allow certain types of microbes to grow while preventing the growth of others. For instance, a medium can be made selective for acid-tolerant microbes by lowering its pH. **Differential media** contain ingredients which make it possible to distinguish one type of microbe from another. A pH indicator is often added to a culture medium to differentiate microbes that produce acidic metabolic products from those that produce neutral or alkaline products. **Enrichment media** contain ingredients that favor the growth of certain types of microbes over others. A medium which contains cellulose, and no other source of carbon, favors the growth of cellulose-decomposing microbes over those that cannot decompose cellulose.

Summary

Bacteria are single-celled, prokaryotic microbes which inhabit soils throughout the world in large numbers. They are much smaller than the eukaryotic microbes and have a much simpler cellular structure. Each bacterial cell consists of a cytoplasmic membrane, cell wall, chromosome, and ribosomes. These structures are regarded as essential because they are present in all bacteria. Different species of bacteria may also possess any of a variety of other structures, such as flagella, fimbriae, pili, capsules, inclusions, or endospores, which provide important functions enabling the bacteria to survive in soils and to adapt to changing environmental conditions.

Despite their small size, bacteria are extremely versatile metabolically. As a group, they can grow in the presence or absence of oxygen, and many species can shift their metabolism from aerobic respiration to anaerobic respiration or fermentation in response to changing oxygen concentrations. Most soil bacteria are heterotrophs, feeding on nonliving organic matter in soils or forming symbiotic associations with plants, insects, or other soil microbes. Many other species are autotrophs, which use light energy or energy obtained from the oxidation of inorganic substrates to synthesize carbohydrates with carbon dioxide as their primary source of carbon.

Through their metabolic activities, bacteria transform soil minerals and organic matter from one form to another and alter the availability of essential nutrients such as nitrogen, sulfur, and phosphorus for plants and other soil organisms to use. As a result, bacteria play central roles in organic matter decomposition, nutrient cycling, and soil formation.

At first glance, soil may appear to be a lifeless collection of weathered rocks and minerals, worn down to fine granules by wind and water and time. Soil is anything but lifeless; it is a habitat filled with "teeming, boisterous life." A single gram of soil may contain hundreds of millions of organisms, some of which can be seen with the naked eye but most of which cannot. It is not an exaggeration to assert that all of the "higher" organisms with which we are more familiar—the plants, mammals, birds, reptiles, amphibians, insects, and others—would cease to exist if not for the activities of the multitudes of "invisible" organisms in soil.

Cited References

Amann, R.I., W. Ludwig, and K.-H. Schleifer. 1995. Phylogenetic identification and in situ detection of individual microbial cells without cultivation. Microbiol. Rev. 59:143–169.

Ames, G. F-L. 1986. Bacterial periplasmic transport systems: Structure, mechanism, and evolution. Annu. Rev. Biochem. 55:397–425.

Clarke, H.R.G., J.A. Leigh, and C.J. Douglas. 1992. Molecular signals in the interactions between plants and microbes. Cell 71:191–199.

Couturier, M., F. Bex, P.L. Bergquist, and W.K. Maas. 1988. Identification and classification of bacterial plasmids. Microbiol. Rev. 52:375–395.

Holt, J.G., N.R. Krieg, P.H.A. Sneath, J.T. Staley, and S.T. Williams. 1994. Bergey's manual of determinative bacteriology. 9th ed. Williams and Wilkins Co., Baltimore.

Madigan, M.T., J.M. Martinko, and J. Parker. 1997. Brock biology of microorganisms. 8th ed. Prentice Hall, Upper Saddle River, N.J.

Ofek, I., R.F. Rest, and N. Sharon. 1992. Nonopsonic phagocytosis of microorganisms. Am. Soc. Microbiol. News 58:429–435.

Pettijohn, D.E. 1988. Histone-like proteins and bacterial chromosome structure. J. Biol. Chem. 263:12793–12796.

Pugsley, A.P. 1993. The complete general secretory pathway in Gram-negative bacteria. Microbiol. Rev. 57:50–108.

Robinow, C., and E. Kellenberger. 1994. The bacterial nucleoid revisited. Microbiol. Rev. 58: 211–232.

Shapiro, L. 1995. The bacterial flagellum: From genetic network to complex architecture. Cell 80:525–527.

Stock, J.B., A.J. Ninfa, and A.M. Stock. 1989. Protein phosphorylation and regulation of adaptive responses in bacteria. Microbiol. Rev. 53:450–490.

Veal, D.A., H.W. Stokes, and G. Daggard. 1992. Genetic exchange in natural microbial communities. Adv. Microbial Ecol. 12:383–430.

Woese, C.R. 1994. There must be a prokaryote somewhere: Microbiology's search for itself. Microbiol. Rev. 58:1–9.

Zuberer, D.A. 1994. Recovery and enumeration of viable bacteria. pp. 119–144. In R.W. Weaver, S. Angle, P. Bottomley, D. Bezdicek, S. Smith, A. Tabatabai, and A. Wollum (eds.) Methods of soil analysis, Part 2. Microbiological and biochemical properties. Soil Science Society of America Book Series, No. 5. Madison, Wis.

General References

Atlas, R.M., and R. Bartha. 1993. Microbial ecology: Fundamentals and applications, 3rd ed. Benjamin Cummings Publishing Co., Redwood City, Calif.

Dawes, I.W., and I.W. Sutherland. 1992. Microbial physiology. 2nd ed. Blackwell Scientific Publications, London.

Pelczar, M.J., Jr., E.C.S. Chan, and N.R. Krieg. 1986. Microbiology. 5th ed. McGraw-Hill, New York.

Schlegel, H.G., and B. Bowien. 1989. Autotrophic bacteria. Science Tech. Publ., Madison, Wis.

Stanier, R.Y., J.L. Ingraham, M.L. Wheelis, and P.R. Painter. 1986. The microbial world. 5th ed. Prentice Hall, Englewood Cliffs, N.J.

Wolfe, S.L. 1993. Molecular and cellular biology. Wadsworth Publishing Co., Belmont, Calif.

Study Questions

1. Identify the common shapes and groupings of bacterial cells, using the correct scientific terms to describe cellular morphology.

2. Describe the structure and function of a bacterial cell, indicating which cellular structures are present in all bacteria and which are present only in certain species.

3. How do bacteria obtain soluble nutrients from the soil environment? How do these small, single-celled organisms break down the large complex polymers in soil organic matter? Do these processes require energy?

4. Contrast the structure and chemical composition of Gram-positive and Gram-negative bacterial cell walls.

5. What are plasmids? What do you suppose might be the ecological significance of plasmids in soil bacteria?

6. List three groups of photoautotrophic and three groups of chemoautotrophic soil bacteria, and explain how they differ from one another. List and contrast the manner in which they obtain carbon and energy for growth.

7. Contrast the types of energy metabolism exhibited by the following groups of bacteria: obligate aerobes, obligate anaerobes, facultative anaerobes, and aerotolerant anaerobes. What type of microenvironment does each group inhabit in soils?

8. Describe the trophic levels of soil bacteria, and outline the roles that bacteria play in nutrient cycling and soil formation.

9. How is the plate count method used to quantify bacterial populations in soils? What are the advantages and limitations of this method?

Chapter 4

♦

Fungi

Joseph B. Morton

♦

When Flora's lovelier tribes give place, the Mushroom's scorned but curious race,
bestood the moist autumnal earth; a quick but perishable birth, inlaid with many
a brilliant dye, of Nature's high-wrought tapestry.

Bishop Mant

Fungi are a diverse group of multicellular organisms with an incredible array of vegetative and reproductive morphologies and diverse life cycles. They are more abundant, on a mass basis, in soils than any other group of microorganisms; their biomass ranges from 500 to 5,000 wet kg ha^{-1} (Metting, 1993). Fungi inhabit almost any niche containing organic substrates, so that they are active participants in ecosystems as degraders of organic matter, agents of disease, beneficial symbionts, agents of soil aggregation, and an important food source for humans and many other organisms. In many cases, they are a vital component of ecosystem function and vitality. Humans depend considerably on fungi for metabolic by-products in food additives and medicines. At least 70,000 species have been described, but at least 20 times that number are estimated to exist worldwide (Hawksworth, 1991).

Fungal Cell Structure

Unlike prokaryotic bacteria discussed in Chapter 3, fungal organisms are eukaryotic. The multitude of membrane-bound organelles present in each cell (Fig. 4–1) are similar to those of insects, plants, and animals, but with some important differences. The membrane surrounding the nucleus constricts during formation of two daughter nuclei, whereas it degenerates and reforms in plant and animal cells. Within the nucleus is a predominantly protein structure called a nucleolus that may persist, become dis-

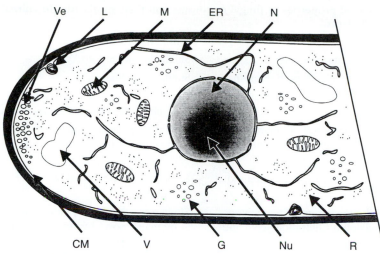

Figure 4–1 General structure and organellar composition of a fungal cell at the apex of a hypha. ER = endoplasmic reticulum, G = glycogen granules, L = lomasome, M = mitochondrion (plural, mitochondria), N = nucleus, Nu = nucleolus, CM = cytoplasmic membrane, R = ribosomes, V = vacuole, Ve = vesicles.

persed within a nucleus or be discharged into the cell cytoplasm. Vacuoles in the cell cytoplasm also usually are smaller than those in plant cells. The network of endoplasmic reticulum is not as extensive, and it has fewer connections with the cytoplasmic membrane compared to that in plants and animals. An extensive membrane system generates many secretory vesicles that substitute for Golgi bodies found in other eukaryotic organisms. These vesicles are uniquely important in filamentous fungi for cellular function and growth because they transport structural and enzymatic molecules to active metabolic regions. These occur at the tips of tubular strands or filaments called **hyphae** (singular, hypha) that elongate and branch indefinitely.

Even though fungi do not contain chlorophyll and thus cannot carry out photosynthesis, they have been considered plantlike because they have cell walls, are generally nonmotile, and reproduce by means of **spores** (microscopic parts that resemble seeds in functioning as vehicles of dissemination and formation of new individuals, but they lack embryos). Accordingly, Latin binomial names of fungi are based on the Botanical Code of Nomenclature. However, other morphological characters and new molecular evidence suggest that fungi are more closely related to animals. We now believe that they originated about one billion years ago as a group equal in rank to plants and animals. Table 4–1 summarizes properties unique to fungi as well as those shared with plants, animals, or both.

Growth and Reproduction, with Reference to Taxonomy

Fungi can be single-celled, but the majority are multicellular organisms with a filamentous vegetative body. In contrast to the relative simplicity of the vegetative body, reproductive structures include various kinds of single-celled spores produced alone or in visible and complex fruiting bodies. Diverse reproductive structures produced

Table 4–1 **General properties of fungal organisms and their similarities to animals and plants.**

Fungal property	Comparison with animals and plants
Eukaryotic organization of cells	Similar to animals and plants
Glycogen is common storage polysaccharide	Similar to animals
Heterotrophic in requiring an external carbon source	Similar to animals, except carbon is obtained by absorption rather than by ingestion
Rigid cell walls composed of chitin and non-cellulosic glucans, except in some Oomycota	Similar to some animal groups Oomycota similar to plants with cellulose in cell walls
Life cycle includes both sexual and asexual stages	Similar to many plant and a few animal groups
Body (or thallus) made up of filamentous hyphae, except in Chytridiomycota and Oomycota	Similar to primitive plants (algae), except hyphae grow by extension from their growing tips
Amino acid, lysine, is produced from a-aminoadipic acid (AAA), except in Oomycota	Similar to the Euglenaceae (Chapter 5), where properties of plants (chlorophyll) and animals (no cell walls) also are mixed
Morphologically distinct asexual and sexual spores produced in variable numbers	Similar to some plant groups, except each spore represents a haploid (1N) germ line, Unlike seeds with preformed embryos
Dominant steroid in the cytoplasm is ergosterol	Unique to Fungi
Nuclei haploid for most of life cycle, except in Oomycota	Unique to Fungi, Oomycota are similar to animals and plants in having diploid nuclei
Hyphae often multinucleate and nuclei migrate in the cytoplasm	Unique to Fungi and Oomycota

during sexual and asexual phases of fungal life cycles provide many of the morphological characters defining species and other groups at higher taxonomic ranks. As technology advances, more characters are becoming available to improve taxonomies based on evolutionary relationships, such as those in sequences of nucleotides and amino acids of DNA and proteins, respectively, and in the chemistry of cell walls, storage carbohydrates, and lipids. Analyses of these characters now indicate organisms traditionally considered to be fungi evolved from separate ancestors; they now are classified in at least two separate Kingdoms (Alexopoulos, et al. 1996; Table 4–2). In this chapter, we will emphasize four **phyla** (singular, phylum) of the Kingdom Fungi and one phylum in the Kingdom Stramenopila because they share common morphological, nutritional, and ecological properties. Members of the Kingdom Fungi are considered a true evolutionarily united group because evidence indicates they originated from a single common ancestor. Members of the phylum Oomycota in the Kingdom Stramenopila closely resemble fungi and have traditionally been treated as such, but most evidence indicates they are more closely related to some algal groups (Barr, 1992).

The vegetative body of a fungus is called a **thallus,** and it generally exists as one of three forms:

Table 4–2 **Sexual and asexual reproductive structures produced by true fungi in four phyla of the Kingdom Fungi and funguslike organisms in the phylum Oomycota of the Kingdom Stramenopila.**

Phylum	Sexual phase (Teleomorph)	Asexual phase (Anamorph)
Chytridiomycota	Resting chytrid cells	Zoospores (from chytrid cells)
Zygomycota	Zygospores (solitary resting spores)	Chlamydospores (from hyphae), sporangiospores (in sporangia)
Ascomycota	Ascospores (in fruiting bodies)	Conidia (on individual hyphae, aggregates of hyphae, or in fruiting bodies)
Basidiomycota	Basidiospores (on individual hyphae or in fruiting bodies)	Conidia, other specialized asexual spores in complex life cycles
Oomycota	Oospores (solitary resting spores)	Sporangia, zoospores (from sporangia)

- **Chytrid cells:** solitary globose cells with or without specialized rootlike filaments called **rhizoids** (Fig. 4–2a). They are unique to organisms of the phylum Chytridiomycota in the Kingdom Fungi and the phylum Hyphochytridiomycota in the Kingdom Stramenopila. The latter group is discussed in more detail by Alexopoulos et al. (1996).

- **Yeast cells:** spherical to ovoid cells are formed by fungi in the phyla Ascomycota and Basidiomycota. They divide by budding or fission (Fig. 4–2b). Some fungi are dimorphic and can change from a mycelium to a yeast form under conditions where penetration of a substrate is not needed to obtain nutrients, such as in aqueous environments or insect cavities.

- **Mycelia:** a filamentous network of hyphae (Fig. 4–2c) that branch and grow only by apical (tip) extension (Fig. 4–3a). This vegetative growth form is the most common of organisms in the Kingdom Fungi, and in soil fungi especially.

A mycelium is incredibly resilient because growth is open-ended for as long as a nutrient source is available. The result is that size is highly variable, ranging from a pinpoint to a mass covering hectares of land. On an **agar** medium in Petri

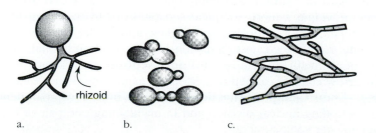

a.   b.   c.

Figure 4–2 Types of cells that make up the vegetative body (thallus) of fungal organisms: (a) chytrid cells, (b) yeast cells, and (c) a mycelium containing compartments separated by crosswalls (septa).

Figure 4–3 Patterns of growth of a mycelium consisting of a branching hyphal network initiated by germination of a spore or a new branch from a hyphal fragment. (a) Radial spread on the surface of a solid agar medium in a Petri dish. (b) Radial spread on the surface of organic litter. (c) Spherical tufts of mycelium submerged in a liquid medium.

dishes or on other flat surfaces, a mycelium forms circular cottony mats (Figs. 4–3a, b). In a liquid broth medium, where no surface boundaries confine growth, a growing mycelium resembles spherical cotton balls (Fig. 4–3c). Soil is more heterogeneous, consisting of zones and microsites that can either benefit or hinder fungal growth. A mycelial strand, or hypha, provides the ideal mechanism to penetrate soil pores and branch in all directions. As a network, hyphal interconnections establish a three-dimensional physical continuum capable of spanning nutrient-deficient zones, navigating around physical barriers or gas pockets in the soil matrix, and penetrating herbaceous and woody organic matter. The branching of hyphae around and within soil particles of all sizes influences soil structure by improving soil aggregation and aeration.

A hypha is essentially one long tube with rigid crosswalls known as **septa** (singular, **septum**). These septa are not the same as cell walls because they contain open pores for continuous flow of cytoplasm and cellular organelles and nuclei. They provide structural support along the length of a hypha and regulate cytoplasmic flow between hyphal compartments. In some fungal groups such as Zygomycota and Oomycota, the hyphae are **coenocyctic** (pronunciation: "seen-o-sitik") in that they rarely form septa in actively growing regions and nuclei are free-floating (Fig. 4–4a). When septa are formed, they are located between cytoplasmically active regions of hyphae and aged or dead regions devoid of cytoplasm. In other fungal groups such as the Ascomycota and Basidiomycota, septa are an integral part of hyphae. Septal pores can be plugged in hyphae of Ascomycota by spherical **Woronin bodies** (Fig. 4–4b), thus providing a more complete separation of hyphal compartments through which movement of cytoplasm, organelles, and nuclei can be blocked. Basidiomycete fungi have the greatest regulation of flow between hyphal compartments by means of a structurally complex **dolipore septum,** with flanges at a central pore and a narrow channel capped by perforated membrane domes called *parenthosomes* (Fig. 4–4c). **Clamp connections** between hyphal compartments, if present, aid in controlling nuclear division and in maintaining one pair of haploid nuclei (a **dikaryon**) per compartment.

Many soil fungi in the Ascomycota and Basidiomycota can organize hyphae into specialized organs such as mycelial strands, rhizomorphs, and sclerotia. Mycelial strands are aggregates of parallel hyphae cemented together by sticky exudates and

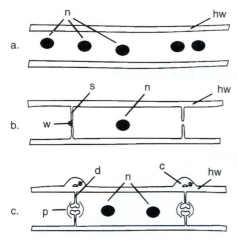

Figure 4–4 Structure of a hypha in (a) Oomycota (wall of cellulose) and Zygomycota (wall of chitin), (b) Ascomycota, and (c) some Basidiomycota. n = nucleus, hw = hyphal wall, s = simple septum, w = Woronin body, c = clamp connection, d = dolipore septum, p = parenthosome.

by fusion of hyphal branches. The cell walls of surface hyphae are thickened and often melanized (containing dark pigments). Mycelial strands of some fungi consist of modified hyphae functioning similar to plant tissues. For example, the wood decay (dry-rot) fungus, *Serpula lacrymans,* produces strands up to 8 mm in diameter, consisting of thin-walled hyphae, thick-walled hyphae for structural support, and wide hyphae lacking cytoplasmic contents for mass flow of water-like plant xylem vessels. These strands are usually in close contact with soil to obtain moisture for translocation to the growing edge of a mycelium as it penetrates new wood tissue (Jennings and Watkinson, 1982).

Rhizomorphs are more complex versions of mycelial strands, with a greater degree of tissue differentiation (Figs. 4–5a, b). They are highly resistant to environmental changes, provide stronger means than hyphae to penetrate soil or organic material, supply oxygen to the leading tips, and transport nutrients throughout the fungal organism. Species of basidiomycete fungi in the genus *Armillaria* are aggressive tree root and crown rot pathogens because of rhizomorph formation. The rhizomorphs resemble gray to black shoestrings, although they are much longer in length. They branch in the soil as much as eight times faster than a typical mycelium and then persist for years. The demise of many orchards started in forests cleared of diseased trees can be attributed to abundant and viable rhizomorph networks in soil even after fumigation.

Sclerotia (singular, **sclerotium**) are spherical to oblong aggregates of interwoven hyphae that resemble plant parenchyma tissue in the center but are thick-walled and melanized on the periphery to form a hard outer rind (Fig. 4–5c, d). Sclerotia often are separated from the parent mycelium and therefore are more readily dispersed than other forms of mycelial aggregates. They usually are 1 to 3 mm in diameter, and dark brown to black in color. Sclerotia are better able to resist drought and temperature fluctuations and, therefore, are much longer-lived than mycelia. They also store lipids, carbohydrates, and proteins until soil conditions are favorable enough to germinate and form either a new mycelium or a fruiting body.

Fungi reproduce by forming distinctive spores or fruiting bodies as a result of either sexual or asexual states designated as teleomorphs and anamorphs, respectively

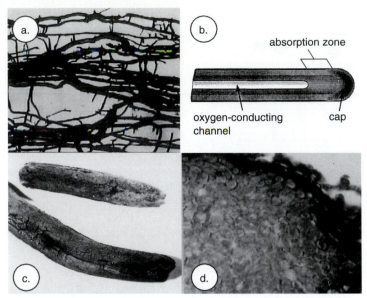

Figure 4–5 Organized structures differentiated by mycelium that enhance survival and other functions. (a) Rhizomorphs of *Armillaria mellea*. (b) Rhizomorph internal structure and function of hyphal tissues. (c) Sclerotia of *Claviceps purpurea* (causing ergot in grasses), each with a hardened (and cracked) brown to black outer rind. (d) Cross-section of a sclerotium showing the cohesive matrix of hyphae and the outer rind with thickened and melanized cell walls.

(Box 4–1). Spores are important in life cycles because they not only encapsulate different nuclei and cytoplasmic factors into numerous discrete packets for survival under stress or for dispersal, but they also increase the probability of contact between opposite mating types.

In general, the sexual cycle requires four sequential processes:

1. **Plasmogamy** is the fusion of cytoplasm of two compatible haploid (1N) sexual organs or **gametangia** (singular, **gametangium**).

2. **Karyogamy** is the fusion of two haploid nuclei to establish a diploid (2N) condition.

3. **Meiosis** is reduction of the diploid nucleus to haploid daughter nuclei, during which genes undergo recombination.

4. **Teleomorph** formation is the encapsulation of each haploid nucleus and its associated cytoplasm in a discrete cell that differentiates into a spore.

Some fungi form morphologically distinctive "male" and "female" gametangia. The gametangia of most fungi are morphologically indistinguishable and, therefore, require chemical methods to recognize each of them. In this case, opposite mating types are designated "+" and "−". In lower fungi (i.e., Chytridiomycota and Zygomycota) and funguslike organisms (i.e., Oomycota), the teleomorph usually

---

Box 4–1

Teleomorphs and Anamorphs. Fungi are ideal opportunists exploiting almost any habitat in part because their life cycles can alternate between sexual and asexual reproduction. Morphologically distinctive spores produced openly or in fruiting bodies develop from the mycelium in the sexual phase (**teleomorph**) and in the asexual phase (**anamorph**). The teleomorph serves as a primary basis for classification of fungi when it is formed (Table 4–2). Many fungi in soil and litter reproduce only asexually under natural conditions, so that Latin binomial names are assigned to the anamorph. Sometimes a sexual phase is discovered in these fungi, at which time it receives a separate name for the teleomorph. This practice has led to dual nomenclature in fungal taxonomy, with separate names for anamorphs and teleomorphs. However, the name of the whole organism (**holomorph**) is the name given to the teleomorph. In the past, Fungi with no known sexual phase were classified in an artificial phylum, Deuteromycota. Today, they are placed in the phylum of their closest sexual relatives if at all possible, based on molecular, biochemical, physiological, and other morphological or ultrastructural characters. Most have proven to be members of the Ascomycota; the remainder are in the Basidiomycota.

---

consists of solitary, thick-walled resting spores. In higher fungi (i.e., Ascomycota and Basidiomycota), spores often are borne on a variety of fruiting bodies like the highly visible mushrooms found in the forest floor, lawns, and flower beds.

The asexual cycle is much simpler than the sexual cycle because it does not require interactions between nuclei or different mating types. The nuclei of one hypha divide repeatedly by mitosis and are partitioned into large numbers of spores called **conidia** (singular, **conidium**). Conidia are formed in a variety of ways, but essentially they arise from hyphal tips. Most soil fungi tend to form conidia freely on a single hypha or hyphae in various degrees of aggregation; a low proportion develop fruiting bodies. Some soil fungi can transform hyphal compartments into thick-walled, darkly pigmented **chlamydospores** (Fig. 4–6d).

The diversity of life cycles found in fungi and funguslike organisms is far too great to be covered in this chapter. Instead, the remainder of this section will illustrate and discuss a representative life cycle of one species from each phylum.

Oomycota

These organisms differ from true fungi in that the mycelium cannot translocate the carbohydrate trehalose, hyphal walls are composed of cellulose instead of chitin, and the nuclei in the mycelium are diploid rather than haploid (Table 4–1). These funguslike organisms are found most commonly in aquatic habitats and moist soils. Many species are saprophytes degrading organic matter but some, such as *Phytophthora infestans* (a pathogen causing late blight of potatoes) and *Plasmopara viticola* (a pathogen causing downy mildew of grape), produce devastating plant diseases. The phylum is characterized by organisms that produce a coenocytic mycelium, thick-walled resting sexual spores called **oospores,** and nonmotile asexual spores called **sporangia** (singular, **sporangium**). Both oospores and sporangia can partition their cytoplasmic contents into clusters of motile **zoospores** under cool, moist conditions. Each zoospore is capable of directional movement in an aqueous environment by the whiplike action of two flagella.

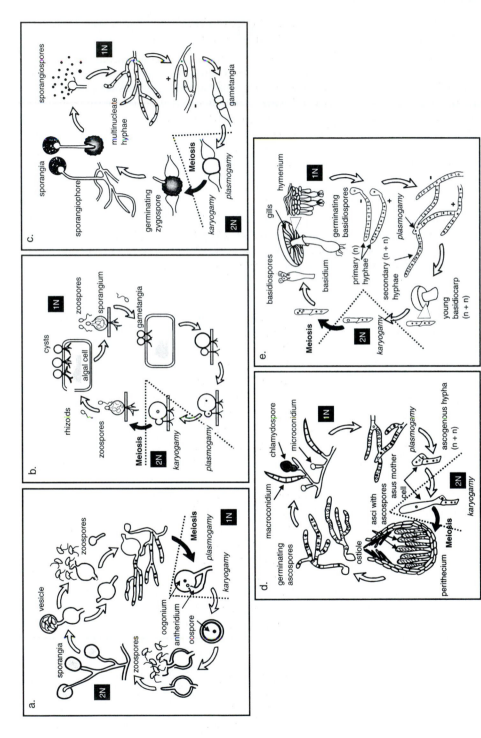

Figure 4-6 Life cycles of fungi in the Kingdom Stramenopila and true fungi in the Kingdom Fungi: (a) Oomycota, *Pythium debaryanum;* (b) Chytridiomycota, *Rhizophydium couchii;* (c) Zygomycota, *Mucor piriformis;* (d) Ascomycota, *Gibberella moniliforme;* and (e) Basidiomycota, *Agaricus campestris.* Filled arrows indicate meiosis and initiation of a sexual (teleomorph) phase; open arrows indicate mitotic divisions of nuclei and asexual (anamorph) phase. The dotted line separates phases where nuclei are haploid (1N) or diploid (2N).

A typical life cycle is exemplified by *Pythium debaryanum,* a soil **saprophyte** (i.e., living on dead organic material) and decay pathogen of seeds or young seedlings. The sexual stage begins with formation of two morphologically distinct gametangia, a female **oogonium** and a male **antheridium** (Fig. 4–6a). The nuclear condition in the mycelium of these organisms is diploid, so meiosis occurs in the gametangia. Each gametangium, once formed, contains a haploid nucleus. Contact between gametangia results in plasmogamy followed by karyogamy and formation of a new diploid nucleus. The fertilized oogonium is transformed into a thick-walled oospore, the main resting stage of the fungus between growing seasons. When a host plant is present, the oospore either germinates directly or it produces a sporangium that further differentiates a vesicle. Within the vesicle, the cytoplasm divides with nuclei to form zoospores that are motile in surface water. Zoospores encyst near their destination nutrient source and either germinate directly when conditions are dry or release zoospores in water. The fungus increases in number rapidly through massive production of asexual sporangia, which either germinate directly or produce and release zoospores in the presence of free water.

Chytridiomycota

These fungi are found mostly in aquatic environments, including wet soils. Many are saprophytes and can be baited from streams and ponds using pollen grains, hemp seed, and other plant parts. Others are parasites of algae and insects. A few are symptomless plant pathogens, such as *Olpidium brassicae,* or produce wart damage on potato tubers, such as *Synchytrium endobioticum.*

These fungi reproduce by forming zoospores with a single whiplash flagellum within a sac-like cell called a sporangium. Most organisms are made up of single cells that vary considerably in how they colonize host cells. Some species form rudimentary coenocytic hyphae, called *rhizomycelia,* which serve as a channel for migration of nuclei much like true hyphae. Some develop completely within the host cell; others develop on the cell surface and are anchored by anucleate rootlike **rhizoids.**

A typical chytrid life cycle is exemplified by *Rhizophydium couchii,* a parasite of green algae (Fig. 4–6b). Sexual reproduction begins with contact between a male gametangium and a larger female gametangium anchored by rhizoids to an algal cell. Karyogamy and meiosis have not been confirmed in this species, but both are presumed to occur. A sporangium develops, within which nuclei and cytoplasm become walled off to form zoospores, which are discharged from exit papillae (i.e., bumps on the cell surface which dissolve). Zoospores swim to new algal cells, penetrate cell walls, and encyst or produce sporangia containing zoospores. Asexual spores consist of zoospores from the mitotic division of nuclei within sporangia.

Zygomycota

These fungi are terrestrial, with many growing as saprophytes in soil and dung. Some are parasites of insects, and others form beneficial endomycorrhizal associations with many land plants. Most are characterized by a coenocytic mycelium, sexual **zygospores** and asexual sporangia. Others, such as endomycorrhizal fungi in the order Glomales, produce diverse asexual spores that share few (if any) morphological similarities with other Zygomycetes. None of the spores produced by zygomycete fungi are motile.

The life cycle of a zygomycete fungus is exemplified by *Mucor piriformis* (Fig. 4–6c), a common degrader of organic matter whose sexual stage occurs in nature and can be induced in **axenic** (single species) culture. Initiation of the sexual stage requires two hyphae with genes of opposite mating type (+ and −). When these mating types make contact, morphologically identical gametangia form that are walled off from the rest of the mycelium. After plasmogamy between gametangia, a zygospore is formed that has an ornamented outer cell wall. The nuclei fuse (karyogamy) to form a single diploid nucleus, which then undergoes meiosis to form daughter haploid nuclei. The zygospore germinates to form a long stalk (called a **sporangiophore**) with a swollen tip on which the sporangium develops. Haploid nuclei in each sporangium become walled off and differentiate into **sporangiospores.** When the sporangium wall decomposes, sporangiospores are released and dispersed on air currents. Asexual sporangia develop from the mycelium of a single mating type and contain sporangiospores with haploid nuclei produced from mitotic divisions.

Ascomycota

Most of these fungi are terrestrial. Life cycles and morphologies of spores and fruiting bodies are more diverse than in any other phylum of the Kingdom Fungi. Many are saprophytes, but some preferentially colonize specific plant tissues or hosts. Notorious plant pathogens include those causing powdery mildew diseases, ergot, chestnut blight, and Dutch elm disease. Ascomycete yeasts are important in bread and alcohol production. Morels and truffles are prized delicacies. Many widespread soil inhabitants are asexually reproducing species, such as those in the genera *Aspergillus, Fusarium,* and *Penicillium* (Box 4–2).

Hyphae are partitioned by septa perforated with one to many small pores. The number of nuclei per compartment is variable, since the movement of nuclei is not tightly controlled. The main features of the sexual stage in fungi of this phylum are formation of **ascospores** enclosed within thin colorless "sacs" called **asci** (singular, **ascus**), which often are clustered within fruiting structures called **ascomata** (singular, **ascoma**). Ascomata vary considerably in shape and organization to accommodate inventive mechanisms for active and passive discharge of ascospores. They include closed spherical **cleistothecia,** flask-shaped **perithecia,** and saucer-shaped **apothecia.** Asexual conidia of various sizes, shapes, and colors are produced on vegetative hyphae. They also can form inside fruiting bodies of similar shapes to those producing asci. Some species, such as *Sordaria fimicola* (Box 4–4), have only a sexual phase whereas many others, especially soil fungi, are strictly asexual.

The life cycle of an ascomycete fungus is exemplified by *Gibberella moniliforme,* a pathogen of corn that produces ascospores in a perithecial fruiting body (Fig. 4–6d). Within a juvenile perithecium, fertile hyphae of opposite mating types anastomose (i.e., fuse) and branch to form hyphae from which asci arise. A stable dikaryon (a pair of nuclei) of each mating type is positioned in the crooked apex of an ascogenous hypha, which then undergoes one mitotic division. The nuclei migrate so that one nucleus of each mating type is paired with a nucleus of the opposite mating type to establish a **heterokaryon,** so named because the pair of nuclei differ in at least one gene (mating type in this case). The heterokaryon is walled off within a dome-shaped ascus mother cell, where it undergoes karyogamy to form a diploid nucleus, meiosis (yielding four daughter nuclei), and one mitotic division (yielding eight nuclei). Each nucleus and its surrounding cytoplasm becomes an ascospore within an elongated ascus. The mature perithecium absorbs water and forces ascospores through the apex pore (ostiole) in spring for subsequent dispersal by splashing or wind-blown rain. Two asexual conidial morphotypes are produced from mitotic division of nuclei in the mycelium of each mating type: one small, globose, and single-celled and the other larger, sickle-shaped, and multicellular.

Basidiomycota

These fungi generally are saprophytes in soil, leaf litter, and decaying wood (including standing trees). Some are plant pathogens that cause tree root and crown rots and destructive flower and foliar diseases, such as smuts and rusts. Others are important beneficial ectomycorrhizal symbionts of tree species. Many of the mushrooms produced by these fungi are prized for their excellent flavor or notorious for their deadly poison.

Hyphae of basidiomycete fungi may be separated into compartments by dolipore septa and clamp connections (Fig. 4–4c), although these features are not present in all species. In the sexual stage, **basidiospores** are produced on the surface of a club-shaped **basidium** (plural, **basidia**) situated within diverse and often complex fruiting bodies such as mushrooms, conks, puffballs, or jellylike masses. Most soil basidiomycete fungi reproduce asexually by producing spores, such as chlamydospores, from the mycelium instead of expending energy on conidia formation. Asexual sclerotia and fragments of mycelia, rhizomorphs, and mycelial strands function somewhat like spores in dispersal, although for much shorter distances.

The life cycle of a basidiomycete fungus is exemplified by *Agaricus campestris,* the common "meadow mushroom" (Fig. 4–6e). Basidiospores released from gills on the underside of the mushroom cap germinate on a suitable substrate, which usually is decaying litter in or on soil. They first produce primary mycelia consisting of hyphae with compartments in which each contains only one haploid nucleus (**monokaryotic**). Primary mycelia of opposite mating types fuse to produce a heterokaryotic secondary mycelium. The heterokaryotic condition is stable, and sexual reproduction begins after a mushroom basidiocarp is formed. The outer layers of the gills on the underside of the mushroom cap are lined with numerous club-shaped basidia in a fertile layer called a **hymenium.** Each basidium contains a heterokaryon that undergoes karyogamy (to form a diploid nucleus), followed by meiosis. Each daughter haploid nucleus is enclosed within a basidiospore positioned at the tip of a pointed projection on the surface of the basidium. That projection is involved in forcible discharge of the basidiospore.

Nutritional Patterns

Fungi are **heterotrophs,** in that they must obtain carbon and other nutrients from organic matter in the external environment. Nutrients are acquired by absorption through chytrid, yeast, or hyphal cell walls. Fungi which degrade nonliving organic matter are called saprophytes, and they are important agents in soil mineralization processes such as ammonification and carbon cycling. In saturated soils or aqueous environments, chytrids and oomycete fungi are the most common saprophytes, whereas fungi in other phyla are more abundant in drier soils. In general, fungi are obligate aerobes in that they cannot grow without an oxygen supply. They tend to be more abundant in acidic soils, where there is less competition from bacteria.

Saprophytic fungi, together with bacteria, are primary agents in the decomposition of cellulose, hemicellulose, and pectin in plant cell walls. Lignin, generally the third most common component of plant residues, can be degraded by fungi, especially some species that decay wood. Enzymes that decompose lignin also contribute to the degradation of organic pollutants and pesticides in soil (Aust, 1990).

Saprophytic fungi are capable of producing many other enzymes as well, including xylanases (to break down xylan, a hemicellulose polymer of plant cell walls) and cutinases (to break down cutin on leaf surfaces). Many of these enzymes are excreted into synthetic media when fungi are cultured in the laboratory. As a result, they are excellent candidates for industrial applications such as food processing, waste treatment, and production of alcoholic beverages (Lowe, 1992).

Numerous fungi are capable of forming specialized positive (**symbiotic**) or negative (**pathogenic**) associations with other living organisms (Fig. 4–7). The relationship is **necrotrophic** when a fungus produces abundant enzymes to degrade and kill tissues of its host. **Biotrophic** relationships are more complex because the fungus establishes close contact with host cell contents via specialized absorptive structures and then induces hormonal changes to channel carbon flow in the host toward those contact sites. Most necrotrophic relationships are **facultative** because the fungi can produce the degradative enzymes to live freely as a saprophyte for a part of its life cycle. Duration of contact between a fungus and its host is highly variable, often depending on the physiology of each associate

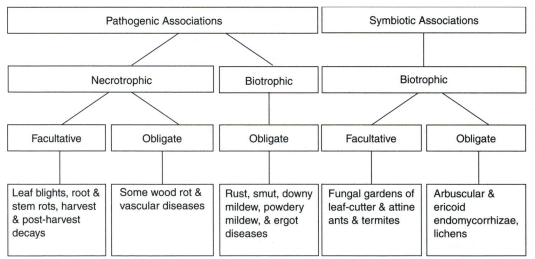

Figure 4–7 Associations between fungi and other living organisms to obtain nutrients for growth under natural ecological conditions. Pathogenic = harmful to one or both associates; symbiotic = beneficial to both associates; necrotrophic = nutrients acquired from decomposition of host cells by enzymes or toxins; biotrophic = no destruction of host cells, with nutrients obtained in most cases from specialized absorptive organs in association with host; facultative = host-associated and also free-living; obligate = only host-associated.

and environmental conditions. Most facultative organisms are readily culturable on synthetic nutrient media. Facultative biotrophic associations are unique in that the fungal partners are saprophytic. For example, some of these fungi digest wood for ingestion by ants, which in turn forage for substrate and provide a competition-free "home" for the fungus. These fungal symbionts are the fungal equivalents of "couch potatoes": they remain confined while their insect partners seek out and transport food resources back to the colony. All other known biotrophic relationships are **obligate** in that the fungi need their hosts to grow and reproduce. Some symbiotic fungi, such as ectomycorrhizal and ericoid mycorrhizal fungi, are obligate symbionts under natural conditions but they still can grow slowly on a synthetic medium containing essential nutrients that can be absorbed by their mycelia.

Some species in Chytridiomycota and Oomycota produce motile zoospores that exhibit chemotaxis and swim along a gradient toward increasing concentrations of simple sugars, amino acids, or organic acids exuded by roots of the target host. The Ascomycota and Basidiomycota do not produce motile spores, but some species have evolved other effective methods to incapacitate their hosts. Some set traps for their hosts. For example, *Arthrobotrys* and *Dactylella* species in Ascomycota have specialized hyphal structures to "capture" their nematode prey, such as adhesive knobs, branches, or nets. Other species have constricting and nonconstricting rings as hyphal outgrowths (see Fig. 6–1). The mycelia of *Pleurotus* spp. in Basidiomycota release toxins into the soil to paralyze nematodes long enough to allow hyphal penetration through the tough host cuticle.

The filamentous nature of the mycelium and the three-dimensional branch architecture of component hyphae afford fungi with a reliable and efficient mechanism to obtain nonuniformly distributed nutrients in soils and translocate those nutrients to distant parts of the organisms or wherever they are most needed for continued growth or reproduction. When nutrient depletion extends over a large area, fungal species often survive by storing available endogenous nutrients in spores and dispersing them to more favorable environments.

Genetic Patterns

The nuclei in hyphae of true fungi are haploid during vegetative growth and asexual reproduction. With only one set of chromosomes, phenotypic traits may be expressed immediately. Hyphae are multinucleate, however, and expression of phenotypes is determined by associations between nuclei and the genetic differences between them. For example, a dikaryon in the hyphal compartment of a septate hypha, as found in many basidiomycete fungi, consists of two physically separated haploid nuclei. They function together as a diploid, and gene products of both nuclei interact in a common cytoplasm. Genetic flexibility is enhanced by independent movement of each haploid nucleus, especially in branches of hyphae that separate to start new "colonies," in repetitive formation of spores from a hyphal branch, or when cytoplasm of two or more fusing hyphae is mixed. The extent to which nuclear migration is controlled varies among phyla, being least in those with coenocytic hyphae and most tightly regulated in basidiomycete fungi with their complex dolipore septa and clamp connections.

Genetic variation can occur rapidly with sexual reproduction as a result of gene recombinations. Many soil fungi have only an asexual phase, so that genetic flexibility and adaptability depends on either mutations (Box 4–2) or a phenomenon known as the **parasexual cycle.** The latter was discovered in laboratory studies, and its significance under field conditions has yet to be clearly established. It entails the following four-step process mimicking a sexual cycle, but without gametangia and meiosis:

1. Anastomosis of vegetatively compatible hyphae and mixing of nuclei until hyphal compartments are heterokaryotic,

2. Fusion of paired nuclei to form a somatic diploid,

3. Mitosis with crossing-over of chromosomes to cause some genetic recombinations, and

4. Spontaneous loss of chromosomes until the stable haploid condition is restored.

Spontaneous mutants can appear at any time, and they are most noticeable in fungal colonies growing in Petri dish cultures. Common manifestations of a mutant are visible differences in mycelium or spore color, architecture of the hyphae (aerial, pressed closely to the agar surface, or submerged within agar), spore abundance, or size (rate and amount of hyphae growing radially over an agar surface).

Aside from hormonal signals, sexuality in fungi is regulated by the physical origin of gametangia and their mating type. A mycelium on which both gametangia develop is called **homothallic.** It is self-compatible in that there are no mating type barriers to plasmogamy. When each gametangium develops on a separate mycelium, then it is **heterothallic.** Plasmogamy occurs only when gametangia are sexually compatible, which requires opposite mating types (+ and −).

In soil environments, some degree of genetic stability is an asset rather than a liability. With the exception of some species in Basidiomycota, soil fungi tend to be homothallic. Genetic variation is not lost, since hyphae still may be heterokaryotic for any number of characters. With karyogamy and meiotic recombination, variation is promoted, but it occurs without too much divergence from the parental phenotype (Raper, 1966). Opposite mating types are more beneficial to fungi producing large numbers of air-dispersed spores which, upon germination, have a greater probability of making contact with each other to begin another sexual cycle (such as *Mucor piriformis,* illustrated in Fig. 4–6c).

Survival Patterns

Fungi are like plants in that they cannot move very far or fast during vegetative growth. Therefore, they must rely on huge numbers of spores for long-range dispersal. Sporulation often is triggered under conditions where growth becomes limiting (e.g., nutrient deprivation or environmental changes), and thick-walled spores often serve as a resting stage. Some fungi inhabiting host tissues do not sporulate, possibly because the niche they occupy affords enough protection. Various mycelial aggregates discussed in the previous section serve as survival structures because they are more resistant than diffuse mycelium in resisting parasitism and adverse environmental conditions.

Spore Production and Mutations. Many soil fungi only reproduce by asexual means so that, in the absence of genetic recombination, mutations provide a means to maintain genetic resiliency. Fungi such as *Aspergillus* (a) or *Penicillium* (b) species produce as many as one million asexual conidia from mycelium in one kilogram of soil or from fruiting bodies like that of a *Phoma* species (c). Assuming one nucleus per spore and a rate of one mutation at a gene in each nucleus, as many as 100 million mutants can occur in the top 15 cm of soil in an area of one hectare. Even if many of the mutations are harmful, they will be detrimental only to that portion of the mycelium where the nuclei are situated. Other parts of the mycelium lacking these mutants will continue to grow normally. Beneficial or neutral mutations may be inherited immediately if they migrate into hyphae or spores that successfully colonize new substrates and reproduce again. Mutants are spread rapidly when spores are dispersed aerially or by water, insects, and soil microfauna.

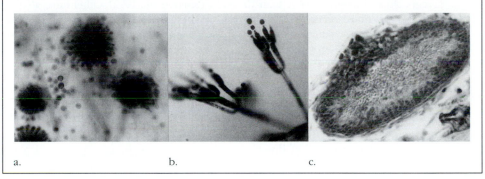

a.   b.   c.

Substrate specialization by saprophytes, or host specialization by parasites or symbionts, are important factors in how long fungi will grow and reproduce in agricultural systems involving crop rotations. A fungus with specialized nutrition is more likely to decline or die out with a nonhost plant (or without a utilizable substrate) in a rotation sequence, whereas a fungus with a wide host range will survive and possibly flourish. Soil fungi often have enough resiliency to convert from one resting structure to another when nutrient levels are too low. Both microconidia and macroconidia of *Fusarium* species (the asexual forms of *Gibberella moniliforme* in Fig. 4–6d) can form dark, thick-walled chlamydospores (Kommedahl, 1966). If conditions are adverse when a chlamydospore germinates, then another chlamydospore is quickly formed. These patterns may be very important to survival in agricultural management systems that involve fallow periods between crop cycles.

The phenomenon of **fungistasis,** a general phenomenon in which natural soils suppress germination of fungal spores or other resting structures, is an important mechanism by which fungi maintain viability when nutrients become limiting. This suppression remains in effect until exogenous nutrients again become available and sometimes even until other environmental conditions are optimal for fungal growth. Fungistasis usually is broken after threshold levels of sugars and nitrogenous compounds are reached in proximity to the dormant spore or resting structure. Plant species-specific stimulants in the rhizosphere also may be involved.

The longevity of fungal organisms has not been measured in many species (Box 4–3), but their genetic plasticity and open-ended growth habit offers unlimited potential.

Fossil evidence, together with measurements of the rate of change in DNA nucleotide sequences, suggest that many fungi have a longevity measured in the millions of years!

Box 4–3

Fungal Longevity. The "individuality" of a single fungal organism is difficult to define because of the diffuse hyphal network and the dispersal of sexual spores, asexual spores, or both. To distinguish individuals, researchers measure genetic homogeneity in mycelium either by testing for vegetative compatibility or comparing random amplified polymorphic DNA (Chapter 9). These methods have been used recently on a mushroom-forming basidiomycete fungus, *Armillaria gallica* (reported as *A. bulbosa*), a saprophyte in the root zone of hardwood trees, present in 20 ha of a forest of Michigan's Upper Peninsula. All of the samples collected were found to be part of one organism estimated to be at least 1,500 years old and weighing more than 10,000 kg (Smith et al., 1992). Other fungi also are long-lived. Fairy rings, which result from extensive fungal spread in the root zone of turfgrass, advance in ever-widening circles, much like growth on a Petri plate, for decades or centuries instead of days or weeks.

Dispersal Patterns

Almost any part of the fungal organism can serve as a vehicle of dispersal and renewed growth at a different location, which is why fungi are found in almost any nook or crevice to exploit diverse nutritional niches. Fragmentation of a mycelium does not necessarily kill a fungal organism; instead, it results in a multitude of new propagules that can initiate new growth on a suitable substrate. Researchers take advantage of this potential by blending intact mycelium (Fig. 4–3) in water to multiply the number of viable units from one to many. Fragmentation occurs naturally in soils from freezing and thawing cycles between seasons, feeding by microfauna, tillage practices, and other natural or human-driven events. Mycelial fragments or discrete resting structures in soil (such as rhizomorph fragments, sclerotia, chlamydospores, oospores, and zygospores) travel short to moderate distances via water runoff or irrigation. Long-range dispersal of these propagules usually requires aerial mobility in wind currents using vectors such as insects, birds' feet, plant debris, soil particles, seeds, and pollen. Airborne sexual spores or asexual conidia usually range in size from 3 to 40 μm, and they are either passively or actively discharged. *Passive discharge* of sporangiospores or conidia produced on individual or aggregated aerial hyphae is caused by external forces such as splashing raindrops or sporadic wind gusts. *Active discharge* involves a variety of unusual and often dramatic mechanisms. In some ascomycete fungi (Box 4–4), spores are forcibly ejected from asci, much like spitting out a fruit pit. Numerous basidiospores in mushrooms are released ballistically (like projectiles) from each basidium of the gill lining. An average-sized mushroom of *Agaricus campestris* discharges up to 16 billion basidiospores within 24 hours.

In aquatic environments or extremely moist soils, chytrids and oomycete fungi are dispersed as zoospores. These thin-walled spores are fragile and not long-lived, but their motility allows them to quickly reach and exploit new food sources. Ascomycete and basidiomycete fungi produce spores with elaborate appendages that function to aid in buoyancy and rapid movement in water.

| Box 4–4 |

Forcible Spore Discharge. *Sordaria fimicola* reproduces only sexually by forming abundant flask-shaped fruiting bodies to promote forcible discharge of ascospores aerially together with phototrophic (i.e., light-seeking) behavior. Both types of behavior can be examined experimentally by placing a small section of a fungal colony on a solid sterile nutrient medium such as potato dextrose agar. The lid of the Petri dish then is covered with black construction paper in which an opening of some shape has been cut (a). The Petri dish is placed under bright light (continuous if possible) for 10–12 days. When the paper is removed, the opening will be darkened due to thousands of dark olive colored ascospores adhering to the underside of the Petri dish lid (b). As each perithecium differentiates (c), ascospores are produced within each ascus (d). The perithecium neck bends in the direction of light passing through the opening in the black paper (e). A mature ascus containing eight ascospores becomes positioned in the ostiole (pore at apex) by extension. When the ascus ruptures, ascospores are released forcibly in a trajectory aimed by the perithecial neck (f). After discharge, the ascus retracts and decomposes while a new ascus takes its place (g). Each ascospore is coated with a mucilaginous material, so that spores often stick together, and they readily adhere to the dish lid or any other object encountered.

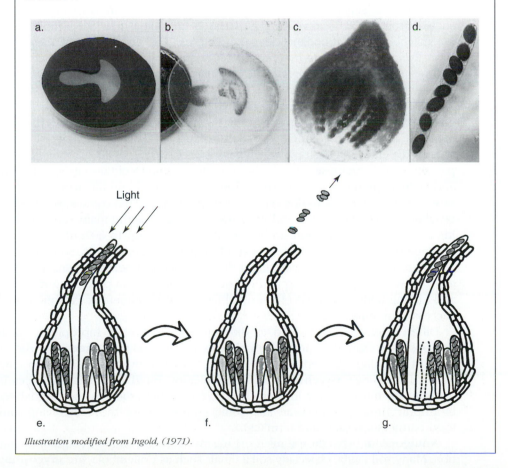

Illustration modified from Ingold, (1971).

Ecological Interactions

Fungal hyphae must be in close contact with a living or dead organic substrate to absorb nutrients, so they rarely grow alone, especially in soils. They compete aggressively among themselves and with other microorganisms for scarce nutrients. Competition and change in available nutrients usually leads to **succession,** or a directional change in composition and abundance of a species. In contrast to plant succession, a stable climax stage is rarely reached with fungi because exploitation of nutrients results in continuous change in nutrient status.

Competitive interactions among fungi are complex because many species can coexist in substrates constantly in a state of flux. For example, Bills and Polishook (1994) carried out extensive culturing procedures on a 5-g sample of leaf litter from a Costa Rican rain forest and recovered between 78 and 134 fungal species in each sample. More than 300 species were found.

Directional taxonomic changes in fungal community structure occur as a plant substrate or a soil undergo shifts in nutrient availability. In general, fungi in the Oomycota, Chytridiomycota, and Zygomycota tend to precede those in the Ascomycota, which in turn precede those in Basidiomycota. Initial colonizers use simple soluble sugars, amino acids, and vitamins in plant protoplasm or plant parts such as fruits, seeds, and vegetables. These "sugar fungi" produce a rapidly spreading mycelium and abundant asexual structures. The dominance of these fungi is short-lived because waste by-products and other conditions cause growth to cease and resting survival structures to be produced. Examples include *Pythium* species (Fig. 4–5a), *Mucor piriformis* (Fig. 4–5c), and common molds like *Rhizopus* species.

Cellulose degraders appear next in this succession, although sugar fungi may grow concurrently by using some of the by-products. Cellulose degraders constitute the most diverse and competitive group because of the heterogeneity of substrates present. Degradation of straw, which has a high C/N ratio of 80/1, requires that fungi parasitize or decompose the mycelium of other fungi to obtain nitrogen for growth and enzyme production. Abundant colonizers of straw in no-till crops include many saprophytes or pathogens that reproduce mainly by asexual conidia, such as species of *Chaetomium, Fusarium,* and *Trichoderma.* Lignin is the main substrate remaining after cellulose depletion. In humus, it constitutes up to 60% of the total mass. The number of fungal species capable of degrading lignin is low, so competition is greatly reduced. Basidiomycete fungi that produce the necessary enzymes tend to grow slowly because they expend considerable biomass in producing rhizomorphs and fruiting bodies. Because of the variables involved, numerous exceptions to these successional trends exist.

Different environments lead to different successions of saprophytic fungi. In compost, for example, successive colonists are increasingly **thermophilic** (heat-tolerant) initially, followed by a decrease in tolerance over time. In frozen environments, increasingly **psychrophilic** (cold-tolerant) fungi are selected. Succession is likely to be much slower in more specialized niches, such as those inhabited by biotrophic fungi (mycorrhizae and some pathogens) or those involving intricate food partnerships (ants and termites).

Antagonism between soil microorganisms also is a common community interaction. Many soil fauna, especially soil animals such as Collembola, are **mycophagous**

(i.e., they feed on fungal mycelia). Fungi also produce antibiotics. Antibiotic production by fungal species in ascomycete genera such as *Aspergillus, Fusarium,* and *Penicillium* provides an ecological advantage in sustaining occupation of a nutrient resource. One of the more clear cut cases of antibiosis in fungi involves the cereal vascular pathogen, *Cephalosporium gramineum.* It produces a broad-spectrum antibiotic that inhibits other organisms from colonizing infested wheat straw when soil pH is less than 6.5 (Bruehl et al., 1969). This antibiotic appears to play more of a role in survival than disease development, since its presence or absence has no effect on disease-causing ability.

Summary

Fungi are highly diverse groups of eukaryotic microorganisms spanning at least two kingdoms. Unicellular chytrids and the flagellated zoospores of oomycete fungi are well adapted to aquatic environments and moist soils. The branching architecture and indeterminate growth of mycelia of terrestrial fungi optimize their ability to explore large soil volumes and exploit heterogeneous organic substrates distributed far and wide. Various mycelial structures optimize this behavior and also provide a hardy resting stage when conditions are adverse. Haploid nuclei are capable of movement in hyphae either singly, as dikaryons, or in greater numbers. Cytoplasmic continuity, even with septa, provides a mosaic of genetic and phenotypic possibilities. Spores are produced sexually, asexually, or both. Spores not only package nuclei of parental or new genotypes (through recombination, mutation, or parasexuality), but they also parcel nutrients, serve as resting structures, and function as vehicles for dispersal.

In the soil environment, fungi are important as food sources, pathogens, beneficial symbionts, saprophytes to degrade crop residues, and biotic agents to improve soil structure and aeration. Saprophytes are relatively unspecialized, producing enzymes that degrade complex organic compounds to simple constituents (e.g., sugars) for absorption. Parasitic or symbiotic associations between fungi and other living host organisms are complex, often producing fungal structures in close contact with host cells to absorb nutrients. Despite the widespread distribution of fungi and their pivotal role in biotic processes, much has yet to be learned about their diversity and the complexity of interactions among each other or with other eukaryotic and prokaryotic organisms.

Cited References

Alexopoulos, C.J., C.W. Mims, and M. Blackwell. 1996. Introductory mycology. 4th ed. John Wiley & Sons, Inc., New York.

Aust, S.D. 1990. Degradation of environmental pollutants by *Phanerochaete chrysosporium.* Microbial Ecol. 20:197–209.

Barr, D.J.S. 1992. Evolution and kingdoms of organisms from the perspective of a mycologist. Mycologia 84:1–11.

Bills, G.F., and J.D. Polishook. 1994. Abundance and diversity of microfungi in leaf litter of a lowland rain forest in Costa Rica. Mycologia 86:187–198.

Bruehl, G.W., R.L. Millar, and B. Cunfer. 1969. Significance of antibiotic production by *Cephalosporium gramineum* to its saprophytic survival. Can. J. Plant Sci. 49:235–246.

Hawkswork, D.L. 1991. The fungal dimension of biodiversity: Magnitude, significance, and conservation. Mycol. Res. 95:641–655.

Ingold, C.T. 1971. Fungal spores: Their liberation and dispersal. Clarendon Press, Oxford, England.

Jennings, L., and S.C. Watkinson. 1982. Structure and development of mycelial strands in *Serpula lacrymans*. Trans. Br. Mycol. Soc. 78:465–474.

Kommedahl, T. 1966. Relation of exudates of pea roots to germination of spores in races of *Fusarium oxysporum* f. sp. *pisi*. Phytopathology 56:721–722.

Lowe, D.A. 1992. Fungal enzymes. pp. 681–706. *In* D.K. Arora, R.P. Elander, and K.G. Mukerji (eds.) Handbook of applied mycology. Vol. 4. Marcel Dekker, Inc., New York.

Metting, F.B., Jr. 1993. Structure and physiological ecology of soil microbial communities. pp. 3–25. *In* F.B. Metting, Jr. (ed.), Soil microbial ecology. Marcel Dekker, Inc., New York.

Raper, J.R. 1966. Life cycles, basic patterns of sexuality, and sexual mechanisms. pp. 473–511. *In* G.C. Ainsworth and A.S. Sussman, (eds.) The fungi: An advanced treatise. Academic Press, New York.

Smith, M.L., J.N. Bruhn, and J.B. Anderson. 1992. The fungus *Armillaria bulbosa* is among the largest and oldest living organisms. Nature 356:428–431.

General References

Carroll, G.C., and D.T. Wicklow (eds.). 1992. The fungal community: Its organization and role in the ecosystem. 2nd ed. Marcel Dekker, Inc., New York.

Harley, J.L. 1971. Fungi in ecosystems. J. Ecology 59:627–642.

Hawksworth, D.L., P.M. Kirk, B.C. Sutton, and D.N. Pegler. 1995. Ainsworth and Bisby's dictionary of the fungi. 8th ed. CAB International, Mycological Institute, Wallingford, England.

Henis, Y. (ed.). 1987. Survival and dormancy of microorganisms. John Wiley & Sons, Inc., New York.

Hudson, H.J. 1986. Fungal biology. Edward Arnold, Ltd., London.

Kendrick, B. 1992. The fifth kingdom. Focus Texts, Newburyport, Mass.

Study Questions

1. What are the unique properties of fungi that distinguish them from prokaryotes, plants, and animals?

2. What are the advantages of a mycelial body plan for (a) growth in soil with patchy nutrient-rich and nutrient-poor zones, (b) reproduction, and (c) survival?

3. Why would a fungal organism occupying a soil pore more likely survive loss of moisture or nutrient substrates than a local population of bacteria?

4. What are the main vegetative structures formed to resist environmental fluxes, and what other, if any, functions do they serve?

5. What are the advantages of soil fungi in producing numerous asexual spores?

6. Besides genetic recombination, what are important functions of sexual spores in each fungal class?

7. Define heterokaryosis, and discuss how this phenomenon benefits genetic plasticity in fungi.

8. In the various specialized symbioses, how do fungi differ in their ability to obtain nutrients?

9. What factors are most important in governing the distribution of fungi in agricultural soils?

10. Define fungistasis and discuss its role in improving survival of fungal resting structures in soil.

Chapter 5

◆

Eukaryotic Algae and Cyanobacteria

Steve L. Albrecht

◆

In any case, prokaryotic organisms held the earth as their exclusive domain during two-thirds to five-sixths of the history of life. With ample justice Schopf labels the Precambrian as the 'age of blue-green algae.'

Stephen Jay Gould

Soil algae occur in nearly all terrestrial environments. They are simple **photoautotrophic** organisms that lack tissue differentiation, but they vary greatly in morphology, physiology, reproduction, and habitat. Although considerable numbers of algae are present at the surface and within the subsurface layers of most soils, there is a general lack of awareness of their presence. Hence, the algae have been much less studied than nonphotosynthetic soil microorganisms. This lack of attention has fostered the impression that they are not an important component of the community of soil microorganisms, even though algae may be the only primary producers present in certain ecosystems.

The study of soil algae can be traced to the beginning of the nineteenth century. Beijerinck first reported the isolation of a soil alga in axenic culture in 1893, and in 1895 Graebner gave the first account of algae as ecological constituents of the soil (Chapman, 1941; Starks et al., 1981).

Classification

There are two major groups broadly defined as algae:

- **Eukaryotic algae** are part of the plant kingdom.

- **Prokaryotic cyanobacteria** (formerly known as blue-green algae) are part of the **Bacteria.**

Table 5–1 **Classes (and common names) for representative genera of the soil algae.**

Chlorophyceae Green	Bacillariophyceae Diatoms	Xanthophyceae Yellow-green	Cyanophyceae "Blue-green"	Euglenophyceae Euglenoids	Rhodophyceae Red
Ankistrodesmus	*Achnanthes*	*Botrydiopsis*	*Anabaena*	*Euglena*	*Cyanidium*
Characium	*Amphora*	*Botrydium*	*Calothrix*	*Peranema*	*Porphyridium*
Chlorella	*Caloneis*	*Bumilleria*	*Gleocapsa*		
Chlorococcum	*Cymbella*	*Bumilleriopsis*	*Lyngbya*		
Hormidium	*Fragilaria*	*Geobotrys*	*Microcoleus*		
Protococcus	*Hantzschia*	*Heterothrix*	*Nostoc*		
Stichococcus	*Navicula*	*Pleurochloris*	*Phormidium*		
Ulothrix	*Stauroneis*	*Vaucheria*	*Stigonema*		

Table 5–2 **Characteristics commonly used to classify algae.**

Class	Pigments	Storage products	Flagellation
Chlorophyceae	Chlorophyll a and b, α- and β-carotene, astaxanthin, lutein, neoxanthin, siphonein, siphonoxanthin, violaxanthin, zeaxanthin	Starch, oils	1, 2, 4 to many, equal apical or subapical insertion
Xanthophyceae	Chlorophyll a and c, β-carotene, lutein, neoxanthin	Chrysolaminarin, lipids	2, (equal or unequal) apical
Cyanophyceae	Chlorophyll a, β-carotene, antheraxanthin, aphanicin, allophycocyanin, aphanizophyll, flavacin, lutenin, myxoxanthin, myxoxanthophyll, oscilloxanthin, phycocyanin, phycoerythrin, zeaxanthin	Cyanophycean starch (the amylopectin portion of starch), proteins	Absent
Bacillariophyceae	Chlorphyll a and c, β-carotene, lutein, diadinoxanthin, fucoxanthin	Chrysolaminarin, oils	1 apical on sperm
Euglenophyceae	Chlorophyll a and b, β-carotene, antheraxanthin, astaxanthin, lutein, neoxanthin	Paramylon, oil	1, (usually) 2, or 3
Rhodophyceae	Chlorophyll a and d, α- and β- carotene, lutenin, taraxanthin, allophycoyanin, phycocyanin, phycoerythrin	Floridean starch, oils	Absent

Adapted from Alexopoulos and Bold (1969)

For convenience, many authors include the **edaphic** (soil dwelling) cyanobacteria as part of the soil algae, a convention that will be followed in this chapter. The most common soil algae are the cyanobacteria (class Cyanophyceae), the green algae (class Chlorophyceae), the diatoms (class Bacillariophyceae), and yellow-green algae (class Xanthophyceae). Euglenoids (class Euglenophyceae) and red algae (class Rhodophyceae) occur less frequently. Several representative genera are presented in Table 5–1 and illustrated in Figure 5–1.

Early classification schemes relied heavily on the presence of specific photosynthetic and accessory pigments. Current systems of classification additionally incorporate information about cell-wall constituents, cellular organization, flagellation, and molecular biology (Table 5–2). Much algal taxonomy, especially that of the Chlorophyceae, remains uncertain. The systematics of soil algae have been reviewed

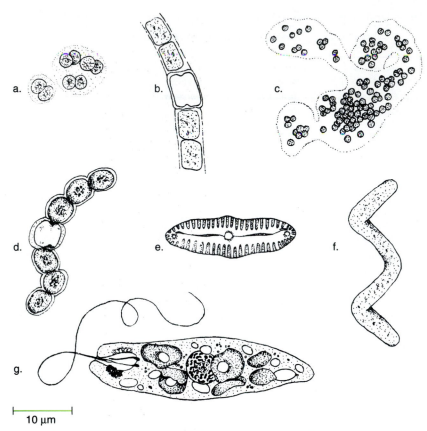

Figure 5–1 Some representative soil cyanobacteria (a–d and f) and eukaryotic algae (e and g). (a) *Gleocapsa,* (b) *Anabaena,* (c) *Microcystis,* (d) *Nostoc,* (e) *Pinnularia,* (f) *Spirulina,* (g) *Euglena. Original drawing by Kim Luoma.*

by Metting (1981), and a recent two-volume survey by Christensen (1980, 1994) provides comprehensive coverage of algal systematics.

Major Groups Found in Soil

Cyanobacteria

The cyanobacteria are widely distributed. In addition to soil, the terrestrial species may also be found on plants, rocks, and even animals. All species belonging to this group are unicellular or filamentous; cells frequently remain together, surrounded by a gelatinous material. At the subcellular level, the cyanobacteria are morphologically and physiologically similar to bacteria. Their cell walls show some chemical similarity to those of bacteria. As would be expected, cyanobacterial DNA is not separated from the rest of the cytoplasm by a nuclear membrane; hence there is no distinct nucleus. However, many species may have a relatively dense

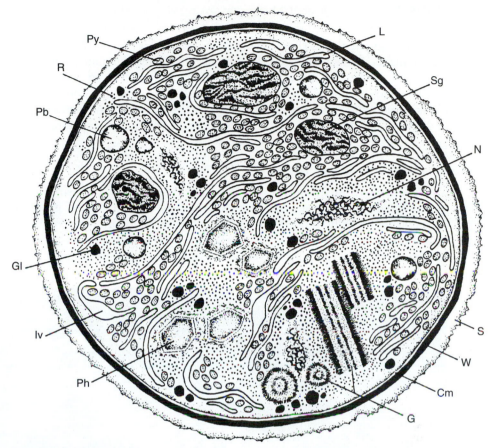

Figure 5–2 Diagram of a cross section of a cell of a cyanobacterium: G = gas vesicles, Gl = glycogen granules, Iv = intralamellar vesicle, L = lamellae, N = nucleoplasm, Pb = polyphosphate body, Ph = polyhedral body, Cm = cytoplasmic membrane, Py = phycobilisomes, R = ribosomes, S = sheath, Sg = structured (cyanophycin) granule, W = wall. *Adapted from Pankratz and Bowen, 1963. Used with permission. Figure redrawn by Kim Luoma.*

mass of material called the *central body*. Cyanobacteria do not have chloroplasts; their photosynthetic pigments are usually associated with membranous layers called **lamellae** and appear uniformly distributed throughout the cytoplasm (Fig. 5–2). The cyanobacteria contain **chlorophyll** a, but not chlorophyll b; however, the dominant pigment is the blue-colored phycocyanin. The storage product in cyanobacteria is starchlike but somewhat different from that of higher plants. Some species store food reserves as oils. Reproduction is by simple cell division without mitosis. The cyanobacteria produce several different types of immobile spores. They are usually found within a filament or trichome, arising from a vegetative cell. They form a thick wall after being filled with food reserves. Many filamentous species can form sporelike cells called **heterocysts,** which are involved in N_2 fixation.

Green Algae

In contrast to the cyanobacteria, considerable cellular organization exists in the eukaryotic green algae. Their organization and physiology closely resemble that of higher plants. In general, the cell walls of eukaryotic algae are similar to those of higher plants; their DNA is localized within minutely perforated nuclear membranes, and the photosynthetic apparatus, including the pigments, is contained in chloroplasts. Green algae possess chlorophyll a as the predominant pigment. In addition to the nucleus, eukaryotic cells may also contain organelles, including vacuoles, flagella, Golgi bodies, and mitochondria.

Diatoms and Yellow-Green Algae

Diatoms live in fresh and salt water and in the soil. Terrestrial diatoms are generally smaller than aquatic species, and most are capable of movement. The diatoms are usually unicellular but may occur in filamentous colonies or as branched or unusual clusters. The cell wall or **frustule** (Fig. 5–3) is composed of two slightly overlapping valves. The cytoplasm contains a single nucleus and

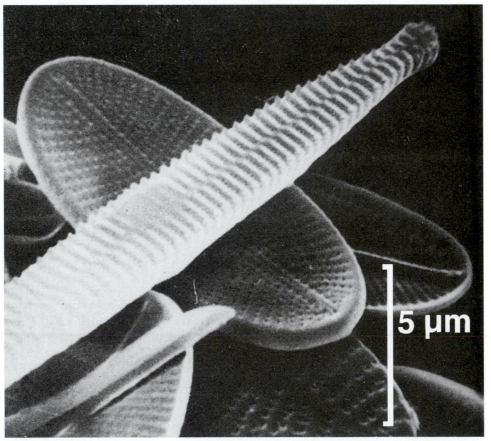

Figure 5–3 Photomicrograph of diatom cell walls, also called frustules. *From the Soil Science Society of America, Division S–3 slide set. Used with permission.*

one to many plastids. The chloroplasts contain chlorophyll a and c. Giving the cells their characteristic color that can vary from green or yellow to golden-brown is the major carotenoid, golden-brown fucoxanthin (Fig. 5–4a). The normal method of reproduction is asexual by division of one cell into two. The storage product in the diatoms is chrysolaminarin, a β-1,3 linked glucan. Diatoms form resting spores with thick, ornamented walls. In addition, they can form resting cells, distinct from the spores, that are morphologically similar to vegetative cells but lack a protective layer.

Yellow-green algae can be motile, with separate forward-directed and posteriorly-directed flagella. Their chloroplasts contain chlorophyll a but lack fucoxanthin; other carotenoid pigments provide the characteristic yellow-green coloration of this group. The cell walls of the yellow-green algae are usually composed of cellulose and consist of two overlapping halves. The principal storage product is probably a β-1,3 linked glucan, but research suggests that lipids are also important for food storage. Members of this class usually multiply asexually by fragmentation and motile or nonmotile spores. Sexual reproduction has been found in *Botrydium* and *Vaucheria*. They have the ability to form resting spores.

Morphology and Reproduction

Soil algae are simple, undifferentiated organisms. As noted previously, they may occur as unicellular types or as cell aggregates, the latter ranging from filamentous to more complex colonial species (Fig. 5–1). Individual algal cells may also differ greatly in size and shape. The cyanobacteria are similar in size to other prokaryotes and commonly have cellular volumes ranging from 5 to 50 μm^3, although volumes from 0.1 to 5,000 μm^3 are known. The unicellular eukaryotic algae are somewhat larger and have a normal range of 5,000 to 15,000 μm^3, with extremes of 5 to 100,000 μm^3. Rudimentary colonies are formed when several generations of cells remain attached after cell division. Other groupings may take the form of branched or unbranched filaments called **trichomes.** Although these organisms are considered simple, they possess great diversity. For example, the euglenoids, a small group often classified as animals rather than plants, lack a cell wall and may ingest food through a gullet. Conversely, the genus *Botrydium* has a balloon-shaped aerial thallus and a mass of branched rhizoids.

Many soil algae are able to form spores. These spores are a means of carrying the species over periods of unfavorable environmental conditions rather than a means of reproduction. They can be formed either sexually or asexually. Some species may form thick-walled resting stages, which are usually morphologically specialized structures. However, formation of resting spores may involve primarily physiological changes, including decreased metabolic activity or the production of protective mucilage.

Algae demonstrate as much variation in reproduction as they do in morphology; vegetative, asexual, and sexual processes are all present. Reproduction in the eukaryotic algae is more varied than in the other groups. Practically all the green algae include some form of sexual reproduction. Reproduction in the diatoms and yellow-green algae is usually asexual by simple cell division. In the euglenoids, the nucleus commonly divides mitotically, and reproduction is by longitudinal cell division.

Physiology

Photosynthesis

Photosynthesis is the most important biochemical reaction on earth. All our food and biological fuels are products of photosynthesis, a complex reaction in which light energy is used to convert carbon dioxide and water to carbohydrates and molecular oxygen (Chapter 10). The biochemistry of both prokaryotic and eukaryotic photosynthesis is very similar to that of higher plants. In fact, the most common pathway of carbon dioxide fixation (the C-3 pathway or Calvin cycle) was first described in the green algae *Chlorella* and *Scenedesmus* by Calvin and co-workers (Bassham and Calvin, 1957).

Photosynthesis in eukaryotic algae and higher plants is carried out in subcellular organelles called **chloroplasts.** The chloroplast, which is bounded by a double membrane, contains both the photosynthetic pigments and the electron-transport systems necessary for photophosphorylation. Chloroplasts also contain the enzymes for carbon assimilation and, therefore, are complete units of photosynthetic function. In contrast, the prokaryotic cyanobacteria do not have chloroplasts. Instead, the photochemical reactions are found in the lamellae containing photosynthetic pigments. This arrangement is structurally and functionally simpler than the chloroplasts of the eukaryotic algae. The enzymes used in the dark reactions of photosynthesis are found in the cytoplasm.

Three main classes of pigments are associated with the photosynthetic apparatus: chlorophylls, carotenoids, and phycobilins (Fig. 5–4). The chlorophylls and carotenoids are always present, but phycobilins are limited to a few groups of algae. The chlorophylls are complex organic molecules structurally related to heme, a component of hemoglobin and some cytochromes (Fig. 5–4c). Chlorophyll a is found in all organisms that carry on oxygen-evolving photosynthesis; one of the remaining chlorophylls (b, c, and d) may also be present. The chlorophylls are the principal light-absorbing pigments of the photosynthetic process.

The carotenoids (Fig. 5–4a, b) are long-chain compounds, characteristically red or yellow in color, which have absorption spectra complementary to those of the chlorophylls. Although they do not play an essential role in photosynthesis, they may contribute indirectly by transferring to chlorophyll part of the light energy they absorb, thus extending the effective spectral range for photosynthesis. In the presence of light, an important and universal function of carotenoids is to protect the photosynthetic apparatus from photochemical oxygen damage.

The phycobilins occur primarily in the cyanobacteria and red algae (Fig. 5–4d). Their distinctive blue or red coloration is conferred by a **prosthetic group,** a cofactor bound tightly to the proteins. As with the carotenoids, they are capable of transferring part of the light energy they absorb to chlorophyll, thus contributing indirectly to photosynthetic energy conversion.

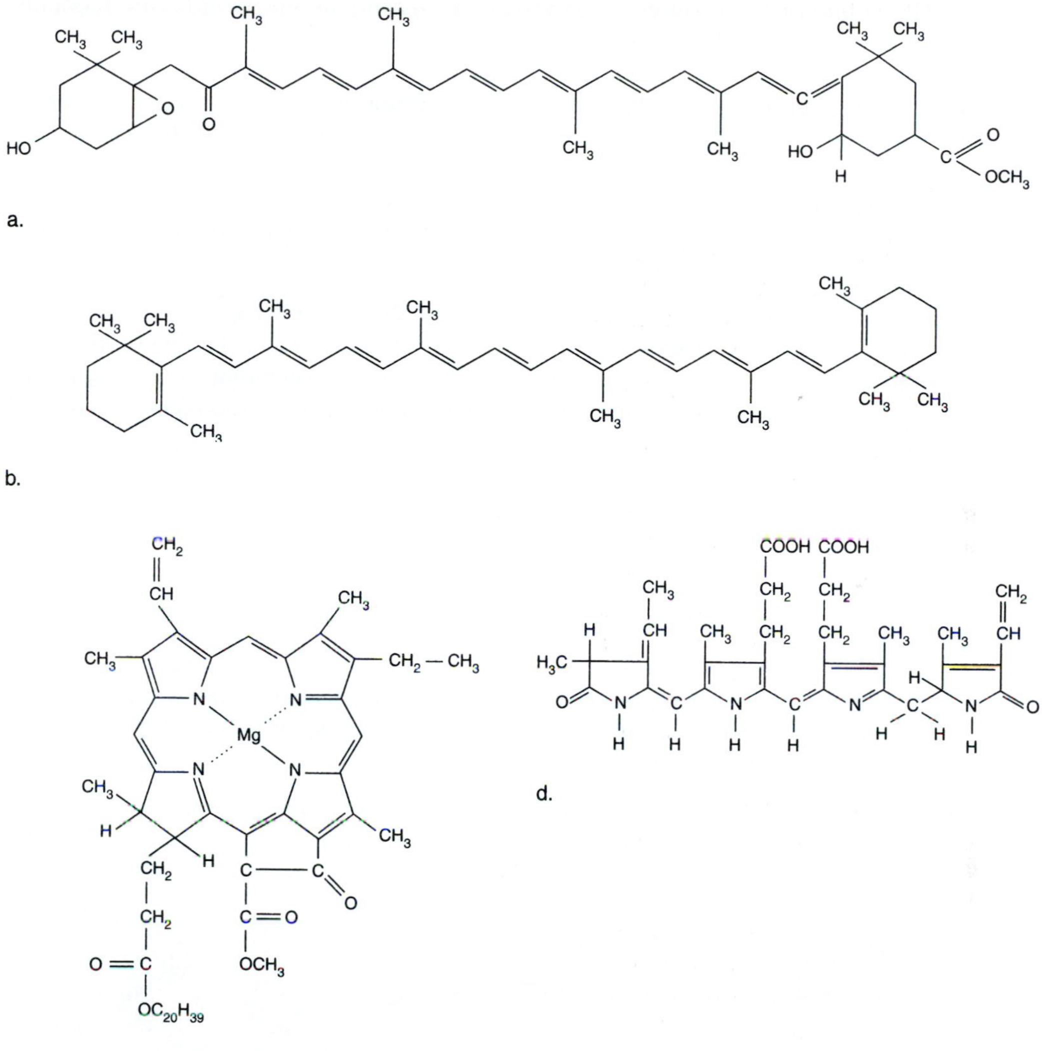

Figure 5–4 The chemical structure of: (a) fucoxanthin, (b) β-carotene, (c) chlorophyll a, (d) phycoerythrobilin. *Drawn by Kim Luoma.*

Dinitrogen Fixation

The reduction or fixation of atmospheric dinitrogen (N_2) to ammonia (NH_3) is an important process in the nitrogen cycle. The largest contribution to biological N_2 fixation from free-living organisms is made by cyanobacteria in the tropics. For example, in rice paddy agroecosystems, the cyanobacteria that grow with rice plants contribute a substantial amount of nitrogen to the crop. This N_2 fixation is probably the reason that rice paddies of many Asian countries have sustained their productivity for hundreds of years (see Chapter 13, Box 13–4).

The ability of N_2-fixing cyanobacteria to withstand adverse conditions accounts, in part, for their wide distribution and emphasizes their ecological importance. Dinitrogen fixation by cyanobacteria is sensitive to pH and decreases markedly below a pH of about 6.6. Unfavorable growth conditions can promote excretion of fixed nitrogen from the cell as inorganic (NH_3 or NH_4^+) or organic nitrogen and can represent a significant nutrient and energy drain.

The enzymes responsible for catalyzing biological N_2 fixation are strongly inhibited by oxygen. The N_2-fixing cyanobacteria have several mechanisms to protect these enzymes from inactivation by oxygen (Gallon, 1981):

- Some filamentous forms (e.g., *Anabaena* and *Nostoc*) have specialized thick-walled cells, or heterocysts, within which N_2 fixation occurs (Fig. 5–5). The thick cell walls of heterocysts block the diffusion of oxygen into the cytoplasm where the N_2-fixing enzymes are located, thus protecting the enzymes from damage.

- The oxygen-evolving portion of the photosynthetic apparatus (photosystem II) is absent from the heterocysts, thus eliminating the biochemical generation of oxygen during photosynthesis. Although heterocysts are the site of N_2 fixation under normal conditions, vegetative cells in the algal trichome may also fix nitrogen under **microaerophilic** (near anaerobic) and very low light conditions.

- Mucilaginous material surrounding the cells can also serve as a barrier to oxygen diffusion. For example, *Nostoc cordubensis* is characterized by mucilaginous colonies; both heterocysts and mucilage must be present in order for N_2 fixation to occur at oxygen levels above 20% (Prosperi, 1994).

- Certain nonheterocystous, filamentous forms, such as *Plectonema* spp., behave like microaerophilic organisms in that they only fix nitrogen at low levels of illumination and oxygen availability.

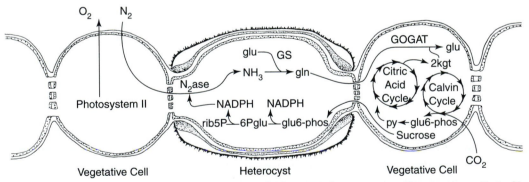

Figure 5–5 Diagram of carbon and nitrogen transfer between heterocysts and vegetative cells in filamentous cyanobacteria. Enzymes: N_2ase = nitrogenase; GS = glutamine synthetase; GOGAT = glutamine-oxoglutarate amidotransferase (glutamate synthase). Metabolites: gln = glutamine; glu = glutamate; glu6-phos = glucose-6-phosphate; NADPH = nicotinamide-adenine dinucleotide phosphate (reduced); py = pyruvate; rib5P = ribose-5-phosphate; 2kgt = 2-ketoglutarate; 6Pglu = 6-phosphogluconate. See Figure 12–5 for additional details. *Adapted from Haselkorn et al., 1978. Used with permission. Figure redrawn by Kim Luoma.*

- Some cyanobacteria fix dinitrogen heterotrophically (i.e., using preformed carbon) in the dark, albeit at low rates.

Some N_2-fixing cyanobacteria form symbiotic associations of considerable economic importance. For example, *Azolla* is a small aquatic fern with a wide distribution that, together with the N_2-fixing cyanobacterium *Anabaena azollae,* can contribute nitrogen to rice production (Chapter 14). Similarly, in the arctic, the group of lichens (defined later in this chapter) comprising reindeer moss is important in the diet of tundra ruminants.

| Box 5–1 |

Algae in Space. Certain species of soil algae are potentially useful for food or energy during space travel. Although the production of photoautotrophic algae on a large scale is uncommon, some have suggested that algae could be a major source of future dietary protein, both in space and on earth. Indeed, some (e.g., *Spirulina*) are already sold as dietary supplements. Hydrogen production by immobilized cyanobacteria also is being investigated as a source of nonpolluting fuel. In addition, algae could remove nutrients from wastewater and replenish oxygen in closed life-support systems.

As humans continue to explore our solar system, and possibly beyond, it is likely that the first carbon-based life form encountered will be a simple photoautotroph. Just as the algae are successful in colonizing inhospitable habitats on earth, there is a chance they, or a life-form similar to them, would be successful pioneer organisms on other worlds. Thus, the algae may serve as model organisms for **exobiology.**

Ecology

Distribution

Soil algae exist from polar to tropical regions, although their abundance is usually greatest in the tropics. Algal communities in Antarctic soils consist of a restricted number of taxa, and a given location may be dominated by only a few species (Davey, 1991). In contrast, hot and dry desert soils may have algae and lichens as the dominant microflora. Terrestrial algae can be found on and in the soil and exposed rocks, mud, sand, and snow. Algae can even be attached to buildings, plants, and animals. There are reports of roof and building discoloration caused by cyanobacteria and green algae. For example, Hermitage Castle, near the English-Scottish border, is being discolored and damaged by filaments of the alga *Trentepohlia,* which are growing into the stonework and turning it orange.

Under both stressful and hospitable conditions, a considerable population of algae is present at the surface and within the subsurface layers of most soils. As many as 10^8 algae g^{-1} of soil have been found, although populations between 10^3 and 10^4 g^{-1} are more common—several orders of magnitude less than for bacteria and fungi, but similar to the levels commonly observed for protozoa. Algal numbers usually decrease with increasing depth and decreasing light intensity. Substantial populations of soil algae may be found in forest soils, although low light intensity may be a limiting factor to algal growth in these habitats. Algal numbers are greatest in the upper 30 cm of soil, even though some algae may occur to a depth of two meters. However, there is no evidence that algae present deep in soils are metabolically active. Rather, they are probably introduced by such mechanisms as cultivation, animals, water seepage, and self-motility.

Spatial distribution of algae can vary widely. Algal populations at sites as little as 10 cm apart have been shown to differ, both quantitatively and qualitatively, due to fluctuations in moisture, pH, nutrient availability, and light intensity. Seasonal changes in the soil algae are generally quantitative and result from fluctuations in soil moisture. In the spring, algae are subjected to diurnal freeze-thaw cycles under conditions of soil saturation, whereas desiccation and water stress commonly occur in the summer. Daily fluctuations of algal populations have been linked to diurnal fluctuations in soil moisture as well as feeding by predators.

Box 5–2

Estimation of Soil Algae and Algal Culture. Assessing algal populations or their diversity in the soil is difficult. Taxonomic problems make comparisons uncertain. Standards for counting soil algae or preparing enrichment media for the isolation of strains are nonexistent. As a result, the estimation of algal populations in soils may differ greatly among studies. Populations of soil algae can be estimated by direct microscopic counts or by various culture-based techniques. However, the microscopes do not differentiate between living and dead organisms, and cultures enumerate only those algae able to grow in the particular medium and environmental conditions provided. Alternatively, the algal biomass may be estimated by the chlorophyll content, but this too has limitations.

Much of our knowledge of the physiology and the ecology of soil algae is based on laboratory cultures. Various types of cultures can be used, depending on the eventual application. Culture types include:

- **maintenance**—where the organisms, either active or dormant, are kept for future use,

- **enrichment**—where selected species are produced without regard to other species in the population,

- **unialgal**—where only one algal species is present, although other organisms, especially bacteria, may be present,

- **axenic**—where only the algal species is present in the culture, and

- **clonal**—where only one species is present and all the individuals are from the same parent.

Terrestrial Colonization

Human activities, along with climatological and geological events, are constantly altering the soil environment. However, no soil remains devoid of algae for much longer than the first gust of wind. Algae are also transported to new environments by water movement, including floods and runoff, and humans and other animals. Algae are pioneer colonizers in barren zones, such as volcanic lava and ash fields, and are important in the early stages of soil formation. They have an important role in plant community succession by providing (Booth, 1941):

- primary production,

- biological weathering of minerals,

- N_2 fixation, and

- stabilization of soil aggregates.

Microscopic algae are found in the atmosphere in numbers up to a few hundred m^{-3} and represent a source of inoculum for terrestrial colonization. Soil algae may contribute to this airborne population when the soil they inhabit is carried into the atmosphere and widely distributed. Most airborne algae have been identified as green algae; cyanobacteria and diatoms are far less numerous. Algae may settle from the air or be removed by rainfall, and numerous algae have been found in rainwater.

Environmental Interactions

Soil algae are subject to many environmental stresses, both biotic and abiotic. Solar radiation, temperature, and particularly water stress are important abiotic factors governing their distribution, metabolism, and life strategies. Algae are not a static soil component. Rather, they are constantly interacting with other biota and their physical and chemical environment. Some soil algae are capable of life in very hostile environments. For example, salt is considered so hostile to life that it is used as a preservative; however, the cyanobacterium *Synechococcus nageli* not only survives in crusts of halite and gypsum but also fixes carbon and nitrogen there (Rothschild et al., 1994).

Moisture. Algae are most abundant when soil is moist or waterlogged for extended periods of time. Although drying conditions are stressful, algae can tolerate desiccation. Many soil algae, especially the cyanobacteria, have well-developed mechanisms for drought tolerance. Tolerance to desiccation often depends on the particular combination of physical factors present and the type of organism undergoing the stress. Most soil algae, especially the green algae and cyanobacteria, possess thick mucilaginous sheaths which protect them against desiccation. Some soil algae form spores, whereas others can tolerate desiccation without apparent special morphological adaptations. Soil diatoms can survive adverse conditions through the storage of energy-rich oils, the buildup of inner plates, and reduction in cell size. To tolerate desiccation and heat, many cyanobacteria produce **akinetes** (thick-walled spores), which are resistant to adverse conditions. The duration of desiccation may influence the size and composition of algal populations. Trainor (1983) studied survival of algae in desiccated soil and found that, of the 30 taxa initially present, 7 were still present after 25 years at a combined population of approximately 2,000 cells g^{-1} of soil. Genera present included *Chlamydomonas, Chlorella, Protosiphon, Tetracystis,* and *Chlorococcum.*

Light. Algae, being phototrophic organisms, are most common in the upper level of the soil. This layer may be referred to as the **photic zone.** It is difficult to assign an exact depth to the photic zone, as light penetration of the soil varies with soil constituents that either reflect (e.g., residue cover) or absorb (e.g., soil organic matter) light. In addition, a physical disruption of the soil, such as tillage, can alter the photic zone. Compared to the eukaryotic soil algae, the cyanobacterial species are repressed by high light intensities. This constraint to their development allows diatoms and unicellular green algae to develop first after a desiccated soil has been rewetted. Cyanobacteria appear when plant cover is sufficiently dense to protect them from high light intensities. Some algae are common in caves, where light intensities are very low. Although there are reports of heterotrophic growth in the absence of light among the green algae, cyanobacteria, and diatoms, it is unlikely that these algae can compete successfully with common soil heterotrophs.

Temperature. Soil algae are metabolically active over a wide range of temperatures. At one extreme are the cyanobacteria in the hot springs of Yellowstone National Park, where temperatures may be as great as 50° to 54°C. Most cyanobacteria prefer temperatures of 30° to 35°C, and their populations may increase relative to eukaryotic soil algae as the temperature increases. At the other extreme are the algae of frozen habitats, which occur characteristically on the surface of snow. For example, *Chlamydomonas nivalis* can cause snow to appear pink during winter surface blooms. Algae in Antarctic soils are subjected to soil temperatures that may exceed 20°C in the summer and below −10°C during winter.

Acidity. The direct effect of pH on the soil algae is difficult to evaluate because pH is correlated with other factors. Nearly neutral or slightly alkaline soils support the most varied algal flora. The cyanobacteria, especially the filamentous forms, are most common in alkaline or neutral soils, but they have also been found in acidic soils. Cyanobacteria do not exist at pH levels below 4, although eukaryotic algae may exist and even grow profusely at that pH (Brock, 1973). The euglenoid, *Euglena mutabilis,* has been observed in acidic coal mine drainage waters at pH values as low as 2.5. The liming of soils may stimulate cyanobacterial growth by increasing the pH. Although the green algae may be found in soils of a wide pH range, they tend to be the most abundant group in acid habitats, probably because of an absence of competition. *Cyanidium caldarium* will grow in hot and extremely acid soils and is the only known photosynthetic organism living at a pH lower than 5 and a temperature greater than 40°C. Diatoms are generally present in neutral or slightly alkaline soils. Some red algae seem to tolerate adverse habitats, such as extremely acid, polluted, or ammonium-rich soils.

Nutrients. The great majority of soil algae are **obligate photoautotrophs;** that is, they use light energy to manufacture all of their organic compounds from inorganic precursors. Thus, they require only water, solar radiation, carbon dioxide, and essential inorganic nutrients for growth—a rather simple recipe to support life. The nutritional requirements for macroelements and microelements are similar to those of higher plants. However, some subtle differences are found; for example, many cyanobacteria have a substantial physiological requirement for phosphorus. The availability of nutrients has an important influence on algal diversity and biomass. The algae are aerobic organisms and require oxygen to support their respiratory functions, including oxidative phosphorylation.

The nutritional relationships of soil algae to other organisms have not been widely explored. The soil algae are primary producers of organic compounds and thus play a central role as the base of the **food chain.** A number of soil algae belong to other nutritional categories, including **photoheterotrophs,** which utilize solar radiation as an energy source but cannot synthesize all necessary organic compounds from inorganic precursors. While heterotrophic activity is rare, some green algae in the soil can use organic complexes produced by the degradative activities of other microorganisms.

Pesticides and pollutants. The soil algae are affected by a wide variety of herbicides, fungicides, pollutants, and soil fumigants. The photosystem-uncoupling herbicides (e.g., diuron, atrazine, simazine, and metribuzin) are generally the most harmful. Various genera react differently to certain pesticides, and the pesticide regimen of cultivated soils may determine microfloral composition. Cyanobacteria appear to be less sensitive than eukaryotic algae and crop plants to many agrochemicals, and their growth is sometimes stimulated because of inhibition of green algal competitors by certain agrochemicals. Algae can be useful indicators of pesticide residues in the soil. Changes in their populations or metabolic activity may provide information on the accumulation of such chemicals and their possible detrimental effects on crop plants.

Biotic Interactions

Interactions of the soil algae with other soil organisms are not well understood but range from antagonistic relations with other soil microbes to symbiotic associations with plants and fungi (Chapter 8). Soil algae are a food source for grazing soil animals, including protozoa and nematodes. Soil algae may affect higher plants through the production of plant growth regulators. Several species produce natural toxins, including *cyanobacterin* and other antibiotics, which inhibit the growth of competing cyanobacteria and eukaryotic algae. For example, *Microcystis* spp. produces the toxin *microcystin* which inhibits the growth of green algae and diatoms. Growth-promoting or inhibitory substances may play an important role in the succession of algal species in the soil. Only rarely are the soil algae considered pathogenic. However, when algae become airborne, they can cause allergic reactions.

Beneficial interactions. Soil algae may enter into relationships with other microorganisms; for example, many soil protozoa contain *endosymbiotic* algae. Algae can also enter into associations with numerous higher plant species. The N_2-fixing cyanobacteria may be *endophytic* within certain liverworts, the aquatic fern *Azolla, Gunnera* spp., and in the roots of cycads. The cyanobacteria provide the plant with fixed nitrogen, while the algae are protected from environmental stresses by the association or symbiosis.

An alga and a fungus can combine to form a **lichen,** a symbiotic association in which the two organisms are intergrown to form a single thallus. The photosynthetic partner (*phycobiont*) may be either a eukaryote, generally a green alga, or a prokaryotic cyanobacterium, usually of the genus *Nostoc*. The fungal member is usually an ascomycete, less often a basidiomycete. There are roughly 18,000 species of lichenized ascomycetes. In the symbiosis, the phycobiont captures light energy for carbon fixation and, if the phycobiont is a cyanobacterium, it may also fix atmospheric N_2. The cyanobacterial partner undergoes changes associated with enhanced N_2 fixation and translocation of fixed nitrogen. These changes can be both morphological, notably an increase in heterocyst numbers and frequency, and physiological. In return for the fixed carbon and nitrogen, the fungal partner (*mycobiont*) is believed to furnish mineral nutrients and help to regulate the water and light environment of its symbionts. The association may allow the survival and even promote the growth of the two partners in severe environments where neither could survive alone. Because they can withstand considerable water stress, lichens play an important role in arid ecosystems.

Detrimental interactions. The soil algae must compete with other photosynthetic organisms for light and with soil microorganisms in general for inorganic nutrients. When present, higher plants also compete with algae for moisture and nutrients. Several cyanobacterial species produce natural toxins which inhibit other cyanobacteria and eukaryotic algae. Algae parasitic on plants are known only among the green algae, and, because they mainly occur on noncultivated plants, they are generally overlooked (Joubert and Rijkenberg, 1971).

All the soil algae are subject to parasitism and predation. The inoculation of soils with N_2-fixing cyanobacteria is rarely successful because their populations are rapidly reduced, apparently because of grazing by soil animals. Viruses are widespread and highly specific in their infection abilities; *cyanophages* are the viruses specific to the cyanobacteria. Viruses can cause declines in algal populations under natural environments. As we better understand soil ecology, the viruses may become tools to manipulate populations of soil algae.

Soil Formation and Quality

Although soil algae may be overlooked in many soil ecosystems, they make significant contributions as geologic agents of soil formation. In addition, they promote the maintenance of soil structure and improve soil quality.

Geological Agents

Algal activity may contribute to geologic weathering. The surfaces of rocks and, to some extent, the interiors of porous rocks, can be inhabited by algae. They can form and enlarge the cavities they occupy. As in the soil environment, the depth to which the algae are found in rock is limited by light penetration. The mechanism by which algae penetrate rock is not completely understood; however, several algal species are capable of secreting acidic substances (e.g., carbonic acid from respiratory carbon dioxide) which may contribute to the weathering process. Soil algae also promote the release of nutrients from insoluble compounds.

The soil algae are important in stabilizing and improving the physical properties of soil. They participate in the formation of soil organic matter through both the exudation of organic compounds and, upon their death, the contribution of their cellular material to soil organic matter. The latter process can be very important in desert ecosystems. Organic matter can help reduce erosion, facilitate water infiltration and root development, and, in agroecosystems, facilitate tillage. The soil algae produce extracellular polysaccharides that stabilize soil aggregates, thus promoting the establishment of higher plants. In particular, the extracellular polysaccharides assist in binding colloidal clay or humic particles through adsorption, adhesion, or cation bridging.

Microbiotic Crusts

Soil algae can form microbiotic *crusts* or *mats,* which are often mistaken for mosses or lichens (Eldridge and Greene, 1994). These surface crusts can retard erosion and reduce the loss of water by evaporation, thus increasing the storage of water. They may also influence nutrient cycling and serve as indicators of pollution. These crusts develop between open shrub and grass communities of arid and semi-arid environments and are distinguishable from crusts formed by chemical or physical processes. Microbiotic crusts are found globally and are common in western North America (Fig. 5–6). Cyanobacteria and algae are the major components of many microbiotic crusts, although mosses, lichens, liverworts, fungi, and an associated microfauna may also be found. Microbiotic crusts are easily damaged by trampling animals and vehicular travel.

| **Box 5–3** |

Diatomaceous Earth. Algae, albeit not soil algae per se, are responsible for the formation of diatomaceous earth. This material consists of diatom frustules, both marine and freshwater, that were deposited during the Tertiary and Quaternary Periods of geologic history. Diatomaceous earth (also called diatomite or Kieselguhr) is an almost chemically inert inorganic material containing about 86% to 88% silica. Today it is used to extract high yields of silica. Diatomite is a filtration aid in sugar refining, the brewing industry, wine making, and antibiotic production. It is also used for insulation and as filler for paints and paper products. Historically, diatomite served as the material for lightweight bricks in the construction of the Cathedral of St. Sofia in Constantinople in 532 A.D. Alfred Nobel used diatomite to absorb nitroglycerin in the manufacture of dynamite (Volesky et al., 1970).

Soil Quality

Soil quality (or soil health) is a critical element in sustainable agriculture (Doran et al., 1996); however, soil quality is hard to define and even more difficult to evaluate. One possible way to evaluate soil quality is to use soil algae as indicators (Pipe and Shubert, 1984). Because of their cell structure, nutrient requirements, biochemical similarity to higher plants, and rapid growth rates, algae may have value as indicator organisms for anticipating crop responses to fertilizers and pesticides.

Not all soil algal activity is beneficial to soil quality. Algal growths can seal the soil surface, reducing water infiltration and soil moisture. Algal crusts may also create a barrier to gas exchange with the atmosphere and inhibit seedling emergence. Conspicuous surface growths of algae on sports turf and ornamental grasses also cause aesthetic problems. In managed turf, algal problems are often caused by unsatisfactory drainage or irrigation practices, but other factors such as fertilizer treatment may be involved. Some evidence suggests that products formed in cyanobacterial surface films on golf putting greens may contribute to the formation of the anaerobic subsurface condition known as *black layer*.

Biofertilizers

Soil algae, especially the cyanobacteria with their ability to fix dinitrogen, have the potential to be biofertilizers; however, this application is not widespread. Cyanobacteria have been used to reclaim saline soil in India and as a source of fertilizer nitrogen in rice production, especially in China and India. Cyanobacteria have

a.

b.

Figure 5–6 Microbiotic crust in southern Utah. Figure (b) is close-up of figure
(a). *Photo courtesy of J.D. Williams. Used with permission.*

contributed as much as 50 kg N ha^{-1} yr^{-1} in tropical rice paddy agroecosystems (Watanabe, 1962). The inability of the inoculum to become established is a primary factor limiting the use of this technology. Although cyanobacteria are primary colonizers, little evidence indicates that introduced strains can succeed in dominating native populations. Genetic engineering may allow the production of cyanobacterial strains that have an enhanced capacity for N$_2$ fixation (Albrecht et al., 1991) or herbicide resistance (thus offering a superior organism for biofertilization). These novel laboratory strains, however, will be of little value if they fail to establish in the field. We presently lack the ecological knowledge necessary to manage these organisms so they can contribute significantly to agricultural practice.

Summary

There are two broad groups of algae: eukaryotic algae and prokaryotic cyanobacteria (formerly known as blue-green algae). On a subcellular level, the eukaryotic algae are morphologically and physiologically similar to higher plants. The cyanobacteria, with the exception of their photosynthetic arrangement, are more comparable to bacteria. Algae are photoautotrophs, and their system of photosynthesis generally resembles that of the higher plants. Some prokaryotic cyanobacteria also can fix dinitrogen. Soil algae vary greatly in form and color, and many have the ability to form spores or resting cells that allow them to withstand harsh environments. Reproduction is both asexual and sexual. They are classified primarily on the basis of morphology, reproduction, pigments, and food storage.

Algae are most abundant in the upper soil layers where light can penetrate. In addition to light, their populations are affected by soil acidity, temperature, and especially moisture. As photoautotrophs, they do not require reduced carbon for growth; however, carbon dioxide or dissolved carbonate is essential. Their requirements for macronutrients and micronutrients are similar to those of higher plants. As primary producers, they are subject to predation by many protozoa, nematodes, and soil insects and are at the base of many soil food chains. Some algae form associations with higher plants. For example, lichens are symbiotic associations formed by algae or cyanobacteria with several types of fungi.

Algae are common pioneers on bare soil and contribute to soil development. In ecosystems such as deserts, they produce ecologically significant amounts of organic matter. The polysaccharides they excrete can bind soil particles together to form aggregates, thereby promoting soil structure. In the future, certain soil algae, especially the N$_2$-fixing cyanobacteria, may be managed for biofertilizers; however, to date, this potential has not fully realized.

Cited References

Albrecht, S.L., C. Latorre, J.T. Baker, and K.T. Shanmugam. 1991. Genetically altered cyanobacteria as nitrogen fertilizer supply for the growth of rice. pp. 163–176. *In* S.K. Duda and C. Sloger (eds.) Biological nitrogen fixation associated with rice production. Oxford & IBH Publishing Co., New Delhi, India.

Alexopoulos, C.J., and H.C. Bold. 1969. Algae and fungi. Collier-Macmillan Co., Toronto.

Bassham, J., and M. Calvin. 1957. The path of carbon in photosynthesis. Prentice-Hall, Inc., Englewood Cliffs, N.J.

Booth, W.E. 1941. Algae as pioneers in plant succession and their importance in erosion control. Ecology 22:38–46.

Brock, T.D. 1973. Lower pH limit for the existence of blue-green algae: Evolution and ecological implications. Science 179:480–483.

Chapman, V.J. 1941. An introduction to the study of algae. Cambridge University Press, Cambridge, England.

Christensen, T. 1980. Algae: A taxonomic study, Fascicle 1. Cyanophyta, Rhodophyta and Chromophyta. AiO Print Ltd., Odense, Denmark.

Christensen, T. 1994. Algae: A taxonomic study, Fascicle 2. Chlorophyta, Euglenophyta and Chlorarachniophyta. AiO Print Ltd., Odense, Denmark.

Davey, M.C. 1991. Effects of physical factors on the survival and growth of Antarctic terrestrial algae. Br. Phycol. J. 26:315–325.

Doran, J.W., M. Sarrantonio, and M.A. Liebig. 1996. Soil health and sustainability. Adv. Agron. 56:1–54.

Eldridge, D.J., and R.S.B. Greene. 1994. Microbiotic soil crusts: A review of their roles in soil and ecological processes in rangelands of Australia. Aust. J. Soil Res. 32:389–415.

Gallon, J. 1981. The oxygen sensitivity of nitrogenase: A problem for biochemists and microorganisms. Trends Biochem. Sci. 6:19-23.

Haselkorn, R., B. Mazur, J. Orr, D. Rice, N. Wood, and R. Rippka. 1978. Heterocyst differentation and nitrogen fixation in cyanobacteria (blue-green algae). pp. 259–278. *In* W.E. Newton and W.H. Orme-Johnson (eds.) Nitrogen Fixation, Vol 2. University Park Press, Baltimore, Md.

Joubert, J.J., and F.H.J. Rijkenberg. 1971. Parasitic green algae. Annu. Rev. Phytopath. 9:45–64.

Metting, B. 1981. The systematics and ecology of soil algae. Bot. Rev. 47:195–312.

Pankratz, H.S., and C.C. Bowen. 1963. Cytology of blue-green algae. I. The Cells of *Symploca muscorum*. Am. J. Bot. 50:387–399.

Pipe, A.E., and L.E. Schubert. 1984. The use of algae as indicators of soil fertility. pp. 214–233. *In* L.E. Shubert (ed.), Algae as ecological indicators. Academic Press, London.

Prosperi, C.H. 1994. A cyanophyte capable of fixing nitrogen under high levels of oxygen. J. Phycol. 30:222–224.

Rothschild, L.J., L.J. Giver, M.R. White, and R.L. Mancinelli. 1994. Metabolic activity of microorganisms in evaporites. J. Phycol. 30:431–438.

Starks, T.L., L.E. Shubert, and F.R. Trainor. 1981. Ecology of soil algae: A review. Phycologia 20:65–80.

Trainor, F.R. 1983. Survival of algae in soil after high temperature treatment. Phycologia 22:210–212.

Volesky, B., J.E. Zajic, and E. Knetting. 1970. Algal products. pp. 49–82. *In* J.E. Zajic (ed.), Properties and products of algae. Plenum Press, New York.

Watanabe, A. 1962. Effect of the nitrogen-fixing blue-green alga *Tolypothrix tenuis* on the nitrogen fertility of paddy soil and on the crop yield of rice plants. J. Gen. Appl. Microbiol. 8:85–91.

General References

Bold, H.C., and M.J. Wynne. 1978. Introduction to the algae. Prentice-Hall, Englewood Cliffs, N.J.

Ehrlich, H.L. 1990. Geomicrobiology. Marcel-Decker, New York.

Gregory, P.H. 1973. The microbiology of the atmosphere. Leonard Hill Books, Aylesbury, England.

Hoffmann, L. 1989. Algae of terrestrial habitats. Bot. Rev. 55:77–105.

Lewin, R.A. 1976. The genetics of algae. Blackwell Scientific, Oxford, England.

Lewin, R.A. 1962. Physiology and biochemistry of algae. Academic Press, New York.

Stein, J.R. 1979. Handbook of phycological methods. Cambridge University Press, Cambridge, England.

Stewart, W.D.P. 1973. Nitrogen fixation by photosynthetic microorganisms. Annu. Rev. Microbiol. 27:283–316.

Study Questions

1. What is the advantage of spore formation? Are there any disadvantages?

2. List the characteristics of algae that allow them to be successful pioneer organisms.

3. Cyanobacterial vegetative cells have the capacity to produce the enzymes that fix dinitrogen. Why are these enzymes localized in the heterocyst?

4. Inoculation of rice paddies with N_2-fixing cyanobacteria rarely improves N_2 fixation in these agroecosystems. What factors mitigate against successful inoculation? What characteristics are necessary for an alga to be a successful inoculant?

5. Soil algae have been suggested as indicator organisms to evaluate soil quality. What characteristics would enable these organisms to be valuable for this purpose? What are be the drawbacks of using soil algae?

6. What benefits do soil algae provide humans? With the use of genetic engineering, can you think of future benefits?

7. What factors affect the depth to which algae will grow in the soil?

8. Sometimes soil algae are a great nuisance. What problems do they cause? Can you think of ways to eliminate these difficulties?

9. How do soil algae contribute to enhanced soil quality?

10. How may algae be disseminated?

Chapter 6

♦

Protozoa and Nematodes

Elaine R. Ingham

♦

*Thus when a soil loses fertility we pour on fertilizer, or at best alter its tame flora
and fauna, without considering the fact that its wild flora and fauna, which built
the soil to begin with, may likewise be important to its maintenance.*

Aldo Leopold

Two components of the wild fauna which Leopold may have never considered are
major players in the important processes that build soil. They are soil protozoa and
nematodes. These organisms are nonphotosynthetic, motile, eukaryotic predators of
bacteria and fungi and each other. Protozoa are considered **microfauna** because
they are usually < 200 μm long, while nematodes are usually considered to be members of the **mesofauna** (200 to 1,000 μm long). Other members of the soil mesofauna (e.g., mites, rotifers, springtails, and tardigrades) and **macrofauna** (> 1,000
μm long; e.g., vertebrates, earthworms, and large arthropods) also play critical roles
in soil processes, particularly with regard to decomposition of organic matter, but are
beyond the scope of this microbiology textbook. For more information on the meso-
faunal and macrofaunal groups, see the *Soil Biology Guide* by Dindal (1990) and
Fundamentals of Soil Ecology by Coleman and Crossley (1996).

Classification

Protozoa

Protozoa are unicellular, eukaryotic organisms which represent a group in which
mitosis and meiosis became established. Certain members of the protozoa may have
been hosts for photosynthetic prokaryotes. These internal "parasites" evolved into

plastids, suggesting an extremely important evolutionary role for protozoa. In the past, scientists generally believed that soil protozoa evolved from aquatic species, but recent evidence shows some unique soil protozoa probably did not evolve from aquatic forms.

Box 6–1

Free-Living Soil Protozoa. Free-living soil protozoa fall into three categories:

- **Flagellates** (Phylum Sarcomastigophora, Subphylum Mastigophora) move by means of one to several flagella.

- **Amoebae** (Phylum Sarcomastigophora, Subphylum Sarcodina) move by protoplasmic extrusions called pseudopodia. Recently, the slime molds (Sporangia) have been classified with amoebae.

- **Ciliates** (Phylum Ciliophora) move by means of cilia on the surfaces of the cells.

Flagellates. These are the smallest members of the protozoa and are divided into two classes based on whether they contain chlorophyll (Phytomastigophorea) or not (Zoomastigophorea). Only nonchlorophyll-bearing flagellates occur in soil; the photosynthetic species are strictly aquatic. Soil flagellates, like *Oicomonas, Scytomonas* and *Peranema,* are morphologically similar to photosynthetic flagellates. Other flagellates are the nonchlorophyll-bearing Zoomastigophorea, such as *Bodo* (Fig. 6–1) or *Pleuromonas.* Flagellates sometimes display amoeboid movements, although they usually have at least one flagellum. Flagellates also have a resting stage, called a cyst, which enables them to survive stressful environmental conditions (Fig. 6–2).

Amoebae. These protozoa move by protoplasmic flow, either with extensions called pseudopodia or by whole-body flow. Within this group:

- Naked amoebae are differentiated by **pseudopodia,** or "false feet" (Fig. 6–3), which can be lobose (rounded), conical (thinner at the tip than the base), filiose (thin filamentous extensions), or reticular (netlike). The number and flow patterns of the pseudopodia also are important taxonomic characteristics for naked amoebae.

- Slime molds have an amoeboid stage that also forms a stage that resembles a slug. This "slug" then differentiates into a sessile stage with a fruiting body containing spores on a stalk.

- Testate amoebae live within a **test** (or shell) constructed of soil particles bound together by secretions. The size and shape of the test, the size and shape of scales and spines or horns on the surface of the test, and the placement and border pattern of the pseudostome or "false mouth" (the opening in the test that allows the amoeba to go in and out of its "mobile home") are important characteristics for identifying species.

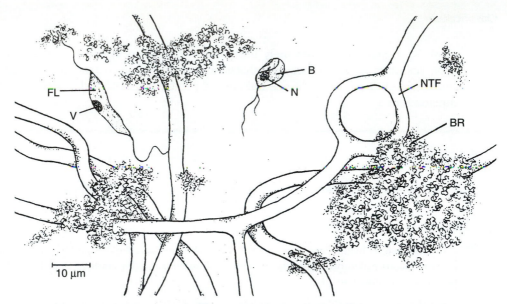

10 μm

Figure 6–1 An example of two flagellates; *Bodo* sp. (B) with a flagellum emerging from anterior end, and a second flagellate (FL) with flagella arising at either end of the flagellate's body. Often the nucleus (N) or vesicles (V) can be seen within the cell. Compare the size of the two flagellates with the matrix of thousands of bacterial cells (BR) on which flagellates feed. A hypha of a large nematode-trapping fungus (NTF) is also visible. This fungus forms rings which produce compounds that attract nematodes. When a nematode contacts the inside of the ring, the ring constricts, holding the nematode so the fungus can grow into the nematode and consume it. *Drawing by Kim Luoma.*

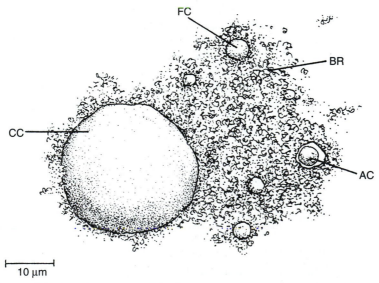

10 μm

Figure 6–2 Flagellate cysts (FC) of the genus *Cercomonas*. Amoebal cysts (AC) have double-walls. Ciliate cysts (CC) are usually much larger. Cysts of all protozoa are usually spheroid, and often enmeshed in a matrix of bacterial cells and organic matter (BR). *Drawing by Kim Luoma.*

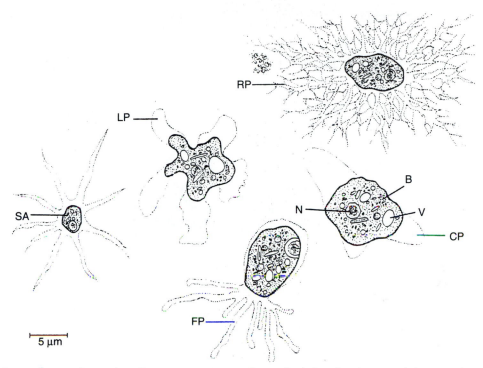

Figure 6–3 Soil amoebae form various types of pseudopods. The shape and the way the pseudopods are formed are important characters for identifying species. Lobose pseudopods (LP) are rounded and form by a smooth, streaming motion or by a more abrupt, "burping" motion. Conical pseudopods (CP) are pointed at the tip, and broader at the base. Filiose pseudopodia (FP) are slender, fingerlike, filamentous structures. Stellate amoebae (SA) form stiff, filamentous pseudopods and so look like floating stars. Reticular pseudopodia (RP) form weblike, fanciful structures. Pseudopods are used for locomotion, to pull or push the amoeba along the soil surface. Within the amoebal body, nuclei (N), vesicles (V), and ingested bacterial cells (B) can be seen. *Drawing by Kim Luoma*.

Ciliates. Ciliates move by beating short, numerous **cilia** (singular, **cilium**) on the surface of their bodies (Fig. 6–4). Distribution of the cilia, fusion of cilia into tufts called *cirri* or into membranelles (undulating membranes), and the placement of the cytostome (mouth) with associated membranelles are important features for species identification.

Sizes of protozoa. Flagellates, naked amoebae, and small testate amoebae normally are barely larger than their bacterial prey, while some forms are much larger (Table 6–1). Giant amoebae, which attack fungal hyphae, spores, or even sclerotia, are up to 1 mm in diameter. Ciliates are larger in size than flagellates, larger than most naked amoebae, and similar in size to large testate amoebae. The large forms are restricted to larger pore spaces and require thicker water films that occur at water potentials of 0 to -0.03 MPa to remain active, generally swimming free in the soil solution.

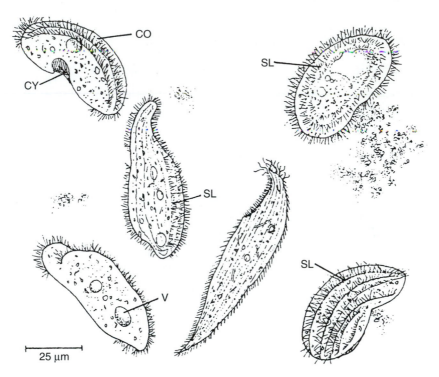

Figure 6–4 A group of soil ciliates. A typical genus is represented by *Colpoda* (CO), about 40 μm in length. The cytostoma (CY) or mouth cavity forms an indentation that makes the organism look like a kidney. Cilia around the cytostoma guide bacteria into the ciliate's mouth, but the major purpose for cilia is locomotion. One important characteristic for identification of ciliates is the pattern made by cilia attachment to the body surface, called "silver lines" (SL). Cilia can run the entire length of the body, part of the body, or occur in patches as shown in this drawing. Vesicles (V) for collecting waste products are also shown. *Drawing by Kim Luoma.*

Table 6–1 **Average length and volume of soil protozoa and nematodes as compared to bacteria.**

Group	Length (μm)	Volume (μm^3)	Shape
Bacteria	<1–5	2.5	Spherical to rod-shaped
Flagellates	2–50	50	Spherical, pear-shaped, banana-shaped
Amoebae			
Naked	2–600	400	Protoplasmic streaming, pseudopodia
Testate	45–200	1,000	Build oval tests or shells made of soil
Giant	6,000	4×10^9	Enormous naked amoebae
Ciliates	50–1,500	3,000	Oval, kidney-shaped, elongated and flattened
Nematodes	250–5,500	5,000	Long, slender, wormlike

Nematodes

Nematodes (Phylum Nematoda) are multicellular, eukaryotic, nonsegmented roundworms. They constitute as much as 90% of all multicellular animals in soil, and their numbers often exceed several million per m^2 of surface soil. Although the total biomass of nematodes may be less than that for other faunal groups (see Table 1–2), nematode metabolic activity is often higher.

Most nematode species are fusiform (tapered ends), while others are vermiform (rounded ends). Nematode digestive, nervous, excretory, and reproductive organs float in a fluid-filled cavity surrounded by a body wall. Nematodes lack circulatory, respiratory, and endocrine systems. The body wall consists of a cuticle, a hypodermis which produces the cuticle, and a layer of muscle. Placement of body openings, such as the stoma (mouth), anus, vulva, excretory openings, and sensory openings, and the structure of these openings are important in identification of nematode species. Also important are cuticular markings, such as striations, ridges, and apparent segmentation. In addition, the position of sensory papillae (small elevations), setae (hair-like projections), and spines help to identify species. The shape and placement of chemoreceptors (called *amphids*) are important in identification of certain species. Prolobae (ornate lip structures) occur in a number of bacterial-feeding species.

Nematodes are grouped into four or five trophic categories based on the nature of their diet, the structure of the stoma and esophagus, and their method of feeding.

- Bacterial-feeding nematodes do not have **stylets** (hard, spear-like structures in the mouth region used to puncture cell walls) but have a simple stoma in the form of a cylinder or cone, terminating in a valvelike apparatus which may bear minute teeth (Fig. 6–5).

- Fungal-feeding nematodes have slender stylets or small teeth used to puncture cell walls; they attack fungi instead of plant roots (Fig. 6–6). After puncturing the hyphal cell wall, the muscular esophagus provides the suction necessary to empty the cytoplasm from the fungus. The morphology of the esophagus in different species varies considerably, from a long, straight esophagus in some genera to those with large cuticularized pumping valves halfway between the mouth and the intestine.

- Root-feeding nematodes usually possess large stylets (Fig. 6–7) and are the most extensively studied group of soil nematodes because of their ability to cause plant disease and reduce crop yield.

- Predatory nematodes possess huge mouths (stoma) armed with powerful teeth (Fig. 6–8).

- Omnivores are sometimes considered as a fifth trophic category of soil nematodes because they may eat many types of food, including roots, fungi, bacteria, algae, and protozoa.

Feeding Behavior

Most soil protozoa are **phagotrophic,** engulfing bacteria, yeasts, and algae. Large protozoa, such as ciliates, may also engulf small protozoa. Some flagellates are

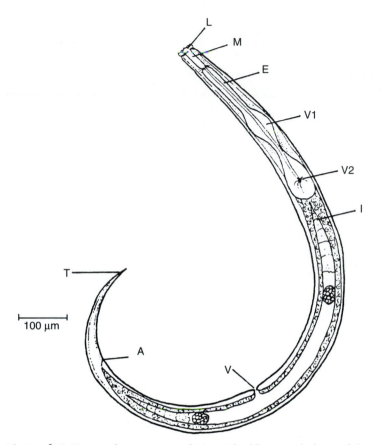

Figure 6–5 Nematode genera are distinguished by morphology of the mouth (M), lips (L), esophagus (E), intestines (I), anus (A), vulva (V), and tail (T). Bacterial-feeding nematodes typically have unique, often elaborate, lip structures; long, straight mouths; and an esophagus with a middle pumping valve (V1) and a second cuticularized pumping valve at the end (V2). *Drawing by Kim Luoma.*

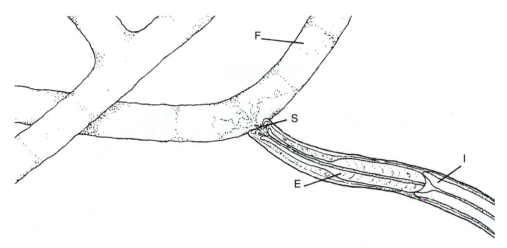

Figure 6–6 Fungal-feeding nematodes feed on the cytoplasm inside fungal hyphae (F). Fungal-feeding nematodes usually have needlelike stylets (S) which are usually only slightly longer than they are wide. The stylet (also called a spear or style) punctures the cell wall of the fungus so the nematode can consume the cytoplasm inside. The nematode's esophagus (E) is usually more slender near the mouth and widens halfway or three-quarters of the way to the intestine (I). *Drawing by Kim Luoma.*

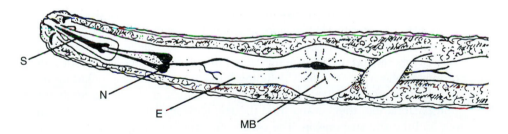

Figure 6–7 Root-feeding nematodes usually have strong, large stylets (S) with knobs (N) at the end. In some cases, root-feeders have small to medium-size cuticularized median bulbs (MB) halfway along the esophagus (E). *Drawing by Kim Luoma.*

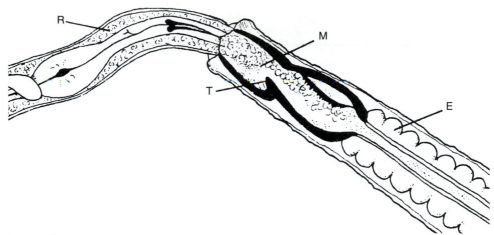

Figure 6–8 Predatory nematodes are easily identified because they have enormous mouths (M), which are armed with one or more large teeth (T) to hold and sometimes crush their prey. This predatory nematode is consuming a root-feeding nematode (R). The esophagus (E) of a predatory nematode is typically long and straight, with no pumping valves or widening of any kind. *Drawing by Kim Luoma.*

saprobic, absorbing nutrients directly from the soil solution. Although rare in most soils, some amoebae are **parasitic,** attacking fungal spores or digesting the cell wall of hyphae to enter the hypha and consume the fungal cytoplasm.

Nematodes eat bacteria, fungi (including yeasts), algae, protozoa, and small invertebrates (including other nematodes). In addition, some nematodes are parasites of invertebrates, vertebrates (including humans), and plants, including roots and all parts above ground. Like "vacuum cleaners of the soil," bacterial-feeding nematodes pull bacteria into their stoma from the surfaces of roots, soil particles, organic matter, and the soil suspension. The ornate lip structures (prolobae) of some bacterial-feeders may serve to sieve the ingested particles. Cuticularized valves within the digestive system may serve to crush particles, aiding in digestion. Fungal-feeding and plant-feeding nematodes search the surfaces of hyphae or roots with their stylets before bracing themselves against soil particles. Then, with a powerful thrust of their stylets, they penetrate hyphal or plant cell walls. The cell cytoplasm is then pumped through the hollow stylet into the nematode's digestive system. Certain nematodes (e.g., *Aphelenchus*) can remove the entire contents of a fungal cell within 15 seconds! Predatory or omnivorous nematodes use teeth to ingest their prey, which include protozoa, rotifers, tardigrades, small worms, and other nematodes.

Different species of bacteria and fungi vary in their palatability and nutritional value for protozoa and nematodes. Highly palatable bacteria and fungi can be overgrazed, while unpalatable bacterial or fungal species can be avoided (Mankau and Mankau, 1963). Some soil fungi avoid being eaten by producing toxins, while others protect themselves by growing in materials that their predators can't penetrate, such as wood.

Protozoan and nematode predation (also called grazing) can limit or prevent the growth of certain fungi and bacteria. Spores of certain fungi have lower viability after passing through nematode digestive systems, and hyphae germinating from spores too large to be ingested whole may be eaten preferentially by nematodes and other soil organisms. Spores of fungi and virus particles can be carried in the digestive system or on the external surfaces of nematodes and possibly protozoa, thus allowing transport of these organisms to infection sites, such as root wounds and necrotic tissue, and increasing the spread of disease.

Selective grazing by protozoa and nematodes may change the bacterial or fungal community structure and alter decomposition rates and nutrient processing. Soil predators might accelerate the succession of bacteria and fungi on litter and alter the decomposition rate by preferentially grazing early successional species and allowing colonization by later successional species. In addition, grazers release waste products that might stimulate decomposition.

An optimal level between predator grazing and prey growth rate seems to exist in mature soil ecosystems. When no predators are present, bacteria and fungi retain most of the nutrients in their biomass, reducing the nutrients available for plant growth. Some bacteria and fungi might eventually autolyse, but rarely at a rate great enough to support rapidly growing plants. When predators first colonize a soil, their grazing on bacteria and fungi increases the percentage of active biomass over ungrazed populations. Although predator grazing is detrimental to the individual bacterium or hypha eaten, nutrients are released from the feeding interaction. These released nutrients are then taken up by roots or nearby bacteria and fungi, which then grow more rapidly. Therefore, the biomass of *active* bacteria and fungi may be greater when some predation occurs, even though total biomass might be the same or greater when predator densities are low.

A different result occurs when predator numbers are so high that the rapidly growing bacteria and fungi are overgrazed. At high predator numbers, fungal and bacterial growth can be slowed or stopped altogether because the predators consume bacteria and fungi more rapidly than they can reproduce. This also reduces decomposition rates of litter and organic matter. Thus, bacterial and fungal growth rates and decomposition rates are optimal when predator densities are at intermediate levels. Low to moderate predation of bacteria or hyphae stimulates microbial growth compared to when no predators are present, while heavy predation consumes too much of the microbial biomass. Lack of predation results in the sequestering of nutrients in bacterial and fungal biomass, while the presence of too many predators eventually results in slow organic matter decomposition rates. In both cases, nutrient cycling is limited, and the growth of plants, bacteria, and fungi can be suboptimal.

Box 6–2

Protozoa and Nematodes Are Not Always the Predators. The fungus *Dactyella* sp. can capture and consume amoebae, thus exploiting a nutrient-rich and abundant resource. Nematode-trapping fungi (e.g., *Arthrobotrys* as shown in Figure 6–1) are also well-known, although the trapping structures are not often found in field-collected soil. Viral and bacterial diseases of protozoa, nematodes, and microarthropods have also been reported.

Distribution

Protozoa and nematodes occur in all ecosystems of the world, from the poles to the tropics, from deserts to humid jungles, and from the tops of mountains to the depths of the ocean. Several thousand species of protozoa and several hundred species of nematodes have been described, although the systematics of any of these soil groups are far from complete.

Both protozoa and nematodes occur in large numbers in soils (Table 6–2 and Table 6–3). A representative picture of the entire community can be "captured" by standard methods such as the *most probable number (MPN)* method (Box 6–3). Nematodes and protozoa occupy the small interstitial spaces in soil and need water films at least several micrometers thick to remain active. Both protozoa and nematodes are usually concentrated near root surfaces that have high densities of bacteria, fungi, or other prey. Compared to aquatic protozoa, soil protozoa tend to be flatter, more flexible, and have less elaborate ornamentation. These modifications are

Table 6–2 **Average numbers of protozoa per gram dry soil in various ecosystems.**

Ecosystem	Flagellates	Amoebae	Ciliates
Agricultural and grassland systems			
Unplanted soil	7,000	30,000	155
Rhizosphere	33,000	100,000	875
Austrian wheat	NE	1,285	NE
Austrian meadows	NE	1,027	NE
Maize			
Noncultivated	348,350	25,190	11,460
nonfertilized	833,000	1,295	2,230
+ farm manure	740,000	144,690	11,265
Bluegrass soil	NE	2,000	800
Grazed pasture	NE	2,000	600
Semi-arid prairie	8,000	5,000	80
"	20,000	18,000	70
Mountain meadow	28,000	24,000	138
Alpine meadow	NE	5,000	2,000
"	NE	1,200	38,000
Forests			
Spruce	NE	2,800	30
Birch	NE	700	120
Pine (min)	NE	1,600	40
Pine (max)	NE	7,000	140
Beech-maple	NE	400	160
Oak-hickory	NE	2,000	200
Lodgepole pine	30,000	25,000	225
Mature Douglas fir	26,532	3,018	74
Subtropical soil			
Soil	1,575,000	419,000	66,000
Rhizosphere	410,000	15,000,000	66,000

Summarized from Ingham (1994a).
NE = not estimated

Table 6–3 **Abundance and biomass estimates for nematode communities in selected ecosystems.**

Ecosystem	Numbers ($\times 10^6$ m^{-2})	Biomass (g fresh mass m^{-2})
Desert (USA)	0.4	0.1
Tundra (Sweden)	4.1	1.1
Beech forest (England)	1.4	0.4
Tropical forest (Africa)	1.7	—
Pine forest (Sweden)	4.1	0.6
Mixed forest (Poland)	7.0	0.7
Pasture (Poland)	3.5	2.2
Grassland (Denmark)	10.0	14.0
Potato field (Poland)	5.5	0.7
Rye field (Poland)	8.6	1.1

Summarized from Ingham (1994b).

likely the result of selection based on surviving in thin water films surrounding soil particles and escaping from larger predators through the small pores in soil. In general, soil nematodes also have less ornamentation than their freshwater or marine counterparts.

Flagellates, smaller amoebae, and ciliates are usually more numerous and more widely distributed than larger ciliates and amoebae and large testate amoebae, which are restricted to continuously moist soil or litter (Table 6–2). Generally, ciliates and testate amoebae reproduce more slowly than flagellates and naked amoebae. In addition, larger organisms require higher prey densities; as a result, ciliate numbers in most soils are lower than the other groups of protozoa. Ciliates number from 1 to as many as 1,000 g^{-1} of soil in temperate systems, although relative numbers appear to be much greater in subtropical forests.

Community Structure in Various Habitats

Both protozoan and nematode communities are distinctive in a wide range of specific habitats. Certain species are cosmopolitan, with a tremendous ability to adapt to environmental fluctuations, while other species are intolerant of even small changes in conditions. A group of strictly terrestrial ciliates, the Grossglockneridae, are uniquely adapted for feeding on fungi and yeasts. These ciliates attach to fungal cells via a feeding tube from the oral cavity, causing the fungal hypha to swell in a distinct manner.

Soil temperature and moisture are two predictors of community composition or numbers of protozoa and nematodes. Equally important predictors are the type of vegetation, the presence (and possibly community composition) of bacteria and fungi, amount and type of soil organic matter, depth and decomposition rate of litter, and type of soil parent material.

Inventories of protozoan community structure in different ecosystems show that some species of protozoa consistently respond to certain environmental stresses. This suggests that protozoa are possibly useful as indicators of soil degradation or restoration and general "soil health" (Doran et al., 1996). For example, deciduous forests in Maryland and Louisiana contain a variety of ciliate species,

Box 6–3

The Most Probable Number (MPN) Method of Enumeration. Like dilution plating, the MPN method is a way to count organisms (Woomer, 1994). It is discussed here because it is the method of choice to enumerate all groups of protozoa at one time and works as well for soil as aquatic samples. The method entails diluting the soil until a dilution is reached where the organisms are no longer present (i.e., dilution to extinction). The problem with diluting to the "point of extinction" is that finding just one organism in a large volume is impossible. The answer is to incubate the sample in a growth medium so that if even one organism is present it will reproduce many, many times, and the presence of the organism will then be easy to detect. To ensure reasonable accuracy, scientists do a number of replications and relate the number of positive replicates to an MPN table to determine the number in the original sample.

By way of example, assume there are four replicates prepared at each 10-fold dilution. After a suitable period of incubation, a sample from each replicate is examined, and the presence of flagellates, ciliates, and amoebae is recorded (Table Box 6–3A). The number of positive replicates for flagellates is 4-4-3 for dilutions of 10^{-2} to 10^{-4} and the MPN is 27,726 individual flagellates in the original sample (Table Box 6–3B).

Table Box 6–3A **Example of most probable number (MPN) data for flagellates (F), amoebae (A), and ciliates (C).**

Replicate	Dilution (if letter is present, that type of protozoan was observed)			
	10^{-1}	10^{-2}	10^{-3}	10^{-4}
1	FAC	FAC	FA	F
2	FAC	FAC	FA	FA
3	FAC	FAC	FC	FC
4	FAC	FC	F	—
Code (F only)	4	4	4	3

Table Box 6–3B **Portion of a most probable number (MPN) table for different presence/absence readings using 10-fold dilutions and four replications.**

Positive replicates at each dilution				MPN
10^{-1}	10^{-2}	10^{-3}	10^{-4}	
4	4	1	1	1.353
4	3	2	1	1.863
4	4	2	1	2.141
4	4	3	1	4.264
4	4	4	1	5.754
4	3	2	2	5.606
4	4	2	2	6.962
4	4	3	2	8.318
4	4	4	2	13.863
4	4	3	3	15.250
4	4	3	4	24.417
4	**4**	**4**	**3**	**27.726***
4	4	4	4	46.060

*Multiplying the MPN value by the inverse of the middle dilution (10^{-3}) provides an estimate of 27,726 organisms in the original sample.

whereas adjoining corn and wheat fields show few species, with their community composition similar to those in desert soils (Bamforth, 1981). In Arizona deserts, the dry climate selects for a low diversity of ciliate species and a predominance of more rounded testate amoebae. The changing quality of food resources appears to be the important factor in determining protozoan responses, and the variability in amount and type of organic matter explains much of the variation in numbers of organisms in soil.

Dormant Stages

Both nematodes and protozoa have resistant stages that enable them to escape temporally stressful conditions. Protozoa form **cysts,** a nonmotile, inactive stage similar in function to fungal spores and nematode eggs, that can survive in soil for considerable lengths of time. Protozoa rapidly encyst when soil dries and rapidly excyst when moisture returns. The process of excystment in protozoa and egg hatch or activation of larval forms of nematodes is species-specific and is influenced by environmental factors such as stress, prey availability, and soil chemical factors. For example, normal seasonal fluctuations cause certain species of protozoa to encyst and become dormant, while others excyst and become active. Normal seasonal dynamics must be understood as the background against which other disturbances select for or select against certain species.

Function

The protozoa and nematodes mediate important functions in soil, including:

- recycling nutrients,

- inhibiting the development of disease-causing organisms in soil,

- controlling bacterial and fungal growth rates and community composition,

- altering litter decomposition rates, and

- causing commercially important diseases of many plants, which may limit the geographical distribution of many other species of organisms.

The most important of these functions is recycling of soil nutrients. Activities of protozoa and nematodes such as predation of bacteria may provide between 25 and 75% of the nitrogen taken up by plants in agricultural systems from their predation of soil bacteria (Box 6–4). Along with nitrogen, protozoa immobilize carbon, phosphorus, and other nutrients in their biomass.

We do not yet understand how to manage predator-prey cycles in soil to maximize nitrogen availability when plants are rapidly growing and to maximize nitrogen storage in soil when plants are not growing. The numbers and species composition of bacteria can be controlled by protozoan predation, but bacterial survival (and thus retention of nitrogen) can be improved by adding clay. The added clay allows the formation of microsites in which protozoa cannot reach their bacterial prey. Both bacterial and fungal pathogens can be suppressed by soil predators in a variety of ways, but managing these interactions for maximum effect has not yet been accomplished. Consistent with classical predator-prey cycles, increases in protozoan numbers lag

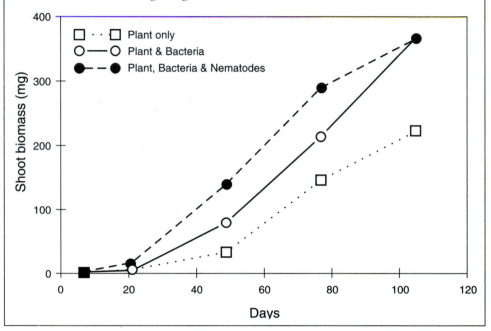

Box 6–4

Role of Predators in Nitrogen Availability. Predators of bacteria and fungi are important to the availability of nitrogen to blue grama (*Bouteloua gracilis*), a typical shortgrass prairie species (Ingham et al., 1985). Plants were nitrogen-limited and grew slowly when grown with only bacteria in the soil, but grew rapidly and contained more nitrogen with the addition of a bacterial-grazing nematode.

slightly behind increases in bacterial numbers, and decreases in protozoan numbers lag behind decreases in bacterial population. Notable peaks in both bacterial and then protozoan numbers have been documented in the spring in semi-arid grasslands and forests, and in both the fall and spring in the Pacific Northwest and California.

Protozoa and nematodes also assist in the dispersal of bacteria and fungi to new microsites and aid their establishment by providing new substrates for exploitation, such as feces and comminuted leaves. Dispersal of propagules by the fauna also ensures that later successional species are distributed to microsites in earlier stages of decomposition.

The factors that determine which predator group is most important in any given system are:

- plant community,

- parent material,

- soil aggregation, and

- relative amounts of bacteria and fungi.

Under some conditions amoebal biomass can comprise the greatest proportion of microbial carbon in soil. This is an example of the "predator paradox" common in marine ecosystems, where the predators of phytoplankton outnumber their prey, and predators of zooplankton can have greater biomass than zooplankton. Protozoa serve as food for predators such as nematodes and arthropods, and larger protozoa prey on smaller protozoa. Ciliates consume flagellates, often in preference to bacterial prey. Nematodes and larger protozoa can be maintained on cultures of flagellates or amoebae in the laboratory, but the importance of these interactions in field situations is unknown. Although protozoa can cause plant disease in certain circumstances (Dollet, 1984), protozoa are more useful as indicators of disturbance, chemical impacts, addition of genetically altered bacteria, and soil degradation (Foissner, 1986).

Nematodes have several important functions in soil ecosystems. Nematodes are generally most abundant in the rhizosphere, where their primary food resources are usually in greatest abundance. Here nematodes interact with other rhizosphere organisms, such as plant pathogens, rhizobia, mycorrhizal fungi, and other nematodes. Nematodes also fill an important link in the soil **food web,** transferring plant- and microbial-based energy sources and nutrients to higher trophic categories. Nematodes are also important pests of agricultural crops, even in native ecosystems. They have been estimated to consume more grass biomass than bison or cattle on the Great Plains of the United States (Ingham and Detling, 1984).

Nematodes can influence the production and accumulation of plant growth factors. Increased hormone concentrations in nematode-infected plants may contribute to greater infection by wilt-causing organisms, because these growth factors make host tissues easier to penetrate by maintaining cells in a rapidly dividing, juvenile state. Plant pathogenic nematodes can also affect nutrients released by roots, which might affect decomposer and mycorrhizal fungi as well as root pathogens. The composition and quantity of exudate compounds can be altered, damaged plant cells can leak nutrients, and pieces of damaged root can become available as substrate for rhizosphere microorganisms. Elements such as calcium, magnesium, sodium, potassium, iron, and copper are exuded in greater quantity from roots infected with *Meloidogyne* sp., a plant-feeding nematode. These alterations might lead to changes in the soil microflora. Plant pathogenic fungi may interact synergistically with plant pathogenic nematodes, where an increase in one usually signals an increase in the other.

Summary

Soil protozoa consist of flagellates, amoebae, and ciliates. The flagellates are the smallest and most abundant members of the protozoa. Amoebae move by protoplasmic extrusions called pseudopodia and have population densities approaching those of soil flagellates. The ciliates move by beating numerous cilia and are larger, but fewer in number, than flagellates and most amoebae. Nematodes, which may constitute as much as 90% of all multicellular animals in soil, are grouped into trophic categories based on the foods they eat and the structure of their feeding organs. Major food sources for nematodes include bacteria, fungi, protozoa, and plants.

Protozoa and nematodes are important in the soil because they regulate bacterial and fungal populations, cycling nutrients from the biomass of these decomposers, competing with and often controlling the populations of pathogenic organisms, and controlling litter decomposition rates. In contrast, protozoa and nematodes negatively impact communities by parasitizing many crop plants.

Cited References

Bamforth, S.S. 1981. Protist biogeography. J. Protozool. 28:2–9.

Coleman, D.C., and D.A. Crossley. 1996. Fundamentals of soil ecology. Academic Press, New York.

Dindal, D. 1990. Soil biology guide. John Wiley and Sons, New York.

Dollet, M. 1984. Plant diseases caused by flagellate protozoa (*Phytomonas*). Ann. Rev. Phytopath. 22:115–132.

Doran, J.W., M. Sarrantonio, and M.A. Liebig. 1996. Soil health and sustainability. pp. 1–54. *In* D.L. Sparks (ed.), Advances in agronomy. Vol. 56. Academic Press, San Diego, Calif.

Foissner, W. 1986. Soil protozoa: Fundamental problems, ecological significance, adaptations, indicators of environmental quality, guide to the literature. Prog. Protistol. 2:69–212.

Ingham, E.R. 1994a. Protozoa. pp. 491–515. *In* R.W. Weaver, S. Angle, P. Bottomley, D. Bezdicek, S. Smith, A. Tabatabai, and A. Wollum (eds.), Methods of soil analysis, Part 2. Microbiological and biochemical properties. Soil Science Society of America Book Series, No. 5. Madison, Wis.

Ingham, R.E. 1994b. Nematodes. pp. 459–490. *In* R.W. Weaver, S. Angle, P. Bottomley, D. Bezdicek, S. Smith, A. Tabatabai, and A. Wollum (eds.), Methods of soil analysis, Part 2. Microbiological and biochemical properties. Soil Science Society of America Book Series, No. 5. Madison, Wis.

Ingham, R.E., and J.K. Detling. 1984. Plant-herbivore interactions in a North American mixed-grass prairie. III. Soil nematode populations and root biomass on *Cynomys ludovicianus* colonies and adjacent uncolonized areas. Oecologia 63:307–313.

Ingham, R.E., J.A. Trofymow, E.R. Ingham, and D.C. Coleman. 1985. Interactions of bacteria, fungi, and their nematode grazers: Effects on nutrient cycling and plant growth. Ecol. Monogr. 55:119–140.

Mankau, R., and S.K. Mankau. 1963. The role of mycophagous nematodes in the soil. I. The relationships of *Aphelenchus avenae* to phytopathogenic soil fungi. pp. 271–280. *In* J. Doeksen and J. van der Drift (eds.), Soil organisms. Proc. Colloquium on Soil Fauna, Soil Microflora and Their Relationships, Oosterbeek, The Netherlands, 10–16 Sept. 1962. North-Holland Publ. Co., Amsterdam, The Netherlands.

Woomer, P.L. 1994. Most probable number counts. pp. 59–79. *In* R.W. Weaver, S. Angle, P. Bottomley, D. Bezdicek, S. Smith, A. Tabatabai, and A. Wollum (eds.), Methods of soil analysis, Part 2. Microbiological and biochemical properties. Soil Science Society of America Book Series, No. 5. Madison, Wis.

General References

Darbyshire, J.F. 1994. Soil protozoa. CAB International, Wallingford, England.

Old, K.M., and S. Chakraborty. 1986. Mycophagous soil amoebae: Their biology and significance in the ecology of soil-borne plant pathogens. Progr. Protistol. 1:163–194.

Study Questions

1. Why do ungrazed bacterial or fungal populations grow less rapidly than grazed populations?

2. The carbon-to-nitrogen ratio of protozoa is about 20/1, and the carbon-to-nitrogen ratio of bacteria is about 5/1. If the average protozoan weighs 10^{-8} grams and half of that weight is carbon, how many grams of bacteria does a single protozoan need to maintain its carbon-to-nitrogen ratio?

3. An average soil bacterium weighs about 1 picogram (10^{-12} g), and half of its weight is carbon. If protozoa eat about 10,000 bacteria per day, how many grams of nitrogen will be released by a single protozoan in one day?

4. What happens to the excess nitrogen consumed by protozoa?

5. Design an experiment to determine protozoan food preferences.

6. In the example of the MPN in Box 6–3, what are the numbers of ciliates and amoebae?

Chapter 7

♦

Viruses

J. Scott Angle and Joel V. Gagliardi

♦

So, naturalists observe, a flea has smaller fleas that on him prey;
and these have smaller still to bite 'em; and so proceed ad infinitum.
Jonathan Swift

Viruses are the most numerous of all soil biological entities, yet we know less about this group than any others in soil. High numbers of viruses in soil (e.g., fertile agricultural soils may contain 10^{11} viruses g^{-1} of soil) suggest that they play a significant role in soil ecosystems, though few viral functions have been elucidated. Viruses require a living host for replication, so the presence and survival of viruses in soil is closely related to the presence of suitable hosts in the environment. Therefore, any discussion of viruses in soil must also consider their host organisms: bacteria, fungi, insects, plants, and animals.

Virions, or complete viral particles, are distinguished from other organisms in that they do not grow in size and their individual components are produced within a living host cell before being assembled into a whole unit. They are usually *submicroscopic* particles less than 0.1 μm in diameter, although some—mainly the poxviruses—are visible with light microscopy. The typical structure of viruses includes a DNA or RNA core that codes for necessary viral components and a protein coat encasing the nucleic acid. Some complex viruses also contain lipid membranes, carbohydrates, and enzymes (e.g., RNA polymerase). Reproduction and assembly are directed by the viral nucleic acid, although some enzymes are not coded by the virus and are taken from the host genome. The metabolic machinery responsible for the production of viral biochemical components is part of the host organism; viruses do not metabolize or respire. Viruses are, therefore, considered *obligate intracellular parasites*. Because they do not grow or metabolize, viruses are not affected by common antibiotics. Viral diseases, however, may be treated with antiviral agents that target virus-specific components.

Box 7–1

Are Viruses Living or Dead? The answer depends on our understanding that viruses can exist in both an extracellular (dormant) state and an intracellular (replicative) state. Outside of their host cells, viruses are essentially dormant particles, protein coats packed with genetic information having no intrinsic metabolism or ability to replicate. Note that our definition of viruses includes the phrase "obligate intracellular parasite." Herein lies the clue to understanding the "reproduction," or replication, of the virus. Viruses "attack" a cell and take over its metabolic machinery. By inserting their own genes into the host cell, viruses manage to use the host's metabolism to replicate all of the parts needed for assembly of new virus particles. Once released from the host cell, the virus cannot replicate until it finds and infects another susceptible cell.

Thus the best response to our original question might simply be, "It depends." It depends on where the virus is and whether we consider piracy of a host cell as "living." One thing is clear: one of the major criteria of living entities, that of self-replication or growth, is accomplished only through the use of the metabolic functions of a living host cell. The only real basis for considering viruses to be alive is their ability to maintain genetic continuity and possibly to mutate.

The Viral Genome

Viruses may contain either DNA or RNA as genetic material in their core, or **genome.** Viral genomes may be double-stranded or single-stranded. In DNA viruses, single strands are named plus (+) strands if they code for mRNA directly, and minus (−) strands if they require a complementary copy made to code for mRNA. In RNA viruses, plus strands function as mRNA while minus strands are a special case. Plus-stranded RNA viruses are translated directly by host cell enzymes. In order for replication to begin in the minus-stranded RNA or double-stranded RNA viruses, a specific viral-encoded enzyme must be packaged with the virus genome. Absence of the enzyme renders the virus noninfectious. Only double-stranded DNA genomes replicate in the usual cellular manner (by DNA polymerase) to make their progeny.

Virus genomes usually contain between 10^3 to 10^6 nucleotides or nucleotide pairs, while bacterial genomes contain 10^6 to 10^7 nucleotide pairs. Plant and animal genomes have 10^8 to upwards of 10^{11} nucleotide pairs. The number of proteins coded by these genomes varies considerably. The smallest known virus, the satellite virus of tobacco necrosis virus, codes only a single protein. Angiosperm-infecting viruses, such as TMV (tobacco mosaic virus), code for four proteins. Bacteriophage T4 codes for over 100 proteins. By comparison, the higher organism-infecting viruses, such as the poxviruses, code for a few hundred proteins (Matthews, 1991; Joklik et al., 1992).

Classification of Viruses

Taxonomic relationships of viruses are partly defined by the hosts they infect, the diseases they cause, and their biochemistry. Viruses are not grouped into a distinct class of organisms and are not included in any of the three domains of living organisms. Viruses are often grouped as animal, plant, insect, fungal, or bacterial parasites. These relationships are often blurred because viruses in the same group or family can infect organisms in different domains. Individual viruses usually infect a limited number of species, although some have broad host ranges. Specificity depends upon the initial attachment to receptor sites or infection method of the virion for its host and the host's metabolic capacity to synthesize necessary viral components.

Bacterial Viruses

Bacterial viruses are called **bacteriophages,** or simply **phages,** which translates literally as "bacteria eater." They are classified into species based primarily on the genera of bacteria they infect (Table 7–1). Viruses have been isolated from nearly every genus of soil bacteria. For classification, a phage must be culturable and have a known morphology. Most bacterial viruses have an icosahedral (12 vertices, 20 faces, and 30 edges) **capsid** (or head) with a tail (Fig. 7–1). Tails vary widely in length but are typically about 0.2 μm long. They may be contractile to inject nucleic acid into the viral genome, flexible or rigid, and may contain fibers, plates, and spikes. The tail is important in attachment to the host and serves as a pathway for the injection of the viral genome through the cell wall of the host (Fig. 7–2). Only 50% of the known tailed phages are classified (Ackerman and DuBow, 1987). Other phage morphologies are cubic, filamentous, or pleomorphic in shape, and several are unique or undetermined.

Table 7–1 **Some agriculturally relevant groups of bacterial viruses.**

Virus group	Example of host	Example of virus species
Tailed phage		
Actinophage	*Streptomyces*	Av–1
Agrobacteria phage	*Agrobacterium*	PIIBNV6
Bacilli phage	*Bacillus*	Phage G (largest known bacterial virus)
Clostridia phage	*Clostridium*	HM2
Cyanophage	*Anabaena*	SM-1
Gram positive cocci phage	*Micrococcus*	Twort
Pseudomonad phage	*Pseudomonas*	M6
Rhizobia phage	*Rhizobium*	XP12
Cubic, filamentous, and pleomorphic phage		
Virus group	Example of host	Example of virus species
Cystoviridae	*Pseudomonas*	Phage 6
Inoviridae	*Xanthomonas*	X
Leviviridae	*Pseudomonas*	7s
Plasmaviridae	*Acholeplasma*	MVL2
Tectiviridae	*Bacillus*	PRD1

Adapted from Ackerman and DuBow (1987).

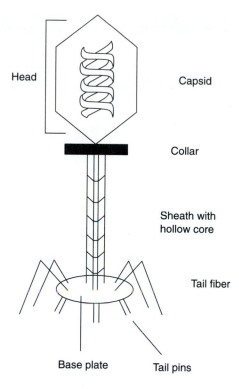

Head

Capsid

Collar

Sheath with
hollow core

Tail fiber

Base plate   Tail pins

Figure 7–1 Simple illustration of a tailed bacteriophage. These bacteriophages are sometimes described as "lunar landers" because they resemble the lunar landers from the U.S. Apollo space missions of the late 1960s and early 1970s. *Drawing by David Zuberer.*

Fungal Viruses

Fungal viruses are called **mycoviruses,** or sometimes mycophages, and are classified according to a taxonomic scheme that accounts for similarities in the virion (e.g., shape, symmetry, serology), nucleic acids (e.g., type of nucleic acid, molecular weight of pieces, percent of G+C (guanine and cytosine content)), proteins (e.g., number, molecular weight, percentages), and lipids (e.g., type, percentages) in the virus. In some cases, mycoviruses resemble bacterial, plant, and animal viruses. The genomes of most mycoviruses are double-stranded RNA, and several are incorporated into the host with no known ill effects. Few mycoviruses, however, have been classified adequately, and most are named for the host they infect or the disease they cause.

Plant Viruses

In addition to the bacteriophages, plant viruses may also be abundant in soils. Most plant viruses contain single, positive-stranded RNA and, with a few exceptions, are of minor economic importance (Table 7–2). There is currently no cohesive taxonomic classification system for plant viruses, mainly because not enough is known about the majority of the plant-associated viruses and because past classification attempts were not based on sound virology. Plant viruses are not indigenous inhabitants of the soil *per se* since they are normally associated with living plant parts. Transmission between plants is typically through an insect vector or by plant-infecting nematodes, although seed transmission, transmission

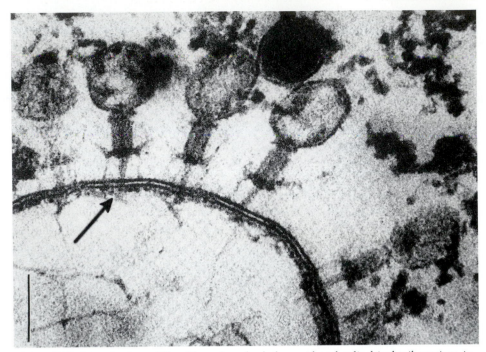

Figure 7–2 Bacteriophage T4, with an icosahedral capsid and cylindrical tail portion, infecting the bacterium *Escherichia coli*. The phages are bound to the bacterial surface by short tail fibers extending directly from their base plates to the cell wall. The needle of one of the phages is penetrating just through the cell wall (arrow). *Micrograph from Simon and Anderson, 1967. Used with permission.*

by other microbes, mechanical means (e.g., planting, spraying, harvesting), and wind and rain splash are also responsible.

In plants, other types of nucleic acid molecules can produce disease much like viruses. *Satellite viruses* are dependent on a helper virus present in the same cell to replicate. They code for their own capsid proteins and can contribute to or amplify the disease caused by the helper virus. *Viroids* are small, circular, single-stranded RNA molecules that do not code for any peptides and replicate without viruses present. They do not have a protein coat and can cause diseases independently. *Satellite RNA* is packaged with a helper virus and contributes to the disease progression, but the RNA is still unique from the helper virus. *Defective-interfering RNA* is identical to that of the helper virus, but it has deletions that can slow the progression of the disease.

Plants can be genetically engineered through the integration of viral coat protein genes and other viral components directly into their genome. The expressed protein can signal infecting viruses that an infection has already occurred and wards off further infections. Classic plant breeding for virus resistance aims to change or eliminate the virus-specific receptor site.

Insect Viruses

Numerous soil insect viruses have been described. Most are classified according to the characteristics of the nuclear envelope, with the majority of them occurring in the family Baculoviridae (Table 7–3). Other common insect viruses belong to the Iridoviridae.

Table 7–2 Major plant viruses. Plant viruses belonging to currently accepted or proposed families of viruses are written in italics.

Group or family	Morphology	Proteins coded	Major modes of transmission	Examples of disease-causing viruses
Positive sense single-stranded RNA genome				
Potyviridae	Helical	8	Insects, fungi, mechanical	
Potexvirus	Helical, flexible	4–5	Mechanical	Papaya mosaic virus
Tobamovirus	Helical	4–5	Numerous	Tobacco mosaic virus
Tymovirus	Icosahedron of pentamers/hexamers	3–4	Mechanical, beetles	European turnip yellow mosaic virus
Comoviridae	Icosahedron	9	Beetles, seed	Cowpea mosaic virus
Bromoviridae	Icosahedron	4	Numerous	Cowpea chlorotic mosaic virus
Tobravirus	Helical	5–6	Numerous	Tobacco rattle virus
Negative sense single-stranded RNA genome				
Rhabdoviridae	Enveloped, bullet shaped, helical	5–6+	Vertebrates	Sonchus yellow net virus
Bunyaviridae	Enveloped helical	6+	Vertebrates, invertebrates	Tomato spotted wilt tosposvirus
Double-stranded RNA genome				
Cryptoviridae	Icosahedron			
Reoviridae	Double Icosahedron	10–12	Leafhoppers	Wound tumor virus
Double-stranded DNA genome				
Caulimovirus	Enveloped Icosahedron	8	Aphids, other	Cauliflower mosaic virus
Single-stranded DNA genome				
Geminiviridae	Icosahedron	5–6+	Leafhoppers, whiteflies	Maize streak virus

Adapted from Mathews (1991).

Table 7–3 Major insect virus families.

Family	Nuclear envelope
DNA genomes	
Baculoviridae	Helical
Poxviridae	Complex, ovoid
Iridoviridae	Icosahedral
Parvoviridae	Icosahedral
RNA genomes	
Reoviridae	Icosahedral
Picornaviridae	Icosahedral
Caliciviridae	Icosahedral
Nodaviridae	Icosahedral
Rhabdoviridae	Bullet-shaped, helical
Nudaurelia B virus group	Icosahedral

Adapted from Joklik et al. (1992).

Animal Viruses

Animal viruses include many of the same families as plant and insect viruses. Most animal viruses are not indigenous to the soil, but they may be added to soil through human activities. For example, the enteroviruses of humans and other animals—like poliomyelitis virus, the cause of polio—replicate in the gastrointestinal tract and are shed through feces in high numbers (Joklik et al., 1992). Nearly 100 types of viruses have been detected in the wastes of humans, of which the most common are the Picornaviridae (enteroviruses, polioviruses, coxsackieviruses, echoviruses), Reoviridae (reoviruses, rotaviruses), Adenoviridae (adenoviruses), and the hepatitis viruses (Bitton et al., 1987). Of these viruses, the Hepatitis A virus is the greatest concern because of its disease severity and its potential to survive for long periods of time in the soil (Sobsey et al., 1986). Vaccination, the administration of attenuated (weakened noninfectious viruses) live viral strains or suspensions of killed virus or virus components, serves to prime the animal immune system for later challenges by infectious viruses. Prevention through vaccination is the best method for treating viral diseases of animals and humans.

Types of Bacterial Virus Life Cycles

Two main types of viruses infect bacteria. The first type, the **lytic** or **virulent** bacteriophage (Fig. 7–3), replicates inside the host cell, releasing progeny by lysing the cell membrane or by *budding,* in which an envelope of host membrane surrounds and remains a part of the virion. Of the lytic soil bacteriophages studied thus far, most lyse their host in the late stage of the infection cycle. Bacterial viruses often encode for a phage **lysozyme** late in the viral replication cycle that dissolves the peptidoglycan layer of the bacterial cell wall, thereby releasing the new virions into the soil. Several hundred viruses may be released from a single host cell in a lytic cycle.

The second type of virus, the **temperate** bacteriophage, enters into an association known as **lysogeny,** in which the host cell is not immediately lysed to release new virions. Instead, the genome of the virion is stably integrated into the genome of the host. The integrated phage genome is then referred to as the **prophage.** Bacterial cells in this condition are called **lysogenic** bacteria. Lysogeny is apparently carried out only by bacteriophages containing double-stranded DNA, though there have been reports of lysogeny with mycoviruses.

Occasionally, external stimuli, such as starvation of the host, cause viral nucleic acid to be excised from the genome of the host, and the lytic phase of the virus life cycle begins. The importance of temperate viruses in soil is difficult to determine because the virions are hidden, or latent, within their host. It has been suggested that temperate viruses are present in most soil bacteria and that they are intimately involved in the survival and evolution of the host, sometimes adding new genes or changing existing ones. In fact, the lysogenized bacterial cell frequently gains new phenotypic traits. The acquisition of new traits due to lysogenization is referred to as *phage conversion* or *lysogenic conversion.* For example, the bacterium *Corynebacterium diphtheriae* exists as toxigenic (pathogenic) and nontoxigenic (nonpathogenic) strains. The production of the diphtheria toxin occurs as a result of the lysogenization of the bacterium with phage β. Thus, a nonpathogenic bacterium is rendered potentially deadly for humans because of the incorporation of

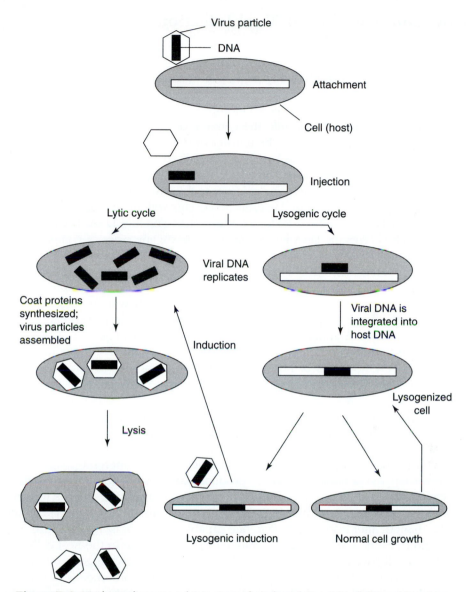

Figure 7–3 Viral attachment and injection of viral nucleic acid is followed by either a lytic cycle (left) leading to the release of new virions or a lysogenic cycle (right) leading to integration of viral nucleic acid into the host genome and subsequent replication. Under some conditions, lysogenized cells can be induced to begin the lytic cycle. Note: the host DNA is shown as a linear molecule only for sake of demonstration. *Original drawing by David Zuberer.*

the viral genome. A similar situation exists with *Clostridium botulinum*. The lysogenized cells of this bacterium produce the deadly botulinus toxin.

Another important consequence of lysogeny is that the infected cell gains immunity to reinfection by the same phage. This may be an important means of protection of the individual in an environment where the threat of viral infection is ever present.

Genetic Consequences of Viral Infection

Transduction, the transfer of host nucleic acid to a new host, is a common phenomenon among bacteriophages in the laboratory (Chapter 3). During transduction, a temperate phage can randomly pick up host nucleic acid during the synthesis of new virions, or a lytic phage may randomly package host nucleic acid into itself instead of viral nucleic acid. The new virions can then infect other hosts, leading to integration of the nucleic acid into the genome of the new host via a virus particle. Transduction is responsible for the transfer of potentially important traits.

Transduction can take place by either generalized or specialized (sometimes called restricted) means. **Generalized transduction** is the packaging of a piece of host nucleic acid (e.g., DNA), which is similar in size to the virus nucleic acid, into a phage particle. An example of generalized transduction occurs with the lysogenic *Escherichia coli* phage, P1, whose genome codes for a nuclease active on host DNA. The random nature of the nuclease cuts the DNA into a variety of pieces; any piece of comparable size to the P1 phage DNA may be packaged into a virion at a rate of approximately 1 in 10^6 viable phage (Freifelder, 1987). Phage P1 is a nonintegrative phage, which means that its DNA is not incorporated into the bacterial chromosome during lysogenization. Rather, it remains as a circular plasmid that replicates simultaneously with the host chromosome.

Specialized, or **restricted, transduction** involves the incorporation of small pieces of host DNA along with viral DNA into the phage (Fig. 7–4). During the beginning of the lytic cycle, imperfect excision of the phage DNA from the host DNA leads to aberrant pieces of host DNA becoming incorporated into the end of the phage genome, often with part of the bacterial virus DNA being left behind. An example is the *E. coli* λ (lambda) phage that integrates into the host chromosome between the genes for galactose (*gal*) and biotin (*bio*) utilization. Either host gene may be excised in full or in part with the phage DNA, leaving a similar-sized piece of phage DNA behind. Any piece of DNA with a length between 79% and 106% of the normal lambda phage genome can be packaged. These aberrant phages are called lambda-*gal* and lambda-*bio* and are produced at a rate of about 1 in 10^6 to 10^7 infected cells (Freifelder, 1987). Screening for these transducing phages requires mutant host cells that cannot metabolize galactose or biotin. If the phenotype of the host cells changes after lysogenization to one capable of metabolizing galactose or biotin, this indicates the transfer and integration of the new gene to the host cells.

Nearly all bacteria isolated from soil are lysogenic. This suggests that the phage enhances the evolution of the organism, although their presence is not necessarily critical to bacterial survival. Transduction may be important for the transfer of traits, such as the ability to metabolize unusual carbon sources or to confer resistance to antibiotics and other environmental toxins. These traits may give the recipient organism a competitive advantage over organisms that are not infected with the virus. An analogy to human disease would be where a bacterium pathogenic to humans acquires a gene for antibiotic resistance to the antibiotic normally used to treat that disease. Wherever that antibiotic is used, natural selection then serves to form a population composed mainly of bacteria with the new antibiotic resistance trait.

The importance of transduction between two indigenous soil bacteria has yet to be resolved (Selander and Levin, 1980). There is evidence that transduction can occur between two bacterial species inoculated into soil (Stotzky, 1989). Resistance to

Specialized transduction

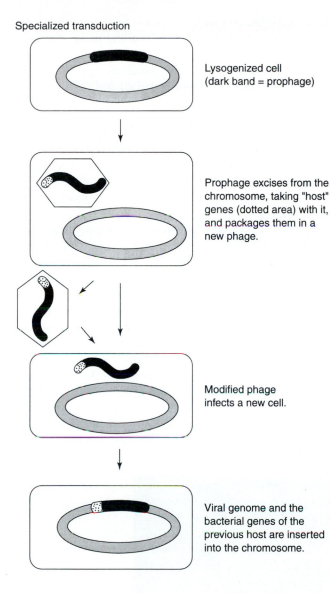

Lysogenized cell
(dark band = prophage)

Prophage excises from the
chromosome, taking "host"
genes (dotted area) with it,
and packages them in a
new phage.

Modified phage
infects a new cell.

Viral genome and the
bacterial genes of the
previous host are inserted
into the chromosome.

Figure 7–4 Genetic transfer
takes place when host DNA
is packaged into the virus either in place of the genome
(generalized transduction) or
when a piece of host DNA is
incorporated with viral DNA
(specialized or restricted
transduction). Specialized
transduction is shown here.
*Original drawing by David
Zuberer.*

environmental toxins (e.g., heavy metals) is an essential trait for survival in some soils,
and the transfer of this resistance may be mediated by a bacteriophage.

Enumeration, Isolation, and Characterization of Viruses

Numerous direct and indirect methods are available for studying and enumerating viruses
in soil (Angle, 1994). The most common indirect method is the soft-agar overlay. Here
the soil is diluted in a buffer and the suspension filtered through a 0.45-μm filter to remove soil particles and microorganisms that are bacteria-sized or larger. Only viruses
pass through the filter. The filtrate and an appropriate host organism are mixed in soft
agar (0.5%) and the mixture is poured over hardened agar (1.5%). The hardened agar

contains nutrients capable of supporting growth of the host organism. After several days, each virion in the overlay will have lysed its original host and those surrounding it to create a clear area, or **plaque,** in the lawn of bacteria (Fig. 7–5). The number of plaques (reported as plaque-forming units ml^{-1}, or PFU ml^{-1}) indicates the number of virions in the filtrate. A similar procedure is used to enumerate plant and animal viruses in soil, except the filtrate is inoculated onto tissue culture plates containing the appropriate host cell line. Plaque formation in the cell culture indicates lysis of host cells. Although the soft-agar overlay is a simple and reliable method, it assays only a fraction of the total number of viruses. This is because of the inefficiency of extraction and the inability to provide suitable hosts for all species of virus present. By some estimates, the soft-agar overlay method enumerates less than 1% of the total population of viruses in soil.

Several direct methods can be used to observe viruses in soil. Electron microscopy is the most universal. Immunofluorescence microscopy is more specific, viewing fluorescently labeled antibodies that fluoresce when excited with specific wavelengths of UV light while bound to the virus (i.e., the antigen). The enzyme-linked immunosorbent assay (ELISA) is similar in that an enzyme that acts on a specific substrate (e.g., a phosphatase) is attached to an antibody, giving a visible color reaction in the well of a microtiter plate when it attaches to the virus particles sorbed to the well and reacts with the substrate for the enzyme. **Monoclonal antibodies** allow the detection of a single trait, often chosen because it is associated with a single type of organism, whereas a polyclonal antibody recognizes many traits and often many different organisms. Although these methods have never gained wide acceptance for studying soil viruses, refinements may make these techniques important in the future. The use of gene probes (Chapter 9) also offers promise as a means for detecting viruses in soil. Because this technique does not require bacterial cells or specific plant or mammalian tissue cultures, it avoids the restriction imposed by limited host range. Unfortunately, the use of gene probes is insufficiently developed for use as a standard method in enumeration

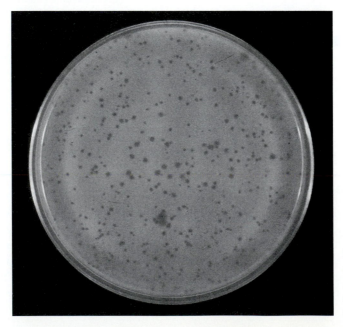

Figure 7–5 Viral plaques (clear zones) in a lawn of *Rhizobium* cells grown in a soft-agar overlay atop a hard agar medium.

of viruses in soil. Use of the polymerase chain reaction (PCR) to amplify low copy numbers of viral DNA or RNA may find application in the future, especially when methods of nucleic acid extraction from soil are improved. Theoretically, the use of PCR will allow the detection of DNA from a single viral particle.

Survival of Viruses in Soil

A variety of factors affect the survival of viruses in soil. First is the type of virus. Some viruses are well adapted for survival in soil and persist for many years. Other viruses, such as human enteric viruses, are ill adapted for survival and may persist in soil for only days to months.

A second factor is the presence of a susceptible host. When populations of the host are high, populations of the homologous virus will increase over time. This shows that continued cycles of infection are occurring. Conversely, when host populations are low, then populations of the homologous virus are low since fewer encounters can take place between the host and virus. As long as an appropriate host is present, a virus can persist indefinitely through continual production of new viruses.

A third factor affecting survival is the collective influence of soil abiotic factors such as temperature, moisture, and pH. Temperature and desiccation are believed to be important factors affecting virus survival: low moisture and high temperature cause viruses in soil to rapidly lose their ability to infect a host (Farrah and Bitton, 1990). Soil pH affects populations of streptomycete phage, with the highest numbers found in neutral to slightly-alkaline soils (Williams and Lanning, 1984).

A fourth factor affecting the long-term viability of phage in soil is adsorption onto organic matter and clays. This adsorption is the result of the electric double layer on clay minerals, cation exchange capacity, electrostatic interactions, hydrophobic interactions, and cation bridging (Lipson and Stotzky, 1987). Presumably, adsorption onto organic and clay surfaces protects virions from predation, degradative enzymes (e.g., extracellular proteases), and other soil biotic factors. For example, reovirus disappeared within 7.5 weeks in distilled water only, but survived for 22 weeks when kaolinite and smectite were added (Lipson and Stotzky, 1987). Tobacco mosaic virus, which causes a necrotic disease on tobacco leaves, is rapidly inactivated in a nonsterile soil, but is stable in an identical sterile soil (Cheo, 1980). This nonspecific adsorption may or may not affect viral infectivity. For example, Williams and Lanning (1984) reported that streptomycete phage retained infectivity despite adsorption onto clay. However, adsorption to soil particulates may also decrease infectivity because the virus can no longer interact with its host. Also, the kind and valence of cations present in the soil solution may affect viral survival. Magnesium and calcium prevent thermal inactivation (degradation of the protein coat) of poliovirus (Wallis and Melnick, 1961).

Because of the interaction of all these factors, it is not possible to define a specific period that a given virus will persist in soil. Under harsh abiotic conditions where no host is present, viruses may survive only a few hours in soil. Alternatively, when soil is cool and moisture is present, viruses may remain infective for many months, even years. Further, if a temperate phage is incorporated into the genome of a host bacterium, its nucleic acid may persist for generations as a stable component of the ecosystem.

Box 7–2

Balance Between Phage and Host Populations. In a sterilized, inoculated soil, the number of viruses in soil was inversely related to the soil population of its host bacterium, *Bradyrhizobium japonicum*. When the phage population was high, the host population was low due to the lytic activity of the virus; when the phage population decreased, the host population increased. Note the number of *B. japonicum* never declined to zero, but rather a tenuous balance between the phage and host populations existed such that neither organism predominated.

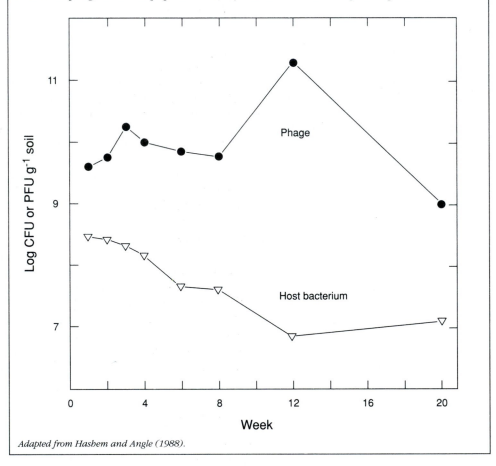

Adapted from Hashem and Angle (1988).

Importance of Viruses

The overall importance of viruses in soil is difficult to determine because methods of studying viruses are imprecise (Farrah and Bitton, 1990). Both the number and diversity of bacterial species in soil are undetermined, so the number of bacteriophages in soil is undetermined as well. Agriculturally important host bacteria for which bacteriophage have been reported include *Rhizobium, Bradyrhizobium, Bacillus, Arthrobacter, Pseudomonas,* and many species of actinomycetes (Angle, 1994). Early reports suggested that the failure of legumes to nodulate was because of the presence of high populations of a homologous bacteriophage in soil, although more recent evidence has failed to confirm these observations (Hashem and Angle, 1988).

Inoculation of soil with bacteriophages has been proposed as a means to reduce soil populations of undesirable and ineffective strains of rhizobia (Basit et al., 1992).

Insect viruses have received increased attention recently because of their potential as biological control agents. Nearly all of the economically important viruses for biocontrol are highly specific and virulent. The nuclear polyhedrosis viruses are most often used for the biological control of insects.

Box 7–3

Biological Control. It is possible to control specific pest populations in soil with viruses. To control an agricultural infestation of celery looper, for instance, the soil could be inoculated with the *Anagrapha falcifera* multiple nuclear polyhedrosis virus (AfMNPV). The virus can potentially infect and kill all susceptible insects. When insect hosts are no longer present, the virus also disappears. The inability of a virus to persist in soil is important from a risk assessment perspective, and this attribute greatly enhances the acceptance of this process.

The presence of human viruses in wastewater is a cause of great concern. Nearly 100 types of viruses have been detected in the wastes of humans, and these may reach the soil through land application of **biosolids** (Box 7–4) and wastewater effluent from treatment plants, septic tanks, and filter fields. Furthermore, the movement of these viruses through soil and the subsequent contamination of groundwater is another troubling possibility (Box 7–5).

Sewage treatment often fails to completely inactivate viruses in wastewater. Primary and secondary treatments typically remove approximately 90% of the viruses contained in the original waste (Gerba, 1981). Although tertiary treatment removes additional viruses from wastewater, removal is never 100% efficient. Because theoretically only a single virion is required to initiate disease in a susceptible individual, even low numbers of viruses in wastewater represent a significant health concern. This is in contrast to bacteria, where hundreds, perhaps even millions, of cells are required to initiate a bacterial infection.

Some Future Prospects for Soil Virology

We still have much to learn about viruses in soil. Research is underway to determine the importance of viruses in the diversity and function of the soil ecosystem. Whether viruses are responsible for controlling specific populations of bacterial species in soil is questionable; however, a fair amount of evidence suggests the possibility. The importance of transduction in the evolution of populations of higher organisms requires further study. Viral transduction may be closely involved in the evolution of bacteria in soil, even to the extent that viruses may control the evolutionary direction of various microorganisms. Until methods are improved that allow us to monitor and assess transduction between two indigenous species of soil bacteria, the importance of transduction will remain unknown.

On a practical basis, many viruses will likely continue to be developed as biological control agents. To date, viruses are used primarily for the control of insect pests. Work is currently underway to develop means whereby viruses reduce

┌─── **Box 7–4** ───┐

A Brief Overview of Wastewater Treatment.

Primary treatment: Large objects are screened out during a pretreatment phase as solid waste for landfill. Flocculated sediment and flotsam are removed for further processing and disposal.

Secondary treatment: Air is bubbled through a second tank containing the effluent from the primary tank. This process encourages microbial growth, and reduces the biological oxygen demand (BOD). Microorganisms form flocs. These flocs carry over to a settling tank (still in secondary treatment) where the flocs settle. Sediment from this tank is called biosolids.

Tertiary treatment: Various methods (e.g., filtration through charcoal) remove chemical and biological components. Chemical coagulation (e.g., alum) can remove greater than 99% of remaining viruses from the wastewater (Berg, 1983).

Biosolids: Biosolids from primary, secondary, and sometimes tertiary treatment can be composted at high temperature, anaerobically digested, lime stabilized, or dewatered.

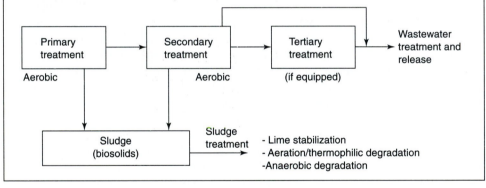

┌─── **Box 7–5** ───┐

Movement of Viruses in Soil. In a study to test the ability of viruses in wastewater to travel horizontally or vertically through soil, the bacteriophages MS-2 and PRD-1 were added to secondarily treated wastewater contained in a storage basin. Water samples were collected from the **vadose zone** (Chapter 21) and wells 3 to 46 meters from the basin. Under conditions of unsaturated flow, the two viruses traveled only 1 meter per day for a total of 4.5 meters. However, under conditions of saturated flow, the two viruses traveled up to 15 m per day for a total of 46 meters (Gerba, 1981).

┌─── **Box 7–6** ───┐

The Viral Threat to Groundwater. The first disease-causing virus isolated during an outbreak of a waterborne illness was poliovirus, taken from a 30.5 m deep well contaminated by a septic tank leach field. The virus had to pass through 5.5 m of clay, 2.5 m of shale, and 22.5 m of limestone to reach the well water (Mack et al., 1972).

the damage caused by fungal, bacterial, and other viral pathogens as well as damage caused by nematodes. Attenuated (viable but not capable of causing disease) viruses may also be inoculated onto or incorporated into plants to protect the crop from subsequent infection by pathogenic viruses. Finally, the rapidly developing field of molecular biology offers new techniques that may soon allow us to answer many of the questions that remain about the importance of viruses in soil.

Summary

Viruses are the most numerous and least understood of all biological entities in soil. Viruses are capable of infecting bacteria, fungi, insects, plants, and animals. Lytic or virulent viruses rapidly kill their hosts while temperate viruses become stably integrated into the genome of the host. Temperate viruses, through the process of transduction, are potentially responsible for the transfer of genetic material from one host bacterium to another. With the possible exception of the animal viruses, there is currently no cohesive classification system for viruses. Virus survival depends on the kind of virus, the presence of a susceptible host, and the appropriate environmental conditions. Viruses may one day find a wide application for the biocontrol of insects, weeds, fungal infections, and other noxious organisms.

Cited References

Ackerman, H.-W., and M. S. DuBow. 1987. Viruses of prokaryotes, Vol. 2. Natural groups of bacteriophages. CRC Press, Boca Raton, Fla.

Angle, J. S. 1994. Viruses. pp. 107–118. *In* R. W. Weaver, S. Angle, P. Bottomley, D. Bezdicek, S. Smith, A. Tabatabai, and A. Wollum (eds.), Methods of soil analysis. Part 2. Microbiological and biochemical properties. Soil Science Society of America, Madison, Wis.

Basit, H. A., J. S. Angle, S. Salem, and E. M. Gewaily. 1992. Phage coating of soybean seed reduces nodulation by indigenous soil bradyrhizobia. Can. J. Microbiol. 38:1264–1269.

Berg, G. 1983. Viral pollution of the environment. CRC Press, Boca Raton, Fla.

Bitton, G., J. E. Maruniak, and F. W. Zettler. 1987. Virus survival in natural ecosystems. pp. 301–332. *In* Y. Henis (ed.), Survival and dormancy of microorganisms. John Wiley, New York.

Cheo, P. C. 1980. Antiviral factors in soil. Soil Sci. Soc. Am. J. 44:62–67.

Freifelder, D. 1987. Microbial genetics. Jones and Bartlett Publishers, Boston.

Gerba, C. P. 1981. Virus survival in wastewater treatment. pp. 39–48. *In* M. Goddard and M. Butler (eds.), Viruses and wastewater treatment. Proc. International Symposium on viruses and wastewater treatment, University of Surrey, Guildford, England, 15–17 Sept. 1980. Pergamon Press, Oxford.

Hashem, F. M., and J. S. Angle. 1988. Rhizobiophage effects on *Bradyrhizobium japonicum*, nodulation and soybean growth. Soil Biol. Biochem. 20:69–73.

Joklik, W. K., H. P. Willett, D. B. Amos, and C. M. Wilfert. 1992. Zinsser microbiology. 20th ed. Appleton & Lange, Norwalk, Conn.

Lipson, S. M., and G. Stotzky. 1987. Interactions between viruses and clay minerals. pp. 197–230. *In* V. C. Rao and J. L. Melnick (eds.), Human viruses in sediments, sludges, and soils. CRC Press, Boca Raton, Fla.

Mack, W. N., Y. Lu, and D. B. Coohon. 1972. Isolation of poliomyelitis virus from a contaminated well. Health Serv. Rep. 87:271.

Matthews, R. E. F. 1991. Plant virology. 3rd ed. Academic Press, San Diego, Calif.

Selander, R. K., and B. R. Levin. 1980. Genetic diversity and structure in *Escherichia coli* populations. Science 210:545–547.

Simon, L. D., and T. F. Anderson. 1967. The infection of *Escherichia coli* by T2 and T4 bacteriophages as seen in the electron microscope. Virology 32:279–297.

Sobsey, M. D., P. A. Shields, F. H. Hauchman, R. L. Hazard, and L. W. Caton, III. 1986. Survival and transport of hepatitis A virus in soils, groundwater and wastewater. Water Sci. Technol. 18:97–106.

Stotzky, G. 1989. Gene transfer among bacteria in soil. pp. 165–222. *In* S. B. Levy and R. V. Miller (eds.), Gene transfer in the environment. McGraw-Hill, New York.

Wallis, C., and J. L. Melnick. 1961. Stabilization of poliovirus by cations. Tex. Rep. Biol. Med. 19:683–687.

Williams, S. T., and S. Lanning. 1984. Studies of the ecology of streptomycete phage in soil. pp. 473–483. *In* L. Ortiz-Ortiz, L. F. Bojalil, and V. Yakoleff (eds.), Biological, biochemical, and biomedical aspects of actinomycetes. Academic Press, New York.

Study Questions

1. What are the differences between lytic and temperate viruses?

2. Why does the enumeration of viruses require a living host?

3. Why are health departments concerned about the contamination of drinking water with viruses?

4. Predict how the soil abiotic factors of temperature, moisture, and pH affect the survival of viruses in soil.

5. Would viruses be adsorbed more tightly to clay at a pH of 5.5 or 7.5?

6. What is the ecological significance of viruses with a broad host range?

7. One ml of a 1/10,000 soil dilution is mixed with a host bacterium in a soft agar overlay. After several days, 79 PFU were observed growing within the host bacterial lawn. How many viruses were present in the original soil sample? Why is this number lower than numbers typically reported in soil?

Chapter 8

♦

Microbial Ecology

Peter J. Bottomley

♦

Everything is everywhere and the milieu selects.
Martinus Beijerinck

Soil microbial ecology is the study of the properties and behavior of microorganisms in their soil environment. We seek to understand how soil physical and chemical properties and organismal interactions influence the microbially mediated processes that drive soil ecosystems. Because of the interplay between the physicochemical and biological properties of soil, many different approaches can be used to study soil microbial ecology. For example, we may wish to identify the soil properties that determine the rate of a specific microbially mediated process in different soil types. Alternatively, microbially mediated processes *per se* may not be our primary concern. We might be more interested in determining if different soil management practices influence the overall composition of the soil microbial community. These changes in community composition might then be correlated with the suitability of the soil for crop production or waste treatment. At the other end of the spectrum of interest, we might wish to identify how one member of a bacterial soil population can prevent a specific plant pathogen from infecting a crop species. The purpose of this chapter is to introduce the subject of soil microbial ecology and to point out the interactions that must occur between microorganisms and macroorganisms if soil ecosystems are to function effectively.

Nature of Soil Organisms and Their Interactions

Many kinds of organisms spend all, or at least a part, of their lives in the soil. These include plants, animals, fungi, bacteria, and viruses. Most soils contain high populations of microorganisms, with nutritional versatility and fast metabolism as the hallmark features of many bacteria and fungi. These nutritional and metabolic characteristics permit microorganisms to play the major role in decomposing organic materials and returning essential plant nutrients to the soil.

Nevertheless, it is difficult for microorganisms to achieve their metabolic potential in the soil matrix. Accessing nutrients is a major problem for soil microorganisms because soil is a discontinuous matrix of vast surface area. Despite the seemingly high densities of microorganisms, it has been estimated that less than 0.4% of soil pore space is occupied by microorganisms. Moreover, because a large percentage of the pore diameters in soil aggregates are less than 2 μm, particulate substrates in these voids are inaccessible to all but the smallest microbes. The bioavailability of hydrophobic substrates is limited by adsorption to soil surfaces, whereas diffusion of soluble substrates is often limited by the lack of continuous water-filled pores. To further complicate matters, most bacteria possess a unicellular growth habit that restricts their ability to efficiently colonize massive, solid, low surface-area-to-volume food sources (e.g., plant residues) and prevents them from moving effectively between spatially separated food sources. Because of their small size, bacteria and many of the smaller fungi occupy micropores that physically exclude the larger organisms that prey upon them.

If decomposition of organic materials is to proceed effectively and if nutrients are to move along the soil **food chain** from microorganisms into the higher **trophic levels,** the limits on substrate availability must be circumvented. The highly-branched nature of a plant's root system and the mycelia of fungi are examples of a spreading growth habit that permits access to spatially separated food sources. Many fungi and soil animals possess the mobility or mass, or both, to physically disrupt both the structure of soil and its massive nutrient sources. These activities circumvent some of the substrate bioavailability problems by permitting bacteria and fungi to gain access to and more efficiently colonize and decompose their nutrient sources. In addition, they enable predators to gain access to their microbial prey.

Influence of Plants on Soil Organisms

Plant life is the primary energy source that drives terrestrial soil ecosystems. An examination of salient features of plant life illustrates many of the interactions among soil organisms. Because soils contain low concentrations of biologically available mineral nutrients, terrestrial plants need large root systems to access a sufficient volume of the soil to obtain an adequate supply of essential minerals and water. As a consequence, plant roots provide a supply of relatively labile organic compounds to microorganisms in their immediate vicinity (i.e., the rhizosphere effect, Chapter 17). The total volume of soil influenced by plant roots may be considerable. An estimated 30 to 90% of fine roots (< 1 mm in diameter) in forest soils are turned over annually and provide at least as much organic material to soil organisms as does leaf fall. The superiority of the rhizosphere for microbial growth is illustrated by the 10- to 100-fold higher population densities of microorganisms closely associated

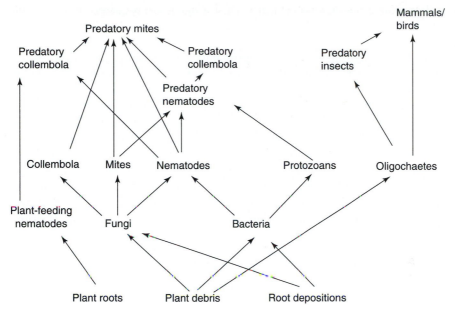

Figure 8–1 Possible interactions among soil organisms displayed in a simplified food web.

with plant roots than in the surrounding soil. As a consequence of the greater microbial productivity near roots, populations of bacterial and fungal predators, such as protozoa and nematodes, also thrive in the rhizosphere, as do their predators, the microarthropods.

The role of plants in driving a soil food chain, the movement of nutrients from one type of organism to another, can also be illustrated at the soil surface. Leaf fall and plant death deposit huge quantities of plant material at the soil surface. The greatest percentage of this plant residue is located in physically resilient, low surface-area-to-volume structures such as leaves, stems, twigs, and branches. Mobile soil animals, such as microarthropods and earthworms, improve the efficiency of microbial decomposition of the large plant debris by increasing its surface area-to-volume ratio through their shredding activities and by distributing the debris throughout the soil volume. In addition, the burrowing activity of moles and other earthworm predators causes further disturbance of the soil environment. These activities result in the redistribution of mineral nutrients and organic matter throughout the soil volume and create more uniform microbial activity and population growth. The interactions of the various members of the soil community and their associated roles in processing essential elements through different chemical and physical forms are often conceptualized schematically in a **food web** (Fig. 8–1).

Growth Characteristics of Soil Microorganisms

Microbial growth is essential for soil ecosystem function. Microbial growth in soil is transitory and depends on the availability of nutrients and adequate physicochemical conditions. Different scenarios of nutrient availability occur in soil and, as a result, different growth strategies are required. For example, under dry surface soil conditions, biodegradable nutrients (fecal material, animal and microbial corpses, plant residues, seeds, and fruits) might accumulate because there is insufficient water to support microbial activity. Upon the return of water, a "flush" of rapid microbial growth occurs. This scenario of transient, luxuriant nutrient availability can be contrasted with the rhizosphere where a more consistent supply of nutrients is available or with the large pool of less labile soil organic matter, or humus, which can be digested only by microorganisms equipped with the enzymes necessary to attack these complex substrates (Stevenson, 1982; Chapter 11). Each of the growth situations described above provides microorganisms with an opportunity to express their individual growth characteristics and to compete for nutrients with whatever unique attributes they can bring to bear on the problem (such as substrate versatility, fast growth potential, high affinity nutrient transport systems, and production of antimicrobial compounds).

Mathematical Concepts of Microbial Growth

This section provides a basic overview of the quantification of bacterial growth. Under favorable environmental conditions of initially nonlimiting nutrients, microbial growth proceeds through various phases shown in Figure 8–2. Briefly, after an adaptive or lag phase (a), the organism enters a period of unlimited growth (b). Eventually, nutrients become limiting, toxic metabolites accumulate, and growth slows down (c). Finally, growth ceases (d) and is followed by cell death (e).

During the unlimited growth phase, cell numbers increase exponentially or double per constant interval of time. This time interval is referred to as the **doubling time,** or **generation time** (often abbreviated to t_d). The number of cells per unit volume (N_t) after a period of growth (t) depends on the number of cells per unit volume in the initial population of cells (N_0) and the number of doublings (n) that have occurred during the time interval (t). Mathematically this can be expressed as:

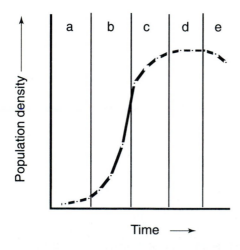

Figure 8–2 Generalized microbial growth curve: (a) adaptive or lag phase, (b) logarithmic or exponential growth, (c) limited growth, (d) stationary phase (growth and death in equilibrium), (e) death phase.

$$N_t = (N_0) \times 2^n \tag{1}$$

Because $n = t/t_d$, equation [1] can be rearranged as:

$$N_t = (N_0) \times 2^{t/t_d} \tag{2}$$

It is more convenient to consider the linear form of equation [2]:

$$\ln N_t = \ln N_0 + \frac{\ln 2 \times t}{t_d} \tag{3}$$

Because it is more convenient to work with logarithms of base 10 than with natural logarithms, the latter are divided throughout by 2.303 to give:

$$\log_{10} N_t = \log_{10} N_0 + \frac{\log_{10} 2 \times t}{t_d} \tag{4}$$

During the unlimited growth phase, a plot of $\log_{10} N_t$ against t will produce a straight line with an intercept on the y axis at $\log_{10} N_0$ and a slope of: $\dfrac{\log_{10} 2}{t_d}$

This type of microbial cell growth is referred to as logarithmic or **exponential growth.** The term $\dfrac{\log_{10} 2}{t_d}$ is a constant, which is called the **growth rate constant** (often abbreviated to "μ"). In some textbooks you will see μ expressed in the natural logarithm form and associated with the denominator 2.303 as follows:

$$\mu = \frac{\ln 2}{2.303 \times t_d} \tag{5}$$

The magnitude of the growth rate constant is inversely proportional to the doubling time. Some bacteria can achieve amazingly short generation times (< 1 hour) when nutrients are nonlimiting. As a result, they can generate populations of astronomical proportions in short periods of time.

As mentioned earlier, soil conditions are rarely conducive for unlimited growth. Most microbial growth in soil probably occurs under substrate-limited conditions, and for long periods of time microorganisms exist in a state of starvation-induced dormancy.

Effect of Substrate Concentration on Growth Rate

Microbial growth is the outcome of hundreds of chemical reactions that occur in the cell and are catalyzed by enzymes. The rate at which an enzyme carries out its reaction is controlled in part by the concentrations of the reactants. The relationship between the rate of an enzyme reaction (v) and the concentration of its reactant or substrate (S) is often expressed by the *Michaelis-Menten equation*.

$$v = \frac{V_{max} \times S}{K_m + S} \tag{6}$$

where v is the reaction velocity when substrate is limited, V_{max} is the maximum velocity when substrate concentration is maximum, and K_m, called the Michaelis

| **Box 8–1** |

Calculating Microbial Growth

Example 1. Using equation [4], we can calculate that if one bacterium has a doubling time of 1 hour, it would produce 16 million progeny in 24 hours.

Rearrange equation [4] to

$$\log_{10}N_t - \log_{10}N_0 = \frac{\log_{10}2 \times t}{t_d}$$

Substitute numerical values into the equation,

$$\text{Log}_{10}N_t - \text{Log}_{10}N_0 = \frac{0.30 \times 24}{1}$$

Rearrange the equation,

$$\log_{10}N_t \times 1 = 7.2 + \log_{10}(1)$$

Therefore, $\log_{10}N_t = 7.2 + 0$ and antilog $7.2 = 1.6 \times 10^7$ cells.

Example 2. Often we know the values of N_t and N_0, and we want to calculate the doubling time. For example, if N_t and N_0 are 10^8 and 10^6 cells ml^{-1} and the length of the unlimited growth interval is 24 hours, we can determine that the doubling time is 3.6 hours:

$$\log_{10}(10^8) - \log_{10}(10^6) = \frac{\log_{10}2 \times 24}{t_d}$$

Therefore, $8 - 6 = \dfrac{0.3 \times 24}{t_d}$ and $2 \times t_d = 7.2$ and $t_d = 3.6$ h

---

constant, represents the concentration of substrate needed to drive the enzyme reaction at one-half of its maximum velocity.

An analogous quantitative relationship exists between the growth rate of a microorganism and substrate concentration. This relationship can be expressed mathematically in an equation referred to as the *Monod equation:*

$$\mu = \frac{\mu_{max} \times S}{K_s + S} \qquad [7]$$

where K_s represents the substrate concentration at which the growth rate of the microorganism is one-half of its maximum rate under specified environmental conditions. The growth response to substrate concentration is illustrated graphically as a hyperbolic curve (Fig. 8–2). When $S \gg K_s$, then $\mu = \mu_{max}$ for all practical purposes. When $S = K_s$ then μ is 0.5 μ_{max}. The lower the value of K_s, the greater the affinity of the microorganism for the substrate and the less the influence of substrate concentration on growth rate. Values of K_s have been determined for a variety of substrates, and generally they are quite low (Table 8–1). When all substrate is consumed, the final microbial biomass is proportional to the product of the initial concentration of the limiting substrate and the **growth yield coefficient** (Y), which is defined as the quantity of biomass carbon formed per unit of substrate carbon consumed. Heterotrophic microorganisms typically exhibit growth yield coefficients of 0.4 to 0.6 g of biomass-carbon g^{-1} of substrate carbon consumed. This

Table 8–1 **Affinity constants (K_s) of the uptake transport systems of bacteria for various substrates.**

Substrate	$K_s(\mu M)$
Glutamate	1–10
Lactate	20–60
Succinate	5–300
Maltose	1–50
Glucose	5–10

value, however, depends on the energy content of the substrate and the nitrogen source available for growth.

Competition for Nutrients

In contrast to pure-culture laboratory conditions, microbial growth in soil occurs under mixed population conditions. When substrates are nonlimiting, the growth rates of two organisms are influenced by their relative maximum growth rates. In contrast, when competitive growth occurs at low substrate concentration, K_s plays a more significant role in the relative success of the two competing microorganisms. Under these conditions, it is theoretically possible that an organism with the smaller K_s will outgrow another organism with a greater μ_{max} (Fig. 8–3).

Classification of Soil Microorganisms Based on Growth Characteristics

The idea that soil microorganisms can be classified on the basis of their growth characteristics, nutritional versatility, and affinity for substrate has surfaced in various forms and at various times throughout the twentieth century. For example, Winogradsky (1924) referred to soil microorganisms that grow rapidly when high

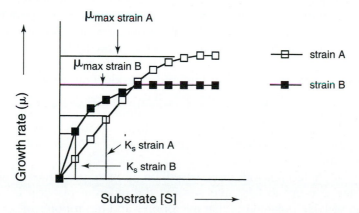

Figure 8–3 Growth as a function of substrate concentration for two bacteria, A and B, with different maximum growth rate potentials (μ_{max}) and with different affinities (K_s) of their substrate uptake.

energy-containing nutrients are readily available as **zymogenous.** In contrast, he referred to organisms that colonized the more recalcitrant material remaining after a primary attack had subsided as **autochthonous** microflora.

Despite the logical intuitiveness of Winogradsky's concept, little experimental evidence has surfaced to indicate that soil bacteria can be concretely separated on these grounds. Another attempt to group soil bacteria based upon growth characteristics arose from the work of researchers in Japan (Ohta and Hattori, 1983). They isolated soil bacteria, referred to as **oligotrophs,** that grow much better at low concentrations of substrates than at high concentrations. These organisms contrast with more typical isolates of soil bacteria that grow better at relatively high concentrations of nutrients, such as those provided in most standard laboratory growth media; these bacteria are often referred to as **copiotrophs.**

In contrast to the ambiguity with soil bacteria, researchers have consistently described a successional colonization of different soil fungal types on plant and animal residues. As decomposition proceeds, the succession is toward organisms with the ability to metabolize residual substrates of increasing chemical complexity and recalcitrance (Griffin, 1972). In addition, fungi that colonize the residues during the later stages of decomposition tend to be slower growing and less competitive at growth on simpler nutrients than the primary fungal colonizers. The latter colonists also tend to produce fewer, and more durable, spores than the fungi that initiate primary colonization.

Andrews (1984) discussed the possibility of linking microbial classification based on growth properties with the r-K theory of colonization and succession. Briefly, the theory suggests that a species needs to adopt a different strategy to colonize an environment in which it is initially present at low density **(r-strategy)** than to persist in an environment in which it is already present near to its carrying capacity **(K-strategy).** The r-strategy species places a large percentage of its available energy into reproduction and attempts to occupy the niche as quickly as possible. Species specializing in this behavior are considered to be poorly adapted in dealing with environmental stress and periods of low nutrient availability. As a consequence, populations of r-strategists are thought to undergo large fluctuations over time. Many soil bacteria, particularly those exhibiting zymogenous or copiotrophic traits, probably belong to the r-strategist category. In the case of K-strategists, growth rate and reproduction are finely tuned to the limited resources available, and these species are adept at dealing with periods of starvation and stress. Many fungal and actinomycete species are considered to be K-strategists because they possess slower growth rates, produce long-lived spores and other survival structures, and metabolize complex nutrients such as lignin and humus.

These theories all reflect classification based on growth properties. Killham (1994) extended the concept to take into account the different physical environments that exist in soil. He pointed out that micropores and macropores of soil can be thought of as distinct microniches from the perspectives of nutrient quality and availability and associated environmental stresses. He hypothesized that microorganisms may need to display different strategies of colonization and persistence in these two pore environments.

Interactions Between Soil Organisms

Given the constraints that soil places on its biota in terms of nutritional limitations and environmental stresses, it is not surprising that many soil organisms interact with each other to circumvent some of these difficulties. Scientists have attempted to define these interactions on the basis of whether one or both organisms gain benefit from the association and if there is any specificity in the association (Table 8–2). However, these interactions cannot be rigidly defined. For example, the specificity of the organisms in a mutualistic association may vary from being highly specific (e.g., legume-*Rhizobium* symbiosis) to promiscuous (e.g., some mycorrhizal symbioses).

Positive Interactions

Commensalism. This term refers to an association in which one population, or microorganism, benefits from the interaction and the second remains unaffected.

Commensalism is illustrated by an organism removing a compound that hinders growth of another organism. For example, the respiratory activity of aerobic soil microorganisms may lower the oxygen concentration sufficiently to allow an anaerobic process to occur in close proximity. This concept is illustrated in Figure 8–4.

Table 8–2 **Types of interactions among soil microorganisms.**

Name of interaction	Effect of interaction on:	
	Population 1	Population 2
Mutualism (Symbiosis)	+	+
Commensalism	+	0
Synergism	+	+ or 0
Amensalism	−	+ or 0
Predation (Parasitism)	−	+
Competition (Antagonism)	−	−

+ = positive effect; − = negative effect; 0 = no effect.

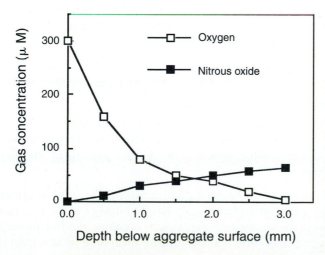

Figure 8–4 Profiles of oxygen and nitrous oxide at different depths in a soil aggregate covered with a decaying clover leaf and wetted with 2 mM nitrate to stimulate microbial activity. Acetylene was added to the incubation so that nitrous oxide became the major end product of denitrification. *Modified from Højberg et al. (1994). Used with permission.*

Microbial respiration was stimulated in the soil aggregate by placing a clover leaf disc in contact with the aggregate surface to act as a nutrient source. As the concentration of oxygen declined below the surface of the aggregate, the concentration of nitrous oxide increased as a product of the anaerobic process of denitrification. In another example of commensalism, ammonia-oxidizing bacteria transform their energy source, ammonia, into a waste product, nitrite, which is a specific substrate of nitrite-oxidizing bacteria. Commensalism is, in fact, the basis for all the nutrient cycles discussed later in this text.

Synergism. Both populations (or microorganisms) benefit from the association. The interaction is not essential for the two microorganisms to exist in that environment, and membership in the association is not usually so specific that one microorganism cannot be replaced by another.

Syntrophy refers to the interaction of two or more populations that supply each other's nutritional needs. Syntrophy often involves the ability of one population to supply growth factors to another population. A specific type of syntrophy is **cross-feeding,** in which one population can metabolize a compound (A), forming a second compound (B), but cannot go beyond that point without the cooperation of a second population. The second population is unable to metabolize compound A, but can use compound B, forming a third compound (C). Both populations can then further metabolize compound C (Atlas and Bartha, 1993).

Examples of syntrophy can be found in flooded soils, where many methane-producing and sulfate-reducing bacteria obtain their energy from acetate, hydrogen gas, or both. Under the same conditions, fermentative bacteria metabolize sugars, fatty acids, and amino acids to generate energy and produce acetate and hydrogen gas as by-products. The methane-producing and sulfate-reducing bacteria benefit from the acetate and hydrogen gas produced by the fermenters, and the fermentative bacteria benefit because the concentration of hydrogen gas is kept low in their environment.

Evidence has accumulated that commensalistic or synergistic associations of microorganisms are required to completely mineralize some soil-applied chemicals, such as pesticides and hazardous waste chemicals. Recently, it has become popular to describe these associations as microbial **consortia.** Two main reasons explain why more than one microorganism is required to effectively carry out these complex transformations.

- One microorganism may provide an essential growth factor (e.g., vitamin or amino acid) needed by the other microorganism, which can metabolize the compound.

  Slater (1978) gives an example of a bacterium that metabolizes cyclohexane to products that a second bacterium can use for growth. In return the second bacterium provides a vitamin, biotin, which is needed for growth of the first bacterium on cyclohexane.

- One microorganism possesses an enzyme that transforms the original chemical into a metabolite that it cannot further metabolize because of steric hindrance, or a lack of the appropriate genes and enzymes to continue the biochemical transformation. However, a second microorganism, incapable of attacking the original parent molecule, can produce an enzyme that attacks the metabolite and thereby continues the degradative process.

Alexander (1994) gives an example of a bacterium that can metabolize polychlorinated biphenyls (PCBs) to chlorobenzoates, which are further metabolized to carbon dioxide and water by another bacterium that cannot attack PCBs.

Mutualism. Also referred to as **symbiosis,** this association usually occurs between specific organisms rather than between populations. Both organisms benefit from the association. The relationships can be highly specialized between the two organisms and allow them to thrive in habitats that neither could occupy alone. The metabolic activities of the partners are often modified in the association, and occasionally the partners cannot grow in the absence of the association.

Many examples of mutualism exist between soil microorganisms and plants and animals. Although plants are the primary source of reduced carbon for the soil microbial community, plant growth is often limited by the availability of nitrogen and phosphorus. Many plants have solved this problem by forming associations with various soil microorganisms. Certain fungi invade plant roots and form an elaborate network of hyphae that penetrate either between or into the cortical cells and also into the soil environment. These associations are collectively referred to as mycorrhizae. The fungus benefits by obtaining a readily-available supply of reduced carbon from the plant. Various mechanisms have been proposed to explain how the fungus benefits the plant, including:

- The mycorrhizal roots enhance the uptake of mineral nutrients and water from soil because the external mycelia explore a greater volume of soil than does a nonmycorrhizal root.

- The fungus sequesters mineral nutrients at low concentrations more efficiently than does the nonmycorrhizal root.

- The fungal hyphae produce various hydrolytic enzymes that release nitrogen and phosphorus from organic compounds that are otherwise unavailable to the plant.

Leguminous plants form mutualistic symbioses with bacteria, primarily of the genera *Bradyrhizobium* and *Rhizobium*. A somewhat random array of plant species form symbiotic associations with the actinomycete genus *Frankia* and are called actinorhizal. These bacteria invade the plant via root hairs or by passage between epidermal and cortical cells of the root. As a result of chemical signals passed between the bacterium and the plant, the latter is stimulated to construct a specialized multicellular structure called a *root nodule*. Inside the nodule the bacteria "fix" atmospheric dinitrogen gas into ammonia, which the plant then assimilates into amino acids. The plant supplies the bacteria with reduced carbon to use as the energy source for metabolic processes including N_2 fixation. As a result of these symbioses, leguminous and actinorhizal plants can grow in environments that are extremely deficient in mineral nitrogen.

Mutualistic associations with microorganisms play an essential role in the well-being of soil animals that feed primarily on plant debris. Some soil animals have protozoa and bacteria in their gut that decompose the cellulose component of plant cells into metabolites that the animal can assimilate. Furthermore, many gut bacteria fix dinitrogen into ammonia, which is then incorporated into protein via amino acids. As a result, the nitrogen-deficient diet of the soil animal is supplemented. In another form of mutualism, various ants, beetles, and termites cultivate specific kinds of fungi. The insects provide the fungi with nutrients in the form of plant debris and feces. By feeding on the fungal biomass, the insects circumvent the problem of digesting cellulose.

Negative Interactions

Predation. This interaction occurs when one population has a negative impact on the size of another population by feeding on it and reducing its numbers.

Soil animals, such as nematodes and protozoa, feed on bacterial and fungal populations and reduce their numbers. As mentioned earlier in this chapter, this activity is an essential part of soil ecosystem function. Although such activity is detrimental to the size of the prey population, the activity of the survivors can be stimulated because the predators release nutrients during their feeding activities.

Competition. In this interaction, both populations compete for the same limited resources and have a negative effect on each other.

For example, heterotrophic microorganisms and ammonia-oxidizing bacteria compete for ammonium. The former needs ammonium to make amino acids and proteins, and the latter needs ammonium primarily as an energy source. If ammonium exists in limiting quantities, then neither population will grow to the extent that it would in the absence of competition. In another example of competition, different strains of a symbiosis-forming species of *Rhizobium* compete with each other to establish a symbiosis with a leguminous plant. In many cases the number of root nodules formed by the competing strains will be less than if only one strain were present in the soil.

Amensalism. In this type of interaction one population gains a competitive advantage over another by producing a growth-inhibiting substance.

Many fungi and bacteria have the potential to produce antimicrobial growth substances that may serve to inhibit other nearby microorganisms. When amensalism occurs in the rhizosphere, the growth of soilborne plant pathogens can be retarded and plant disease lowered (Table 8–3). This phenomenon, known as **biological control,** is described in Chapter 19. Amensalism can also be caused by more general effects of metabolism. For example, sulfur-oxidizing microorganisms can cause soil acidification, which may adversely affect acid-sensitive *Streptomyces* species in their immediate vicinity (Chapter 15).

Table 8–3 **Relationship between the production of the antibiotic phenazine-1-carboxylic acid (PCA) by *Pseudomonas aureofaciens* and control of take-all root rot disease of wheat.**

Seed Treatment	ng of PCA g^{-1} of root	Disease rating[a]
Phenazine-producing strain	133	2.08
Phenazine-deficient mutant	0	4.22
No bacterial treatment	0	4.42

[a]The lower the number, the less the disease incidence
Adapted from Thomashow et al. (1990). Used with permission.

Microbial Succession

A fundamental principle of ecosystem development is that of **succession,** a sequence of changes in populations leading to the establishment of a so-called climax community. Succession arises because as organisms (from microbes to trees) grow, they alter the environment around them. Just as large trees shade out smaller trees on the forest floor, so too do microbes alter their environment as they grow, thereby causing changes that pave the way for development of different populations. However, microbial ecosystems generally lack the stability that so many macro-ecosystems possess.

We sometimes use the term *primary microflora* to describe the early populations that develop in response to the addition of fresh carbonaceous substrates, such as crop residues or leaf litter, to the soil. As these microbes grow, they alter their environment and the substrate and, thereby, change plant-derived products into microbial biomass and metabolic waste products. Soon a *secondary microflora* develops and feeds on the dead cells of the primary flora and its products. Thus, we see a wavelike progression of change in microbial populations as the substrates are metabolized and the abiotic environment (e.g., pH, oxygen in the soil atmosphere, redox potential) changes. As a simple example, consider how aerobic microbes consume oxygen from waterlogged soil environments, permitting the development of anaerobes which cannot function unless the oxygen in the soil atmosphere is depleted (Box 8–2). Upon reaeration, another set of successional changes begins.

Diversity in Soil Microbial Populations

Generally, ecologists feel that a stable ecosystem should have high genetic diversity to tolerate environmental fluctuations. Nevertheless, it is by no means clear what level of diversity is required to maintain the stability of a soil ecosystem, or what degree of disturbance will trigger a significant decline in diversity. Although microbial diversity in soils is generally assumed to be high, some soil management practices (e.g., conventional tillage operations) can bring about both a decline in soil microbial biomass and deterioration of soil physical properties. It would be extremely useful to know if changes in microbial population composition and genetic diversity could serve as an early warning of deterioration in soil properties. Conceptually, high genetic diversity exists in most soils because immigration rates are high and environmental stresses are neither of sufficient intensity nor sustained for sufficient time to selectively eliminate large numbers of genotypes. Furthermore, nutrients are rarely present in excess for a sufficient period of time to allow selective enrichment of the more adapted genotypes.

Measuring Microbial Diversity

The measurement of diversity in microbial communities is not a trivial matter. Until recently, microbiologists depended on culturing methods to recover and examine representative microorganisms from an environment. Unfortunately, many microorganisms are not easily cultured, and diversity measurements are invariably underestimated with this approach. To circumvent this problem, researchers have resorted to genetic and biochemical methods that do not require culturing of the organisms as the first step. Brief descriptions of four such methods follow.

Box 8–2

Winogradsky Column. A simple method for demonstrating microbial succession in a wetland soil is to set up a Winogradsky column. This is also a good tool for demonstrating the concept of using **microcosms** to mimic or model large-scale ecosystems. To prepare a Winogradsky column, pack a transparent plastic or glass cylinder (5 cm diameter by 45 cm is convenient, but dimensions are not critical) with soil or with sediment from an aquatic (freshwater or marine) habitat. Add a small amount of calcium carbonate and calcium sulfate to about 3 cm of mud in the bottom of the column to provide carbonate (CO_2) for autotrophic microbes and sulfate as a source of sulfur for growth and an electron acceptor for the sulfate-reducing bacteria. A carbon source, such as grass clippings or shredded newspaper, can be added to the remaining soil or sediment, which is then mixed with water to form a slurry with the consistency of heavy cream. Pour the preparation into the column, which should be maintained under flooded conditions in the presence of light.

As the column comes to a new equilibrium following preparation, visual changes in the flooded soil begin to signal various microbial processes. Most notable is the development of sulfate-reducing bacteria, which cause formation of blackened areas in the lower portion of the column and may even blacken the entire sediment profile if the aerobic bacteria deplete the oxygen supply through their respiratory activities. Later, the black areas give way to development of green, purple, or rust-colored patches that represent the anaerobic green and purple photosynthetic bacteria using the sulfide produced by the sulfate reducers. At the sediment-water interface, one may observe a lighter-colored layer indicating the depth to which oxygen has penetrated and limited the activities of the anaerobic organisms. Later this zone may deepen and spread as the cyanobacteria take hold and begin to aerate the upper column through their oxygenic photosynthesis.

These are but a few of the changes that can be observed with this very simple method. These columns are teaching tools if students observe them regularly and record visual changes that occur over the course of a semester or longer. These columns can last for years if they are kept flooded and left near a window where they can receive sunlight.

DNA extraction and reassociation. By extracting DNA from a soil sample and determining the rate of reassociation of the DNA (see Chapter 9), researchers have discovered that an estimated 4,000 genetically different kinds of bacteria exist in a single soil sample (Torsvik et al., 1990). Unfortunately, it is not a trivial matter to consistently recover soil DNA of the quality necessary to conduct reassociation analyses. As a result, other methods have been tested for their ability to provide accurate information about community composition in different soils.

16S and 23S ribosomal DNA sequencing. The 16S and 23S rRNAs are major components of prokaryotic ribosomes, which function as the sites of protein synthesis in living cells. Modern molecular techniques allow researchers to recover the genes (rDNA) encoding for these rRNA molecules directly from environmental samples and to circumvent the culturing step. With these technologies, researchers have examined the microbial community composition of various environments such as open ocean water, hot springs, hydrothermal vents, and soils. Briefly, DNA is extracted directly from soil, and rRNA genes are recovered and sequenced. The sequences are compared with sequences already deposited in gene data banks. As a result of this technique, microorganisms have been detected in soil for which no close relatives are known to exist in culture. For example, Liesack and Stackebrandt

Box 8–2 cont'd

The Winogradsky Column

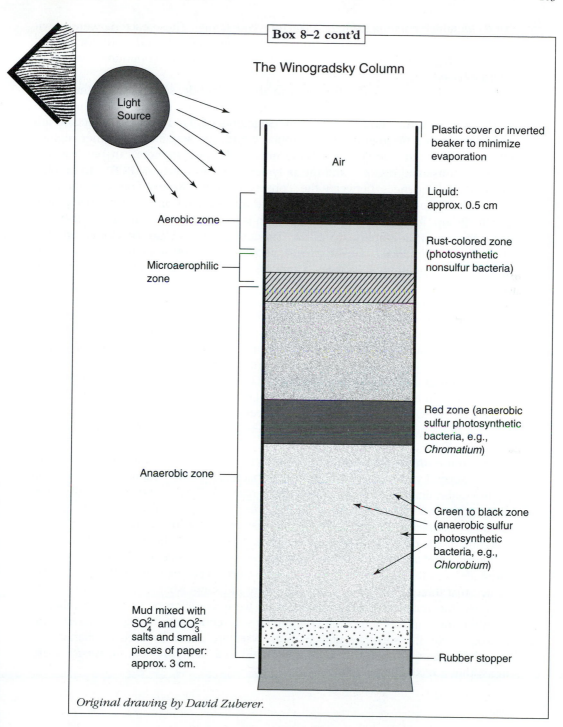

Light Source

Air

Aerobic zone

Microaerophilic zone

Anaerobic zone

Mud mixed with SO_4^{2-} and CO_3^{2-} salts and small pieces of paper: approx. 3 cm.

Plastic cover or inverted beaker to minimize evaporation

Liquid: approx. 0.5 cm

Rust-colored zone (photosynthetic nonsulfur bacteria)

Red zone (anaerobic sulfur photosynthetic bacteria, e.g., *Chromatium*)

Green to black zone (anaerobic sulfur photosynthetic bacteria, e.g., *Chlorobium*)

Rubber stopper

Original drawing by David Zuberer.

(1992) identified bacteria from the Planctomycete family. These bacteria were previously thought to inhabit primarily unpolluted aquatic habitats and were not considered to be typical soil bacteria. Their role in the soil microbial community has yet to be established.

Relative similarities and differences in the nucleotide sequences of the rDNA molecules have been used by microbiologists as a means to classify the microbial world. Although the sequences in some segments of the 16S rDNA molecule are highly conserved throughout the microbial world, the sequences in other regions of the molecule are unique to specific microbial types. Primers can be designed from the conserved regions, and the polymerase chain reaction (PCR) can amplify the unique sequences between the conserved areas. Researchers can then sequence the resulting PCR product and place the organism into a particular phylogenetic group. A recent alternative approach is to extract soil community DNA and use primers to PCR amplify 16S rDNA. The amplified DNA can be cloned and the DNA inserts sequenced. This approach allows 16S rDNA sequences to be determined from nonculturable isolates, a process which was unobtainable prior to PCR analysis.

Lipid composition. The lipid composition of pure cultures of microorganisms has been used for several years as a tool for taxonomic classification of microorganisms. Lipids are easily extracted from natural samples and analyzed. Some fatty acids are associated with a class of lipid found in cell membranes called phospholipids because they contain phosphate. Some of the fatty acids in phospholipids are only found in certain microorganisms. These are referred to as *signature fatty acids,* and representatives have been recognized in fungi, Gram-positive actinomycetes, and Gram-negative bacteria. Recently, researchers have analyzed the fatty acids found in microbial lipids recovered directly from differently managed soils (Fig. 8–5).

Substrate utilization patterns. Researchers have used commercially available kits for several years to tentatively identify bacterial species. Identification is based on the determination of a phenotypic "fingerprint" reflecting the specific substrates that support microbial growth or on the presence of particular enzymes and biochemical pathways. Recently, researchers examined the microbial communities in soil under six different plant communities from the Chihuahan desert in New Mexico (Zak et al., 1994). Kits were inoculated with soil recovered from these communities, and the patterns of substrate utilization were examined. These data revealed that three of the six communities were indistinguishable from each other. In contrast, the remaining three communities were as different from each other as from the three indistinguishable communities. In contrast to the lipid analyses discussed above, the substrate utilization analyses are generally used to characterize the *physiological diversity,* sometimes referred to as the *functional diversity,* existing within a mixed population.

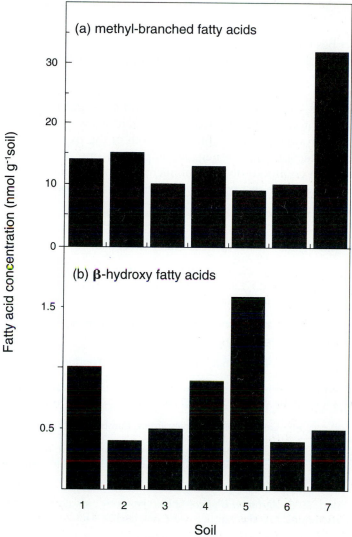

Figure 8–5 Concentrations of (a) methyl-branched and (b) β-hy-droxy fatty acids in cropped soils (1–4), hopyard soils (5 and 6), and an undisturbed grassland soil (7). The highest concentration of methyl-branched fatty acids was found in lipids recovered from undisturbed grassland soil, indicating an abundance of actinomycetes. A copper fungicide-treated soil (5) had the highest concentration of β-hydroxy fatty acids, indicating a greater abundance of Gram-negative bacteria. Neither the causes nor the impact of these community changes upon the soil ecology are known. *Modified from Zelles et al. (1994). Used with permission.*

Summary

This chapter places soil microorganisms in the context of their soil environment and conceptually describes their roles in soil ecosystems. Soil organisms play crucial and often complementary roles in nutrient cycling. However, the complexity of soil structure and fluctuations in environmental conditions and nutrient availability combine to make it difficult for soil microorganisms to achieve their metabolic potential. Organismal interactions circumvent some of these problems and help to maintain crop productivity and promote the degradation of hazardous wastes. Some interactions between soil microorganisms benefit one or both of the participating populations, while other interactions have a negative impact on one of the participants. Clearly, soil contains an enormous diversity of microbial populations, yet we remain woefully ignorant of the forces that control population composition.

Cited References

Alexander, M. 1994. Biodegradation and bioremediation. Academic Press, San Diego, Calif.

Andrews, J.H. 1984. Relevance of r- and K-theory to the ecology of plant pathogens. pp.1–7. *In* M.J. Klug and C.A. Reddy (ed.), Current perspectives in microbial ecology. American Society of Microbiology, Washington, DC.

Atlas, R.M., and R. Bartha. 1993. Microbial ecology: Fundamentals and applications. 3rd ed. Benjamin/Cummings Publishing Co., New York.

Griffin, D.M. 1972. The ecology of soil fungi. Chapman and Hall, London.

Højberg, O., N.P. Revsbeck, and J.M. Tiedje. 1994. Denitrification in soil aggregates analyzed with microsensors for nitrous oxide and oxygen. Soil Sci. Soc. Am. J. 58:1691–1698.

Killham, K. 1994. Soil ecology. Cambridge University Press, Cambridge, England.

Liesack, W., and E. Stackebrandt. 1992. Occurrence of novel groups of the domain bacteria as revealed by analysis of genetic material isolated from an Australian terrestrial environment. J. Bacteriol. 174:5072–5078.

Ohta, H., and T. Hattori. 1983. *Agromonas oligotrophica* gen. nov., sp. nov., a nitrogen-fixing oligotrophic bacterium. Antonio van Leeuwenhoek 49:429–446.

Slater, J.H. 1978. Microbial communities in the natural environment. pp.137–154. *In* K.W.A. Chater and H.S. Somerville (ed.), The oil industry and microbial ecosystems. Heyden and Sons, London.

Stevenson, F.J. 1982. Humus chemistry. Wiley-Interscience, New York.

Thomashow, L.S., D.M. Weller, R.F. Bonsall, and L.S. Pierson. 1990. Production of the antibiotic phenazine-1-carboxylic acid by fluorescent *Pseudomonas* species in the rhizosphere of wheat. Appl. Environ. Microbiol. 56:908–912.

Torsvik, V., J. Goksoyr, and F.L. Daae. 1990. High diversity in DNA of soil bacteria. Appl. Environ. Microbiol. 56:782–787.

Winogradsky, S. 1924. Sur la microflore autochthone de la terre arable. C.R. Acad. Sci., Paris 178:1236–1239.

Zak, J.C., M.R. Willig, D.L. Moorhead, and H.G. Wildman. 1994. Functional diversity of microbial communities: A quantitative approach. Soil Biol. Biochem. 26:1101–1108.

Zelles, L., Q.Y. Bai, R.X. Ma, R. Rackwitz, K. Winter, and F. Beese. 1994. Microbial biomass, metabolic activity, and nutritional status determined from fatty acid patterns and polyhydroxybutyrate in agriculturally managed soils. Soil Biol. Biochem. 26:439–446.

Study Questions

1. List the complementary roles of the bacteria and eukaryotes that contribute to efficient nutrient cycling in soil.

2. Describe the characteristics of soil which influence substrate inaccessibility.

3. If 54 bacterial cells were to divide every 45 min for a period of 20 h, what would be the final number of cells in the population?

4. List the benefits of organismal interactions in ecosystem function.

5. Describe an interaction between two microorganisms and an interaction between one macroorganism and one microorganism. Explain how the organisms immediately benefit from the interaction, and how the soil ecosystem ultimately benefits from it.

6. Why are soil microbial populations so diverse in composition?

Chapter 9

◆

Molecular Genetic Analyses in Soil Ecology

Ian L. Pepper and Karen L. Josephson

◆

It has not escaped our notice that the specific pairing we have postulated immediately suggests a possible copying mechanism for the genetic material.
F. Crick and J.D. Watson

In the past decade or so there has been an explosion of molecular genetic methodologies to examine soil microorganisms at the molecular level. These techniques are conceptually simple, but breathtakingly powerful. For example, one can detect minute concentrations of microbial nucleic acids from a particular species or use these methods to classify particular organisms into phylogenetic groups. One can even differentiate between genetic potential and actual microbial activity in environmental samples.

Specific molecular analyses are based on the structure of the nucleic acids and the intricate mechanisms that synthesize specific microbial compounds at the genetic level. Nucleic acids are either **deoxyribonucleic acid (DNA) or ribonucleic acids (RNA).** DNA is made up of four deoxynucleotide bases: guanine (G), cytosine (C), adenine (A), and thymine (T). DNA consists of these bases linked to the sugar deoxyribose and a phosphate radical. Structurally, DNA consists of two strands of material combined to form a double helix. These two strands are linked by hydrogen bonds between corresponding pairs of bases. Specifically, G bonds only to C and vice versa (i.e., G–C), and A only binds to T and vice versa (i.e., A–T). Each G–C pairing has three hydrogen bonds, whereas the A–T pairing has only two hydrogen bonds. Because these bases are specific in their ability to bind together, they are **complementary** to each other. The assumption is that single-stranded DNA (ss DNA) will only form double-stranded DNA (ds DNA) if the sequence of one is complementary to the sequence of the other. In other words, a DNA strand containing the nine bases A–T–T–C–G–G–A–A–T will only pair up with the complementary strand T–A–A–G–C–C–T–T–A with the resulting ds DNA being nine base pairs (bp) in length.

Box 9–1

Hybridization and Denaturation. When two strands combine because they are complementary, the process is known as DNA-DNA **hybridization** to reflect that the resulting ds DNA is a hybrid of the two separate strands. The reverse process, in which ds DNA unwinds into two single strands, is called **denaturation.** DNA can be unwound chemically or simply by heating the DNA to 94°C. Upon cooling, the two single strands automatically hybridize back into a single, double-stranded molecule, a process known as **reannealing.** This complementary nature of DNA is important because if one strand sequence is known, the sequence of the other strand is easily deduced.

Double-stranded DNA is the basis for two extremely important molecular analyses—**gene probes** and **polymerase chain reactions (PCR).** Hybridization techniques can also be applied to RNA sequences. Except for two important differences, RNA is structurally similar to DNA. The two differences are that the base uracil (U) replaces thymine (T) and that the sugar in the molecule is ribose rather than deoxyribose. In RNA, G still binds to C, but A binds to U. All organisms have three forms of RNA: **ribosomal RNA (rRNA), messenger RNA (mRNA),** and **transfer RNA (tRNA).** Two types, rRNA and mRNA, are used extensively in molecular genetic analyses, including phylogenetic studies and estimates of metabolic activity. In all cases, the base sequence of RNA is transcribed from conserved DNA sequences. For more in-depth information on the structure and function of DNA and RNA, review Watson et al. (1983, 1987).

Methodologies

Gene Probes and Probing

Considerable research skills and a strong background in molecular biology are needed to construct gene probes, and specific methodologies are presented in manuals such as Sambrook et al. (1989). Here we present the general concepts involved in the construction of gene probes, along with some of their practical applications. These concepts are also critical to the understanding of more sophisticated techniques that are discussed later.

Gene probes take advantage of the ability of DNA to be denatured and reannealed. To make a gene probe, the DNA sequence of the gene of interest must be known. If the gene is unique to a particular microbial species, the sequence may be useful for the specific detection of that organism. Alternatively, the gene may encode for the production of an enzyme unique to some metabolic pathway; a gene probe constructed from such a sequence may indicate the potential activity of a group of bacteria in a soil or water sample. For example, gene probes can be made from sequences that code for enzymes involved in N_2 fixation, and such a probe could then estimate whether a particular soil contained any bacteria having the genes for N_2 fixation. Additional probes could later be constructed to determine whether such genes were in N_2-fixing *Rhizobium, Azospirillum,* or cyanobacteria. Many gene sequences are now known. Computer software packages enable researchers to search for sequences that are unique to a particular bacterium or, conversely, sequences that are conserved within particular groups of bacteria. The strategy that we use to select a particular sequence will depend on the intended use of the gene probe.

After a target sequence has been identified, the next step is to construct a gene probe, which is the complementary strand for that particular sequence. The size of the probe can range from 30 to 40 bp to as many as several hundred base pairs. The basic strategy is to obtain a ds DNA molecule composed of the target sequence and to select a portion of this sequence for use as a probe, with specific nucleotides marked or labeled in some way in order to make the reaction detectable. The label may involve a radioactive chemical, such as ^{32}P incorporated into the sugar-phosphate backbone of DNA, or it can consist of a nonradioactive label such as biotin-streptavidin.

Consider a ^{32}P gene probe in which the DNA sequence is only found in the intestinal bacterium *Escherichia coli.* After obtaining a pure culture of a known isolate of *E. coli,* the cells can be filtered onto a membrane and lysed before the released bacterial DNA is denatured into two single strands. If the gene probe is similarly denatured into single strands and is subsequently added to the membrane with the lysed cells, reannealing can occur between the single strands of the gene probe and the single complementary sequences associated with the denatured DNA from the *E. coli* cells. After washing the membrane free of unhybridized probe, the paper is incubated with X-ray film, and the radioactivity of the ^{32}P associated with the hybridized DNA results in a photographic image (Fig. 9–1). This process is called **autoradiography.**

Box 9–2

Autoradiography. During autoradiography a photon of light, beta particle, or gamma ray emitted by the filter activates silver bromide crystals on the film. Through film development, the silver bromide is reduced to silver metal and forms a visible grain or black spot on the film. Film images obtained by this technique may be quantified using microdensitometry.

A positive image implies that ^{32}P was in fact in the sample, and that hybridization between the probe and the sample occurred. Such hybridization and, hence, detection only occurred because a gene probe with a specific sequence for *E. coli* reannealed with target DNA from *E. coli* cells. If the probe had been added to the DNA obtained from lysed *Bacillus* cells, no strands complementary to the probe would have been obtained, no hybridization would have occurred, and no positive image

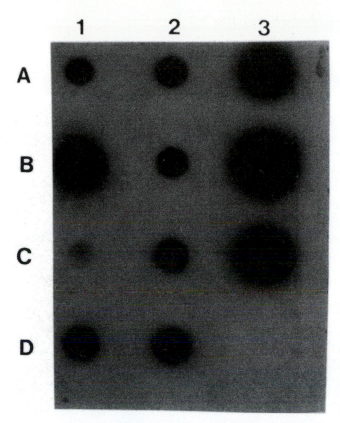

Figure 9–1 Detection of *Escherichia coli* DNA by a ^{32}P-labeled gene probe. The different intensities of the image correspond to different amounts of DNA. Lane 1C is the negative control that contained no *E. coli* DNA.

would have been obtained on the film. Thus, this particular probe would allow specific detection of *E. coli*. The probe could similarly check whether unknown cultures contained cells of *E. coli*.

Colony lifts/hybridization. The principles of gene probe detection of specific DNA sequences can also detect a specific gene sequence within bacterial colonies on a Petri plate containing a mixed population of bacteria by a process called *colony hybridization* or *colony lifts.* A colony hybridization entails lightly pressing a piece of filter paper onto the Petri plate so that some bacterial cells from each colony adhere to the paper. The cells are lysed directly on the filter paper and probed as described above. After this procedure, only those colonies that contained the specific DNA sequence would give a radioactive signal. Referring to the original Petri plate, the viable colony of interest can now be isolated and retained for further study (Fig. 9–2).

Southern blots/hybridization. This analysis further identifies the location of the target sequence of interest. For example, it may be important to know whether a gene is within a particular plasmid contained within a bacterial strain. To determine this, all the plasmids within the strain can be extracted and separated by gel electrophoresis. The different-sized plasmids on the gel can be transferred to a special membrane by blotting and subsequently probed. Once again, only those DNA molecules that contain the target sequence will hybridize with the probe, thus allowing detection of those plasmids containing the target sequence (Fig. 9–3).

| Box 9–3 |

Agarose Gel Electrophoresis. Agarose gel electrophoresis is a fundamental tool in nucleic acid analysis. It is a simple and effective technique for viewing and sizing DNA molecules, such as plasmids or DNA fragments. The DNA samples are loaded into wells in the agarose gel medium. Voltage is applied to the gel, so that the negatively charged phosphates along the DNA backbone cause the DNA to migrate toward the anode. The gel is stained with a dye, such as ethidium bromide, allowing for visualization of the DNA when viewed under UV light. The molecular weight in base pairs (bp) of the DNA determines the rate of migration through the gel and is estimated from standards of known size that are run in parallel on the gel.

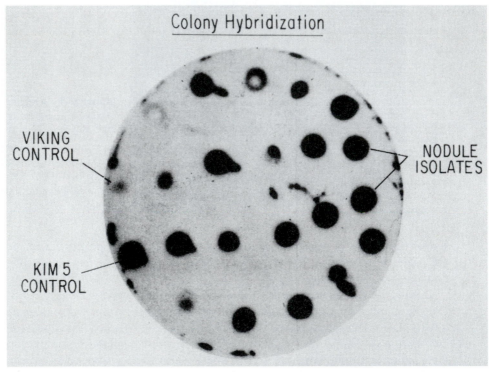

Figure 9–2 Colony hybridization of rhizobia isolated from root nodules. Dark spots are colonies that hybridized to a gene probe specific for *Rhizobium phaseoli* strain Kim-5. The Viking 1 control strain of *Rhizobium* acts as a negative control that does not contain Kim-5 DNA.

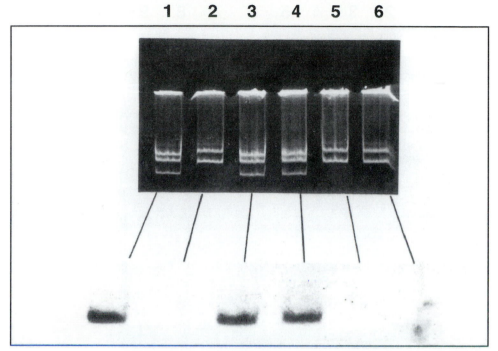

Figure 9–3 Plasmid profile of rhizobia isolated from root nodules (top) and corresponding southern hybridization (bottom). The gene probe, originally constructed from a Kim-5 plasmid, confirmed the presence of strain Kim-5 in Lanes 1, 3, and 4. Lanes 2, 5, and 6 did not contain any Kim-5 DNA.

Overall, the principles of gene probe analysis form the basis of many more sophisticated molecular genetic techniques. These analyses have combined with deductive reasoning to allow considerable progress in understanding the molecular ecology of soil organisms.

Polymerase Chain Reactions

The polymerase chain reaction (PCR) has revolutionized molecular biology methodologies. First discovered in 1985, PCR has become a key protocol in many biological laboratories, including those concerned with soil and environmental microbiology. PCR is an enzymatic reaction that allows amplification of DNA through a repetitive *in vitro* process. During each cycle of PCR, a specified piece of the total DNA is copied, thus doubling the amount of DNA. In practice, 25 cycles result in approximately a million-fold increase in the amount of DNA present. The PCR product is visualized by agarose gel electrophoresis and the size estimated by comparison to DNA standards (or ladders) of known size (Fig. 9–4). PCR has many applications, but the simplest to understand is probably the sensitive detection of a specific gene of interest.

Theory of PCR. A typical cycle of PCR has three steps. The first step involves denaturing DNA that contains a gene sequence of interest into two single strands of target or template DNA. Added to the reaction mixture are two different short pieces of single stranded DNA or **oligonucleotides,** called **primers,** that have been commercially

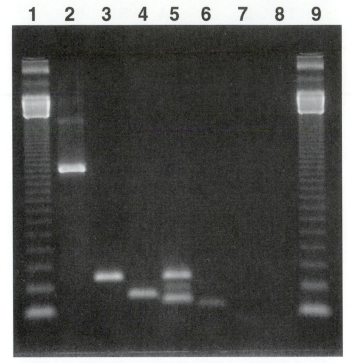

Figure 9–4 Agarose gel electrophoresis showing PCR amplification of specific DNA sequences. Lanes 1 and 9 contain DNA size markers; Lane 2, 1,500-bp 16S ribosomal DNA from *Alcaligenes eutrophus;* Lane 3, 306-bp DNA from adenovirus; Lane 4, 205-bp DNA from *tfdB* gene of pJP4 plasmid of *Alcaligenes eutrophus;* Lane 5, 299/236/173-bp DNA from *Salmonella* spp.; Lane 6, 179-bp DNA from *E. coli;* Lane 7 and 8, negative control (no DNA added).

synthesized. If the primers have a complementary sequence to the ss DNA template, they can hybridize or anneal to this DNA. Thus the second step within a cycle of PCR is primer annealing. The third and final step is extension. Here a DNA polymerase synthesizes a complementary strand to the original ss DNA by the addition of appropriate bases between the primers. The net result at the end of a cycle is two double-stranded molecules of DNA identical to the original double-stranded molecule of DNA (Fig. 9–5). Repeating the process results in PCR amplification of the DNA and an exponential increase or amplification of the copies of the original DNA present.

Box 9–4

PCR, Step by Step. The three steps in a PCR amplification are:

- template DNA denaturation,

- primer annealing, and

- primer extension.

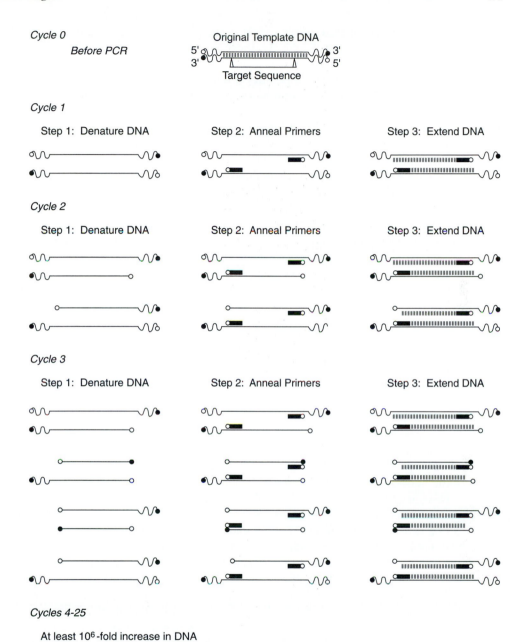

Cycles 4-25

At least 10^6-fold increase in DNA

Figure 9–5 Schematic representation of PCR.

All three steps of PCR amplification occur at different but defined temperatures for specific time intervals. Generally, these three steps are repeated 25 to 40 times to obtain sufficient amplification of the target DNA. These cycles are conducted in an automated, self-contained temperature cycler or thermal cycler. These instruments are relatively inexpensive, but allow precise temperature control for each step and provide for a substantial reduction in labor. The actual amplifications take place in small (approximately 1 ml) centrifuge tubes, and commercial kits are available that provide all the nucleotides, enzymes, and buffers necessary for the reaction.

Temperature is a critical part of PCR. Template denaturation occurs at a temperature greater than the melting temperature of the DNA; for most PCR reactions, this is standardized at 94°C for 1.5 min because it guarantees complete denaturation of all DNA molecules. Primer annealing occurs at a lower temperature, typically 50 to 70°C for 1 min. It is possible for a primer to anneal to a DNA sequence that is similar to the correct target sequence but contains a few incorrect bases. This will result in the incorrect amplification of the DNA, which is termed *nonspecific amplification,* and can result in a false positive identification of the DNA sequence of interest. The higher the temperature of annealing, the more specific the annealing is, and thus the extent of primer and template mismatching is reduced. However, as the annealing temperature increases, PCR sensitivity normally decreases.

The final step of PCR is extension. The essential component of this reaction is the enzyme Taq polymerase, which adds bases to the primers. This unique enzyme was obtained from the thermophilic bacterium *Thermus aquaticus;* it is heat stable, withstanding temperatures up to 98°C, allowing reuse for many cycles. The *Taq* polymerase is needed in the extension step, where the complementary strand of DNA is synthesized at 72°C for 1 min in most reactions. A 25-cycle PCR reaction may take about three hours, including ramp times between each step, although the time varies depending on the type of thermal cycler. For each primer pair, the researcher must initially optimize the conditions of temperature, incubation times, and concentration of the various reaction components to obtain the desired results.

Design of primers. The choice of the primer sequences is critical for successful amplification of a specific DNA sequence. As in the case of gene probes, primer sequences can be deduced from known DNA sequences. The overall choice of primers is guided by the objectives of the researcher. If detection of a target DNA that is specific to a given species or genus of a bacterium is required, then only sequences unique to that bacterium are appropriate for the design of the primers. For example, the *lamB* gene encodes for the production of an outer membrane protein in *E. coli;* primers designed from this gene sequence will specifically detect *E. coli* (Josephson et al., 1991). However, some objectives may require primers from conserved sequences. Conserved sequences are those that are similar and can be found in many different bacterial species. Some conserved sequences are universal and found in all known species. For example, primers designed from common nod gene sequences (*nod abc*) could detect all species of *Rhizobium* (Chapter 14). Therefore, the experimental objective is an important consideration in the choice of the sequence from which the primers will be designed.

In addition, other criteria contribute to the ultimate choice of primer sequences. In general, most primers are 17 to 30 bp in length and are usually separated by as few as a 100 bp or as many as several thousand. The distance between the primers determines the size of the amplification product. Because the location of the primers within the genome defines the size of the amplification product, this theoretical size can be compared to the actual size of the product obtained on the gel. Further proof that the amplified product is the DNA sequence of interest can be obtained by the use of a gene probe that targets an internal sequence of the PCR product. It is also critical that the primers contain sequences that are different from each other, so that they anneal at different sites on the chromosome. If primers contained complementary sequences, they could hybridize to each other producing a *primer dimer*. Most primers should also contain high G–C content (more than 50%) because the melting temperature of the three hydrogen bonds between G–C pairs is higher than that of the two hydrogen bonds between A and T.

Restriction Fragment Length Polymorphism (RFLP) Analysis

Unique DNA sequences also can be identified by the use of **restriction enzymes.** These enzymes recognize specific 4 to 6 bp sequences and will cut the DNA at these sites. Thus the original piece of DNA is split into several smaller fragments. Through gel electrophoresis and ethidium bromide staining, the DNA fragments show up as a characteristic fingerprint of the original piece of DNA. This process is known as **restriction fragment length polymorphism (RFLP) analysis.** RFLP can be performed on a specific piece of DNA, such as a PCR product (Fig. 9–6). RFLP analysis can also be conducted on whole bacterial chromosomes.

Many Gram-negative enteric bacteria contain repetitive DNA sequences that are located at different points within the bacterial chromosome. Because these sequences are evolutionarily conserved, gene probes utilizing portions of these sequences can be designed. If a genomic DNA preparation of a particular isolate is cut with specific restriction enzymes, some DNA fragments will contain the conserved repetitive DNA sequence. A gene probe can identify these fragments, and two different strains of the same species will produce different patterns or fingerprints, depending on the original location of the sequences. RFLP analysis, coupled to specific gene probe detection of specific sequences within DNA fragments on the gel, has successfully fingerprinted bacteria (Fig. 9–7).

Plasmid Profiles

Although most of the genetic information in a bacterial cell is contained in the chromosome, some bacteria contain genetic material called **plasmids** (Chapter 3). Methods are available to extract plasmids from bacterial cells. Through agarose gel electrophoresis and ethidium bromide staining, the plasmids show up as one or more bands on the gel. This type of characterization is known as a *plasmid profile*. A typical plasmid profile is shown in Figure 9–3. Gene probe detection of a specific sequence within the plasmid, in conjunction with plasmid profiles, can confirm a particular sequence. Because the size of multiple plasmids within a given bacterium are often unique, analysis of these plasmids provides useful information about the organism's identity or capabilities.

Reporter Genes

Nucleic acid sequences cannot only detect specific organisms, but can also demonstrate specific activity. One method involves the use of **reporter genes.**

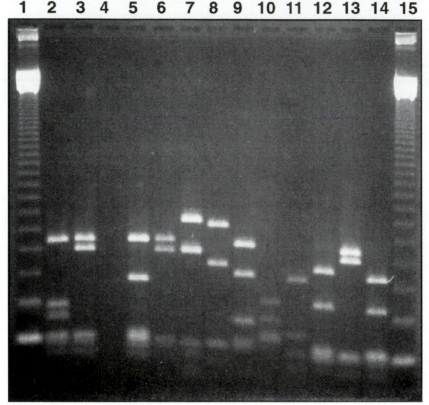

Figure 9–6 RFLP analysis of bacterial 16S ribosomal DNA. Lanes 1 and 15 contain DNA size markers. The DNA fingerprints in all other lanes result from different bacterial isolates. For each isolate, 16S ribosomal DNA has been cut with a restriction enzyme.

Reporter genes are genes that signal the activity of an associated gene of interest. Reporter genes are typically inserted into a specific bacterial **operon** containing the gene that codes the activity of interest, creating a gene fusion. These gene fusions are often created using **transposons** to randomly insert reporter genes into the genome of the bacterium. The process produces mutants with many different gene fusions. This collection of mutants must be screened to ensure that the mutant still has the activity of interest and that the activity of interest occurs concomitantly with the expression of the reporter gene. Reporter gene expression typically gives some easy-to-detect signal, such as production of β-galactosidase (*lac z* gene) (Box 9-5). This enzyme can be easily detected via a colorimetric assay on solid growth media. Luminescence has also been used as a reporter gene. In this case, the *lux* gene emits light via luciferase activity on the substrate luciferin; that activity can be measured on solid or liquid growth media or even in environmental soil and water samples.

Other reporter genes include: the *uidA* gene coding for glucuronidase, the *phoA* gene coding for alkaline phosphatase, and the *tfdA* gene coding for 2,4-D monoxygenase.

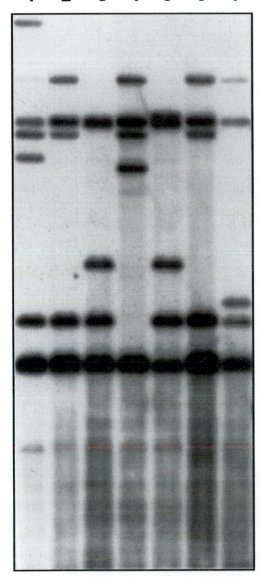

Figure 9–7 RFLP analysis of a *Salmonella* strain isolated from sewage sludge, coupled to gene probe detection using a ^{32}P-labeled 16S ribosomal DNA sequence. Lane 1, *Salmonella typhimurium* positive control. Lanes 2 to 7, various *Salmonella* environmental isolates.

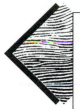

Box 9–5

The *lac z* Gene. One widely used reporter gene approach is to use the Tn5 transposon containing a constitutively expressed *lac z* gene which codes for β-galactosidase to create insertion mutants. Isolates that receive the insertion grow on an agar medium containing the chromogenic substrate X-gal and lactose. When the *lac z* gene is active, a blue colony results as compared to the wild type (*lac−*), which will produce a white colony. Blue colonies are thus isolated and tested for (A) the activity of interest, e.g., 2,4-dichlorophenoxyacetic acid degradation via a monoxygenase, and (B) activation of the *lac z* gene along with the monoxygenase activity. Once a suitable mutant has been identified, β-galactosidase activity can be inferred from the *lac z* activity.

Detection of Specific DNA Sequences in Soil

Viable but Nonculturable Soil Organisms

Comparisons of direct microscopic enumerations and viable plate counts reveal that the majority of soil organisms cannot be cultured on common laboratory media. In addition, many bacteria introduced into soil, including pathogens from animal or human wastes, may be sublethally injured. When bacteria are injured in some way, the stress of isolating them on a selective medium may kill them. These organisms are considered **viable but nonculturable** (Roszak and Colwell, 1987). Thus, detection of soil bacteria based on growth in a culture medium is tenuous, and estimates of the prevalence of organisms based on such methodology, or the prevalence of gene sequences in a soil based only on examination of cultured isolates, may be incorrect by several orders of magnitude. Any method that allows detection without the culturing of organisms is welcome. Such methods include using gene probes or PCR to analyze the nucleic acids of the entire soil community.

Obtaining Intact Cells From Soil

One way to obtain a representative portion of the nucleic acids of the entire soil community is to separate all cells from the soil and subsequently lyse the cells and extract the nucleic acids. Differential centrifugation based on density gradient separation can also separate cells from soil. One problem with isolating microorganisms from soil is that they bind to clay and organic matter. Obtaining all the cells within the soil matrix is difficult if not impossible. To overcome this, we add surfactants to remove cells that are sorbed to colloidal or humic substances, allowing subsequent removal of microbes from the soil with a solvent. Extracting solutions include phosphate buffers or soluble sodium and calcium solutions.

In practice, the percentage recovery of cells from soil depends on the nature of the soil and the nature of the organisms. Typically, more microorganisms are extracted from coarse-textured than fine-textured soils and from soils high in organic matter. Oligotrophic cells that have been present in soil for long periods of time are normally sorbed more strongly than young rapidly growing zymogenous cells. Because some cells secrete polysaccharides, these cells may be intrinsically more sticky than others. All of these factors can result in the differential extraction of some bacterial populations over others. However, despite these problems, cell extraction followed by DNA analysis is still likely to result in a more representative estimate of the total bacterial community than that provided by sampling only those soil bacteria that are culturable on agar growth media.

Once the cells are extracted from soil, the total cell biomass can be lysed and subjected to PCR analysis. Intact cells may be lysed by heat, chemical, or enzymatic treatment. It is essential that the sample undergo one or more purification steps because PCR is inhibited by humic materials. The particular type of PCR analysis conducted can be any of the techniques described later in this chapter.

In some instances it may be desirable to enumerate soil organisms using culture methods. PCR, in conjunction with culture methods, can identify the resulting bacterial colonies. In this method, intact cells are obtained from the soil with an extracting solution, and the sample is diluted and plated on a growth medium. The medium may be specific for a certain bacterium, or it may support the growth of many different culturable bacteria. Isolated colonies are resuspended in a buffer or grown in pure culture in a liquid medium before the cells are heat lysed and subjected to PCR. Purification of the lysed cells prior to PCR is not usually required for this method.

Extraction of Bacterial Community DNA From Soil

A different approach to obtain a more representative estimate of the total bacterial community in soil is to obtain community DNA directly. With this methodology, bacterial cells are lysed *in situ* within the soil matrix, and the DNA is subsequently extracted and purified. Several protocols for the extraction of community DNA have recently been published (Moré et al., 1994); most rely on the use of lysozyme or sodium dodecyl sulfate as the lysing agent. Once extracted, the DNA is purified by cesium chloride density centrifugation or the use of commercial kits. An alternative approach is to purify DNA with a phenol extraction followed by ethanol precipitation. Once the DNA has been purified, it can be subjected to PCR analysis. Although extraction of DNA from soil is an attractive approach, it is not without limitations. Soils high in clay adsorb DNA so that the efficiency of extraction varies with soil type and organic matter content. Also, the efficiency of lysis is not always high and varies with the species composition within the soil. DNA is lost during the purification process, and even purified DNA may still contain humic residues from the soil, which interferes with the PCR reaction. Another potential problem is that community DNA extracts may contain eukaryotic DNA as well as bacterial DNA.

Overall, the community-based extraction methods (cells or DNA) probably obtain a higher proportion of the total bacterial population than dilution plating alone. They also are likely to represent a different assemblage of bacterial populations than that obtained with culture methods.

Applications of Molecular Analyses in Soil Microbiology

Specific Detection of Soil Bacteria

As we have seen, PCR can detect specific soil bacteria with great sensitivity in community DNA extracted from soil or in DNA obtained from extracted cells. Fecal coliforms in soil have been detected by PCR, with even greater sensitivity achieved when gene probes were used to visualize the amplified DNA (Josephson et al., 1991). In addition, PCR can be coupled to culture techniques to quickly identify isolated colonies from soil. For example, *Salmonella* isolated from animal feed lots and identified by conventional methods can be confirmed using primers specific for *Salmonella* spp. (Way et al., 1993).

Specific Detection of Microbial Populations

The judicious design of primers can detect more conserved gene sequences that allow broader estimates of microbial populations. Examples might include detection of rhizobia or total or fecal coliforms. PCR can also provide estimates of potential bacterial activity. In ecological studies of nitrification, primers have been designed to detect the nitrifying bacteria (Navarro et al., 1992). PCR has also been used in biodegradation studies. Neilson et al. (1992) used primers specific to the *tfdB* gene that is involved in the degradation of 2,4-dichlorophenoxyacetic acid. These primers are also useful in gene transfer studies, since the *tfdB* gene is located on a catabolic plasmid. It is important to note that currently most PCR assays only tell the researcher whether or not the DNA sequence of interest is present in the soil. They give no indication as to the amount of the target DNA that is present. New approaches utilizing high-performance liquid chromatography (HPLC) or detection of fluorescently labeled primers are now evolving that allow for the quantification of PCR product rather than just a presence/absence test. (Marlow et al., 1997).

Detection of Viruses in Soil

Because PCR amplifies DNA, it can also detect organisms other than bacteria. These include viruses that contain DNA or RNA. Protocols are available for the extraction of viruses from soil and subsequent lysis and purification. Gel filtration with Sephadex columns has been one successful approach in purifying viral extracts. Straub et al. (1994) detected enteroviruses in sludge-amended soil with PCR. Because these are RNA viruses, an extra step is needed in the protocol. RNA must first be transcribed into DNA using the enzyme **reverse transcriptase.** Following this step, PCR proceeds normally with primers specific to the viral nucleic acid. Care must be taken, however, in interpreting results, because noninfectious virus can also be detected if the viral nucleic acids are intact (Ma et al., 1994). Analogous results have also shown PCR detection of nonviable bacteria (Josephson et al., 1993). However, it is important to note that dead bacterial cells are normally degraded quickly in soil, so that a successful PCR amplification implies that the bacteria were viable in the soil, or at least recently viable. This same implication does not necessarily apply to viruses, which even if inactivated can remain in soil for several months if the viral capsid is intact.

DNA Fingerprinting

Because organisms contain unique DNA sequences, these sequences differentiate or fingerprint organisms at the strain level. PCR-based fingerprinting is becoming standard procedure in many laboratories. Arbitrarily primed PCR (AP-PCR) is one such method. Here a single 10-bp primer of random sequence is used. Because it is smaller than the normal 20- to 30-bp primers used in standard PCR, it can anneal at several locations within the bacterial genome. This in turn results in the produc-

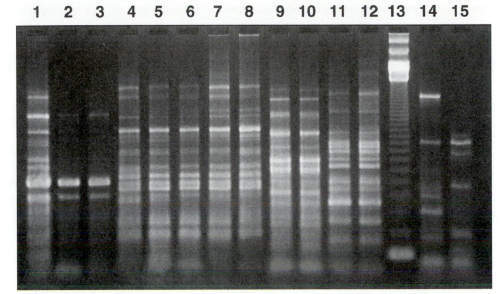

Figure 9–8 ERIC-PRC on bacteria isolated from soil. Different fingerprint patterns are seen in Lane 1, Lanes 2-3, Lanes 4-8, Lanes 9-10, and Lanes 11-12. Lanes 13 = DNA size marker; Lane 14 = known *Alcaligenes eutrophus* control; Lane 5 = known *Pseudomonas* spp. control.

tion of multiple amplification products. These multiple products act as a unique fingerprint of the target organism. AP-PCR has been used to fingerprint bacterial isolates involved in hydrocarbon degradation in soil.

Two other PCR analyses based on this methodology are REP-PCR and ERIC-PCR. REP (Repetitive Extragenic Palindromic) and ERIC (Enterobacterial Repetitive Intergenic Consensus) are repetitive, conserved sequences at different genomic points for a particular bacterium. REP- and ERIC-PCR differ from AP-PCR in that two 20 bp primers are used. Primers designed from these sequences result in multiple products of variable size, which again result in a specific fingerprint for the target organism. ERIC-PCR can distinguish different soil isolates (Fig. 9–8).

Detection of Expressed Microbial Activity

Messenger RNA (mRNA) is necessary for the translation of DNA gene sequences prior to specific gene product formation. Because translation necessarily proceeds enzyme production, detection of mRNA gives an estimate of actual expressed metabolic activity. This is particularly true of mRNA because it has a half-life of only a few minutes and is only produced immediately prior to protein synthesis. Tsai and Olson (1991) employed guanidine isothiocyanate to extract mRNA from soil organisms and estimate the activity of toluene degradation. Such mRNA determinations in ecological studies can estimate specific rates of microbial transformations or, in bioremediation studies with probes, can detect rates of biodegradation of xenobiotics.

Reporter genes also can monitor specific bacterial activity in environmental samples. Lam et al. (1990) incorporated a *lac z* reporter gene to study rhizosphere colonization in response to root exudates. In soil, *lux* gene insertions detected *E. coli,* although 10^2 to 10^3 cells g^{-1} of soil were required for detection (Rattray et al., 1990). Reporter genes can show fungal degradation of lignin (Godfrey et al., 1994). Clearly, reporter genes are a valid approach for quantifying microbial activity *in situ.*

Summary

Sequence information of nucleic acids associated with many microbial organisms is being gathered rapidly. These sequences, both DNA and RNA, are frequently analyzed by gene probe and PCR technologies. These analyses are now being applied to detect specific viruses, bacteria, and fungi in environmental samples and the activity of specific enzymes associated with particular organisms. Gene probes and PCR can also detect organisms that previously could not be isolated or cultured. Amplification of mRNA or use of reporter genes now allow for estimates of microbial activity. Phylogenetic studies are also now being conducted in many laboratories. However, while many of the analyses are fairly routine when applied to pure culture isolates, these analyses are much more complex for soil samples because of the presence of PCR-inhibitory substances. New procedures that purify nucleic acids extracted from soil without subsequent loss of PCR sensitivity are necessary and will allow for exciting breakthroughs in the area of soil molecular ecology.

Cited References

Godfrey, B.J., L. Akileswaran, and M.H. Gold. 1994. A reporter gene construct for studying the regulation of manganese peroxidase gene expression. Appl. Environ. Microbiol. 60:1353–1358.

Josephson, K.L., C.P. Gerba, and I.L. Pepper. 1993. Polymerase chain reaction detection of nonviable bacterial pathogens. Appl. Environ. Microbiol. 59:3513–3515.

Josephson, K.L., S.D. Pillai, J. Way, C.P. Gerba, and I.L. Pepper. 1991. Fecal coliforms in soil detected by polymerase chain reaction and DNA-DNA hybridizations. Soil Sci. Soc. Am. J. 55:1326–1332.

Lam, S.T., D.M. Ellis, and J.M. Ligon. 1990. Genetic approaches for studying rhizosphere colonization. Plant Soil 129:11–18.

Ma, J.F., T.M. Straub, I.L. Pepper, and C.P. Gerba. 1994. Cell culture and PCR determination of poliovirus inactivation by disinfectants. Appl. Environ. Microbiol. 60:4203–4206.

Marlowe, E.M., K.L. Josephson, B. Zenner, R.M. Miller, and I.L. Pepper. 1997. A method for the detection and quantification of PCR template in environmental samples by high performance liquid chromatography. J. Microbiol. Methods: 28:45-53.

Moré, M.I., J.B. Herrick, M.C. Silva, W.G. Ghiorse, and E.L. Madsen. 1994. Quantitative cell lysis of indigenous microorganisms and rapid extraction of microbial DNA from sediment. Appl. Environ. Microbiol. 60:1572–1580.

Navarro, E., P. Simonet, P. Normand, and R. Bardin. 1992. Characterization of natural populations of *Nitrobacter* spp. using PCR/RFLP analysis of the ribosomal intergenic spacer. Arch. Microbiol. 157:107–115.

Neilson, J.W., K.L. Josephson, S.D. Pillai, and I.L. Pepper. 1992. Polymerase chain reaction and gene probe detection of the 2,4-dichlorophenoxyacetic acid degradation plasmid, pJP4. Appl. Environ. Microbiol. 58:1271–1275.

Rattray, E.A.S., J.I. Prosser, K. Killham, and L.A. Glover. 1990. Luminescence-based nonextractive technique for in situ detection of *Escherichia coli* in soil. Appl. Environ. Microbiol. 56:3368–3374.

Roszak, D.B., and R.R. Colwell. 1987. Survival strategies of bacteria in the natural environment. Microbiol. Rev. 51:365–379.

Sambrook, J., E.F. Fritsch, and T. Maniatis. 1989. Molecular cloning: A laboratory manual. Cold Spring Harbor Press, New York.

Straub, T.M., I.L. Pepper, M. Abbaszadegan, and C.P. Gerba. 1994. A method to detect enteroviruses in sewage sludge-amended soil using the PCR. Appl. Environ. Microbiol. 60:1014–1017.

Tsai, Y.L., and B.H. Olson. 1991. Rapid method for direct extraction of DNA from soil and sediments. Appl. Environ. Microbiol. 57:1070–1074.

Watson, J.D., J. Tooze, and D.T. Kurtz. 1983. Recombinant DNA — A short course. Scientific American Books. W.H. Freeman and Company, New York.

Watson, J.D., N.H. Hopkins, J.W. Roberts, J.A. Steitz, and A.M. Weiner. 1987. Molecular biology of the gene. 4th ed. Benjamin/Cummings Publishing Company Inc., Menlo Park, Calif.

Way, J.S., K.L. Josephson, S.D. Pillai, M. Abbaszadegan, C.P. Gerba, and I.L. Pepper. 1993. Specific detection of *Salmonella* spp. by multiplex polymerase chain reaction. Appl. Environ. Microbiol. 59:1473–1479.

Study Questions

1. Why is the concept of DNA complementarity important in the study of nucleic acids?

2. What specific technique would you use to determine if a bacterial isolate contained a certain gene?

3. How would you determine if a bacterial plasmid contained a specific gene?

4. Describe the three key steps of PCR.

5. Why does the analysis of community DNA probably give a better estimate of microbial diversity than dilution plating alone?

6. What is the major difference between PCR amplification of bacterial and viral nucleic acids?

7. Discuss the different ways that bacterial DNA can be fingerprinted.

8. In what ways do studies of mRNA and 16S rDNA differ (i.e., what can we learn from analysis of both of these nucleic acid sequences)?

Part 2

♦

Microbially Mediated Transformations in Soil

Chapter 10

♦

Microbial Metabolism

Jeffry J. Fuhrmann

♦

Organic chemistry is the chemistry of carbon compounds.
Biochemistry is the study of carbon compounds that crawl.

Anonymous

The preceding chapters documented the high degree of morphological and structural diversity within and among populations of soil microorganisms. This chapter, however, will consider microbial diversity from another perspective—metabolic diversity among microorganisms. The term **metabolism** refers to the sum total of the biochemical reactions that occur within living cells.

Although many of the topics and concepts discussed in this chapter are familiar, they warrant repetition nevertheless because of the important role that metabolism plays in defining populations of soil microorganisms and in understanding their interactions with each other and the environment. It is simply impossible to appreciate fully the discipline of soil microbiology without a fundamental understanding of the underlying biochemical principles. In discussing metabolic diversity, a central concept is that the overall goal of all forms of metabolism is essentially the same: to obtain energy or carbon in the production of cellular constituents necessary for growth, survival, and reproduction. Thus, the metabolic diversity among microorganisms described in this chapter comes down to differences in strategies rather than the goals of these strategies.

An Overview of Microbial Metabolism

Biochemical reactions in metabolism either yield energy (**exergonic**) or consume energy (**endergonic**). Exergonic reactions result in the production of energy needed

189

to support cellular processes. They are often referred to as "spontaneous" because, given the proper conditions, they can be sustained without the addition of external energy. Endergonic activities result in the biosynthesis of cellular components and microbial biomass. These reactions are not spontaneous and require a supply of energy from exergonic reactions to proceed. Although such a separation of exergonic and endergonic reactions is helpful for purposes of discussion, metabolism involves both types of reactions and is a highly integrated and interdependent system.

Exergonic Reactions

Microbes obtain energy from a myriad of sources, including organic compounds, inorganic substances, and light. Furthermore, microorganisms as a group, and sometimes as individuals as well, can use these sources under both aerobic and anaerobic conditions. Thus, to a large extent, the ubiquity of microorganisms reflects their diverse strategies for obtaining energy.

Two general types of energy-rich compounds are derived from the exergonic reactions of metabolism:

• high-energy phosphate compounds and

• stored electrons associated with specialized carrier molecules.

Adenosine triphosphate (ATP) is the dominant high-energy phosphate compound in cells (Fig. 10–1). The potential energy stored in the "high-energy phosphate bonds" of ATP is useful in a wide variety of biosynthetic reactions. Electrons released during exergonic metabolism are generally captured by oxidized electron carriers such as **nicotinamide adenine dinucleotide** (NAD^+), NAD phosphate ($NADP^+$) or related compounds (Fig. 10–2). The reduced forms of these carriers will be referred to as NADH and NADPH, respectively. An additional but less energy-rich electron

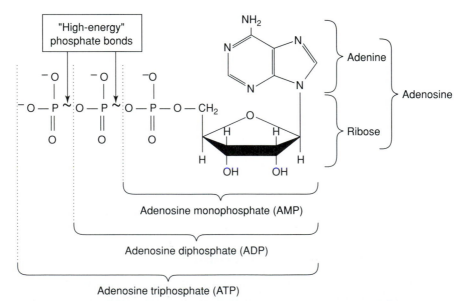

Figure 10–1 Structure of adenosine triphosphate (ATP).

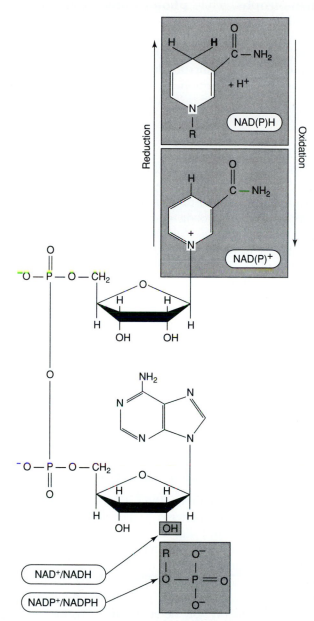

Figure 10–2 Structure of oxidized nicotinamide adenine dinucleotide (NAD^+) and nicotinamide adenine dinucleotide phosphate ($NADP^+$). The reduced forms (NADH and NADPH) are also shown.

carrier is flavin adenine dinucleotide ($FAD/FADH_2$). The electrons stored in these compounds are often called **reducing equivalents** or **reducing power.**

Most organisms obtain cellular energy from the biodegradation of energy-rich organic compounds, such as carbohydrates, proteins, and lipids; this "dismantling" of organic substances is **catabolism.** Alternatively, some organisms obtain energy from transformations of various inorganic compounds, whereas others use light as an energy source. Organisms that derive energy from organic or inorganic substances are called **chemotrophs,** while **phototrophs** convert light energy into chemical energy to support metabolic processes. Organisms that obtain reducing equivalents from organic compounds are referred to as **organotrophs,** and those that use inorganic substances are called **lithotrophs.** The vast majority of microorganisms encountered by soil microbiologists fit into the three metabolic groups shown in Table 3–4, i.e., chemoheterotrophs, photoautotrophs, and chemoautotrophs.

Endergonic Reactions

A portion of the energy derived from exergonic reactions is subsequently used for the endergonic biosynthesis of cellular components needed for cell growth, biomass production, and reproduction. This biosynthetic process is called **anabolism,** which is the functional opposite of catabolism.

One fundamental source of diversity among microorganisms relates to carbon nutrition. Most microorganisms obtain the carbon intermediates, or "building blocks," needed for biosynthesis from the degradation of organic compounds; these organisms are **heterotrophs.** The remaining organisms (algae and certain bacteria) can convert inorganic carbon from carbon dioxide (or carbonates) or other one-carbon substances (e.g., methane) into organic compounds by the endergonic process known as *carbon fixation;* these organisms are called **autotrophs** (Fig. 10–3).

A Note on Terminology

As indicated in the previous two sections, microorganisms can be categorized with respect to three important metabolic requirements (Fig. 10–3):

* source of energy (chemotrophs or phototrophs),

* source of reducing equivalents (organotrophs or lithotrophs), and

* source of carbon for anabolism (heterotrophs or autotrophs).

According to this system (summarized in Table 10–1), a filamentous fungus might be called a "chemoorganoheterotroph" or perhaps a "heterotrophic chemoorganotroph," whereas a photosynthetically active alga could be referred to as an "photolithoautotroph." In practice, however, many soil microbiologists have informally adopted an abbreviated system for referring to these metabolic categories that recognizes that certain combinations of these traits nearly always occur together. For example, the description of the fungus given above is typically reduced to simply "heterotroph" because the vast majority of heterotrophic organisms are also chemoorganotrophs.

The algal cell presents a somewhat more complicated case in that autotrophy commonly occurs with both phototrophy (e.g., eukaryotic algae and cyanobacteria) and chemotrophy (e.g., nitrifying bacteria). Given this ambiguity, the label "photoautotroph" would probably be used by most soil microbiologists. Lithotrophic organisms are usually also autotrophic, and this association can be assumed unless otherwise noted. Used alone, the terms "lithotroph" and "autotroph" generally refer

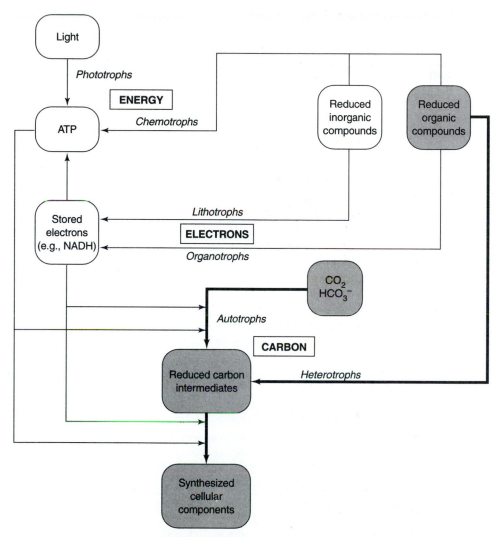

Figure 10–3 Overview of different metabolic pathways used by soil microorganisms for acquiring cellular energy (primarily ATP) for anabolic processes, storing electrons (reducing equivalents) such as NADH, and reducing carbon intermediates for the synthesis of other cellular compounds and polymers. Bolded arrows and shaded boxes show flow of carbon; other arrows and boxes represent energy and electron flow.

to chemolithoautotrophs (commonly referred to simply as chemoautotrophs), such as the nitrifying bacteria, although a much less ambiguous label would also indicate whether the energy source is light or chemical compounds, as in photoautotroph or chemoautotroph.

Table 10–1 **Summary of metabolic classifications of microorganisms based on their source of energy, reducing equivalents, and carbon.**

Metabolic requirement	Alternative metabolic strategies	Definition and comments	Representative microorganisms
Energy	Phototrophy	Light is used as the primary source of energy for metabolism and growth.	Algae, cyanobacteria
	Chemotrophy	Light-independent chemical reactions are the source of metabolic energy.	All nonphotosynthetic cellular organisms
Reducing equivalents (electrons)	Lithotrophy	Inorganic substances are used as electron donors; lithotrophic organisms may be either phototrophs or chemotrophs.	Algae and cyanobacteria (photolithotrophs); nitrifying and sulfur-oxidizing bacteria (chemolithotrophs)
	Organotrophy	Organic substances are used as electron donors; normally associated with chemotrophy, although also exhibited by certain photosynthetic bacteria (phototrophs).	Fungi, most bacteria, and protozoa
Carbon	Autotrophy	Cellular carbon is derived mostly or entirely from carbon dioxide; associated primarily with lithotrophy.	Algae, cyanobacteria, nitrifying and sulfur-oxidizing bacteria
	Heterotrophy	Cellular carbon is derived from preformed organic compounds; generally associated with chemoorganotrophy.	Fungi, most bacteria and protozoa

Summarized from Singleton and Sainsbury, 1987.

The Chemical Foundations of Metabolism

Discussion of metabolism requires a basic understanding of certain chemical principles. This section is provided for those readers wishing to review the more important concepts.

Oxidation/Reduction Reactions

Metabolism involves **oxidation-reduction (redox) reactions.** *Oxidation* refers to the loss of electrons by a substance. Oxygen may or may not be involved in the reaction. The substance oxidized is transformed into a second substance with a lower level of potential energy. Conversely, *reduction* refers to the acceptance of electrons by a substance. This represents a net gain in energy by the substance being reduced. Hydrogen is often transferred to the reduced compound simultaneously with the electrons. A critical point here is that oxidation and reduction are inseparable; the oxidation of one compound necessarily implies the reduction of a second compound.

Oxidation State

Redox reactions result in changes in the **oxidation state,** or valence, of elements. In a reaction, it is often helpful to determine which elements are being reduced and oxidized. In many cases, the oxidation state of an element can be determined by applying the following general rules:

- Any element in the free state is assigned an oxidation number of zero; e.g., elemental sulfur (S) has an oxidation state of zero; each oxygen atom in O_2 has an oxidation state of zero.

- Elements present as monoatomic ions have an oxidation state equal to the ionic charge; e.g., magnesium in Mg^{2+} has an oxidation state of $+2$.

- When combined with other elements, hydrogen and oxygen typically have oxidation states of $+1$ and -2, respectively. (The exceptions are hydrides for hydrogen and peroxides for oxygen; in both cases, the oxidation state is -1.)

- The sum of the oxidation states of the atoms comprising a compound or polyatomic ion must equal the overall net charge of the substance; e.g., the N in NO_3^- has an oxidation state of $+5$; the C in CH_4 has an oxidation state of -4; the S in H_2SO_4 has an oxidation state of $+6$.

In general, a relatively high ratio of hydrogen to oxygen atoms in a given substance indicates that the associated elements are in a comparatively high state of reduction (see Box 10–1).

Box 10–1

Example Redox Reactions. The numbers above the compounds indicate the oxidation state of the constituent elements.

$$\overset{0}{2H_2} + \overset{0}{O_2} \rightarrow \overset{+1\ -2}{2H_2O}$$

Each hydrogen loses one electron and is oxidized from 0 to $+1$ oxidation state; each oxygen gains two electrons and is reduced from 0 to -2 oxidation state.

$$\overset{0}{2S} + \overset{0}{3\,O_2} + \overset{+1\ -2}{2\,H_2O} \rightarrow \overset{+1\ +6\ -2}{2H_2SO_4}$$

Sulfur is oxidized from 0 to $+6$; each oxygen from O_2 is reduced from 0 to -2.

$$\overset{-3\ +1}{2\,NH_4^+} + \overset{0}{3\,O_2} \rightarrow \overset{+3\ -2}{2\,NO_2^-} + \overset{+1}{4\,H^+} + \overset{+1\ -2}{2\,H_2O}$$

Nitrogen is oxidized from -3 to $+3$; oxygen is reduced from 0 to -2.

$$\overset{+4\ -2}{CO_2} + \overset{0}{4\,H_2} \rightarrow \overset{-4\ +1}{CH_4} + \overset{+1\ -2}{2\,H_2O}$$

Carbon is reduced from $+4$ to -4; hydrogen is oxidized from 0 to $+1$.

Reduction Potentials

Although oxidation states are useful for monitoring changes to the specific elements comprising a redox reaction, they do not indicate the direction in which the reaction will tend to proceed. Although nearly all metabolic reactions are reversible, a given reaction will proceed spontaneously only in the direction yielding a net release of energy. For redox reactions, this will reflect the relative concentrations of the substances comprising a reaction and their tendencies to donate or accept electrons.

For example, the overall reaction:

$$H_2 + \tfrac{1}{2} O_2 \rightarrow H_2O \tag{1}$$

has two half-reactions:

an oxidation: $$H_2 \rightarrow 2\,H^+ + 2e^- \tag{2}$$

and a reduction: $$\tfrac{1}{2}O_2 + 2\,H^+ + 2e^- \rightarrow H_2O \tag{3}$$

Therefore, assuming equal concentrations of the reaction components, the tendency for reaction [1] to result in the formation of H_2O versus H_2 and O_2 will be determined by the relative tendencies of reactions [2] and [3] to proceed in the direction indicated.

By convention, the tendency for a half reaction to occur is determined by measuring the electrical potential (E_o) it generates relative to a reference substance (H_2) under standardized conditions of temperature, acidity, and pressure (25°C, pH = 0, and 1 atm). The resulting potentials are then corrected to pH = 7 (and designated E'_o) to reflect that metabolic reactions generally occur under near neutral conditions. Finally, to facilitate comparison among reactions, the half reactions are written as reductions and the sign of the corresponding electrical potential is adjusted accordingly. These standardized values are referred to as **reduction potentials.**

Some reduction potentials are given in Table 10–2. A relatively low (more negative) reduction potential indicates a relatively low tendency for the reduction to occur, i.e., low tendency for the half reaction to proceed in the direction written. Conversely, it indicates a relatively high tendency for the opposite oxidation reaction to occur. Thus, for our example half reactions (both written here as reductions):

Table 10–2 **Standard redox potentials for selected redox pairs that have metabolic importance to soil microorganisms.**

Redox pair	E'_o (V)
O_2/H_2O	+0.82
Fe^{3+}/Fe^{2+}	+0.77
NO_3^-/NO_2^-	+0.43
NO_2^-/NO	+0.36
$FAD/FADH_2$	−0.22
CO_2/CH_4	−0.24
SO_4^{2-}/HS^-	−0.23
NAD/NADH	−0.32
H^+/H_2	−0.42

$$2 H^+ + 2e^- \rightarrow H_2 \qquad\qquad E'_o = -0.42 \text{ V} \qquad\qquad [4]$$

$$\tfrac{1}{2}O_2 + 2 H^+ + 2e^- \rightarrow H_2O \qquad\qquad E'_o = +0.82 \text{ V} \qquad\qquad [5]$$

the E'_o for equation [4] is more negative than that for equation [5], indicating that equation [5] has the greater tendency to proceed as written. Therefore, under the stated conditions, the combined reaction [1] can proceed spontaneously to form H_2O. The reverse reaction can also occur but only with the input of energy from an external source.

Free Energy Change

Perhaps the most useful measure of a metabolic reaction is its corresponding change in **free energy** (also called **Gibb's free energy**). Free energy ($G^{0'}$) refers to the intrinsic energy contained in a given substance. For the reaction $X \rightarrow Y$, the change in free energy ($\Delta G^{0'}$) refers to the difference between the free energy of product Y and the reactant X, i.e., Gy − Gx. The $\Delta G^{0'}$ value is important because it indicates the amount of energy that is released by or required for a particular reaction. A negative or positive $\Delta G^{0'}$ indicates that the corresponding reaction is exergonic or endergonic, respectively; equilibrium conditions are indicated by $\Delta G^{0'}=0$. The value of $\Delta G^{0'}$ for a redox reaction can be calculated by using the following equation:

$$\Delta G^{0'} = -n F \Delta E'_o \qquad\qquad [6]$$

where   n = the number of electrons transferred,

F = Faraday's constant ($96.5 \text{ kJ V}^{-1} \text{ mol}^{-1}$ [$=23.1 \text{ kcal V}^{-1} \text{ mol}^{-1}$]), and

$\Delta E'_o$ = the difference in reduction potentials for the two half reactions comprising the combined reaction.

For the reaction: $H_2 + \tfrac{1}{2} O_2 \rightarrow H_2O$,

$$\Delta E'_o = +0.82 -(-0.42) = +1.24 \text{ V and} \qquad\qquad [7]$$

$$\Delta G^{0'} = -(2)(23.1)(1.24) = -239 \text{ kJ mol}^{-1} \text{ of water produced.} \qquad\qquad [8]$$

This highly negative $\Delta G^{0'}$ value indicates that a large amount of free energy is released by the reaction. According to our definition of $\Delta G^{0'}$, this value is also the amount of energy that must be expended to convert one mole of water into its component elements.

Essential Concepts

Metabolic processes in cells involve oxidation-reduction (redox) reactions. All redox reactions consist of two parts, an oxidation half reaction and a reduction half reaction. When a substance is oxidized, it donates electrons to a second substance which thereby becomes reduced. The tendency to be reduced or oxidized differs among substances and is commonly expressed as a standard reduction potential (E'_o). A combined redox reaction can proceed spontaneously in the direction written provided the numeric difference of its corresponding reduction potentials is positive. The free energy change ($\Delta G^{0\prime}$) associated with a reaction gives a measure of the energy released by ($\Delta G^{0\prime} < 0$) or required for ($\Delta G^{0\prime} > 0$) the reaction. The value of $\Delta G^{0\prime}$ is a function of the number of electrons transferred in a reaction and the corresponding difference in reduction potentials for the half reactions.

The Role of Enzymes in Metabolism

Although redox chemistry and free energy principles are useful in predicting energy relationships in metabolic processes, they provide no information concerning the rate at which reactions proceed. Certain reactions may release energy too rapidly for cellular processes and are incompatible with living cells, generally because of heat production. Conversely, many chemical reactions do not occur at appreciable rates even though the corresponding free energy change is highly favorable. This results from the need to satisfy the **activation energy** for the reaction—the energy input necessary to break the existing chemical bonds within each reactant. For a reaction to proceed, its activation energy must either be met by an external energy source or reduced in magnitude by the presence of a catalyst (Fig. 10–4).

A **catalyst** is a substance that promotes a reaction by reducing the required energy of activation without itself being altered by the ensuing reaction. This catalytic role in living cells is provided by specialized proteins called **enzymes.** Unlike most nonbiological catalysts, enzymes are highly specific in that they generally promote only one particular reaction or class of reactions. Enzymes promote efficient metabolism not only by facilitating the initiation of biochemical reactions but also by controlling the rate and extent of these reactions. Enzymes do this by temporarily

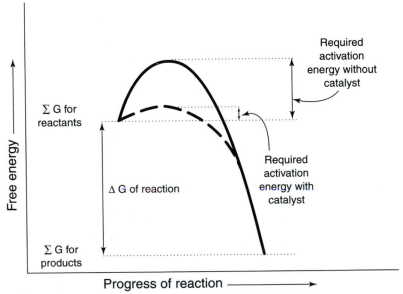

Figure 10–4 Effect of a catalyst on the activation energy required for initiating an exergonic chemical reaction.

binding the reactants at the enzyme's **active site** and thereby physically orienting them so as to both promote and control their interaction (Fig. 10–5). In this way, enzymes allow reactions to proceed in a step-by-step fashion with minimum energy input.

Enzymes often require the presence of various nonprotein substances called **coenzymes** to carry out their catalytic functions. The noncomplexed protein component is referred to as the **apoenzyme,** whereas the apoenzyme-coenzyme complex is called the **holoenzyme.** Coenzymes may bind either covalently or noncovalently to the apoenzyme. Covalently bound coenzymes are called **prosthetic groups.** For example, the heme group in many enzymes is usually present as a prosthetic group (Fig. 10–6). Noncovalently bound coenzymes (e.g., NADH) are only transiently associated with the enzyme and may be thought of as cosubstrates. Many substances serve as coenzymes, including

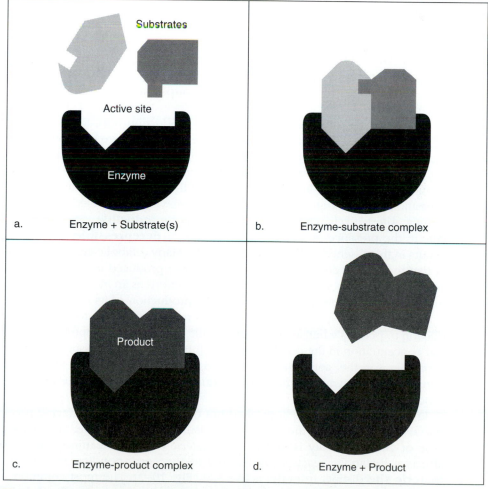

Figure 10–5 Major steps in an enzymatic reaction.

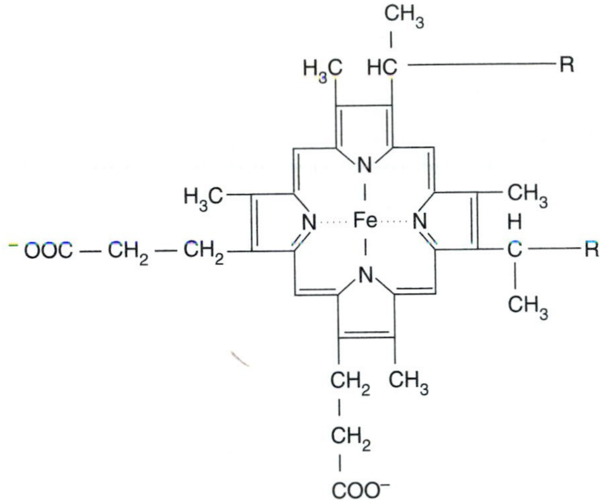

Figure 10–6 Structure of heme. Associated enzymes are attached at the letter R. The coordination iron at the center is the active site for oxidation and reduction.

metal ions (e.g., iron, molybdenum, and magnesium) and carrier molecules (e.g., NADH, $FADH_2$, and various vitamins). The carrier molecules function to supply substances required by the reaction, such as electrons and hydrogens, or to aid in the transfer of chemical groups, such as phosphate, between molecules.

Microorganisms need control of enzyme production both to promote balanced cell growth and to minimize unnecessary energy expenditure for protein production. Therefore, various strategies have evolved to regulate the production and activity of cellular enzymes. Enzymes needed for fundamental cellular processes are generally maintained at constant levels in the cell, and their production is unaffected by concentrations of substrates or products; such enzymes are called **constitutive enzymes**. Alternatively, enzyme production may be **inducible**, when production occurs in the presence of an *inducer* molecule, or **derepressible**, when production occurs in the absence of a *repressor* molecule. Many catabolic enzymes are inducible in that the proper degradative enzymes are only produced in response to the presence of the corresponding substrate, which functions as an inducer; for example, lactose (milk sugar) acts as an inducer for the production of its degradative enzymes. Conversely, some anabolic enzymes are derepressible so as to limit enzyme production to periods when a specific enzymatic product has dropped below some critical concentration in the cell; that product functions as a repressor. The enzymes responsible for the conversion of dinitrogen gas to ammonia are examples of derepressible enzymes in that their production is inhibited by the presence of the product.

Enzyme activity can be controlled not only at the level of enzyme production, but also by modifying the activity of existing enzymes in the cell. This general type of altered activity is known as *allosteric control,* meaning that a molecule affects a metabolic reaction by binding to the corresponding enzyme at an **allosteric site** distinct from the active site (Fig. 10–7). The most common form of allosteric control is allosteric inhibition, also referred to as **feedback inhibition**

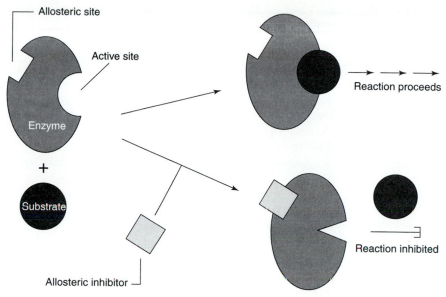

Figure 10–7 Allosteric inhibitors decrease the rate of enzymatic reactions by binding to a separate allosteric site on the enzyme and thereby indirectly altering the configuration of the active site to prevent substrate binding.

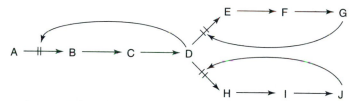

Figure 10–8 Schematic of feedback inhibition. The enzymatic products D, G, and J inhibit early steps in their respective metabolic pathways.

or **end product inhibition.** Allosteric inhibitors are generally metabolic end products which attenuate the metabolic pathway that produced them by inhibiting an early step in the pathway (Fig. 10–8). Attachment of the inhibitor at the allosteric site reduces enzymatic activity by altering the conformation of the active site, thereby preventing attachment of the substrate(s). In other cases, the attachment of a substance at the allosteric site may result in increased enzyme activity.

Enzymes promote many types of reactions within living cells and are generally named for the particular reaction catalyzed. Enzyme names are further designated by the suffix "-ase." Table 10–3 summarizes the kinds of reactions carried out by different classes of enzymes. In many respects, the metabolic diversity discussed in the remainder of this chapter is a direct reflection of the enzymatic diversity that exists among soil microorganisms. Specific examples of enzymes important in soil microbiology include:

Table 10–3 **Major classes and subclasses of enzymes and the corresponding types of reactions catalyzed.**

Class	Representative subclasses	Types of reactions catalyzed by enzyme class
Oxidoreductases	Dehydrogenases Oxidases Reductases Oxygenases Peroxidases Catalases	Catalyze oxidation-reduction reactions. Important in fermentation and respiration pathways.
Transferases	Aminotransferases Kinases	Catalyze the transfer of molecular substituents among molecules.
Hydrolases	Glycosidases Peptidases Phosphatases Ribonucleases	Catalyze the hydrolytic cleavage of chemical bonds.
Lyases	Decarboxylases Synthases Lyases	Catalyze the addition or removal of chemical groups such as carbon dioxide, ammonia, and water.
Isomerases	Racemases Isomerases	Catalyze inversions at asymmetric carbon atoms and the intramolecular transfer of molecular substituents.
Ligases	Synthetases Carboxylases	Catalyze the binding of two molecules with the expenditure of ATP. Important in anabolic pathways.

Box 10–2

Common Enzyme Groupings. Soil microbiologists often find it useful to place enzymes into broad groupings based on certain functional characteristics. One common grouping refers to the intended location of enzymatic activity:

- **Intracellular enzyme**—an enzyme that catalyzes biochemical reactions occurring within cells; many intracellular enzymes can be found in soil apart from cells and are thought to be released to the environment by cell lysis.

- **Extracellular enzyme**—an enzyme purposefully released exterior to cells, generally to catalyze the degradation of a polymeric substance too large to cross the cellular membrane.

Another common grouping relates to the position within a polymer at which an enzyme is active:

- **Exoenzyme**—an enzyme, typically extracellular, that catalyzes the removal of terminal monomers of a polymeric substance; some microbiologists use this term synonymously with extracellular enzyme.

- **Endoenzyme**—an enzyme, typically extracellular, that degrades polymers at positions other than the termini, thereby producing oligomers, chains containing more than two monomers, which can be further degraded by exoenzymes.

- cellulases, which degrade the polymer cellulose into smaller constituents,

- nitrogenase, which converts dinitrogen gas into biologically available ammonia,

- sulfatases, which release sulfate from protein and certain other organic compounds, and

- phosphatases, which remove phosphate groups from organic compounds such as nucleic acids.

Not all enzymes are intracellular, meaning that they are produced and active within the cell. Enzymes that catalyze the degradation of polymeric substances added to soil, such as cellulose from crop residues, are necessarily extracellular because the polymer is too large to be transported across the cellular membrane. However, once the polymer has been reduced to its smaller subunits, subsequent catabolism may proceed intracellularly. Because many enzymes are important in nutrient cycling and other soil-related processes, standardized assays have been developed for measuring their activities. For instance, an assay for phosphatase activity in soil is described in Chapter 16.

Production of ATP

Adenosine triphosphate (ATP) is synthesized in cells by one of two mechanisms: substrate-level phosphorylation or oxidative phosphorylation. **Substrate-level phosphorylation** is the direct production of ATP during the enzymatic oxidation of a substance and involves the transfer of a phosphate group from a substrate to an ADP molecule. This type of phosphorylation is reaction specific and closely linked to the corresponding oxidative reaction. In contrast, **oxidative phosphorylation,** sometimes called electron-transport-chain phosphorylation, involves membrane-localized reactions that are physically removed from the initial energy-yielding reaction. Oxidative phosphorylation is implied to be active in respiration, which is discussed later in this chapter. In the case of oxidative phosphorylation, electrons released during oxidation are transferred to electron carriers such as NAD^+ and FAD and transported to an **electron-transport chain** rather than being directly used for ATP production. These electron-transport chains consist of membrane-bound electron carriers that function as a unit to accept electrons from NADH and $FADH_2$, thereby releasing NAD^+ and FAD back to the cell (Fig. 10–9). Within the electron-transport chain, the electrons participate in a carefully controlled series of redox reactions that capture a portion of the energy released as ATP.

In comparison with substrate-level phosphorylation, oxidative phosphorylation is a more generalized means of ATP production because reducing equivalents derived from diverse oxidative reactions may be processed by the same electron-transport chain. It is also a much more efficient means of producing ATP. However, oxidative phosphorylation requires that an external oxidant be present to accept the electrons as they exit from the electron-transport chain. This **terminal electron acceptor** is generally oxygen, but some microorganisms can use alternative compounds. Substrate-level phosphorylation can also take advantage of its ability to occur under more diverse conditions than can oxidative phosphorylation because it does not require specific external electron acceptors.

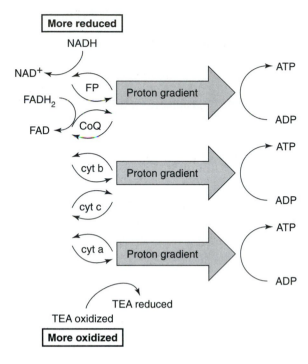

Figure 10–9 Schematic of oxidative phosphorylation via an electron-transport chain. Reduced electron carriers are oxidized by a membrane-bound electron transport system which results in the eventual reduction of a terminal electron acceptor (TEA). These oxidations produce a membrane-contained proton gradient which, in turn, can produce ATP. Note that 3 and 2 ATP molecules are produced by the oxidation of each NADH and FADH$_2$ molecule, respectively (FP, flavoprotein; CoQ, coenzyme Q; cyt b, cytochrome b; cyt c, cytochrome c; cyt a, cytochrome a).

| Box 10–3 |

Measurement of ATP in Soils. Soil microbiologists who study general soil processes such as nutrient cycling often wish to know the biomass of the resident microbial population. Although several different approaches for the measurement of soil microbial biomass have been proposed, many share the principle of measuring some component of biomass that is common to the entire microbial community. Because ATP is inherent to all forms of metabolism and persists only in living cells, it has received considerable attention as an index of soil biomass.

Briefly, the general procedure for the measurement of soil ATP is to treat a soil sample so as to rupture, or lyse, the cells of the indigenous microorganisms and to then measure the ATP released by means of the *luciferin-luciferase reaction*. Luciferin is an aromatic compound found naturally in fireflies that reacts, via the enzyme luciferase, to produce light in the presence of ATP:

$$\text{reduced luciferin + ATP + O}_2 \xrightarrow{\textit{luciferase}} \text{oxidized luciferin + CO}_2 \text{ + ADP + Pi + light.}$$

Light production is measured in a photometer or scintillation counter. Given the presence of excess luciferin, the amount of light produced will be proportional to the amount of ATP present in the sample of soil extract analyzed. Although useful under certain well-defined conditions, the method has significant limitations (see Horwath and Paul, 1994).

Glycolysis

The **glycolytic,** or **Embden-Meyerhof, pathway** is the most common pathway for the catabolism of glucose and other simple sugars to pyruvate (Fig. 10–10). The products of glycolysis are pyruvate, ATP, and NADH. The initial six-carbon portion of glycolysis involves the utilization of two molecules of ATP to produce fructose-1,6-bisphosphate. This activated sugar is then split into two three-carbon molecules, dihydroxyacetone phosphate and glyceraldehyde-3-phosphate, which are interconvertible by enzymatic action. Therefore, subsequent metabolic steps occur twice for each molecule of glucose catabolized. During several energy-yielding steps, each molecule of glyceraldehyde-3-phosphate is converted to pyruvate with the coproduction of one molecule of NADH and two molecules of ATP, the latter by substrate-level phosphorylation. This results in a total net gain of two NADH and two ATP molecules per molecule of glucose.

Two additional points should be noted. The end product of glycolysis, pyruvate, is incompletely oxidized and constitutes a rich potential source of metabolic energy.

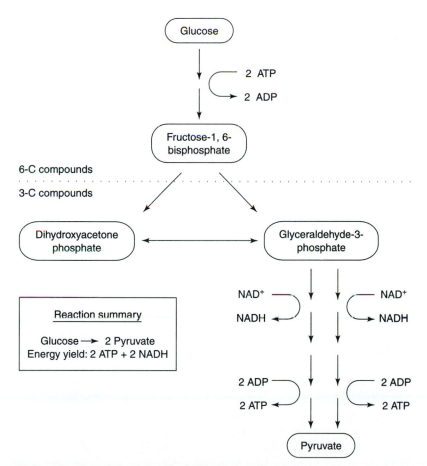

Figure 10–10 Embden-Meyerhof glycolytic pathway.

Second, the NAD^+ needed to accept electrons from glyceraldehyde-3-phosphate has not been regenerated, a situation that must be corrected if energy production is to continue. As the following discussions illustrate, the alternative metabolic pathways subsequently employed by microorganisms differ primarily in the degree to which the energy contained in pyruvate is used and the strategy employed to recover NAD^+ for cellular processes.

Fermentation

Following glycolysis, some microorganisms recover NAD^+ by using NADH to reduce pyruvate or one of its derivatives to various organic compounds via any of several anaerobic pathways. The overall reaction of glucose to organic end products, including glycolysis, is called **fermentation** and is characterized by its use of internally produced (endogenous) organic electron donors and acceptors. Depending on the particular fermentative pathway used by an organism, compounds produced may include various alcohols (e.g., ethanol, butanol, and isopropanol), organic acids (e.g., lactic, acetic, propionic, and butyric acid) and, in some cases, carbon dioxide and hydrogen (Fig. 10–11).

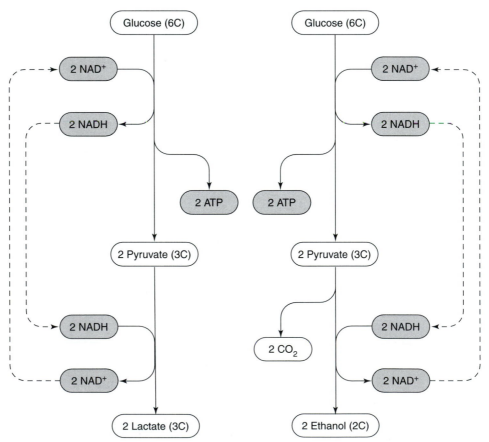

Figure 10–11 Generalized pathways for fermenting glucose to lactic acid (left) or ethanol (right).

The ATP produced during fermentation is derived from substrate-level phosphorylation and is generally limited to that which is produced during glycolysis. As a result, the metabolic steps proceeding from pyruvate function almost exclusively as a means of regenerating NAD^+ for use in the glycolytic pathway. Fermentation is inherently energy inefficient because the final organic compounds produced are still a rich source of energy. Conversely, because it uses an internally generated terminal electron acceptor, fermentation has the potential advantage of its ability to proceed under a wider range of environments than can alternative metabolic strategies, such as respiration, which require the presence of an external electron acceptor.

Fermentative processes are often associated primarily with the catabolism of glucose and other sugars, but other compounds, such as amino acids, can serve as fermentation substrates for certain microorganisms. Fermentative processes are important commercially in the production of chemicals and in baking and brewing some foods and beverages. They are also significant ecologically as sources of carbon and energy for other microorganisms, metal chelators, and agents of mineral weathering in soils.

Respiration

Besides fermentation, the second possible catabolic fate of pyruvate is **respiration.** The defining characteristic of respiration is its use of an external terminal electron acceptor (rather than the internally produced acceptors found in fermentation). The overriding benefit derived from this change in metabolic strategy is that respiratory organisms have the potential to maximize their recovery of energy contained in reduced compounds such as pyruvate; these compounds are no longer needed as terminal electron acceptors themselves. A disadvantage, of course, is that the proper external electron acceptor must be present in the environment and available to the organism for respiration to proceed.

Respiratory catabolism of pyruvate is perhaps best thought of as a two-step process:

- Pyruvate is oxidized to carbon dioxide in a series of reactions called the **tricarboxylic acid (TCA) cycle,** also known as the *Krebs cycle* and the *citric acid cycle* (Fig. 10–12). As a result of these oxidative steps, the TCA cycle produces NADH, $FADH_2$, and a small amount of GTP, a high-energy compound functionally equivalent to ATP. GTP is produced by substrate-level phosphorylation. Although a detailed discussion of the TCA cycle is beyond the scope of this chapter, it is important to note that the only other products of the cycle are carbon dioxide and water.

- The reduced electron carriers are further processed by electron-transport chains and oxidative phosphorylation to produce additional ATP and simultaneously regenerate NAD^+ and FAD for use during glycolysis and the TCA cycle (Fig. 10–13).

Acceptable external electron acceptors vary among organisms and may be either oxygen (aerobic respiration) or another highly oxidized compound (anaerobic respiration). Furthermore, although most aerobic microorganisms use the respiratory catabolism of organic compounds such as pyruvate, certain microorganisms are capable of obtaining energy by the respiration of reduced inorganic compounds.

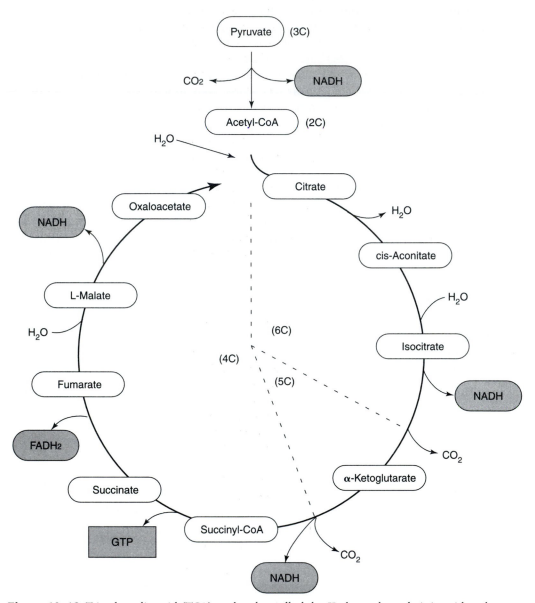

Figure 10–12 Tricarboxylic acid (TCA) cycle, also called the Krebs cycle and citric acid cycle.

Aerobic Respiration of Organic Compounds

The most widely studied soil microorganisms are those that rely on organic compounds as their source of energy and reducing equivalents. These microorganisms are classified metabolically as chemoorganotrophs. As noted previously, chemoorganotrophs are also nearly always heterotrophic (i.e., they obtain carbon from organic compounds). These organisms are essential to soil productivity and environmental quality because they have the primary responsibility for the degradation of organic substances added to soil, including plant residues, dead soil organisms, and natural and synthetic compounds from agricultural and industrial sources. Here we discuss the general metabolic characteristics of those chemoorganotrophs that carry out respiration in the presence of oxygen.

With glucose as an example substrate, the overall reaction for its oxidation to carbon dioxide under aerobic conditions via glycolysis and the TCA cycle can be given as:

$$C_6H_{12}O_6 + 6O_2 \rightarrow 6CO_2 + 6H_2O \qquad (\Delta G^{0\prime} = -2870 \text{ kJ mol}^{-1} \text{ glucose}) \qquad [9]$$

The highly negative change in free energy for this reaction indicates a high energy yield associated with the aerobic respiration of glucose. This high potential energy yield is clearly evident in Figure 10–13, which emphasizes the energy yielding reactions shown in Figures 10–9, 10–10, and 10–12. Assuming that all of the energy contained in a molecule of glucose is allocated to ATP production, the effective energy yield from aerobic respiration can be summarized as shown in Table 10–4. Clearly, the theoretical production maximum of 38 moles of ATP per mole of glucose oxidized is far in excess of the 2 moles of ATP produced by the major fermentative pathways discussed earlier.

Anaerobic Respiration of Organic Compounds

Some microorganisms, almost all of them bacteria, can substitute certain other oxidized inorganic compounds for oxygen as a terminal electron acceptor in a process known as **anaerobic respiration.** The most common alternative electron acceptors are nitrate, sulfate, and carbon dioxide. All of these substances accept electrons less readily than oxygen, because they have less positive reduction potentials (Table 10–2). Thus the resulting energy yields are correspondingly reduced relative to aerobic respiration. Nevertheless, anaerobic respiration is more efficient than fermentation as a mode of metabolism and may confer a significant ecological advantage to those microorganisms possessing this ability, provided the proper electron acceptors are available.

Bacteria that can use nitrate as a terminal electron acceptor are generally facultative anaerobes, meaning that they carry out normal aerobic respiration in the presence of oxygen but can switch to nitrate in its absence. Oxygen is the preferred acceptor for these organisms, and its presence strongly inhibits nitrate reduction. Two types of dissimilatory nitrate reduction are common among soil microorganisms.

- **Denitrification** is the reduction of nitrate to gaseous products, primarily nitrous oxide and nitrogen (Chapter 12). This process returns biologically available or "fixed" nitrogen to the atmosphere and is effectively the opposite of dinitrogen fixation (Chapter 13). Denitrification is extremely important because, depending on one's perspective, it alternatively represents a loss of soil fertility for plant growth, a means of reducing nitrate contamination of ground and surface waters, and a source of gaseous pollutants that contribute to the destruction of atmospheric ozone.

- Various other bacteria carry out a less complete reduction to produce nitrite in a process called **nitrate respiration** (Chapter 12). This process is less efficient than denitrification in that fewer electrons are accepted per nitrate molecule. It may also lead to the accumulation of nitrite in the soil environment, which can be toxic to plants and microorganisms.

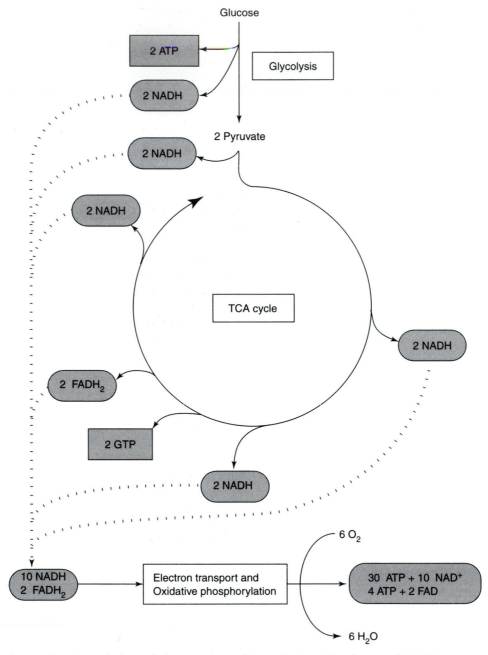

Figure 10–13 Catabolism of glucose via aerobic respiration. Note how each FADH$_2$ produces only 2 ATP while NADH produces 3 ATP molecules.

Table 10–4 Energy yield (expressed as ATP) for the aerobic respiration of glucose.

	Reducing equivalents	ATP yield
Glycolysis		
Substrate-level phosphorylation	not applicable	2 ATP
Oxidative phosphorylation	2 NADH	6 ATP
Tricarboxylic acid cycle		
Substrate-level phosphorylation (GTP)	not applicable	2 ATP
Oxidative phosphorylation	8 NADH	24 ATP
	2 FADH$_2$	4 ATP
Total energy yield		38 ATP

A rather diverse group of bacteria can use sulfate and other oxidized forms of sulfur as terminal electron acceptors in respiration (Chapter 15). Unlike the nitrate reducers, the **sulfate reducers** are generally obligate anaerobes, also called *strict anaerobes,* that are only active in the absence of oxygen. The final product of microbial sulfate reduction is hydrogen sulfide, which is partly responsible for the disagreeable "rotten egg" or "sulfur" smell associated with waterlogged soils.

The utilization of carbon dioxide as a terminal electron acceptor to produce methane (CH_4) is called **methanogenesis.** Methanogenesis is of environmental concern because methane is one of several gases implicated in global warming. The methanogenic bacteria belong to the Archaea, which are noted for their unique cell wall composition, distinctive ribosomal characteristics, and certain other attributes. All of the methanogenic bacteria are strict anaerobes that can use hydrogen as a source of energy and reducing equivalents (chemolithotrophy) and carbon dioxide as a source of carbon (autotrophy).

Respiration of Inorganic Compounds

A restricted but extremely important group of bacteria can obtain energy and reducing equivalents necessary for oxidative phosphorylation from reduced inorganic compounds. These bacteria are called **chemolithotrophs** according to our previous discussion of metabolic groupings of organisms (Table 10–2). However, because these organisms generally also obtain carbon from carbon dioxide, they are often called **chemoautotrophs.** A number of reduced inorganic substances can be used as energy sources by this group, including hydrogen gas, methane, ferrous (divalent) iron, and certain reduced nitrogen and sulfur compounds. Most chemolithotrophs are aerobes, but a few can use alternative electron acceptors such as nitrate and sulfate.

A given microorganism generally can oxidize only a single compound, and its common name often reflects this ability. For example, particularly important in normal agricultural soils are the aerobic nitrifying bacteria, which derive energy from the oxidation of ammonium to nitrate (Chapter 12). In reality, these bacteria are composed of two groups: the ammonium oxidizers, which convert ammonium to nitrite, and the nitrite oxidizers, which complete the conversion to nitrate.

Table 10–5 Standard free energy of reactions utilized by chemolithotrophic microorganisms.

Reaction	$\Delta G^{0\prime}$ (kJ mol^{-1})
$HS^- + H^+ + \frac{1}{2}O_2 \rightarrow S + H_2O$	−209
$S + 1\frac{1}{2}O_2 + H_2O \rightarrow SO_4^{2-} + 2H^+$	−587
$NH_4^+ + 1\frac{1}{2}O_2 \rightarrow NO_2^- + 2H^+ + H_2O$	−275
$NO_2^- + \frac{1}{2}O_2 \rightarrow NO_3^-$	−76
$H_2 + \frac{1}{2}O_2 \rightarrow H_2O$	−237
$2Fe^{2+} + 2H^+ + \frac{1}{2}O_2 \rightarrow 2Fe^{3+} + H_2O$	−31

Values for Δ^{0t} are at pH = 7 except Fe^{2+}/Fe^{3+}, which is at pH = 2 to 3.

$$NH_4^+ + 1\frac{1}{2}O_2 \rightarrow NO_2^- + 2H^+ + H_2O \quad (\Delta G^{0\prime} = -275 \text{ kJ mol}^{-1}) \quad \textbf{[10]}$$

$$NO_2^- + \frac{1}{2}O_2 \rightarrow NO_3^- \quad (\Delta G^{0\prime} = -76 \text{ kJ mol}^{-1}) \quad \textbf{[11]}$$

The nitrifying bacteria are important both agriculturally and environmentally. Nutritionally, many crop plants prefer nitrate over ammonium as a nitrogen source. However, by converting nitrogen from a cation to an anion, nitrifiers promote nitrogen contamination of ground and surface waters because the anion exchange capacity of soil is usually low and anions are relatively free to leach through the soil profile with percolating water. Additionally, the transformation of ammonium to nitrate completes the nutrient cycling link between the opposing processes of nitrogen fixation and denitrification.

Other important groups of chemolithotrophic soil microorganisms are the *sulfur-oxidizing bacteria, iron-oxidizing bacteria, hydrogen-oxidizing bacteria,* and *methane-oxidizing bacteria,* or methanotrophs. Values for the change in free energy associated with some of the reactions carried out by chemolithotrophs are given in Table 10–5. Note that these reactions yield significantly less energy for cellular processes than the aerobic respiration of glucose, for example. Balanced against these low energy yields, however, is the ability of chemolithotrophs to use energy sources that are unavailable to other microorganisms, thereby reducing competitive constraints on their growth.

Phototrophy

The metabolic process of obtaining chemical energy from light is known as **phototrophy** (Fig. 10–3, Table 10–2, and Chapter 5). Nearly all phototrophs are autotrophic and use energy captured from light to fix carbon dioxide by means of photosynthesis. Similarly, most phototrophs are lithotrophic because they obtain reducing equivalents from water or another inorganic substance. Photosynthesis is conveniently divided into its energy-gathering *light reactions* and the subsequent *dark reactions* that are involved in carbon fixation. Only the light reactions will be discussed here; the dark reactions are covered later in this chapter under carbon nutrition.

The light reactions of photosynthesis (Fig. 10–14) require specialized light-gathering pigments called **chlorophylls.** Among microorganisms, chlorophylls are found in algae and certain bacteria. Several different types of chlorophylls exist but all are planer molecules that contain a magnesium-bearing core. A large number of

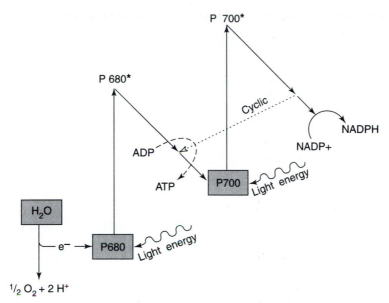

Figure 10–14 Light reactions of photosynthesis. Asterisks show the high energy forms of chlorophyll.

chlorophyll molecules function together to absorb light energy over their combined surface area and then funnel the absorbed energy to a reaction-center chlorophyll. Within this latter chlorophyll molecule, the captured light energy causes electrons to be elevated in energy and ejected from the pigment in a process called *photooxidation*. These ejected electrons then reduce specific acceptors to form high-energy compounds. The high-energy electrons within these acceptor compounds then return to a lower energy state by traveling through an electron-transport system, thereby producing ATP by **photophosphorylation** in a manner analogous to that for oxidative phosphorylation during respiration.

Depending on the organism involved, electrons involved in photophosphorylation may simply reduce the original chlorophyll molecule (i.e., return it to its original state) and simultaneously produce ATP in a process known as *cyclic photophosphorylation*. This set of reactions involves a reaction-center chlorophyll molecule designated P700 to indicate that its maximum light absorption occurs at a wavelength of 700 nm; P700 and its associated chlorophylls and electron-transport components are referred to as *Photosystem I*. Alternatively, many phototrophs carry out *noncyclic photophosphorylation* in which electrons from the initial photooxidation do not cycle back to P700 but rather reduce NADP to NADPH. In this case, the electrons needed to reduce P700 arise from the light-mediated oxidation of water to oxygen occurring at a second reaction-center chlorophyll designated P680, which has an absorption maximum at 680 nm (*Photosystem II*). As in the case of cyclic photophosphorylation, ATP is also produced. Thus, compared with cyclic photophosphorylation, noncyclic photophosphorylation additionally produces NADPH and molecular oxygen. Noncyclic photophosphorylation is associated with green plants, algae, and cyanobacteria and is sometimes referred to as *oxygenic photophosphorylation.* Conversely, because cyclic photophosphorylation does not produce oxygen, it is often called *anoxygenic photophosphorylation;* it is limited to the green and purple bacteria.

Carbon Nutrition

One remaining major source of diversity among microorganisms is how they obtain carbon. As mentioned previously, certain organisms called autotrophs have the capacity to transform carbon dioxide or other inorganic carbon compounds, such as methane, into organic molecules in a process known as **carbon fixation.** Autotrophic organisms are ultimately responsible for fixing essentially all of the carbon used by heterotrophs, those organisms which require preformed organic compounds for growth. Carbon nutrition is a fundamental attribute of microbial metabolism because it determines the degree to which a microorganism relies on other organisms to fix carbon needed for biosynthetic processes. Carbon is the major element comprising organisms on a dry mass basis—for example, bacteria are approximately 50% carbon—and its acquisition is correspondingly critical to cellular processes.

Autotrophs are defined as organisms that obtain essentially all of their cellular carbon from carbon dioxide. Implicit in this definition is the realization that, under certain circumstances, otherwise autotrophic organisms may obtain some carbon from organic sources. Autotrophy occurs in green plants, algae, and certain bacteria. Carbon fixation is an endergonic process and, therefore, requires energy captured by the cell by means of exergonic reactions. Autotrophic organisms obtain this energy by either phototrophy or chemolithotrophy. The pathway utilized by nearly all autotrophs to fix carbon is the **Calvin cycle** in which carbon dioxide is converted to a C_3 compound (glyceraldehyde 3-phosphate) with the expenditure of 9 ATP and 6 NADPH (Fig. 10–15). As noted previously, glyceraldehyde 3-phosphate is one of the 3-carbon products of glycolysis. This newly synthesized glyceraldehyde 3-phosphate is a potential precursor in the biosynthesis of needed cellular constituents. In the case of photosynthetic organisms, the Calvin cycle is often referred to as the "dark reactions" because it can proceed independent of light as long as sufficient captured light energy is available in the form of ATP and NADPH. Assuming carbon fixation is occurring in concert with noncyclic photophosphorylation, the production of glucose from carbon dioxide can be summarized as:

$$6\ CO_2 + 6\ H_2O \rightarrow C_6H_{12}O_6 + 6\ O_2 \qquad (\Delta G^{0\prime} = 478\ kJ\ mol^{-1}\ CO_2) \qquad \textbf{[12]}$$

This reaction is the opposite of equation [9].

Heterotrophs, which include animals, protozoa, fungi, and the majority of bacteria, obtain essentially all of their carbon from preformed organic compounds. Although nearly all heterotrophs are chemoorganotrophs, some bacteria combine heterotrophy with either phototrophy or chemolithotrophy. Whereas autotrophs obtain biosynthetic intermediates by means of anabolic reactions, heterotrophs generally obtain these compounds via catabolic processes. Similarly, while autotrophs must produce and maintain the enzymes necessary

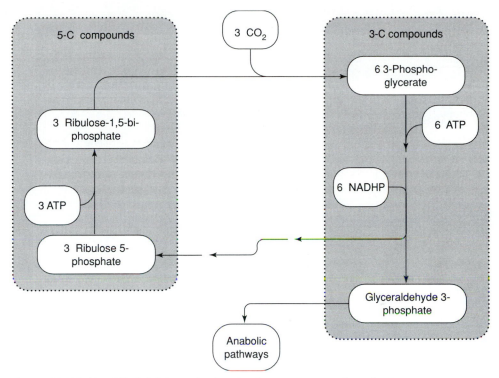

Figure 10–15 Simplified Calvin cycle showing key reactants and products.

for carbon fixation, heterotrophs often have evolved the capacity to enzymatically degrade large or complex organic compounds that are inaccessible to autotrophs. Thus, the two groups of organisms may be thought of as having the same metabolic goal (biosynthetic intermediates) but different means of obtaining this goal.

Integration of Metabolic Pathways

Although it is useful for purposes of discussion to introduce the various metabolic processes carried out by microorganisms as separate entities, metabolic pathways are actually highly integrated and interdependent. In particular, many of the compounds comprising glycolysis or the TCA cycle play an important role in both catabolic and anabolic reactions in cells. Central pathways such as these that serve in both energy-yielding and biosynthetic roles are called *amphibolic pathways*. Figure 10–16 summarizes some of the general pathways by which glycolysis and the TCA cycle are linked to various other pathways, both catabolic and anabolic, in many microorganisms. Such integration serves to increase metabolic efficiency of cells by allowing them to allocate most of their enzymatic resources to a limited number of central pathways.

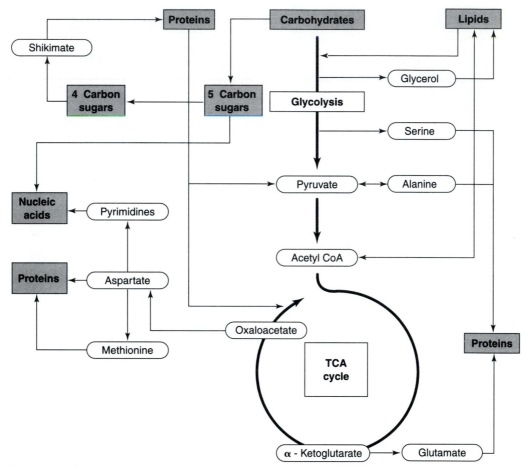

Figure 10–16 Some of the major relationships of glycolysis and the TCA cycle to biosynthesis and catabolism of various cellular constituents.

Summary

The different strategies by which soil microorganisms obtain carbon and energy are impressively diverse and innovative. To many soil microbiologists (this author included) this range of metabolic types and abilities, and the corresponding complexity of the interactions among the populations, is in large part the fascination of the discipline. It is hardly an overstatement to say that essentially all forms of metabolism and life strategies can be found among the microorganisms that inhabit soil environments. Hopefully, it is also evident that underlying this diversity is the unifying principle that life processes require that organisms acquire energy and nutrients to meet their needs of growth, survival, and reproduction. This marriage of metabolic diversity and uniformity of purpose makes the study of soil microorganisms a formidable but stimulating challenge.

Cited References

Horwath, W.R., and E.A. Paul. 1994. Microbial biomass. pp. 753–773. *In* R.W. Weaver, S. Angle, P. Bottomley, D. Bezdicek, S. Smith, A. Tabatabai, and A. Wollum (eds.), Methods of soil analysis, Part 2. Microbiological and biochemical properties. Soil Science Society of America Book Series, No. 5. Madison, Wis.

Singleton, P., and D. Sainsbury. 1994. Dictionary of microbiology and molecular biology. 2nd ed. Wiley, New York.

General References

Hamilton, W.A. 1988. Microbial energetics and metabolism. pp. 75–100. *In* J.M. Lynch and J.E. Hobbie (eds.), Micro-organisms in action: Concepts and applications in microbial ecology. Blackwell Scientific Publishing, Oxford, England.

Lehninger, A.L., D.L. Nelson, and M.M. Cox. 1993. Principles of biochemistry. 2nd ed. Worth Publishing, New York.

Madigan, M.T., J.M. Martinko, and J. Parker. 1997. Brock biology of microorganisms. 8th ed. Prentice Hall, Upper Saddle River, N.J.

Prescott, L.M., J.P. Harley, and D.A. Klein. 1993. Microbiology. 2nd ed. Wm. C. Brown, Dubuque, Iowa.

Voet, D., and J.G. Voet. 1995. Biochemistry. 2nd ed. John Wiley & Sons, New York.

Study Questions

1. In Chapter 2, Figure 2–4 is a drawing of a microaggregate. The interior of the aggregate has an E_h of -400 mV; the exterior has an E_h of $+100$ mV. What changes in bacterial metabolism would you expect between the interior and exterior of the aggregate?

2. For each of the following types of microorganisms, specify which kind of culture medium and environmental conditions would be needed to support growth: (a) nitrifying bacteria, (b) cyanobacteria, (c) actinomycetes, (d) methanogens.

3. Determine the oxidation state of sulfur in each of the following compounds. Which among those listed is the richest potential source of energy for chemolithotrophic bacteria? S (elemental), $S_2O_4^{2-}$, SO_4^{2-}, H_2S, SO_3^{2-}, $S_2O_6^{2-}$.

4. Many sources of groundwater are aerobic despite being both water saturated and colonized by large numbers of heterotrophic bacteria. Some also contain high concentrations of nitrate, which may be of environmental and health concern. One means suggested to remediate such aquifers is to encourage nitrate removal by denitrification. What conditions must exist for denitrification to occur? How might these requirements be met in the field? Briefly describe the sequence of events that would lead to denitrification.

5. Explain why it is necessary that certain degradative enzymes be produced extracellularly (e.g., those involved in starch and cellulose degradation).

Chapter 11

♦

Carbon Transformations and Soil Organic Matter Formation

George H. Wagner and Duane C. Wolf

♦

"Die Zertrümmerung der organischen Reste ist die für den Natur-Haushalt wichtigste Leistung der Mikroorganismen." (translation: The disintegration of organic residues in nature is the most important function of microorganisms).

F. Löhnis

Concern about high levels of carbon dioxide in the earth's atmosphere constitutes the public's pivotal interest in the global carbon cycle. For soil scientists, other pools and fluxes of carbon are of equally high interest. Particularly important is the balance between autotrophic fixation, primarily by green plants, which brings carbon from the atmosphere into the biosphere, and respiration, which releases carbon dioxide back to the atmosphere. Much of the respiratory activity on earth occurs in the soil. Soil respiration represents decomposition of organic residues, root respiration, and slow decay of **soil organic matter (SOM).** Currently soil scientists are refocusing considerable effort toward quantifying soil respiration. Beginning in the eighteenth century, land clearing with accompanying cultivation accelerated soil respiration and was the largest contributor to carbon dioxide buildup in the atmosphere from that time until 1950 (Schlesinger, 1986; Post et al., 1990). We now recognize that accelerated soil respiration occurs during the initial 50+ years of cultivation of virgin soils; during that period, topsoils may lose 30 to 50% of their organic carbon. Nevertheless, SOM in mineral soils remains a major storehouse or sink of global carbon, containing about twice that in the atmosphere (Table 11–1). Thus, any change in this pool can significantly influence the global carbon dioxide level.

Each of the two major biological fluxes of carbon dioxide in nature, namely photosynthetic fixation and global respiration, transfers about 7% of atmospheric carbon annually, but in opposite directions (Bolin, 1983). Put another way, 15 years of photo-

Table 11–1 **Soil carbon pools in relation to atmospheric carbon pools and annual fluxes.**

	Carbon (10^{15} g)*
Pool	
Soil organic matter	1,200 to 1,600
Living biomass	420 to 830
CO_2-C in atmosphere	740
Annual flux	
Net primary production	40 to 80
(net photosynthesis – plant respiration)	
Terrestrial soil respiration	50 to 70
(release of CO_2-C to atmosphere)	

*All values are 1980 estimates.
Adapted from Post et al (1990) Used with permission.

synthetic fixation without renewal by respiration would exhaust the atmosphere of its carbon dioxide. Some quantitative relationships of the global carbon cycle are shown in Table 11–1. This global perspective provides a framework for understanding the cycling of carbon in smaller, more easily defined ecosystems. This chapter focuses on carbon cycling within discrete ecosystems with a major focus on agricultural ecosystems.

In an agricultural ecosystem, crop residues constitute the primary carbon source that furnishes substrate for the soil biota and gives rise to the respiratory release of carbon dioxide into the local atmosphere. Decomposition of organic residues is a major function of the vast soil microbial population. These residues, the bulk of which are of plant origin, constitute a reservoir of energy stored in carbon compounds. Heterotrophic soil microorganisms need the energy in carbon sources derived through photosynthesis of green plants to grow, multiply, and survive.

Significance of Microbes to Carbon Cycling

A basic activity of soil microbes, like that of other life forms, is survival through reproduction. Soil microbes use residue components as substrates for energy and also as carbon sources in the synthesis of new cells. Energy is furnished to the microbial cells through the oxidation of the organic compounds. A major end product is carbon dioxide, which is released back into the atmosphere.

| Box 11–1 |

Definitions of related terms.

- **Decomposition**—chemical breakdown of a compound into simpler compounds, often accomplished by microbial metabolism.

- **Mineralization**—Conversion of an organic form of an element to an inorganic form as a result of microbial decomposition.

- **Respiration**—catabolic reactions producing ATP in which either organic or inorganic compounds are primary electron donors and exogenous compounds are the ultimate electron acceptors.

Overall, decomposition (also referred to as microbial respiration or mineralization, Box 11–1) can be viewed as one part of the carbon cycle (Fig. 11–1). From the atmospheric pool of carbon dioxide, green plants and other autotrophs (e.g., photosynthetic and chemoautotrophic bacteria) fix carbon into various organic compounds. This annual fixation is balanced by heterotrophic decomposition carried out primarily by microorganisms.

Microorganisms are generally short-lived and may themselves be decomposed by successive populations that find the dying cells to be more suitable substrates than the initial residues. The overall process of decomposition generally involves a broad spectrum of complementary microbes that act in concert on a substrate. They differ in the kinds of enzymes they produce to disassemble the array of organic compounds present. By their combined efforts, carbon present as various organic compounds is progressively transformed to its most oxidized form, that of carbon dioxide.

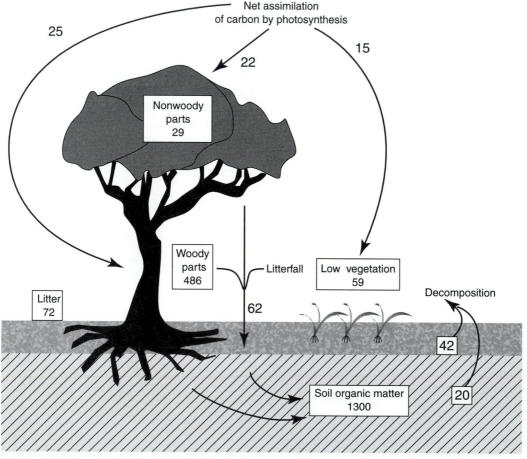

Figure 11–1 Terrestrial carbon cycle showing fluxes and reservoirs of the various components. Photosynthesis results in net assimilation of atmospheric CO_2-C which is balanced by the decomposition of plant residue and litter, and soil organic matter. The sizes of the compartments are indicated in units of 10^{15} g. *Adapted from Post et al. (1990). Used with permission.*

It is important to realize that the flow of carbon through the soil depends on the efficiency with which the microbes use the residues as substrates for growth. Later we shall examine the decay process in field situations on an annual basis to obtain a broader perspective of the flow of carbon from residue incorporation to its release as carbon dioxide.

One of the more stable transformation products that forms in the course of residue decay is SOM, also referred to as humus. Soil organic matter has long been recognized as a valuable constituent because it imparts many desirable biological, chemical, and physical properties to soil. The nature of organic residues and the subsequent microbial decomposition of their components results in the SOM that is usually related to enhanced **soil quality** (Doran et al., 1996).

Nature of Organic Materials Added to Soil

The first step in our discussion of carbon transformations in soil is to focus on the nature of organic materials added to soil each year, beginning with the kinds, amounts, and composition of plant residues from cultivated crops or native vegetation.

Plant Residues

In agricultural ecosystems, the residues that decompose in soil are generally derived from crop plants and consist primarily of leaf, stem, and root tissues that remain after harvest. The residues may comprise an annual production of several Mg ha^{-1} that, with time, become fully incorporated into the soil. Because of their location in the soil, roots are very susceptible to decay. Even during growth, roots supply exudates and sloughed-off cell materials that serve as carbon substrates for the microbial population in the same manner as post-harvest root residues.

In natural ecosystems, the residues that undergo decomposition are derived primarily from native prairie and forest vegetation. In the native prairie, the above-ground plant debris accumulates year after year on the soil surface. In addition, considerable below-ground residue is deposited each year from root systems that die back periodically during periods of cold- or drought-induced dormancy. In the forest system, leaf litter and other plant debris fall to the soil surface and accumulate as layers in which decay occurs.

Incorporation of residues. Decay of above-ground residues begins and progresses in a manner characteristic to the particular agricultural management system. Conventional tillage operations, such as plowing and disking, incorporate straw and stalks within the soil where contact with moisture and microbes hastens decay. Minimum and no-till operations result in residues on the soil surface, and the decay process progresses in a manner similar to that for natural ecosystems. At the interface between residue litter and moist soil, the microbial population becomes very active. The decay progresses from beneath this litter layer, and in moist environments, the bulk of the residue may be actively under attack by soil microbes. During the decay process, some of the litter may be transported beneath the soil surface by the soil fauna, where active decay occurs as it becomes mixed into the moist upper mineral horizon of the soil.

Table 11–2 **Annual production (g m^{-2})of residue biomass for three agronomic crops.**

	Corn	Soybean	Wheat
Stems and leaves	1,262	400	590
Cobs or pod hulls	158	136	—
Roots	1,304	478	521
Total	2,724	1,014	1,111

From Buyanovsky and Wagner (1986). Used with permission.

Variations in moisture, temperature, nutrient supply, and pH modify the rate of decay. Limitations imposed by these influences may cause greater fluctuations in the rate of decay for surface residues than for residues residing in the bulk soil.

Residue quantities. Major crops furnishing residues, such as corn, wheat, and soybean, are widely grown, and considerable knowledge about the decay of their residues exists. The quantities of residues produced each year vary for different regions as modified by soil, climate, and agricultural management. Here, we will limit our consideration to mean annual values that are representative of subhumid agriculture in the central United States. The values illustrate the quantitative relationships involved in the decomposition process and serve as reference points for the study of specific cases in other regions and materials.

Residue yields for some Missouri crops, fairly representative of central U.S. agriculture, are given in Table 11–2. For wheat and soybean, the annual residues approximate 1,000 g m^{-2}, of which the root contribution is no more than one-half (note that g m^{-2} × 10 = kg ha^{-1}). This amount of net annual production is similar to that for native prairie in this region. Corn production is considerably greater, with a mean annual value of about 2,700 g m^{-2}. The more productive forests of the region would show an annual litter drop similar to this value. These residues, in combination with root exudates, supply the total annual substrate input for the soil organisms to a depth of approximately 20 cm. The corresponding microbial biomass is around 150 g m^{-2} (1,500 kg ha^{-1}), and the SOM content ranges from 2 to 2.5%. A portion of the substrate carbon is used by the microbial community for reproduction, but a larger share is liberated as carbon dioxide derived from energy-yielding metabolic processes. Soil organisms work to decompose the residue substrates at a rate that generally maintains an equilibrium level of SOM year after year for a given crop production system.

Residue composition effects. The crop residues added to and decaying in soil have various compositions and are comprised of complex polymers such as **cellulose** and **lignin.** Representative compositions of above-ground residues of corn, soybean, and wheat are shown in Table 11–3. Composition influences decay because residues do not decompose as whole units. Rather, the various groups of organic compounds comprising plant tissues are selectively attacked by a range of soil microbes, each of which produces a particular suite of degradative enzymes active on the insoluble polymers. Generally speaking, the more resistant a particular compound is to degradation, the lower the frequency of the corresponding enzyme within the population.

The general processes involved in decomposition of residues may occur under aerobic or anaerobic conditions (Fig. 11–2). For most soils, the aerobic pathway is of greatest importance. Residue decay progresses through the interrelated activities of many different microorganisms. Simple substrates that are readily assimilated may have a community of microorganisms feeding in competition with one another as the material comes under attack. More complex and resistant substrates, such as carbohydrate polymers, may initially attract one group of microbes that breaks down the polymer into simpler components. This is followed by activity of other groups of microorganisms that can use the simpler components. In nearly all cases, the final step is the assimilation of decay products by diverse microbes that oxidize the compounds to obtain energy and carbon for the production of new cell tissue. A generalized summary of the microbial decomposition process under

Table 11–3 Representative compositions of above-ground corn, soybean, and wheat residues (g kg^{-1} dry mass).

Component	Corn	Soybean	Wheat
Soluble components	293	557	288
Hemicellulose	268	90	184
Cellulose	284	222	361
Lignin	56	119	141
Ash*	93	64	84
Nitrogen	10	22	9

*Substance remaining after ignition of the plant material.
From Broder and Wagner (1988). Used with permission.

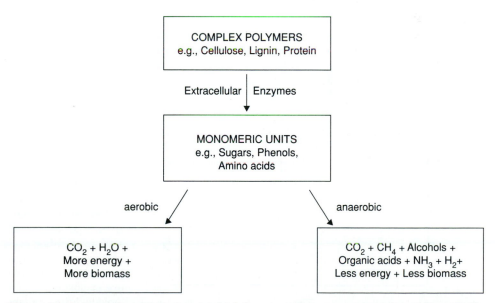

Figure 11–2 General reactions in microbial decomposition processes. *Adapted from Wolf and Legg (1984). Used with permission.*

aerobic conditions is presented in Figure 11–3. This figure shows a progressive decrease in substrate accompanied by an increase in evolved carbon dioxide along with a temporary increase in microbial numbers.

Although agricultural residues consist primarily of various complex polymers, they also contain small amounts of fairly simple, often soluble, organic constituents that can be utilized by a vast array of soil microbes. In favorable environments, these simple constituents may be completely converted to new cell structures and carbon dioxide within a few days. For example, studies with a common simple sugar like glucose show that this substrate is completely metabolized by soil microorganisms in one or two days. Free amino acids also are readily utilized by soil microbes. In a similar manner, other microbes readily decompose many of the cytoplasmic constituents of lysed and dead cells. Other microbial cell components can be very resistant to decay. An

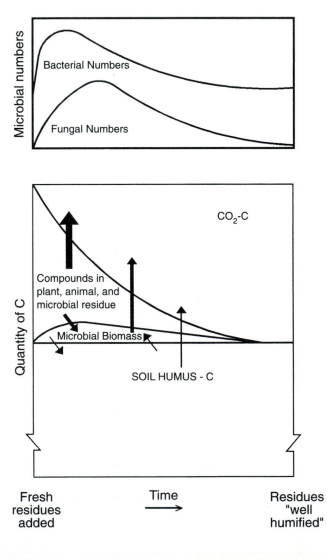

Figure 11–3 General reactions in microbial decomposition of plant, animal, and microbial residues. When residue is added to the soil and undergoes decomposition, the amount of original residue decreases with time. The microbial biomass increases as does the carbon evolved as carbon dioxide. Note the initial increase in microbial numbers and then the steady decline. As the soil returns to a steady state condition, the microbial populations and soil organic matter return to approximately the same levels prior to adding residues. Width of the arrow indicates the relative amount of carbon exchanged between carbon pools. Long upward arrows indicate relative magnitudes of carbon dioxide fluxes from the soil. *Adapted from Brady and Weil (1996). Used with permission.*

example is melanized (darkly pigmented) fungal cell walls whose resistance leads to their eventual appearance in SOM. By contrast, hyaline (nonpigmented) fungal cell wall tissues, such as that from the fungus *Aspergillus niger,* consist predominantly of carbohydrate polymers, which are easily decomposed.

As the decay of residue progresses, the more resistant components tend to accumulate and reactive aromatic compounds are generated, some by microbial modification of decomposing plant constituents and others from synthesis by microorganisms. The reactive aromatics, such as phenolic compounds, enter into condensation reactions to form new polymeric materials, some of which are considerably more resistant than the original plant tissues. The resulting pool of complex carbonaceous materials is highly resistant to decomposition and constitutes the SOM. This process of **humification** of residues occurs concomitantly with microbial decay of residues. The residue carbon that enters the SOM pool may be only 10 to 20% of the carbon in the original residue.

The selective action of microbes in attacking the various components of the residue was demonstrated by Broder and Wagner (1988) who studied the decay of wheat straw buried in nylon bags in a field environment. Their experimental system allowed periodic removal of one or more samples to examine the progression of decay. The results shown in Figure 11–4 demonstrate the rapid rate at which at least a part of any crop residue can decompose. Of the various polymeric carbohydrates

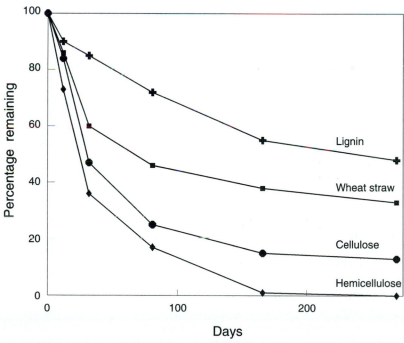

Figure 11–4 Decomposition of wheat straw and its major constituents in a silt loam with the residue buried 12 June in nylon bags to facilitate periodic removal and analysis. The initial composition of wheat straw was 50% cellulose, 25% hemicellulose, and 20% lignin. *Adapted from Broder (1985). Used with permission.*

(polysaccharides), hemicelluloses were degraded with relative ease by numerous soil microorganisms, and as a result, this component was essentially exhausted within 160 days. Cellulose, the most abundant polysaccharide, also decomposed more rapidly than the residue as a whole, and approximately 20% of this component remained at the end of the study. Lignin decayed markedly slower than the other wheat straw components; this stability is a reflection of its polymerized aromatic character.

Although this study did not examine degradation of root residues, other studies have shown that wheat roots decompose more rapidly than straw. This most likely occurs because the roots are already in intimate contact with an active soil population, and their decomposition may begin quickly once plant senescence begins. An active heterotrophic population surrounding the roots would have already developed and been nurtured by simple organic compounds exuded by the roots and other cellular debris that is sloughed off as the plant root grows. It should be reiterated that residues derived from above-ground crop tissues, such as straw and stubble, may begin to decompose at the soil surface if soil moisture is adequate, but their decay will accelerate following soil incorporation by tillage or animal activities.

Seasonal influences. The annual cycle of climate greatly influences the progression of residue decay for various crops, because microbial activity may be restricted by cold temperatures in winter or by dry conditions in summer. In the case of wheat in the central United States, about 50% of the original residue mass carries forward through winter, and the warming that occurs in spring marks the beginning of a second phase of decay. Corn and soybean crops, which mature later in the fall, may have a greater proportion of the total residue carried into the next spring, at which time a flush of decomposition activity occurs. This carryover for soybean is largely stem residues, because leaf materials drop to the ground in the summer as the plant matures and degrade almost completely by the end of the fall season.

The overall process of decomposition in soil, therefore, should be looked upon as an annual cycle. New residues are produced each year, a large percentage of which are completely mineralized to carbon dioxide during that year. However, residues generally persist into the second year so that most materials have a residence time of two or more years. Some of the residue carbon is eventually incorporated into SOM, which, due to its highly resistant nature, can persist in the soil for hundreds of years. Nevertheless, even this highly resistant humic material is slowly mineralized at rates of 2 to 5% yr^{-1} by enzymes produced by some members of the heterotrophic population.

Temperature and soil moisture strongly modify residue decay within the framework of the seasonal pattern described above. Wheat residues decay rapidly in late summer and fall when soil temperatures in the central United States approach 24°C, with soil moisture at a satisfactory level at least for some part of that period. The rate of microbial response to increasing temperature is greatest over the range of 3 to 19°C, but more modest responses occur if temperature increases from 20 to 25°C.

The soil water content for maximum decomposition is about 60% of the water holding capacity of the soil or −20 to −50 kPa moisture potential. This level helps to ensure good aeration for oxidative processes and provides sufficient moisture for all groups of heterotrophic microbes that are most active in the decomposition process.

Microorganisms

In addition to plant residues, the cells of dead soil microbes are also substrates for living microorganisms. Fungal cell walls composed of cellulose, chitin, and chitosan can be degraded. Bacterial cell walls containing N-acetylglucosamine and N-acetylmuramic acid in peptidoglycans, along with other polysaccharide materials, are also substrates. Representative decomposition curves for a number of different microbial cells are presented in Figure 11–5.

Biosolids and Animal Manure

Land application of biosolids (previously called sewage sludge) and animal manure has increased substantially in the past 25 years. Land application allows for efficient disposal of the waste material and recycling of nutrients. As a result of the magnitude of waste disposal on and in soil, waste materials have become important substrates for soil microbes in certain soil systems.

Environmental concerns have led to a reduction in ocean dumping and incineration of biosolids in recent years. Therefore, land application is becoming more common for disposal and recycling of nutrients. As with manures, biosolids and industrial and municipal wastewaters contain organic materials similar to those in plant residue; thus decomposition follows the same general pathways. However, the presence of

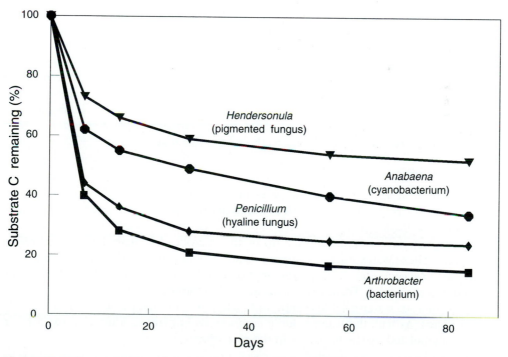

Figure 11–5 Decomposition of representative microbial cells added to soil and incubated under aerobic conditions. The percentage substrate carbon remaining in the soil does not consider formation of microbial biomass or humus. *Adapted from Burns and Martin (1986). Used with permission.*

excessive levels of toxic materials, such as heavy metals and xenobiotics, can inhibit microbial decomposition of the organic carbon substrates (Chapters 16 and 20). Pathogenic viruses, bacteria, and protozoa may also represent a public health hazard if appropriate regulations are not followed in land application programs.

Box 11–2

Animal Manure. Application of animal manure to soil has been an accepted method of recycling nutrients since the beginning of agriculture. However, today's confined animal feeding operations have resulted in extensive generation of animal manure in concentrated areas. Thus, the manure is often applied to soil at rates that provide a substantial amount of organic material for the soil microbes to decompose. The primary concern in most manure application studies has been the potential excess nitrogen applied and the subsequent contamination of surface and groundwater with nitrate. Survival and transport of microbial pathogens and pathogen indicators are also of concern. Recent interest also relates to the excessive amounts of phosphorus applied in biosolids and manures and the potential for phosphate in runoff to contaminate surface water and result in eutrophication. The organic carbon materials in manure are like plant residue in nature and are thus similar in decomposition characteristics.

Decomposition of Naturally Occurring Organic Materials

Many of the naturally occurring compounds degraded by soil microorganisms are similar in structure and modes of decomposition to man-made compounds discussed in Chapters 20 and 21. Specific microbial metabolic pathways are presented in Chapter 10. Here we use a generalized composition for plant residue to illustrate the relative importance of each component in the decomposition process. Residues with differing compositions exhibit differing decomposition rates. In soils amended with high rates of biosolids and animal waste, the importance of plant residues as microbial substrates may be secondary to the added material.

Soluble Components

Plant residue contains an appreciable amount of water-soluble organic compounds (Table 11–3), such as free amino acids, organic acids, and sugars that are readily available for microbial decomposition by the vast majority of soil microbes. These materials are rapidly taken up by the microbes for catabolic and anabolic activities. These water-soluble compounds are generally utilized by bacteria and "sugar fungi" (Zygomycetes such as *Mucor* spp. and *Rhizopus* spp.), all of which exhibit rapid growth rates. These organisms are called **zymogenous.**

Numerous water-soluble compounds also pass from the plant root into the soil as exudates. As with the water-soluble compounds in plant residue, root exudates are rapidly utilized by the microbes in the rhizosphere.

Proteins

Proteins are generally a small but significant component of plant residue added to soil. However, proteins can be a sizable portion of plant material used as a green

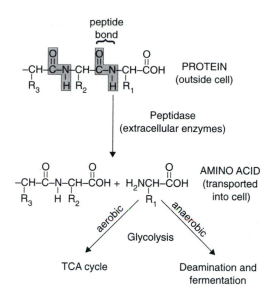

Figure 11–6 Degradation of protein by hydrolysis of the peptide bond.

manure. As polymers of amino acids linked by peptide bonds, they are readily decomposed in soil. A wide variety of microbes produce proteolytic enzymes (protease, peptidase) that hydrolyze proteins into individual amino acids. These products are then transported into the cell for further catabolism or for synthesis of new proteins required by the microbe (Fig. 11–6).

Structural Components of Plant Cell Walls

Cellulose. In most plant residues, the greatest amount of carbon is in the form of complex carbohydrates, such as the structural polysaccharides. **Structural polysaccharides** are so named because they are generally the constituents responsible for conferring structural rigidity to cell walls, thereby allowing plants to grow as erect structures. The most common structural polysaccharide in residues is cellulose, which is a linear chain of glucose units joined by β 1-4 linkages (Fig. 11–7). Each cellulose molecule may contain up to 10,000 glucose units. The cellulose content of plant material generally increases as the plant matures; it may be as low as 15% on a dry mass basis for young plants, but may be greater than 50% in wood, straw, and leaves. Cellulose is also an important component of some fungal and algal cell walls.

Because cellulose is a large polymer that is insoluble in the soil and too large to enter the microbial cell, it must first be cleaved by extracellular enzymes into smaller units that can be transported into the microbial cell for subsequent catabolism (Fig. 11–2 and Box 11–3). Cellulose decomposition is initiated by a diverse group of enzyme complexes known as cellulases. The cellulase enzyme complex generally contains three types of enzymes: a β 1,4-endoglucanase, a β 1,4-exoglucanase, and a β 1,4-glucosidase, also called cellobiase. The first step in cellulose mineralization involves the loss of crystalline-like structure, followed by depolymerization. The resulting linear chains of two or three glucose units are called cellobiose and cellotriose, respectively. These chains can enter the microbial cell. Before the carbon

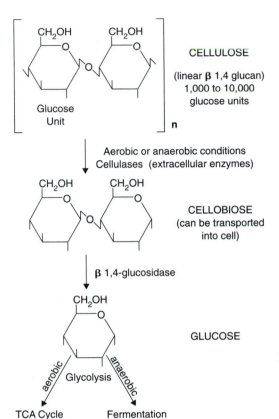

Figure 11–7 Decomposition of cellulose under aerobic and anaerobic soil conditions. Note that the lines attached to the rings are positions where hydroxyl groups (-OH) are attached.

can be metabolized by the cell for energy production, cellobiose and cellotriose must be hydrolyzed to single glucose units by the β 1,4-glucosidase. The individual glucose molecules can then be metabolized for energy and biomass production.

Soil fungi, such as *Trichoderma, Aspergillus, Penicillium,* and *Fusarium,* and bacteria, such as *Streptomyces, Pseudomonas,* and *Bacillus,* are important in the initial extracellular depolymerization of cellulose. Once cellulose is hydrolyzed into smaller units that can be transported into the cell, the entire glucose-metabolizing microbial population of the soil can participate in the decomposition process. Particularly noteworthy are the **"brown rot" fungi,** because they decompose the cellulose and hemicellulose in wood, leaving lignin and phenolic materials behind. Rotting wood under attack by these organisms takes on a brown color. In contrast, the **"white rot" fungi** can decompose all of the wood including the lignin, but a greater quantity of cellulose accumulates, leading to a white residue. Recent interest in the white rot fungi, notably *Phanerochaete chrysosporium,* has been stimulated by the observation that they may be important in bioremediation of soils contaminated with toxic organic chemicals (Chapters 20 and 21).

Hemicelluloses. Hemicelluloses are the second most common carbohydrate constituents in plant residues and consist of polymers containing hexoses (6-C sugars), pentoses (5-C sugars), and uronic acids. They constitute a diverse group of

Box 11–3

Extracellular Enzymes Determine the Rate of Decomposition of Naturally Occurring, Large-Molecular-Weight Materials. As is generally the case with microbial decomposition of polymeric materials, the rate-limiting process is the extracellular enzymatic steps required to convert the large molecular weight material into smaller units that can be transported into the microbial cell. To carry out the decomposition of complex molecules to usable substrates, extracellular enzymes, which operate outside the cell, are required. **Endoenzymes** operate along the internal portions of the polymer chain, while **exoenzymes** act at the end of the chain, cleaving off monomers and dimers and sometimes larger chain fragments. An appropriate glycosidase cleaves these into monomers that can be transported into the cell and used as carbon and energy sources. At least three theories aim to explain how extracellular enzymes are formed and function:

- The enzymes are synthesized inside the cell and excreted into the soil.

- Cells that are lysed, and thus lose membrane integrity, release their enzymes into soil, and the enzymes carry out reactions necessary for survival of the remaining viable cells.

- Enzymes are attached to the microbial cell wall, and the active site is free to function outside cells.

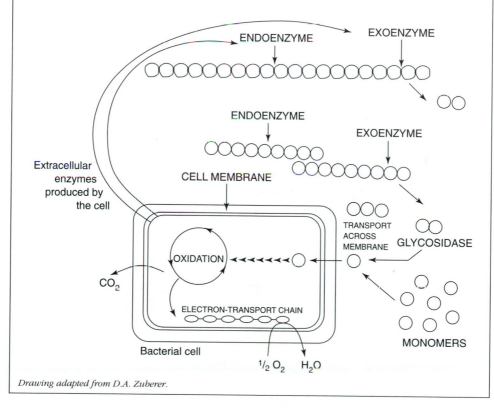

Drawing adapted from D.A. Zuberer.

structural polysaccharides that comprise up to 30% of the dry mass of plant residue. Most frequently, hemicelluloses in plant residues contain from 50 to 200 sugar units. The polymers are generally branched and contain more than one type of sugar or uronic acid unit as opposed to the linear arrangement of glucose, which is characteristic of cellulose. The presence of uronic acid indicates that the $-CH_2OH$ group of the 6-C sugar has been oxidized to $-COOH$.

Pectic substances, an important component of plant cell walls, are chains of galacturonic acid, which is often esterified with a methyl group (i.e., methylgalacturonic acid). They occur primarily in the middle lamella, where they help to "cement" the plant cell walls together. An example of pectin decomposition is shown in Figure 11–8. The decomposition process involves several enzymes often referred to collectively as pectinases. Many of the same fungi and bacteria that degrade cellulose are active in pectic substance decomposition. Plant root invasion by plant pathogens and symbionts, such as the N_2-fixing *Bradyrhizobium* and *Rhizobium* and mycorrhizal fungi such as *Glomus,* has been related to production of pectinases.

Decomposition of hemicellulose is generally rapid and exceeds the rate for cellulose (Fig. 11–4). Similarly, high levels of available simple sugars reduce hemicellulose decomposition, as the microbes will preferentially use more readily available substrates before attacking more complex materials. As with cellulose, the initial hy-

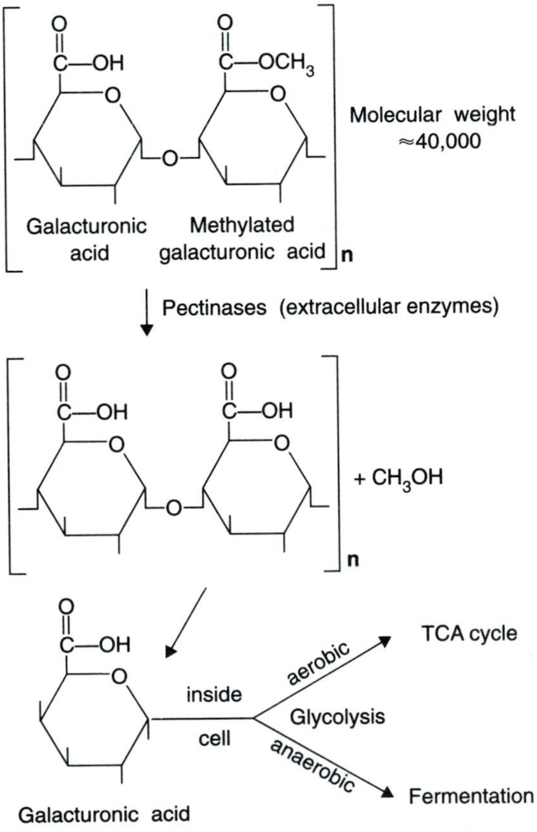

Figure 11–8 Decomposition of the hemicellulose, pectin, in soil.

drolytic disintegration of the polymer occurs outside the cell; when the monomeric sugar or uronic acid residues are produced, they can be transported into the cell for subsequent metabolism.

Lignin. Lignin is generally the third most common component of plant residue. It is especially prominent in woody tissues and is thus particularly important in forest ecosystems. Lignin is an integral component of the cell walls of plants, and its content increases as the plant ages. The lignin content of young plants is often less than 5% by mass, while mature plants may contain 15% and the wood in trees may have levels near 35%. Because the chemical composition of lignin varies among plant species, it is not possible to give a specific structure that is representative of all lignins. The basic building block of lignin is the phenylpropene unit that consists of a hydroxylated 6-C aromatic benzene ring (phenol) and a 3-C linear side chain. Three common phenylalcohols used by plants in lignin synthesis are shown in Figure 11–9.

Lignin is typically composed of 500 to 600 phenylpropene units with randomly condensed units of substantial cross-linkage. The linkage can involve ether linkages (C-O-C) or C-C bonds and may occur between benzene rings, side chains, or rings to side chains. Figure 11–10 illustrates an example of possible degradation pathways for lignin.

Because of the size and complexity of lignin, its decomposition rate is slow compared to that of cellulose and hemicellulose (Fig. 11–4). The decomposition process generally begins with oxidation and removal of the exposed side chains.

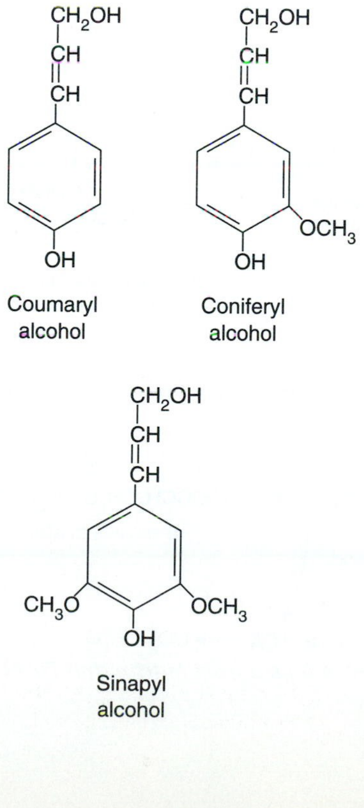

Coumaryl
alcohol

Coniferyl
alcohol

Sinapyl
alcohol

Figure 11–9 Examples of phenylpropene units that are the basic building blocks of lignin.

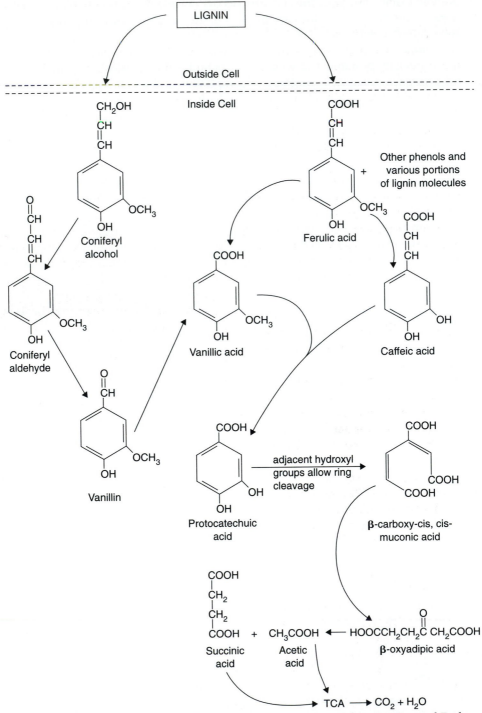

Figure 11–10 One possible pathway of lignin degradation. *Adapted from Martin and Focht (1977). Used with permission.*

The methoxyl (CH_3O-) group is oxidized to carbon dioxide and a hydroxyl (−OH) group is left on the ring. The second step in lignin degradation is depolymerization, which results in the liberation of individual phenolic units with side chains. Once these units are released, they are available to much of the soil microbial population and may be transported inside microbial cells for subsequent metabolism. The side chains are then removed and, if hydroxyl groups exist on adjacent carbons on the phenolic ring, the ring can be cleaved. Ring cleavage via ortho or meta (position on the benzene ring) fission occurs under aerobic conditions. Degradation of lignin can occur under anaerobic conditions, but the process is limited and lignin tends to accumulate in these environments (e.g., peat bogs).

Fungi, especially basidiomycetes, are generally recognized as the major microbial group responsible for lignin degradation. However, actinomycetes, such as *Streptomyces* spp., and bacteria are also involved. Research suggests that the efficiency of carbon utilization for microbial biomass production is low with most of the carbon being either evolved as carbon dioxide or incorporated into soil organic matter.

Starch

Starch is a common plant **storage polysaccharide** that serves as an energy storage product in cereal grains (seeds), stems, roots, rhizomes, and tubers. Starch exists as a linear and as a branched polymer of glucose. When glucose is linked in the α1-4 position, the linear polymer is known as amylose. The α1-4 linkage facilitates a more rapid breakdown rate than the β1-4 linkage found in cellulose. As a branched polymer, the glucose is linked in the α1-4 and α1-6 positions to produce a material known as amylopectin (Fig. 11–11).

Extracellular enzymes known as amylases (α1-4 exo- and endoglucanases) are produced by numerous bacteria and fungi and hydrolyze amylose and amylopectin to the disaccharide (two-glucose units) maltose. Subsequent hydrolysis of maltose by an α1-4 glucosidase (maltase) yields monomeric glucose units that can enter the glycolytic pathway. Inulin, a linear 1,2 fructosan (fructose polymer), is another common storage polysaccharide. It is degraded by inulinase, an extracellular fructosanase.

Decomposition Activity of Microorganisms

Magnitude of the Microbial Population

The dynamic population of heterotrophic microbes engaged in residue decomposition is regulated primarily by aeration, moisture, temperature, pH, and nutrient status in the soil. Conditions favoring rapid decomposition of plant residues and multiplication of microorganisms include:

- plant residue with a low lignin content and small particle size,

- adequate available nitrogen or residue with a low carbon-to-nitrogen ratio,

- near neutral soil pH to allow diverse microbial populations to be active,

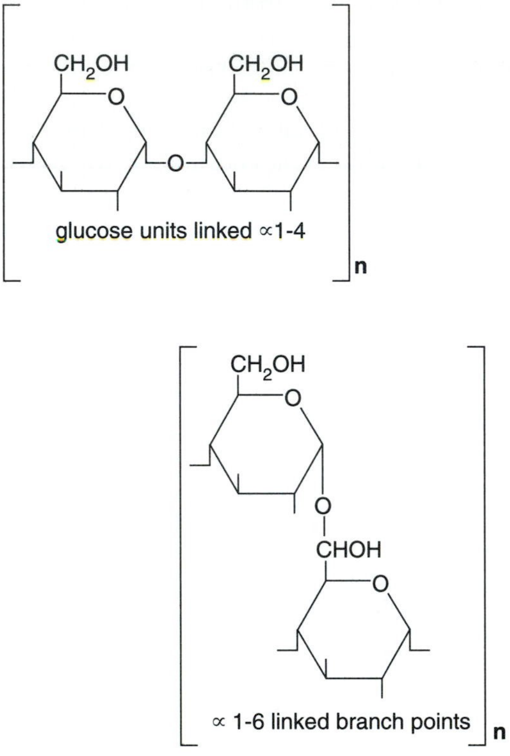

Figure 11–11 Structural formulas for starch showing fragments from linear and branched polymers.

- adequate soil moisture and aeration, and

- warm soil temperature with an optimum of 30 to 45°C.

Population increases following addition of fresh residues may result in a doubling of the microbial biomass. Subsequent declines in populations result from the onset of unfavorable environmental conditions, such as cold temperatures or dry conditions, or from depletion of the more readily available substrates. Typical soil **microbial biomass** levels approximate 150 g m^{-2} (or 1,500 kg ha^{-1}) and are influenced by the method used in the determination. A frequently used method involves fumigating the soil with chloroform and measuring the amount of carbon released upon subsequent incubation (Box 11–4).

The magnitude of the **microbial population** engaged in decomposition activity has also been estimated by monitoring carbon dioxide evolution from the soil. Broder and Wagner (1988) monitored the actual invasion by soil microbes of residues for corn, soybean, and wheat. When wheat straw was incorporated into soil, the microbial population was high during late summer and fall, reflecting active decomposition (Fig. 11–12). Because of the large indigenous population found in soil, the population differences due to stimulation of

Box 11–4

Soil Fumigation-Incubation Method. A current method of determining microbial biomass involves a short laboratory incubation of a soil sample in a closed container with chloroform to lyse the microbial population (below). Subsequently the chloroform vapor is removed and the soil is reinoculated with a small amount of fresh soil. A flush of carbon dioxide evolution follows as the readily available carbon compounds released from the original microbial population by lysis are decomposed. The carbon liberated as carbon dioxide during a 10-day incubation is quantitatively related to the initial microbial biomass. Generally the flush of carbon dioxide accounts for 41% of the mineralized microbial biomass carbon.

For example, following chloroform fumigation, vapor removal, and soil reinoculation, the amount of CO_2-C evolved during a 10-day incubation was determined to be 35.8 mg CO_2-C 100 g^{-1} dry soil. A nonfumigated subsample of the same soil evolved 23.5 mg CO_2-C 100 g^{-1} dry soil during the 10-day incubation.

The amount of microbial biomass is calculated as:

$$\text{Microbial Biomass} = \frac{A - B}{k}$$

where A = CO_2–C evolved from the fumigated soil,
     B = CO_2–C evolved from the nonfumigated soil,
and k = fraction of biomass carbon mineralized to carbon dioxide = 0.41 (at 22°C);

$$\text{therefore, microbial biomass} = \frac{35.8 - 23.5 \text{ mg } CO_2 - C \text{ 100 } g^{-1} \text{ dry soil}}{0.41}$$
(mg C 100 g^{-1} dry soil)

$$= 30.0 \text{ mg C 100 } g^{-1} \text{ dry soil or about 670 kg C ha}^{-1}$$

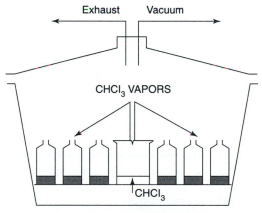

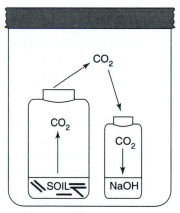

Soil samples are fumigated under vacuum in a desiccator for 24 hours in the dark. After fumigation, the chloroform is removed by repeated evacuation of the chamber. Individual samples are then transferred to gas-tight jars containing an alkali trap to collect CO_2 and incubated for 10 days. Unfumigated samples are incubated in a like manner.

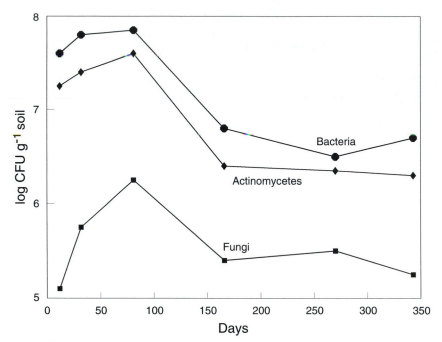

Figure 11–12 Populations of soil microorganisms at various sampling dates after incorporating wheat straw in June into a silt loam (CFU = colony forming units). Mean annual microbial biomass values in the surface 10 cm of soil were: bacteria, 300 kg ha^{-1}; actinomycetes, 130 kg ha^{-1}; fungi, 290 kg ha^{-1}. *Adapted from Broder and Wagner (1988). Used with permission.*

added residue were not dramatic. More dramatic changes in the microbial population over time were demonstrated by isolating the microorganisms present within the plant residue where active decomposition was occurring. Following residue addition in June and as decomposition progressed into the cold fall and winter conditions, the microbial population declined sharply. When active decay resumed the following spring, the number of bacteria and actinomycetes in the residue was similar to that of the previous fall. Fungi, in contrast, had a large increase in biomass during the spring, and the larger biomass levels existed for several months before declining. This is further evidence that the populations change within succession (particularly fungi) following seasonal cycles and indicates that decomposition activity extends into a second year.

Progression of Decay in Stages: Microbial Succession

Over time, the residue-degrading microbial population responds to the variety of substrate components added to the soil. These components differ in ease of decomposition. Simple compounds, such as water-soluble sugars and amino acids, are metabolized first because they serve as excellent growth substrates for a wide spectrum of soil microbes that exhibit short generation times: these are the r-strategists, or copiotrophs, described in Chapter 8. Successive stages of decay involve the metabolism of progressively more complex components for which the number of microorganisms able to break down the components becomes progressively fewer. These are the K-strategists, or oligotrophs. The decomposition rates for the remaining residues also decline. Throughout the decay process, carbon dioxide and new microbial cells are being formed. The newly produced microbial biomass, however, constitutes additional substrate that, in general, is relatively easy to decompose. The overall decay process might be conceptualized as proceeding in various stages as outlined in Figure 11–13.

In the first stage of decay, the easily mineralized components of residues are utilized by the soil microorganisms, with about one-half of the carbon liberated as carbon dioxide and the remainder assimilated in the production of new microbial biomass. This stage may occur during a period of days or weeks after residues are incorporated into soil.

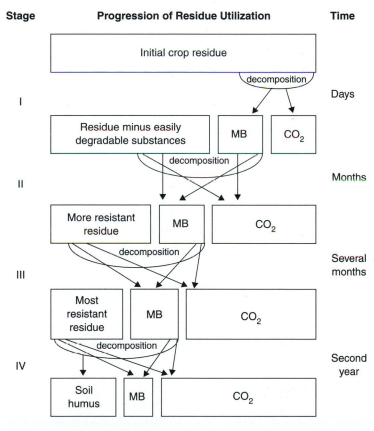

MB = Microbial Biomass

Figure 11–13 Stages of residue decomposition in soil. Changing box sizes indicate the relative quantities of carbon in the various fractions as decomposition progresses toward completion. *Adapted from Stevenson (1986). Used with permission.*

In a second stage of decomposition, the fresh microbial biomass is decomposed to produce additional carbon dioxide and a new generation of microbial biomass. At the same time, additional residue components, particularly cellulose and other carbohydrates, are attacked. The microbial utilization of these substrates releases about 50% of the carbon as carbon dioxide with the remainder passing into new microbial biomass.

A third stage further involves decay of the recently synthesized microbial biomass and continued decay of more resistant components of the remaining residues.

The final stage involves the decay of the most resistant residue components along with the previously produced microbial tissues. The remaining substrate is high in lignin content and aromatic decomposition may be accompanied by synthesis of complex humic materials. Thus, the overall efficiency of microbes is reduced so that as much as two-thirds of the residue passes off as carbon dioxide and one-third remains in the soil as resistant carbon complexes and microbial tissues.

Efficiency of Microbial Population Development

In addition to the production of microbial biomass, a second major product of residue decomposition in soil is carbon dioxide. This gas, which is generally released into the atmosphere, should be viewed as the end product of decay. However, various intermediate products may be produced due to incomplete oxidation of the substrate, and these intermediates may subsequently undergo further oxidation. Oxidations furnish energy to heterotrophic organisms via oxidative phosphorylation, with the maximum energy yield obtained by complete oxidation to carbon dioxide.

The proportion of the total-substrate carbon that is synthesized into new cell material is commonly taken as a measure of the efficiency of the microorganism in metabolizing substrate for new cell production (Wagner, 1975). By measuring carbon dioxide that is liberated over a period of weeks or months, microbial metabolism in aerobic soils has been found to have an efficiency of around 35%; based on the amount of carbon dioxide liberated, this represents 65 to 70% of the substrate utilized. During this time, numerous generations of microorganisms probably have participated in the decay process. Many of the microbes feeding on the more readily available constituents have generation times of a day or a few days at most. Thus, the apparent efficiency is really an overall display of the activity of numerous successive generations of microbes, many of which are attacking and obtaining energy from previously active members of the decomposer population.

For those microbes feeding on simple substrates like soluble carbohydrates or amino compounds, the efficiency of utilization of substrate carbon in synthesis of new cell material may be as high as 65% (Gilmour and Gilmour, 1985). However, because of the rapid recycling of this biomass carbon, the overall relationship between carbon dioxide liberated and microbial carbon produced approximates the much lower values we have noted earlier. Many current models of the carbon cycle use efficiencies of 30 to 50%. Measured values of about 60% efficiency have been reported only rarely and, in any case, would not be expected to exceed 70% based on thermodynamic consideration of the process of cell synthesis (Wagner, 1975).

Different members of the microbial population display broad differences in their efficiency of substrate assimilation. Overall efficiency may be lowest for the bacteria because many may operate under anaerobic as well as aerobic environments. In anaerobes, the oxidation of the substrate is incomplete and produces a variety of partially oxidized compounds, fermentation products such as alcohols, organic acids, and methane. As a result, the energy available for new cell synthesis is lower, and the efficiency of cell production relative to substrate utilized will be less.

Substrate Utilization by Microbes in Relation to Carbon Dioxide Production

The liberation of carbon dioxide occurs during residue decay with the simultaneous production of new microbial tissues. This carbon dioxide evolution is widely accepted as a quantitative expression of active decay processes. Despite this convention, the actual rate of carbon utilization by microbes in a decaying substrate is not quantitatively the same as the rate of liberation of CO_2-C.

To define decay quantitatively requires familiarity with alternative methods of expressing decomposition that are based on carbon remaining in the soil. One useful representation of decomposition is a curve that represents the amount or proportion of substrate carbon remaining in the soil as a function of time. This value would include carbon in undecomposed residue, synthesized resistant components, and microbial biomass. An alternative approach is to plot the proportion of carbon that has not been utilized, or consumed, by microbes against time. This latter curve would fall below that which reports the total carbon remaining (Fig. 11–14). The difference

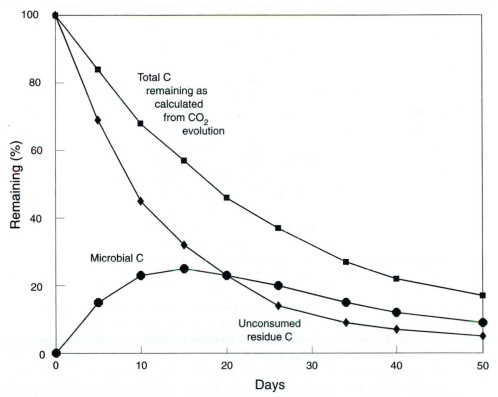

Figure 11–14 Total carbon remaining in soil and microbial biomass carbon produced during residue decomposition. The values were calculated using a microbial efficiency of 60%. *Adapted with modifications from Paul and van Veen (1978). Used with permission.*

between these two curves represents the portion of the decomposing residue carbon that has been incorporated into microbial biomass. Paul and van Veen (1978) have shown that, if the amount of carbon incorporated into microbial biomass and the efficiency of decomposition are not considered, the actual decomposition rate could be substantially underestimated for short-term incubation studies with a reasonable efficiency of biomass production. An example of the problem is presented in Box 11–5.

The substrate-utilization efficiencies that we have discussed are directly related to the rate at which the biomass curve increases relative to the rate of decrease in total carbon remaining in the soil. Equal rates of change indicate an efficiency of 50%. As Figure 11–14 illustrates, the substrate utilization efficiency is initially high and approximates 50%. This efficiency is indicated by the fact that the carbon remaining (not liberated as carbon dioxide) declines at about the same rate that microbial biomass increases. With time the relationship becomes more complex, because microbial biomass production levels off and eventually decreases. Even though new microbes are being generated, others are dying and being decomposed. A concomitant change is that the remaining residue on which the microbes are feeding is becoming progressively more resistant to degradation and less suitable as an energy substrate for new growth. Eventually there is a progressive loss of microbial biomass which represents a more readily utilizable substrate than does the remaining residue components. Under these conditions, total carbon remaining is comprised of carbon not yet utilized along with microbial tissues that are resistant to decay. From the interrelationships of the values for the three curves, we see that total microbial carbon generally decreases once less than one-half of the total carbon remains.

Microorganisms feeding on crop residues require nitrogen and other nutrients for the synthesis of new microbial cells. For a given amount of residue carbon incorporated into microbial biomass, the nitrogen requirement of bacteria is approximately twice that for fungi. In either case, the new microbial cells are rich in nitrogen relative to that for most residues being decomposed. For residues that are typically low in nitrogen, the decay process does not yield sufficient nitrogen from the residue itself to meet the nitrogen requirement of the developing microbial biomass. The inorganic nitrogen pool of the soil is then drawn upon to meet the nitrogen needs of microorganisms. This removal of available nitrogen from the soil and its assimilation into an organic form in microbial tissues that renders it not readily available is called **immobilization.**

Determining Decomposition Rates

Production of radiolabeled carbon dioxide. The carbon dioxide released from soils reflects the net metabolic activity of microorganisms, plants, and animals. Thus, in order to measure carbon dioxide evolution from only decomposition processes, it is often necessary to label the material in question with radioactive [14]C. The labeling of the added carbon distinguishes the heterotrophic decomposition of the added substrate from other sources of carbon that contribute to total soil respiration. The latter includes some nonlabeled carbon dioxide arising from root respiration and from the breakdown of SOM. The [14]CO_2 that is evolved over time traces

Box 11–5

Three Ways to Measure Decomposition. Decomposition of wheat straw in a laboratory study can be determined by (A) measuring the amount of carbon evolved as carbon dioxide and (B) calculated using first-order kinetics (described on page 245). Microbial biomass can also be estimated using microbial efficiency (C).

In this example, a flask containing 100 g of Captina silt loam was amended with 500 mg wheat straw (45% C). The amount of CO_2–C evolved after 14 days was 94.6 mg. A control sample with no straw evolved 18.1 mg CO_2–C under the same conditions.

(A) Decomposition based upon carbon dioxide evolution data.

$$\% \text{ decomposition} = \frac{[CO_2\text{–C evolved from wheat straw amended soil]} - [CO_2\text{–C from control]}}{\text{amount of wheat straw C added}} \times 100$$

$$= \frac{94.6 \text{ mg } CO_2\text{–C} - 18.1 \text{ mg } CO_2\text{–C}}{(500 \text{ mg wheat straw}) (45\% \text{ C in straw})} \times 100$$

$$= \frac{76.5 \text{ mg C}}{225 \text{ mg C}} \times 100 = 34\%$$

(B) Decomposition based upon first-order kinetics.

Assume wheat straw is 15% sugars and amino acids, 65% cellulose and hemicellulose, and 20% lignin with first-order rate constants of 0.2, 0.08, and 0.01 days^{-1}, respectively. The respective microbial efficiencies are 60%, 40%, and 10%.

$$A_t = 15 \, e^{-0.2(14)} + 65 \, e^{-0.08(14)} + 20 \, e^{-0.01(14)}$$

$$= 15 \, (0.06) + 65 \, (0.32) + 20 \, (0.87)$$

$$= 0.9 + 21.2 + 17.4$$

$$= 39.5\% \text{ of the C from the wheat straw remaining in the original form.}$$

The amount of carbon in the microbial biomass and SOM is calculated as 100% − [CO_2–C evolved + C remaining in the original form] or 100% − (34% + 39.5%) = 26.5%.

(C) Estimation of microbial biomass formed using microbial efficiency.

$$\text{Microbial efficiency (E)} = \frac{\text{mg biomass–C}}{\text{mg biomass–C} + \text{mg } CO_2\text{–C}}$$

$$\text{Solving, mg biomass–C} = \frac{E}{1 - E} (\text{mg } CO_2\text{–C})$$

| Box 11–5 cont'd |

In Step A, 34% of the wheat residue-C, or 76.5 mg C, was evolved as CO_2–C. At a microbial efficiency, E, of 0.4 or 40%,

$$\text{mg biomass-C} = \frac{0.4}{1 - 0.4} (76.5 \text{ mg } CO_2\text{–C})$$

$$= 0.67 (76.5 \text{ mg } CO_2\text{–C})$$

$$= 51 \text{ mg C}$$

Thus, 22.7% of the wheat straw carbon exists in the form of microbial biomass. The amount of wheat straw carbon remaining in the unaltered form and SOM is 100% − [% evolved as CO_2–C + % as microbial biomass] or 100 − [34 + 22.7] = 43%. The value for carbon remaining as unaltered wheat straw as calculated in Part B using first-order kinetics is similar to the value calculated using microbial efficiency.

---

the progressive decay of the particular substrate being studied. The decomposition of many types of substrates ranging from simple sugars to complex aromatic polymers and even xenobiotics has been examined by this approach.

When applying the technique to the study of an agricultural ecosystem, the entire mass of crop residues needs to be labeled. Ideally, the growing crop is exposed to an atmosphere of $^{14}CO_2$ in the field to ensure that both above-ground parts and the root system carry the ^{14}C label. It is important to include labeled roots in such studies because they may constitute as much as half of the residue biomass undergoing decomposition.

Studies reporting residue decomposition as $^{14}CO_2$ evolution over time have frequently found that the evolution rate decreases as the substrate is consumed. Furthermore, the total carbon liberated as $^{14}CO_2$ generally does not reach 100% of that in the decomposing substrate, even though the study is prolonged for several years. This is because a portion of the substrate carbon is assimilated as microbial tissue and remains in the soil microbial biomass. Additionally, certain components of the substrate, particularly aromatic compounds, may be only partly degraded to produce intermediates that become repolymerized by humification reactions and pass into the resistant SOM pool.

A representative curve showing cumulative $^{14}CO_2$ liberation from labeled wheat residues whose decay in soil was monitored for 28 months is shown in Figure 11–15. The repeating pattern is one of accelerated decay followed by limited activity due to cold temperature in winter. The rate measured during the acceleration phase diminished each successive year. By the end of 28 months, the amount of ^{14}C evolved as $^{14}CO_2$ was approximately 85% of the total initially added to the soil. The slope of the curve at that time suggests that the remaining carbon had acquired a resistance to decay approaching that of soil humus.

For late-maturing, summer-grown crops, such as corn in the central United States, the fall season remaining after harvest is short and does not allow decomposition to proceed to the extent recorded for wheat and similar crops. In corn, only about a third of the residue carbon is liberated in the fall, and much highly degradable

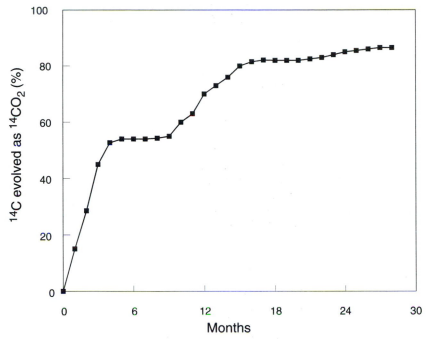

Figure 11–15 Cumulative curve of $^{14}CO_2$ evolution from labeled wheat residues following a June harvest with residues disked into soil. *Adapted from Buyanovsky and Wagner (1987). Used with permission.*

residue overwinters in the soil. As a result, the primary flush of carbon dioxide evolution occurs in the following spring and summer months.

Kinetics of residue decay. Decomposition of carbonaceous residues in soil over time does not plot as a linear relationship. Rather, the decay rate progressively declines as the microbial population utilizes the substrate and slowly, or only partially, breaks down the more resistant materials.

The pattern of declining rate with time suggests that decomposition may be a *first-order reaction;* that is, the natural logarithm of the quantity of substrate remaining at any time is proportional to the period of time. The proportionality constant in such a true first-order reaction is k. Figure 11–16 shows the decay rate curve for a typical first-order reaction. In this example, the natural logarithm (ln) of substrate concentration is plotted on the vertical axis against time on the horizontal axis. The slope of the curve is the *rate constant* (k) at which the substrate is disappearing. If we know the original concentration of substrate (A_0) and that remaining (A_t) at a later time (t), we can calculate the rate constant from the relationship:

$$\ln (A_t/A_0) = -kt$$

In practice, the decomposition of a complex residue does not strictly obey first-order kinetics. However, it does appear that the decomposition of a simple substrate can be characterized by one rate constant depicting the rapid decay phase and that

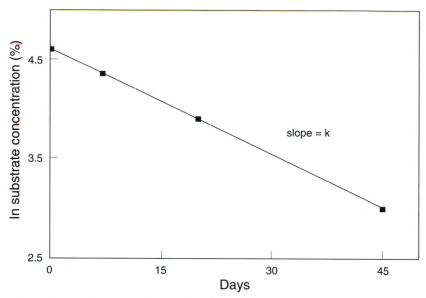

Figure 11–16 First-order function for decay. The y-axis is the natural or base e logarithm (ln) of substrate remaining (A_t). The slope is the rate constant (k) which has units of inverse time or $time^{-1}$.

decomposition of the more recalcitrant components can be described with another rate constant or several other constants. Many scientists have chosen to describe decay by two rate constants defining two phases of decomposition of the residue itself and often a third constant to characterize the very slow decomposition of soil humus. This approach recognizes the presence of several substrate pools, each having a different rate constant.

In the central United States under typical agricultural conditions, the rate constant, or k, for the rapid decay of simple carbohydrates in soil may be about 0.03 day^{-1}. Highly complex and degradation-resistant materials such as lignin have a k value of 0.0003 day^{-1}. In comparing these decay rates with the rates used in Box 11–5, it should be noted that the experiment described in Box 11–5 was a laboratory incubation, yielding more rapid rates, while the above rates are for field conditions.

As the fresh residues of a particular year are decomposed and become incorporated into the resistant humic fraction of the soil after as many as five years, the material is highly resistant to further microbial attack. In reality, studies of carbon turnover have shown that the humic fraction of SOM is comprised of several pools rather than a single pool. Some carbon in the humic pool may be characterized by a turnover time of tens of years, while other parts have turnover extending to hundreds and thousands of years.

Soil Organic Matter (SOM)

Plant residue serves as the major substrate for SOM formation. In this section the beneficial impact and the formation, chemical properties, and decomposition of SOM are considered.

Beneficial Properties Imparted to Soil

When soil microorganisms decompose organic material, the majority of the carbon is evolved as carbon dioxide or incorporated into biomass. However, a small portion of the carbon is biochemically altered and remains in the soil as organic matter (Box 11–6). Soil organic matter imparts many desirable biological, chemical, and physical properties to soil (Box 11–7). The amount of SOM accounts for up to 5% of the total soil volume, but it exerts a disproportionally large influence on soil properties.

The major chemical effect of SOM in most mineral soils is that it contributes 20 to 80% of the cation exchange capacity (CEC). The source of the charge is pH dependent so that as the pH increases, the CEC also increases. The SOM also influences soil physical properties by promoting **aggregation.** Polysaccharides produced by soil bacteria and humic substances produced by soil fungi improve aggregation. A soil in which the smaller soil particles are bound into water-stable aggregates is highly desirable because it allows:

- higher infiltration rates that reduce runoff and erosion,

- more rapid gas exchange with the atmosphere and thus better soil aeration, and

- easier cultivation and penetration by plant roots because the soil offers less resistance.

In general, organic residues that have the highest levels of readily available carbon exert the largest and most rapid increase in soil aggregation. However, the effect

Box 11–6

Definitions of Various Fractions of Soil Organic Matter.

- **Soil organic matter**—nonliving organic fraction of the soil exclusive of undecayed macroanimal and plant residues. It consists of humic and nonhumic substances.

- **Humic substances**—series of brown to black substances of relatively high molecular weight formed by secondary synthesis reactions.

- **Nonhumic substances**—unaltered remains of animals, microbes, and plants. This fraction contains compounds belonging to known biochemical classes (e.g., amino acids, carbohydrates, fats, waxes, resins, organic acids).

- **Humic acid**—dark brown to black organic material that can be extracted from soil by various reagents (e.g., dilute alkali such as 0.5 M NaOH) and that is precipitated by acidification to pH 1 to 2.

- **Fulvic acid**—yellow material that remains in solution after removal of humic acid by acidification.

- **Humin**—fraction of soil organic matter that cannot be extracted from soil with dilute alkali.

Adapted from Soil Science Society of America (1997) and Stevenson (1994). Used with permission.

Box 11–7

Beneficial Properties Imparted to Soil by Organic Matter. This list illustrates the importance of adopting soil management systems, such as no-tillage, crop rotation, and manure addition, which maintain, if not augment, the level of organic matter in soil.

Biological properties

- Provides a slowly available carbon and energy source to support a large, diverse, metabolically active microbial population

- Source of certain compounds that may exert plant growth-promoting effects

Chemical properties

- Increases cation exchange capacity of soil (often 20 to 80% of the total CEC is due to organic matter)

- Buffers pH change

- Provides a slow release supply of organically bound nutrients such as nitrogen, phosphorus, and sulfur (soil organic matter decomposes at 2 to 5% yr^{-1})

- Enhances chelation and thus bioavailability of trace elements to plants

- Accelerates mineral weathering and aids in solubilization of plant nutrients (such as phosphorus) from otherwise insoluble minerals

- Has a high adsorptive capacity for organic compounds and thus reduces the bioavailability of toxic xenobiotics

Physical properties

- Contributes to improving soil structure and aggregation

- Decreases bulk density and thus increases percentage pore space

- Increases total water holding capacity, but may also increase the strength with which water is held

- Increases heat absorption in early spring due to dark color, but soil often holds more water which requires more energy to heat

---

is generally of short duration. Organic residues that are more resistant to decomposition require more time to express maximum aggregation, but the effect persists for a longer period. Lower soil temperatures prolong the effect, and higher temperatures decrease the aggregating effect. Better aggregation increases water infiltration into the soil, especially during rainstorms of short duration and high intensity.

Soil organic matter can be divided into nonhumic substances and humic substances. The nonhumic material, or organic residue, consists of biochemically unaltered remains of plants, animals, and microorganisms and can represent up to 20% of the soil organic matter. Before the organic matter content of soil is determined, the recognizable plant roots and litter are removed, usually by sieving. However,

Table 11–4 **Comparison of chemical properties of humic materials and lignin.**

Characteristic	Humic material	Lignin
Color	Black	Light brown
$-OCH_3$ (methoxyl) content	Low	High
N content	3–6%	0%
Carboxyl and phenolic hydroxyl	High	Low
Total exchangeable acidity (cmol kg^{-1})	≥150	≤0.5
α amino N	Present	0
Vanillin content	<1%	15–25%

small fragments of plant residues that are no longer recognizable as to their origin remain in the soil and are inadvertently included in the soil organic matter determination.

The remaining 80 to 100% of the organic matter is called humic material and is characterized as being high in molecular weight, brown to black in color, highly resistant to decomposition, and formed by secondary synthesis reactions. The differences between humic materials and various biopolymers, such as lignin and cellulose which are commonly produced by plants and soil microbes, are numerous (Table 11–4). The common term for the humic fraction is humus.

Formation of SOM

Plant, animal, and microbial residues are the substrates for the synthesis of humus. Several hypotheses have been presented to explain the mechanism of humus formation (Box 11–8). Current information suggests that, as the residues undergo biochemical alteration and additional compounds are synthesized by microbes, some of the compounds polymerize or condense either through chemical reactions or enzymatic reactions (Fig. 11–17). Figure 11–17 also shows how the decomposition of plant residues and the synthesis of soil humus are intimately associated. The major mechanism of humus formation appears to be related to the ability of phenolic compounds to undergo enzymatic or autooxidative polymerization reactions. Certain orthodihydroxy- and trihydroxy-phenols readily autooxidize at pH ≥6 and form polymers. Monophenol monooxygenases (phenolases) and peroxidases produced by soil microbes are important in enzymatic polymerization of compounds.

An example of oxidation of phenolic compounds and possible polymerization reactions is given in Figure 11–18. The result is random condensation products of hydroxyphenols, hydroxybenzoic acids, and other aromatic compounds that are chemically linked to partially degraded proteins, amino acids, amino sugars, carbohydrates, and other materials derived from residue decomposition or microbial synthesis. Within the bound amino acid, the nitrogen closest to the aromatic ring is least mineralizable, whereas the nitrogen farthest removed from the aromatic ring is readily mineralizable. The energy of bond formation and steric considerations most likely reduce the availability of the nitrogen adjacent to the aromatic ring. Because various phenolic compounds, amino acids, and polysaccharides are involved, the condensation or polymerization reactions result in a wide variety of polymers with different structures but somewhat similar elemental composition.

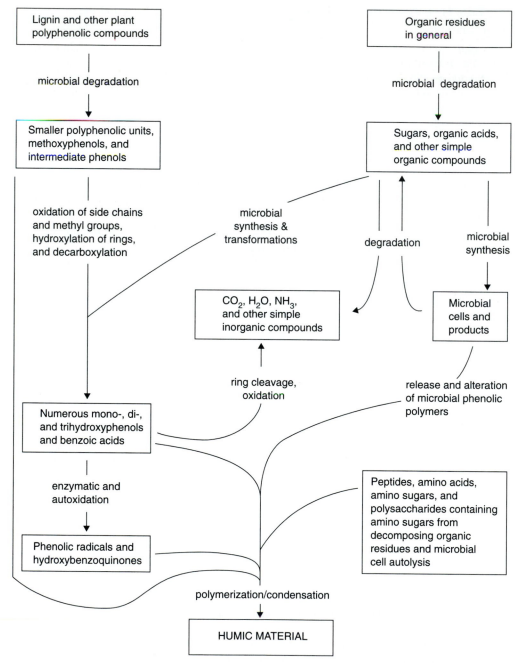

Figure 11–17 Possible pathways showing the relationship of residue decomposition and microbial synthesis of soil humus. *Adapted from Martin and Focht (1977). Used with permission.*

Box 11–8

Theories of Humus Formation. Several theories exist to explain soil humus formation. Those included here are in a more or less chronological order.

- Plant Alteration Theory or Waksman's Lignin-Protein Theory
  Lignin is incompletely utilized by microbes, and the residuum becomes part of the soil humus. Lignin is modified by hydroxylation and oxidation of $-OCH_3$ to $-COOH$, which react with amino compounds to yield humic acids that are oxidized to fulvic acid.

- Reducing Sugar Theory (Browning Reaction)
  Reducing sugars and amino acids are formed as by-products of microbial metabolism and undergo nonenzymatic condensation to yield humus.

- Lignin-Derived Quinone Theory (Flaig)
  Phenolic compounds released during lignin decomposition undergo enzymatic conversion to quinones (Fig. 11–18), which condense with amino compounds to yield humus.

- Microbial Synthesis Theory (Martin and Haider; Kononova)
  Polyphenols are synthesized by fungi from nonlignin carbon sources such as cellulose. The polyphenols are then enzymatically oxidized to quinones that condense with amino compounds to yield humus.

  The current thinking is that the latter two theories involving polyphenols, and likely a combination of the two, is correct in aerobic soils. In poorly drained soil where peat is formed, plant alteration is likely the dominant pathway.
  Adapted from Stevenson (1994). Used with permission.

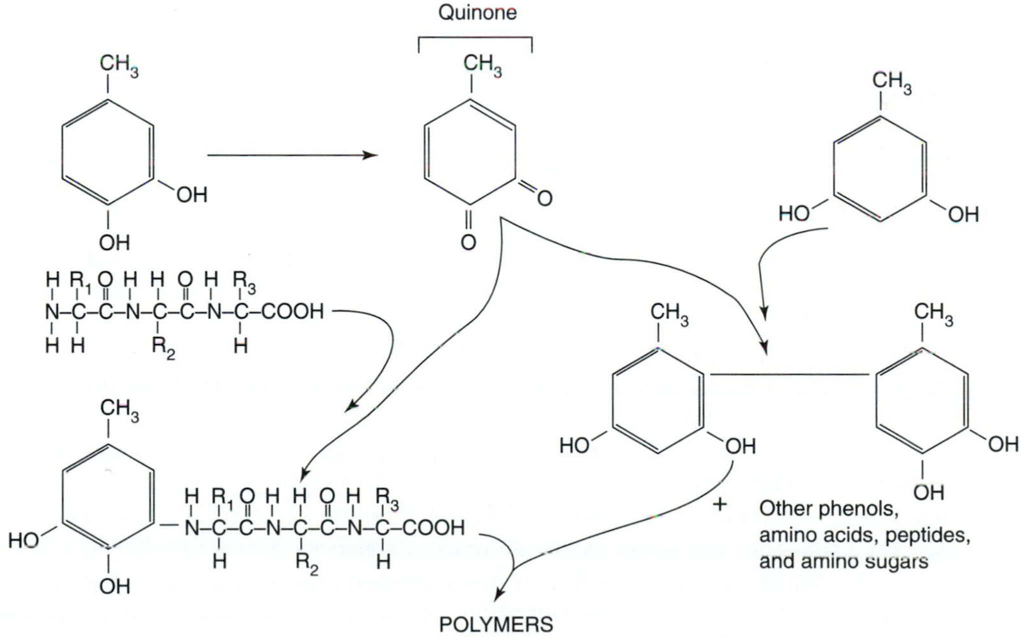

Figure 11–18 Possible condensation or polymerization reactions of phenols involving quinone formation that result in the production of humic polymers. *Adapted from Burns and Martin (1986). Used with permission.*

Characterization of SOM

The exact chemical composition of humus is not known, but it is characterized as a complex mixture of condensation products of phenolic units, polysaccharides, and proteins. The carbon content of humus is generally given as 58% (1% organic-C = 1.72% SOM), whereas nitrogen values range from 3 to 6%. The carbon/nitrogen/phosphorus/sulfur ratio for humus is approximately 100/10/1/1. The SOM is generally thought to have a netlike, three-dimensional structure that coats mineral particles and can be electrochemically bound to clay and metal oxides in soil. A hypothetical model of a soil humus-clay complex is illustrated in Figure 11–19. A structural formula of humic acid has been presented by Shevchenko and Bailey (1996).

Currently, less than 50% of the structural composition of the organic carbon, about 50% of the organic nitrogen, and 40 to 70% of the organic phosphorus forms have been elucidated and then only as products of specific degradative procedures. A typical analysis of SOM shows that approximately 50% of the carbon is present as polyphenolics, 20% as polysaccharides, 20% as various nitrogen complexes, and 10% in miscellaneous forms. Recent findings suggest that the aliphatic component of SOM may be greater and more important than previously recognized. Based upon the ^{14}C dating technique, the average age of SOM ranges from 150 to over 1,500 years.

Chemical fractionation. Soil organic matter can be chemically separated into three general components known as fulvic acid, humic acid, and humin (Box 11–6). The classical chemical extraction and fractionation procedure for soil organic matter involves:

- washing the soil with 0.1 M HCl to remove cations and iron and aluminum hydroxides to increase extraction efficiency,

- extraction with 0.5 M NaOH under a nitrogen atmosphere,

- acidification to pH 2 with HCl, and

- separation of flocculated and soluble material by centrifugation.

The material that precipitates upon acidification is known as humic acid, and the material remaining in solution is known as fulvic acid (Box 11–6). The organic material that remains with the soil mineral fraction following the NaOH extraction is known as humin and may represent 50 to 70% of the SOM carbon. Selected chemical properties of humic acid and fulvic acid are given in Table 11–5. Because of its nonextractability, less information is available about humin.

Functional groups. The CEC of SOM ranges from 150 to 300 cmol kg^{-1} and is derived from such functional groups as carboxyls and phenolic hydroxyls. As noted in Chapter 2, this range is considerably higher than the exchange capacity of many clay minerals. Sites for anion exchange reactions also exist and are due in part to protonated amino groups ($-NH_3^+$). A noted property of organic functional groups is the influence of pH on their ionization. As the pH increases, more negatively charged sites ($-COO^-$ and $-O^-$) are available for cation exchange and fewer sites exist for anion exchange.

Complexes. Many organic compounds found in the soil form complexes with various metal ions in soil and water. When a metal ion such as Zn^{2+} combines with an electron donor (certain organic compounds), the substance formed is

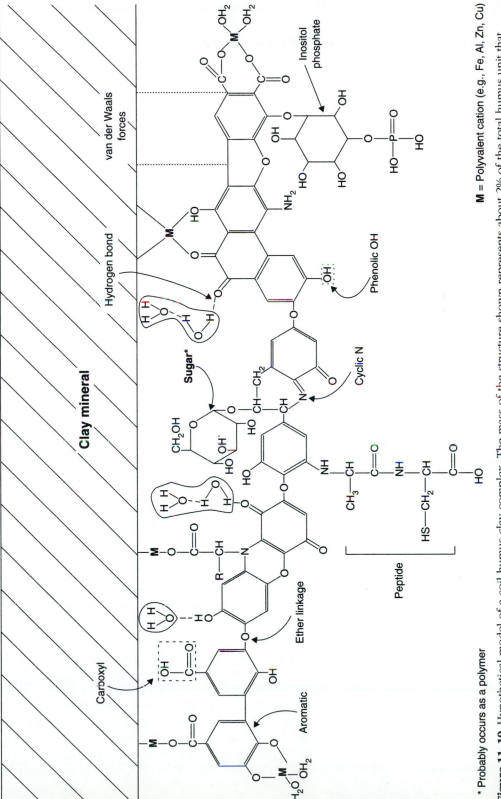

* Probably occurs as a polymer

M = Polyvalent cation (e.g., Fe, Al, Zn, Cu)

Figure 11–19 Hypothetical model of a soil humus-clay complex. The mass of the structure shown represents about 2% of the total humus unit that would have a molecular weight of >10⁵. *Adapted with modifications from Stevenson and Ardakani (1972). Used with permission.*

Table 11–5 **Chemical characteristics of humic acid and fulvic acid fractions of soil humus.**

Property	Humic acid	Fulvic acid
Color	black	yellow
Molecular weight (daltons)	80% $>10^5$	80% $<10^3$
Elemental analysis (%)		
Carbon	57	49
Oxygen	34	45
Hydrogen	4	5
Nitrogen	3	2
Sulfur	1	2
Phosphorus	0.3	0.3
Functional groups		
Total acidity (cmol kg^{-1} dry ash free)	700	1,200
Carboxyl (cmol kg^{-1} dry ash free)	400	800
Phenolic hydroxyl (cmol kg^{-1} dry ash free)	300	400
Polyphenolics	high	low
Polysaccharides	low	high

Adapted from Stevenson (1994). Used with permission.

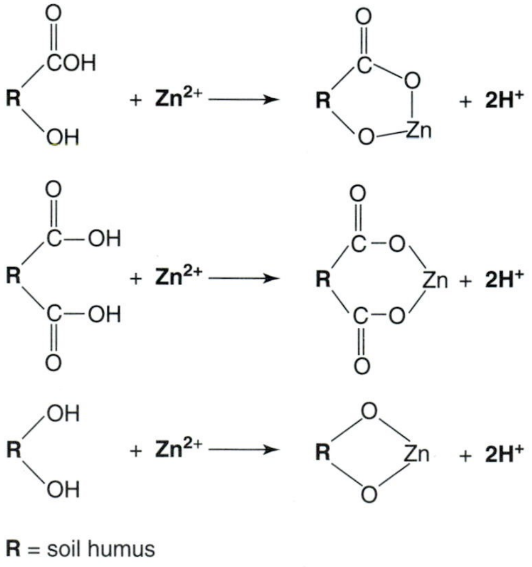

Figure 11–20 Possible Zn^{2+} chelates formed with soil organic matter functional groups.

R = soil humus

known as a complex or **chelate.** Examples of Zn^{2+} chelates are shown in Figure 11–20. The bonds between the electron-accepting metal (Zn^{2+}) and the electron-donating group may be ionic or covalent. Metal complexes are important in micronutrient and heavy metal availability to plants. It is generally thought that most metal complexes are more bioavailable to plants and other organisms and more mobile in soil than the corresponding ionized form. Micronutrient availability is facilitated by chelate formation, and it has been demonstrated that up to 75% of the copper in soil solution is in a chelated form. However, chelate formation by heavy metals such as nickel and cadmium would be undesirable if it resulted in increased plant uptake.

Certain organic compounds in SOM form chelates with iron and aluminum. The chelates facilitate the downward movement of iron and aluminum through the soil profile in a process known as *cheluviation*. The chelates are subsequently deposited in the lower soil horizons, and the resulting redistribution of iron and aluminum is important in soil profile development.

Adsorption. Adsorption of pesticides by SOM plays a major role in determining pesticide activity, effectiveness, and leachability (Stevenson, 1994). The mechanism of adsorption varies depending upon the chemical form of the pesticide. A common observation is that the application rate of certain herbicides must be increased as the SOM level increases to maintain adequate weed control. Higher SOM levels adsorb larger amounts of the herbicide and result in a lower concentration of the herbicide available for uptake by the weed. Similarly, adsorption of toxic organic contaminants can influence transformation and transport of xenobiotics in soil.

Decomposition of SOM

Aerobic conditions. Humus represents a stable material with a decomposition rate of only 2 to 5% per year depending upon climatic conditions. Increasing soil temperatures will result in increasing humus decomposition up to about 45°C. Tropical conditions favor rapid SOM decomposition, and nitrogen deficiencies commonly result following SOM depletion. In some tropical sites, one approach for introducing cultivated agriculture has been "slash and burn." In this practice, the forest is burned and the cleared land brought into production for three to four years. When most of the nutrients in the soil are depleted, the land is abandoned and new land cleared. Given sufficient time—more than 20 years—vegetation reestablishes on the land and fertility is restored. Rapid SOM decomposition at the high temperatures found in the tropics is at least partly responsible for this practice.

Adequate soil aeration favors oxidative decomposition and results in a faster decomposition rate than anaerobic conditions. The presence of adequate available nitrogen will increase the rate of decomposition. Cultivation or mechanical mixing of the soil exposes new surfaces of organic matter to microbial oxidation and thus increases the rate of SOM decomposition. The practice of no-till or minimum-till crop production increases SOM levels compared to conventional tillage management due to less soil cultivation and inherently lower levels of residue incorporation.

Anaerobic conditions. When plant residue is added to an anaerobic environment (e.g., wetland, swamp, or marsh), decomposition is greatly reduced and organic residue typically accumulates, often in layers. These layers represent various successions of vegetation types and various degrees of decomposition of the plant residue. Eventually an organic soil, or Histosol, is formed. In the past, Histosols were called peats or bogs to indicate little plant residue decomposition or mucks to indicate advanced plant residue decomposition.

Oftentimes, Histosols were drained for vegetable production. A unique management problem that arises when Histosols are drained is the resulting rapid decomposition of the SOM, which results in a drop in land elevation. This decline in soil volume, known as *subsidence,* occurs at rates up to 10 cm yr^{-1}.

Summary

Decomposition of plant residue, microbial biomass, and SOM is a major function of soil microorganisms. In agricultural ecosystems, crop residue is the primary carbon substrate for the microbial population and is often added at rates of 1 to 3 kg m^{-2}. The residue consists of readily decomposable fractions such as soluble components and protein, less available components such as cellulose and hemicellulose, and the slowly available fraction typified by lignin. Decomposition rates differ for each of the various components, and the rates are influenced by soil abiotic factors of temperature, moisture, aeration, and pH. The levels of nitrogen in the soil and in the residue are critical. First-order kinetics often describe the rate of residue decomposition.

During the decomposition of the plant residue, a portion of the carbon is evolved as carbon dioxide and an additional fraction is incorporated into microbial biomass. The efficiency with which the microbes incorporate the residue carbon into biomass may be as high as 65% for soluble components, but may be only 10% for lignin. The newly formed biomass also serves as substrate for other microorganisms.

As the plant residue and microbial biomass undergo biochemical alteration and additional compounds are synthesized by soil microorganisms, some of the compounds react and form SOM. The formation of SOM and the decomposition of residue are intimately associated. Soil organic matter accounts for a small percentage of the soil volume, but it exerts a large influence on the biological, chemical, and physical properties of soil. This decomposition of organic materials in the soil also releases carbon dioxide into the atmosphere and is a major factor in the carbon cycle.

Cited References

Bolin, B. 1983. The carbon cycle. pp. 41–45. *In* B. Bolin & R.B. Cook (eds.), The major biochemical cycles and their interactions. John Wiley & Sons, New York.

Brady, N.C., and R.R. Weil. 1996. The nature and properties of soils. 11th ed. Prentice Hall, Upper Saddle River, N.J.

Broder, M.W. 1985. Changes in the chemical composition and microbial population on corn, wheat, and soybean residue decomposing in Sanborn Field. Ph.D. diss. University of Missouri, Columbia (Diss. Abstr. 86-07891).

Broder, M.W., and G.H. Wagner. 1988. Microbial colonization and decomposition of corn, wheat, and soybean residue. Soil Sci. Soc. Am. J. 52:112–117.

Burns, R.G., and J.P. Martin. 1986. Biodegradation of organic residues in soil. pp. 137–202. *In* M.J. Mitchell and J.P. Nakas (eds.), Microfloral and faunal interactions in natural and agroecosystems. Martinus Nijhoff/Dr. W. Junk Publications, Boston.

Buyanovsky, G.A., and G.H. Wagner. 1986. Post-harvest residue input to cropland. Plant Soil 93:57–65.

Buyanovsky, G.A., and G.H. Wagner. 1987. Carbon transfer in a winter wheat (*Triticum aestivum*) ecosystem. Biol. Fert. Soils 5:76–82.

Doran, J.W., M. Sarrantonio, and M.A. Liebig. 1996. Soil health and sustainability. Adv. Agron. 56:1–54.

Gilmour, C.M., and J.T. Gilmour. 1985. Assimilation of carbon by the soil biomass. Plant Soil 86:101–112.

Martin, J.P., and D.D. Focht. 1977. Biological properties of soils. pp. 115–169. *In* L.F. Elliott and F.J. Stevenson (eds.), Soils for management of organic wastes and waste waters. Soil Science Society of America, Madison, Wis.

Paul, E.A., and J.A. van Veen. 1978. The use of tracers to determine the dynamic nature of organic matter. Trans. 11th Int. Congress Soil Sci. 3:61–102.

Post, W.M., T.H. Peng, W.R. Emanuel, A.W. King, V.H. Dale, and D.L. DeAngelis. 1990. The global carbon cycle. Am. Scientist 78:310–326.

Schlesinger, W.H. 1986. Changes in soil carbon storage and associated properties with disturbance and recovery. pp. 194–220. *In* J.R. Trabalka and D.E. Reichle (eds.), The changing carbon cycle: A global analysis. Springer-Verlag, New York.

Shevchenko, S.M., and G.W. Bailey. 1996. Life after death: Lignin-humic relationships reexamined. Critical Rev. Environ. Sci. Tech. 26:95–153.

Soil Science Society of America. 1997. Glossary of soil science terms. Soil Science Society of America, Madison, Wis.

Stevenson, F.J. 1986. The carbon cycle. pp. 1–44. *In* Cycles of soil: Carbon, nitrogen, phosphorus, sulfur, micronutrients. John Wiley and Sons, New York.

Stevenson, F.J. 1994. Humus chemistry: Genesis, composition, reactions. 2nd ed. John Wiley and Sons, New York.

Stevenson, F.J., and M.S. Ardakani. 1972. Organic matter reactions involving micronutrients in soils. pp. 79–114. *In* J.J. Mortvedt (ed.), Micronutrients in agriculture. Soil Science Society of America, Madison, Wis.

Wagner, G.H. 1975. Microbial growth and carbon turnover. pp. 269–305. *In* E.A. Paul and A.D McLaren (eds.), Soil biochemistry. Vol. 3. Marcel Dekker, Inc., New York.

Wolf, D.C., and J.O. Legg. 1984. Soil microbiology. pp. 99–139. *In* M.F. L'Annunziata and J.O. Legg (eds.), Isotopes and radiation in agricultural sciences. Academic Press, London.

General References

Allison, F.E. 1973. Soil organic matter and its role in crop production. Elsevier Science Publishing, New York.

Jenkinson, D.S. 1988. Soil organic matter and its dynamics. pp. 564–607. *In* A. Wild (ed.), Russell's soil conditions and plant growth. 11th ed. John Wiley and Sons, New York.

Kononova, M.M. 1966. Soil organic matter. 2nd ed. Pergamon Press, New York.

Tate, R.L. 1987. Soil organic matter: Biological and ecological effects. John Wiley and Sons, New York.

Waksman, S.A. 1936. Humus: Origin, chemical composition, and importance in nature. Williams and Wilkins Co., Baltimore, Md.

Study Questions

1. What are the relative quantities of carbon in the earth's atmosphere and in soil organic matter compared to annual photosynthetic fixation by terrestrial plant life?

2. What are the major organic constituents of crop residues and how do they differ in ease of decomposition by the soil microbial population?

3. In the decomposition of wheat straw in soil over time, is the loss of carbon from the soil as carbon dioxide more or less than the utilization of carbon in the residue as substrate for microorganisms? Why?

4. Draw, on a linear scale, a graph that describes the decay of crop residue in soil if it progresses according to first-order kinetics.

5. In a laboratory study, a Captina silt loam was amended with poultry litter at a rate of 1 g carbon 100 g^{-1} soil, adjusted to optimal soil moisture, and incubated at 30°C for 50 days. The litter contained 15% easily decomposable, 75% intermediately decomposable, and 10% slowly decomposable fractions with first-order rate constants of 0.2, 0.08, and 0.01 $days^{-1}$, respectively. Calculate the percentage of the carbon in the poultry litter that remains after 50 days.

6. One gram of ^{14}C glucose is added to 100 g of soil, and approximately 80% of the glucose carbon is evolved as $^{14}CO_2$ during four weeks of soil incubation. After four weeks of soil incubation, in what general forms would you expect to find the ^{14}C remaining in the soil?

7. Write the names of the specific organic matter fractions isolated with the following procedure.

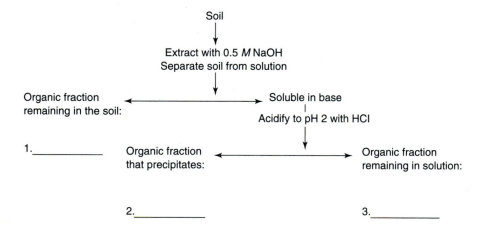

Chapter 12

♦

Transformations of Nitrogen

David D. Myrold

♦

Nature works only in cycles, there are no straight lines.
The forward movement is provided by time. Everything within it must revolve.

Anonymous

Perhaps more time and effort have been invested in studying the nitrogen cycle than any other topic in soil microbiology. Nitrogen is an essential nutrient for all life on earth. Thus its fixation into usable forms by bacteria and subsequent transformations and recycling through organic and inorganic forms are of great practical interest. Indeed, nitrogen is the nutrient most often limiting plant growth in terrestrial ecosystems. The nitrogen cycle affects the environment as well. Current concerns include high concentrations of nitrate in ground and surface waters and the contribution of gaseous nitrogen oxides, such as NO and N_2O, to large-scale environmental problems of acid rain, ozone depletion, and greenhouse warming (Chapter 23). The large diversity of nitrogen-containing compounds, which exist in numerous oxidation states, and the wide array of microbial transformations makes the nitrogen cycle an extremely interesting intellectual challenge.

The Nitrogen Cycle

An overview of the nitrogen cycle is presented in Figure 12–1. Nitrogen is present in various forms (Table 12–1)—primarily as dinitrogen gas (N_2), organic nitrogen (in plants, animals, microbial biomass, and soil organic matter), and ammonium (NH_4^+) and nitrate (NO_3^-) ions. Microbially mediated processes transform nitrogen

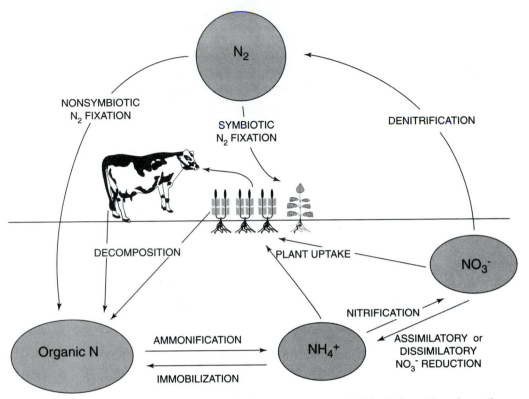

Figure 12–1 Overview of the nitrogen cycle, showing major pools (shaded areas) and transformations (lines) of N.

from one form to another. Certain bacteria can transform dinitrogen to ammonia (NH_3) by a process known as N_2 fixation. The processes of ammonification/immobilization, nitrification, and denitrification are responsible for moving the fixed nitrogen from one form to another in the soil and will be discussed in turn in this chapter.

The nitrogen cycle can be divided into three subcycles (Fig. 12–2):

- **Elemental**—emphasizing the biological oxidation-reduction reactions that interconvert nitrogen and dinitrogen into various chemical forms,

- **Phototrophic**—driven by plant nitrogen uptake, which is fueled by photosynthesis and converts inorganic nitrogen (NH_4^+ and NO_3^-) into organic, nitrogen-containing plant constituents,

- **Heterotrophic**—linked to decomposition processes and driven by the need of heterotrophic organisms for preformed carbon.

These three subcycles function in concert but they are also in competition for one or more of the pools of nitrogen. The outcome of the competition determines which subcycle dominates. This competition is focused at two main control points: the ammonium and the nitrate pools, as discussed later in this chapter. First, how-

Table 12–1 **Pool sizes of terrestrial nitrogen based on soil to a depth of one meter.**

Pool	Typical size (range) (g N m^{-2})	Remarks
N_2 (Dinitrogen)	1,150 (230–27,500)	Minimum based on 0.25 m^3 air-filled pore space in the soil; maximum based on soil air plus a 30-m tall cylinder of air above the soil surface, e.g., a tall forest stand.
Organic N	725 (100–3,000)	From Post et al. (1985); typical value is median of reported soil N contents. Histosols are not included and would likely contain 3,000–8,000 g N m^{-2}.
Plant N	25 (1–240)	Minimum based on desert regions, maximum based on agricultural crops (Olson and Kurtz, 1982) and forest systems (Waring and Schlesinger, 1985; Anderson and Spencer, 1991).
NH_4^+ (Ammonium)	1 (0.1–10)	Assumes 1 m^3 soil at a bulk density of 1.25 Mg m^{-3} and typical NH_4^+ concentrations for soil extracts.
NO_3^- (Nitrate)	5 (0.1–30)	Assumes 1 m^3 soil at a bulk density of 1.25 Mg m^{-3} and typical NO_3^- concentrations for soil extracts.

ever, we need to review the various forms of nitrogen and their characteristics, because these important fundamentals influence the microbial transformations of nitrogen in soil.

Forms of Nitrogen

The sizes of the nitrogen pools vary over several orders of magnitude (Table 12–1). Although we tend to ignore the relatively inert dinitrogen pool, probably because it is an invisible gas, it actually represents the largest pool of biologically active nitrogen in terrestrial ecosystems. Soil organic nitrogen makes up the next largest pool of nitrogen and varies widely among soil types. The variation in soil organic nitrogen is determined largely by the factors of soil formation, particularly temperature and precipitation. The amount of nitrogen tied up in plant biomass is of intermediate size and varies as a function of vegetation type (forests versus grasslands), climate, and soil nitrogen availability. Soil inorganic-nitrogen pools are usually small, generally just a few mg N kg^{-1} in natural ecosystems and rarely exceeding 100 mg N kg^{-1} in the plow layer of recently fertilized agricultural soils.

Larger pools tend to be the less reactive (i.e., they turn over more slowly) and the smaller pools usually are more dynamic. For example, the atmospheric dinitrogen pool is the largest pool of nitrogen and has a mean residence time on the order of thousands to millions of years. Decades are required to turn over the organic-nitrogen pool. Nitrogen in plant biomass often turns over annually, whereas inorganic-nitrogen pools are so dynamic that they may turn over more than once a day.

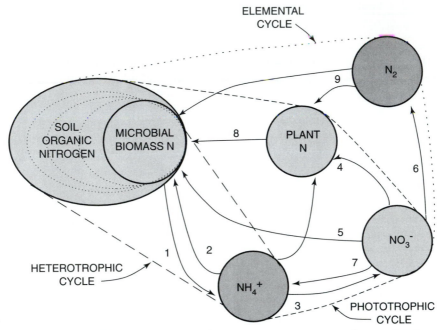

Figure 12–2 Detailed nitrogen cycle showing major processes and control points. The nitrogen cycle consists of three overlapping subcycles. The demand of heterotrophic organisms for organic carbon drives one subcycle, the heterotrophic cycle (long dashed lines), where ammonium is both consumed and produced. The demand of plants for inorganic nitrogen drives the second subcycle, the phototrophic cycle (short dashed lines). Finally, the oxidation and reduction of nitrogen by microorganisms drives the third subcycle, the elemental cycle (dotted lines), where nitrogen is converted into various forms. All the subcycles contain the soil organic nitrogen and ammonium pools. The soil organic nitrogen pool contains several nitrogen fractions of differing biological availability as well as a separate microbial biomass nitrogen pool. Important biological transformations of nitrogen include: 1, ammonification; 2, immobilization; 3, autotrophic nitrification; 4, plant uptake; 5, nitrate immobilization; 6, denitrification; 7, dissimilatory nitrate reduction to ammonium; 8, decomposition; and 9, N_2 fixation. *Based on Jansson and Persson (1982). Used with permission.*

Soil organic nitrogen. The nitrogen contained in soil organic matter occurs in a wide range of compounds, of which only about half can be definitively identified. Naturally occurring organic-nitrogen compounds isolated from soils include: proteins and amino acids, microbial cell-wall polymers and amino sugars, nucleic acids, and a variety of vitamins, antibiotics, and metabolic intermediates (Figs. 12–3 and 12–4). Because much of the organic nitrogen in soil is of unknown composition, a fractionation procedure based on acid hydrolysis has been used to characterize soil organic nitrogen (Table 12–2). It is interesting to note that the range given for amino-sugar nitrogen, which is found mainly in microbial cell walls (Fig. 12–4), is similar to that often found for microbial biomass nitrogen, which is typically about 5% of total soil nitrogen.

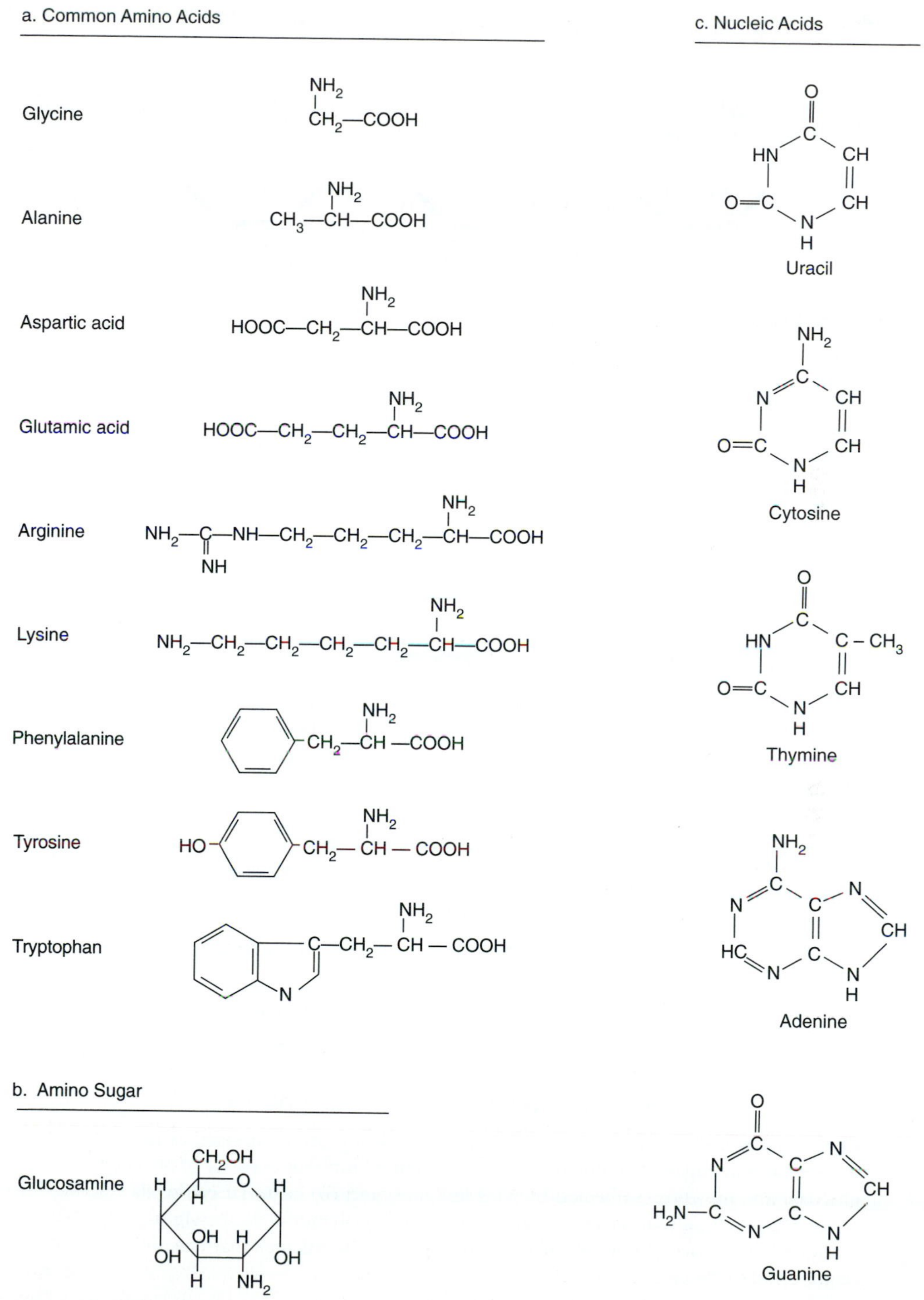

a. Common Amino Acids

Glycine

$$\overset{\displaystyle \overset{NH_2}{|}}{CH_2-COOH}$$

Alanine

$$CH_3-\overset{\displaystyle \overset{NH_2}{|}}{CH}-COOH$$

Aspartic acid

$$HOOC-CH_2-\overset{\displaystyle \overset{NH_2}{|}}{CH}-COOH$$

Glutamic acid

$$HOOC-CH_2-CH_2-\overset{\displaystyle \overset{NH_2}{|}}{CH}-COOH$$

Arginine

$$NH_2-\overset{\displaystyle \underset{NH}{\overset{}{C}}}{}-NH-CH_2-CH_2-CH_2-\overset{\displaystyle \overset{NH_2}{|}}{CH}-COOH$$

Lysine

$$NH_2-CH_2-CH_2-CH_2-CH_2-\overset{\displaystyle \overset{NH_2}{|}}{CH}-COOH$$

Phenylalanine

$$\text{(benzene ring)}-CH_2-\overset{\displaystyle \overset{NH_2}{|}}{CH}-COOH$$

Tyrosine

$$HO-\text{(benzene ring)}-CH_2-\overset{\displaystyle \overset{NH_2}{|}}{CH}-COOH$$

Tryptophan

$$\text{(indole ring)}-C-CH_2-\overset{\displaystyle \overset{NH_2}{|}}{CH}-COOH$$

b. Amino Sugar

Glucosamine

c. Nucleic Acids

Uracil

Cytosine

Thymine

Adenine

Guanine

Figure 12–3 Examples of important organic nitrogen compounds in soil: (a) common amino acids, (b) an amino sugar, and (c) common nucleic acids.

a.

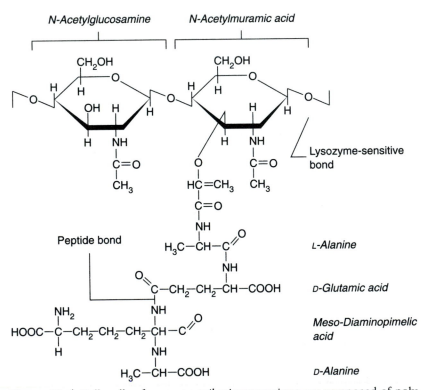

b.

Figure 12–4 Cell walls of common soil microorganisms are composed of polymers of amino sugars: (a) fungal cell walls contain chitin that consists of N-acetylglucosamine moieties connected by β1-4 linkages, and (b) bacterial cell walls contain a peptidoglycan layer whose backbone is a polymer of N-acetylglucosamine and N-acetylmuramic acid connected by β1-4 linkages that has an oligopeptide side chain.

Table 12–2 **Classic fractionation of soil nitrogen based on acid hydrolysis.**

Form of nitrogen	Definition and method	Typical range (% of soil N)
Acid insoluble-N	Largely aromatic N. Nitrogen remaining in soil residue following acid hydrolysis (6 M HCl).	10–20
Ammonia-N	Exchangeable NH_4^+ plus amide N. Ammonia recovered from hydrolysate by steam distillation with MgO.	20–35
Amino acid-N	Protein, peptide, and free amino acid N. Determined by ninhydrin reaction of hydrolysate.	30–45
Amino sugar-N	Microbial cell walls. Ammonia recovered from hydrolysate by steam distillation with phosphate-borate buffer at pH 11.2 minus the ammonia-N fraction.	5–10
Hydrolyzable unknown-N	Largely unknown but contains non-α-amino-N of arginine, tryptophan, lysine, and proline. The hydrolyzable N not accounted for as ammonia, amino acids, or amino sugars.	10–20

Based on Stevenson (1982).

One purpose of soil organic matter fractionation schemes is to determine which fractions are most active in nutrient turnover. One method of measuring turnover is to add inorganic nitrogen labeled with [15]N (a stable, heavy isotope) and measure how it is partitioned among the various organic-nitrogen fractions. Such studies of recently immobilized nitrogen have shown that the amino acid and unknown hydrolyzable nitrogen fractions are often relatively enriched in ^{15}N whereas the acid insoluble fraction shows very little incorporation of the labeled nitrogen.

Soil inorganic nitrogen. Unlike soil organic nitrogen, the important inorganic forms of nitrogen in soil ecosystems are well characterized, primarily because most inorganic-nitrogen compounds can be readily separated and measured. Inorganic-nitrogen pools in soil are usually small compared to organic nitrogen, but are nevertheless important because they serve as substrates, metabolic intermediates, alternate electron acceptors, or products of the many biological nitrogen transformations. Some key inorganic forms of nitrogen and their characteristics are shown in Table 12–3.

Nitrogen Mineralization (Ammonification)/Immobilization

Nitrogen mineralization has several meanings. It is sometimes used in a generic sense for the production of inorganic nitrogen, both ammonium and nitrate, and sometimes more narrowly for the production of ammonium. The increase (or sometimes decrease) in inorganic nitrogen is most often called *net nitrogen mineralization* because it represents the sum of the concurrent ammonium production and consumption processes. It is more correct to use **ammonification,** or *gross nitrogen mineralization,* to describe the biological transformation of organic nitrogen to ammonium.

Table 12–3 **Important inorganic nitrogen compounds found or produced in soil.**

Compound	Formula	Oxidation state	Form in soil	Major attributes
Ammonium	NH_4^+	−3	Fixed in clay lattice, dissolved, as gaseous ammonia (NH_3)	Cationic, rather immobile, volatilizes as NH_3 at high pH, assimilated by plants and microbes, substrate for autotrophic nitrification (NH_3 oxidation)
Hydroxylamine	NH_2OH	−1	Not detected	Intermediate in NH_3 oxidation
Dinitrogen	N_2	0	Gas	Largest pool of N, relatively insoluble, substrate for N_2 fixation, end product of denitrification
Nitrous oxide	N_2O	+1	Gas, dissolved	Greenhouse gas and implicated in ozone destruction, very soluble, an intermediate in denitrification, by-product of nitrification
Nitric oxide	NO	+2	Gas	Chemically reactive, an intermediate in denitrification, by-product of nitrification
Nitrite	NO_2^-	+3	Dissolved	Normally present at very low concentrations, toxic, product of NH_3 oxidation, substrate for NO_2^- oxidation, an intermediate in denitrification
Nitrate	NO_3^-	+5	Dissolved	Anionic, mobile, readily leached, assimilated by plants and microbes, end product of nitrification, substrate for denitrification

Less confusion surrounds the term **immobilization** because it almost always describes the conversion of ammonium to organic nitrogen, primarily as a result of the assimilation of ammonium by the microbial biomass, a process which temporarily renders the nitrogen unavailable for plants or microbes. Less frequently, immobilization may refer to the assimilation of both ammonium and nitrate. The assimilation of nitrate by the microbial biomass is usually specified explicitly as nitrate immobilization. It is important to remember, however, that nitrate assimilation requires that nitrate be reduced to ammonium before the nitrogen can be incorporated into cell constituents.

Ammonification

The conversion of organic-nitrogen compounds to ammonium is mediated by enzymes produced by microbes and soil animals. Production of ammonium often involves several steps. Extracellular enzymes first break down organic-nitrogen polymers, and the resulting monomers pass across the cell membrane and are fur-

Table 12–4 **Extracellular enzymes involved in microbial nitrogen mineralization.**

Substrates	Enzymes	Products
Proteins	Proteinases, proteases	Peptides, amino acids
Peptides	Peptidases	Amino acids
Chitin	Chitinase	Chitobiose
Chitobiose	Chitobiase	N-acetylglucosamine
Peptidoglycan	Lysozyme	N-acetylglucosamine and N-acetylmuramic acid
DNA and RNA	Endonucleases and exonucleases	Nucleotides
Urea	Urease	NH_3 and CO_2

Based on Ladd and Jackson (1982).

ther metabolized, with the resulting production of ammonium, which is released into the soil solution.

Extracellular enzymes important in nitrogen transformations. The major extracellular enzymes produced by microorganisms depolymerize proteins, aminopolysaccharides (microbial cell walls), and nucleic acids and hydrolyze urea (Table 12–4).

Proteins are broken down by a wide variety of proteinases, also called proteases, and peptidases. Proteinases work on large proteins whereas peptidases may cleave dipeptides or split off an individual amino acid. These enzymes are classified according to their active site and substrate specificity, but all hydrolytically cleave peptide bonds to ultimately produce individual amino acids. Examples of proteolytic enzymes isolated from soil microbes include subtilisin, clostripain, and thermolysin.

Although microbial cell walls are thought to be relatively recalcitrant in soils, several common extracellular enzymes will degrade these polymers. Chitin (Fig. 12–4), which forms the cell wall of many fungi and is also part of insect exoskeletons, is degraded by the combined activities of chitinase and chitobiase. Chitinase breaks chitin, a polymer of N-acetylglucosamine, into dimers (chitobiose), which are subsequently cleaved to two molecules of N-acetylglucosamine by chitobiase. This process is analogous to the enzymatic degradation of cellulose (Chapter 11). Several enzymes work to degrade the peptidoglycan portion of bacterial cell walls. Lysozyme is perhaps the most well known. It breaks the β 1,4 linkage between N-acetylmuramic acid and N-acetylglucosamine. Individual aminosugar monomers are the end products of the extracellular enzymes that degrade microbial cell walls.

Nucleic acids are degraded by ribonucleases (RNases) and deoxyribonucleases (DNases), which hydrolyze the ester bonds between the phosphate groups and pentose sugars of nucleic acids. The known types of RNases and DNases are divided into exonucleases, which split off a single nucleotide from the end of the nucleic acid polymer, or endonucleases, which cleave within the nucleic acid polymer. Individual nucleotides are the ultimate product of the nucleases.

Urease is another important extracellular enzyme involved in ammonification. Ureases hydrolyze urea into carbon dioxide and ammonia. Nickel is the cofactor associated with the active site of at least some ureases. Ureases function in the utilization of natural sources of urea (e.g., animal wastes) but perhaps most importantly in making the nitrogen in urea fertilizer available to plants.

Considerable research has focused on the interactions between extracellular enzymes and soil mineral and organic constituents. These interactions are complex. For example, both the enzyme, which is a protein, and the substrate may be adsorbed onto clay surfaces. This may act to stabilize the enzyme or substrate and protect it from degradation. This type of stabilization provides one explanation for the presence of free DNA in soils. If the active conformation of an extracellular enzyme is altered by adsorption, this will likely inactivate the enzyme, but if the catalytic site is not affected, the enzyme may remain active. In the latter case, the protected, active enzyme may be an important catalyst as long as its substrate is accessible.

Intracellular enzymes important in nitrogen transformations. In most cases, the final production of ammonium occurs within microbial cells through the action of intracellular enzymes. Of course, some of these intracellular enzymes may become "extracellular" when microbial cells are lysed.

Two types of nitrogen are found in amino acids: the amine (NH_2-CR_3) and amide (NH_2-$CR=O$) functional groups. The amide groups of asparagine and glutamine are cleaved by asparaginase and glutaminase. Amino nitrogen is released primarily by amino-acid dehydrogenases and amino-acid oxidases in a process known as *deamination*. Dehydrogenases, such as glutamate dehydrogenase, use NAD as a cofactor to accept electrons.

Amino sugars are metabolized in two steps. First, the amino sugar is phosphorylated by a *kinase* and then ammonia is released through a deamination reaction.

Degradation of nucleotides and the release of ammonium typically requires several steps. First nucleotides are hydrolyzed to produce nucleosides and PO_4^{3-}. Following the dephosphorylation, the nucleosides are further hydrolyzed to purine or pyrimidine bases and pentose sugar moieties. Normal metabolic pathways then release ammonium during the catabolism of purines and pyrimidines, with urea as a prominent intermediate.

In most instances the microbial degradation of amino acids, amino sugars, and nucleic acids is driven by the need of heterotrophic microbes for energy and carbon. Thus, the ammonium released as a result of ammonification can be considered a byproduct of catabolism. At least in pure culture studies, microbes grow better with a carbohydrate as a carbon and energy source and ammonium or nitrate as a source of nitrogen than if grown on organic-nitrogen compounds alone.

Immobilization (Assimilation)

Microbes and other organisms assimilate ammonium by two primary pathways (Fig. 12–5): glutamate dehydrogenase and glutamine synthetase-glutamate synthase (GOGAT). When ammonium is present in relatively high concentrations (> 0.1 mM or about 0.5 mg N kg^{-1} soil), glutamate dehydrogenase, acting with $NADPH_2$ as a coenzyme, can add ammonium to α-ketoglutarate to form glutamate.

In most soils, ammonium is present at rather low concentrations, which results in low intracellular ammonium concentrations. Under these conditions, the second ammonium assimilation system is operable. The GOGAT pathway is complex. The first step requires ATP to add ammonium to glutamate to form glutamine. The second step transfers the ammonium from glutamine to α-ketoglutarate to form two glutamates. Once ammonium has been incorporated into glutamate, it can then be transferred to other carbon skeletons by various transaminase reactions to form additional amino acids.

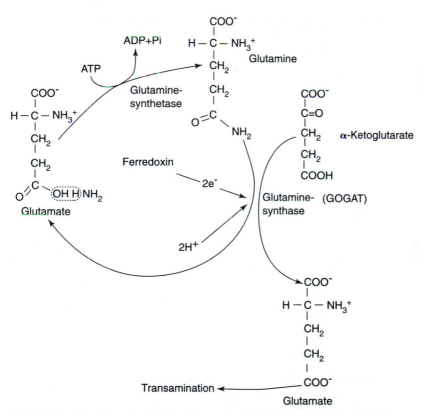

Figure 12–5 Two main pathways of ammonium assimilation: (a) glutamate dehydrogenase, which is a reversible reaction that works at higher ammonium concentrations and (b) glutamine synthase-glutamate synthetase (GOGAT), which is induced and functions at low ammonium concentrations. This uptake system requires energy.

Net Ammonium Production or Consumption

Several factors influence whether there is net production or consumption of ammonium by microorganisms in soil. The general principle is that net immobilization of ammonium occurs if the availability of nitrogen is limiting; otherwise, net production occurs. In most soils, the growth and activity of heterotrophic microorganisms are limited primarily by the amount of available carbon; net nitrogen mineralization is the norm in these soils. This is actually not surprising, because plants, which require inorganic nitrogen, grow well in most soils.

When plant residues are returned to the soil or organic amendments are added, we can predict what effect this addition has on nitrogen availability. Adding a material high in carbon, such as sawdust, will likely immobilize inorganic nitrogen, whereas adding a material relatively high in nitrogen, such as manure, will provide available nitrogen. Most organic materials are about 45% carbon by mass; consequently, the ratio of carbon to nitrogen is determined largely by the concentration of nitrogen in the material. Decades of research have shown that when organic amendments with **C/N ratios** below 20/1 are added to soils, net ammonium production results; at wider C/N ratios, inorganic nitrogen from the soil is immobilized. This critical, or "break-even," C/N ratio can be calculated from basic principles based on knowledge of the C/N ratio of the heterotrophic microorganisms and the **yield coefficient,** the amount of substrate carbon converted to microbial biomass (Box 12–1).

The C/N ratio of soil microorganisms ranges from 4 or 5 for bacteria to as high as 15 for fungi. Given that fungal biomass is often about twice that of bacteria in many soils, a typical C/N ratio for soil microbial biomass is about 8. Yield coefficients vary widely, depending on the type of organic matter (substrate quality), type of microorganism, and environmental conditions. Readily degradable organic compounds, such as simple sugars, may have yield coefficients as high as 0.6, whereas assimilation efficiencies for complex, recalcitrant compounds, such as lignin, may be less than 0.1. A reasonable average yield coefficient for plant-derived substrates is about 0.4. Fungi are typically more efficient than bacteria and thus have higher yield coefficients, perhaps 0.5 versus 0.4. Environmental conditions that stress microorganisms would generally decrease the yield coefficient because more energy is required for cell maintenance than for growth. Another example of the effect of environmental factors is that yield coefficients are often much lower under anaerobic conditions, primarily because anaerobic metabolism normally produces less energy per mole of substrate.

Net production of ammonium is influenced not only by environmental factors and the C/N ratio of substrates and microbes but also by other biotic factors. Most important of these is the role that soil animals play as predators of the primary decomposers, bacteria and fungi. About 30% of the yearly net nitrogen mineralization is directly released by the activities of soil animals, such as protozoa and nematodes. When soil animals prey upon microorganisms, ammonium is often released as a waste product because the predators have a C/N ratio similar to their prey, which results in an excess of nitrogen due to loss of carbon as carbon dioxide during metabolism. This phenomenon may be important in the rhizosphere where "grazing" by soil protozoa facilitates liberation of available nitrogen for plant growth.

| **Box 12–1** |

Calculating the Critical C/N Ratio that Determines Whether Nitrogen is Mineralized or Immobilized.

Start with these basics:

- Fungi typically make up about two-thirds of the total microbial biomass; bacteria make up about one-third.

- Fungi typically convert about 44% of the carbon of readily decomposable organic matter into cell biomass; thus their yield coefficient (Y) is 0.44.

- Bacteria typically convert about 32% of the carbon of readily decomposable organic matter into cell biomass; thus their yield coefficient (Y) is 0.32.

- Fungal cells commonly have a C/N ratio of about 10.

- Bacterial cells commonly have a C/N ratio of about 4.

We can calculate that the decomposing organic substrate must have a C/N ratio of 20 or less in order to meet the nitrogen needs of the microbial decomposers by the following steps:

Step 1. Calculate the average microbial yield coefficients and C/N ratios;

$$Y = (\tfrac{2}{3})0.44 + (\tfrac{1}{3})0.32 = 0.4$$

$$C/N = (\tfrac{2}{3})10 + (\tfrac{1}{3})4 = 8$$

Step 2. Calculate how much microbial biomass carbon is produced;

100 g substrate C → 60 g CO_2−C + 40 g microbial biomass C

Step 3. Calculate how much microbial biomass nitrogen is produced;

40 g microbial biomass C ÷ C/N ratio of 8 = 5 g microbial biomass N

Step 4. Calculate the substrate C/N ratio that would be needed;

substrate C/N ratio = 100 g substrate C ÷ 5 g substrate N = 20

Fate of Ammonium in Soil

In addition to the mineralization/immobilization cycle, ammonium has several other fates in soil. It can be chemically held on cation exchange sites or become fixed in the lattice of clay minerals (ammonium fixation), such as illite and vermiculite. Ammonium may react chemically with organic compounds, such as quinones, or it may be volatilized at high pH. Major biological fates are plant uptake, microbial assimilation, or oxidation to nitrate by nitrifying microorganisms.

Nitrification

Nitrification is the microbial production of nitrate from the oxidation of reduced nitrogen compounds. Most often we think of autotrophic nitrification, the two-step, two-organism process of oxidizing ammonium to nitrate, in which the inorganic nitrogen serves as the energy source for the nitrifying bacteria. The first step of autotrophic nitrification is ammonia oxidation, the conversion of ammonium (actually, ammonia at the enzyme level) to nitrite by the ammonia-oxidizing bacteria of the "Nitroso-" genera (Table 12–5). Nitrite is then oxidized to nitrate by the nitrite—oxidizing bacteria of the "Nitro-" genera.

In addition to the oxidations by the autotrophic nitrifying bacteria, other microbes can produce nitrite and nitrate by enzymatic oxidation processes that are not linked to microbial growth. For example, the many genera of methane-oxidizing bacteria contain a membrane-bound methane monooxygenase enzyme that will oxidize ammonia as well as methane, which is an interesting linkage between carbon and nitrogen cycling. Perhaps more widespread is heterotrophic nitrification, the oxidation of ammonium or organic-nitrogen compounds by a variety of heterotrophic bacteria and fungi.

Table 12–5 **Chemoautotrophic nitrifying bacteria.**

Genus	Species	Characteristics of genera
NH$_3$ oxidizers		
Nitrosomonas	*europeae* *eutrophus* *marina*	Ellipsoidal to rod-shaped cells; intracellular membranes as peripheral flattened vesicles; 46–54% G+C; some have urease; obligate chemoautotrophs; soil, freshwater, marine, sewage
Nitrosococcus	*nitrosus* *mobilis* *oceanus*	Spherical to ellipsoidal cells; intracellular membranes as peripheral or central flattened vesicles; 48–50% G+C; some have urease; obligate chemoautotrophs; marine, soil
Nitrosospira	*briensis*	Spiral cells; intracytoplasmic membrane system, invagination of protoplasm; 53–55% G+C; some have urease; obligate chemoautotroph; freshwater and soil
Nitrosolobus	*multiformis*	Pleomorphic, lobate cells; compartmentalized protoplasm; 53–56% G+C; some have urease; obligate chemoautotroph; soil
Nitrosovibrio	*tenuis*	Curved, slender rods; extensive intracytoplasmic membrane lacking, invagination of protoplasm; 54% G+C; some have urease; obligate chemoautotroph; soil
NO$_2^-$ oxidizers		
Nitrobacter	*winogradskyi* *hamburgensis* *vulgaris*	Pleomorphic, rod- to pear-shaped cells; no separate peptidoglycan wall; intracytoplasmic membranes as polar, flattened vesicles; 59–60% G+C; facultative chemoautotrophs and mixotrophs; soil
Nitrospina	*gracilis*	Slender rods; nonmotile; no intracytoplasmic membrane system; 58% G+C; obligate chemoautotroph; obligate halophile; marine
Nitrococcus	*mobilis*	Cocci; motile; intracytoplasmic membranes as tubules; 61% G+C; obligate chemoautotroph; obligate halophile; marine
Nitrospira	*marina*	Spiral or vibrioid cells; nonmotile; no intracytoplasmic membrane system; 50% G+C; chemoautotroph and mixotroph; marine and soil

Based on Prosser (1989) and Holt et al. (1994).

Ammonia Oxidation

Ammonia-oxidizing bacteria are thought to be a relatively defined and coherent group. Recent phylogenetic work based on 16S rRNA sequences has largely confirmed this: except for two strains of *Nitrosococcus oceanus,* all autotrophic ammonia oxidizers are tightly clustered together phylogenetically. Representatives from each of the five genera of ammonia oxidizers have been isolated from soil. Although *Nitrosomonas* has been the best characterized and most well studied ammonia oxidizer, particularly with respect to its enzymology and the biochemistry of ammonia oxidation, *Nitrosolobus* is thought to be the dominant ammonia oxidizer in many soils. *Nitrosospira* is associated with acid soils.

The overall reaction for the conversion of ammonia to nitrite is:

$$NH_3 + 1.5O_2 \rightarrow NO_2^- + H^+ + H_2O$$

This oxidation is a $6e^-$ transfer that yields 271 kJ (65 kcal) mol^{-1} NH_3. The first step in the reaction is the conversion of NH_3 to NH_2OH (hydroxylamine) by the membrane-bound *ammonia monooxygenase* enzyme:

$$NH_3 + O_2 + 2H^+ + 2e^- \rightarrow NH_2OH + H_2O$$

This reaction is endergonic and requires a small amount of energy. It is not coupled to ATP synthesis. Like many monooxygenase enzymes, ammonia monooxygenase has broad substrate specificity. It can oxidize methane but does so at much lower rates than methane-oxidizing bacteria. Ammonia monooxygenase has been shown to **cometabolize** several other small organic compounds, including some halogenated organics such as trichloroethylene, chlorinated ethanes, and chloroform. A practical characteristic of the broad substrate specificity of ammonia monooxygenase is that it will bind irreversibly to acetylene. Thus, acetylene is a useful inhibitor of ammonia oxidation. Several other inhibitors of ammonia oxidation have been developed and used in agriculture (Box 12–2).

Hydroxylamine is converted through several undefined steps to nitrite with an overall reaction of:

$$NH_2OH + H_2O \rightarrow NO_2^- + 5H^+ + 4e^-$$

This is an energy-yielding reaction, with two of the electrons produced passing down the electron-transport chain to oxygen while the other two are used in the ammonia monooxygenase reaction. The initial step of this reaction is catalyzed by hydroxylamine oxidoreductase, a soluble enzyme. The nitroxyl radical (HNO) is thought to be produced from the oxidation of NH_2OH and may be the source of some of the nitric oxide (NO) that is released as a by-product of nitrification. The final step(s) in the production of nitrite are not well-defined.

Two other products of ammonia oxidation are nitrous oxide (N_2O) and acidity. Ammonia oxidizers contain a nitrite reductase, which is capable of reducing NO_2^- to N_2O. Under aerobic conditions, the production of nitrous oxide by this mechanism is small, less than 1% of the ammonia oxidized. As oxygen availability decreases, however, relatively more nitrous oxide is produced as nitrite is used as the electron acceptor. In some habitats, nitrification may be a major source of gaseous nitrogen oxides.

Box 12–2

Nitrification Inhibitors. For several decades scientists have attempted to find specific inhibitors of ammonia oxidation with the ultimate goal of commercializing these compounds for use in agriculture. The initial motivation was to increase the efficiency of fertilizer nitrogen (typically ammonium compounds or urea) use by crop plants to maximize economic yield. More recently the goal has expanded to include minimizing environmental consequences of nitrate in excess of plant needs. Some of the more successful or commercially available nitrification inhibitors are listed in the following table.

Effectiveness of nitrification inhibitors expressed as average percent inhibition of nitrification in three soils treated with 5 mg active ingredient kg^{-1} soil that had been amended with 200 mg $(NH_4)_2SO_4$-N kg^{-1} soil and incubated at 25°C.

Common name(s)	Chemical	Inhibition (%)	
		14 d	28 d
	2-Ethynylpyridine	97	87
	Phenylacetylene*	92	55
Dwell	Etridiazole	90	75
N-serve, nitrapyrin	2-Chloro-6-(trichloromethyl)pyridine	85	65
ATC	4-Amino-1,2,4-triazole	87	60
	2,4-Diamino-6-trichloromethyl triazine	76	41
DCD, dicyan	Dicyandiamide	61	15
AM	2-Amino-4-chloro-6-methylpyrimidine	60	37
ST	Sulfathiazole	52	17
Tu	Thiourea	2	0

*Average of two soils after 10 and 30 days incubation.
Data from McCarty and Bremner (1986, 1990).

Currently the most widely used nitrification inhibitors are probably N-serve and DCD, although Dwell, a more recent product, would seem to be a more effective choice. The acetylenic compounds 2-ethynylpyridine and phenylacetylene also seem to show promise, along with wax-coated calcium carbide. As the wax coat surrounding the calcium carbide granules breaks down, the calcium carbide reacts with water to form acetylene, a potent inhibitor of ammonia oxidation. Field studies with irrigated cotton and flooded rice have shown these acetylenic compounds to be at least as good as N-serve in increasing fertilizer nitrogen recovery (Freney et al., 1993; Keerthisinghe et al., 1993).

Ammonia oxidation acidifies soils by releasing one mole of H^+ for every mole of ammonia oxidized. This presents a paradox, as nitrifying bacteria generally grow best at neutral pH and their activity is often inhibited by low pH. Nevertheless, the production of acidity by ammonia oxidizers has been shown to be responsible for lowering the pH of natural and agricultural soils.

Nitrite Oxidation

Nitrite-oxidizing bacteria are phylogenetically more diverse than the ammonia oxidizers. Most soil isolates are *Nitrobacter* spp., although a *Nitrospira* strain has also been isolated from soil.

The oxidation of nitrite to nitrate is a one-step reaction, with the following stoichiometry:

$$NO_2^- + 1/2O_2 \rightarrow NO_3^-$$

Nitrite is oxidized to nitrate by a membrane-bound *nitrite oxidoreductase*, which transfers an oxygen from water and transfers a pair of electrons to the electron-transport chain for the production of ATP via oxidative phosphorylation:

$$NO_2^- + H_2O \rightarrow NO_3^- + 2H^+ + 2e^-$$

This reaction yields 77 kJ (18 kcal) mol^{-1} nitrite utilized, about one-third that of ammonia oxidation. Nitrite oxidation can be competitively inhibited by chlorate (ClO_4^-), which is useful in experimental studies to determine rates of autotrophic versus heterotrophic nitrification.

Unlike ammonia oxidizers, which are strict autotrophs, nitrite oxidizers are capable of heterotrophic growth under some limited conditions. Even anaerobic heterotrophic growth may be possible, with nitrite oxidoreductase reducing nitrate to nitrite. Heterotrophic growth by nitrite oxidizers is much slower than other heterotrophic bacteria and slower than when nitrite oxidizers grow autotrophically.

Heterotrophic Nitrification

Several heterotrophic microorganisms oxidize either ammonium or organic nitrogen to nitrite or nitrate. Heterotrophic nitrifiers include both fungi (e.g., *Aspergillus*) and bacteria (e.g., *Alcaligenes, Arthrobacter* spp., and some actinomycetes). A particularly interesting bacterium is *Thiosphaera pantotropha,* which is a heterotrophic nitrifier that can also denitrify under aerobic conditions. Unlike the autotrophic nitrifiers, heterotrophic nitrifiers gain no energy through this activity. In fact, it is uncertain what benefit heterotrophic nitrifiers gain by oxidizing organic nitrogen, although hydroxamic acids, which act as **siderophores,** compounds involved in iron acquisition, are one type of oxidized nitrogen product.

The relative importance of heterotrophic versus autotrophic nitrification is still debated. In pure cultures, the highest rates of nitrite or nitrate production are just one-tenth that of autotrophic nitrifiers, which would suggest that heterotrophic nitrifiers are of minor importance. The case is not as clear-cut in soils, however, where the relative rates of the two nitrification processes have been assessed with inhibitors (e.g., nitrapyrin and acetylene are thought to only block autotrophic nitrification) or the use of ^{15}N. For example, in one study most of the nitrate produced in a coniferous forest soil was from heterotrophic nitrification.

Factors Affecting Nitrification in the Environment

Many interacting factors control nitrification in soils. The decision tree shown in Figure 12–6 is one way of assessing the relative importance of these factors. The most important, or most commonly limiting, factors are listed at the top of the decision tree. If all factors are favorable, then nitrification is possible; if any factor is unfavorable, then significant rates of nitrification are unlikely. Implied by this organization is that the factors affecting nitrification rates are multiplicative (i.e., they interact). The dashed line shows that alleviating a limiting factor has the potential to increase the growth of nitrifiers, hence increasing their populations in soil.

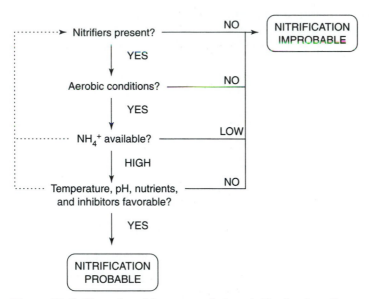

Figure 12–6 Hierarchy of factors regulating nitrification in soil.
The dotted lines suggest that these factors may limit nitrifier populations.

Nitrifier populations. For nitrification to occur, either autotrophic or heterotrophic nitrifiers must be present. Nitrifiers are present in most soils, however,
they may be present in populations too low to be of much importance in producing
nitrate. For example, if we extrapolate from the activity of pure cultures of autotrophic nitrifying bacteria, we can calculate that about 3×10^5 nitrifiers g^{-1} soil
are required for a nitrification rate of 1 mg N kg^{-1} day^{-1} (Box 12–3). Unfertilized
soils contain far fewer nitrifiers than this, often 10^3 to 10^4 g^{-1}, but upon nitrogen fertilization, nitrifier populations have been observed to increase to more than 10^6 g^{-1}.
A similar response is often seen when wildland soils are disturbed. In their natural
state, many wildland soils have very low concentrations of nitrate and small populations of nitrifiers. If disturbance increases the availability of ammonium, nitrifier
populations and rates of nitrification often increase gradually until they reach a new,
higher steady state.

Soil aeration. Because nitrifiers are almost exclusively aerobic microorganisms,
soils must have sufficiently high concentrations or fluxes of oxygen for nitrification
to occur. Like general aerobic heterotrophic activity, nitrification is typically optimal
when bulk soils are near field capacity or at about 60% water-filled pore space. It
should be noted, however, that even flooded soils and sediments normally have a
narrow aerobic zone where nitrification occurs. As oxygen becomes more limiting,
autotrophic nitrifiers produce relatively more nitric oxide and nitrous oxide.

Substrate availability. Provided that aerobic conditions exist, the most important regulating factor for nitrification is substrate availability, particularly ammonium
availability. Studies have shown that ammonium and nitrite oxidation follow
Michaelis-Menten kinetics (Chapter 8). The saturation constants for these oxidations

Box 12–3

Using Theoretical Nitrifier Activity to Estimate the Size of Soil Nitrifier Populations.
  Given that:

- Actively growing ammonia oxidizers in pure culture oxidize ammonium at a rate of about 10 fmol* NH_4^+ $cell^{-1}$ $hour^{-1}$.

- A reasonable net nitrification rate in soil is about 1 mg N kg^{-1} day^{-1}.

We can estimate the population of ammonia oxidizers to be about 3×10^5 g^{-1} soil;

$$\frac{1 \text{ mg N}}{\text{kg d}} \times \frac{1 \text{ d}}{24 \text{ h}} \times \frac{1 \text{ mmol N}}{14 \text{ mg N}} \times \frac{\text{kg}}{10^3 \text{ g}} \times \frac{10^{12} \text{ fmol}}{1 \text{ mmol N}} \times \frac{\text{cell h}}{10 \text{ fmol}} = 2.98 \times 10^5 \text{ cell g}^{-1}$$

*f = femto = 10^{-15}

---

are in the same range as typical soil concentrations of ammonium and nitrite, which suggests that substrate availability is often limiting to growth and activity. This is consistent with the previous description of nitrifier populations being limited by available substrate.

Because nitrification activity is often dominated by autotrophic nitrifiers, it is possible that carbon dioxide concentrations may also influence the growth of nitrifiers. The higher carbon dioxide concentration found in soils compared to the atmosphere may be beneficial to nitrifying bacteria, as long as oxygen does not become limiting. Carbonate equilibrium may also help to poise the soil pH at a level more favorable for nitrifiers.

Soil pH. Nitrification rates are often low in soils below pH 4.5, particularly in agricultural soils. At high pH, nitrite can accumulate because of greater inhibition of nitrite oxidizers relative to ammonia oxidizers. These observations, along with the fact that most isolates of autotrophic nitrifying bacteria grow best at neutral pH, support the generalization that autotrophic nitrifiers are neutrophilic. Nevertheless, high rates of nitrification or high concentrations of nitrate have been observed in many acid (pH < 4.5) soils. Several explanations for this apparent paradox include acidophilic autotrophic nitrifiers, heterotrophic nitrifiers, and alkaline microsites.

Several ammonia-oxidizing and nitrite-oxidizing bacteria have been isolated from low pH soils, with *Nitrosospira* and *Nitrobacter* being the most common genera. With the exception of a few strains of *Nitrobacter* isolated from acid forest soil (Hankinson and Schmidt, 1984), however, nitrifier isolates are neutrophilic or alkalophilic in pure culture. More recently co-cultures of *Nitrosospira* and *Nitrobacter* have been isolated that hydrolyze urea and produce nitrate (De Boer and Laanbroek, 1989). Adding urea stimulates nitrification above the small increase in pH associated with urea hydrolysis. It appears that acidophilic nitrifiers may exist, or at least operate, as consortia in acid soils.

Perhaps the most common explanation given for nitrate production in acid soils is the presence of heterotrophic nitrifiers. As mentioned previously, studies with either inhibitors that are thought to be specific for autotrophic nitrifiers or $^{15}NH_4^+$ have shown significant heterotrophic nitrification in some acid soils. Most isolated heterotrophic nitrifiers are not acidophilic, however.

In a habitat as diverse as soil, it would not be surprising to find microsites of higher pH. These might be associated with surfaces of minerals, organic matter, or even roots; however, conclusive evidence is lacking and empirical evidence is mixed. It is perhaps more likely that microsites are created by the activity of microorganisms themselves. For example, ammonia released during mineralization of organic nitrogen by heterotrophs or from urea hydrolysis may alter microsite pH. Such an effect would be difficult to distinguish from enhanced substrate availability, however. Thus we see that the puzzle of nitrification in acid soils remains to be solved.

Miscellaneous soil and environmental factors. Clearly factors such as temperature, water potential, salinity, and availability of nutrients other than nitrogen all have the potential to affect the activity of nitrifiers. Because of their slow growth rate and relatively inefficient metabolism, nitrifiers are thought to be more sensitive to temperature, particularly low temperatures, than common heterotrophs. Psychrophilic nitrifiers have been identified, however. There have also been indications that phosphorus availability may limit nitrification rates in some soils. Some scientists have suggested that nitrification is a sensitive indicator of alterations in the soil environment.

Allelochemical inhibitors. The observation of low nitrate concentrations in soils of climax communities in natural ecosystems led to a theory of specific inhibition of nitrification by allelochemicals produced by the climax vegetation. This theory suggests that tannins and polyphenols are among the more important allelochemical agents. Subsequent research has demonstrated that this allelochemical theory is probably not the major reason for the low soil nitrate concentrations. Active competition for nitrate by plant uptake and microbial immobilization are probably dominant (Box 12–4). In fact, several studies that examined a wide suite of proposed allelochemicals have failed to show any direct effect on nitrification rate or nitrifier populations. That is not to say that naturally occurring inhibitors of nitrification do not occur, rather that such compounds do not seem to be the primary reason for low soil nitrate concentrations in climax communities.

Fate of Nitrate in the Soil Environment

Like ammonium, nitrate has many competing fates in the soil ecosystem (Fig. 12–1). Because it is an anion, nitrate is easily leached. Removal of nitrate from the soil by leaching has several consequences. Obviously, nitrate leaching represents a loss of available nitrate from the plant-soil system. When nitrate is leached, it must be accompanied by an equivalent amount of cations to maintain charge balance. Thus soils are also depleted of cations when nitrate is leached. The leaching of basic cations, such as K^+ and Ca^{2+}, reduces the base saturation of a soil and increases exchangeable acidity. Nitrate that leaches eventually enters ground and surface waters, where it may have potentially adverse environmental effects. High concentrations of nitrate in surface waters can lead to **eutrophication** (the sudden enrichment of natural waters with excess nutrients which can lead to the development of algal blooms and other vegetation). Current federal regulations require that drinking

Box 12–4

Is Nitrification Really Inhibited in Soils of Mature Forests?   Observations of nitrate pool sizes and net production of nitrate during incubations of undisturbed forest soils led to the suggestion that nitrate is a relatively unimportant pool of available nitrogen in these soils and that nitrification was not an important process. Various reasons were given for this, including low populations of autotrophic nitrifiers, limited substrate (ammonium) availability, and allelochemical inhibition. When mature forests were disturbed by clear-cutting, for example, a subsequent increase in soil and streamwater nitrate concentrations and net nitrification rates often resulted. This change was in part explained by a reduced amount of plant competition for ammonium. Collectively, these observations led to the dogma that as ecosystems mature, losses of nitrogen are reduced and nitrogen cycling becomes more conservative, primarily because nitrification is turned off.

Work using ^{15}N isotope dilution methods to measure the gross rates of ammonification and immobilization and nitrification and nitrate immobilization has revised our view of how nitrogen is conserved in mature forest ecosystems. An excellent example is the study by Davidson et al. (1992), who applied these methods to forest soils from an old-growth (more than 100 years) mixed-conifer forest and a 10-year-old mixed-conifer plantation in northern California and found the following:

Nitrogen cycling characteristic*	Young forest	Old-growth forest
Inorganic N (mg N m^{-2})[†]		
NH$_4^+$	340	210
NO$_3^-$	300	80
N mineralization (mg N m^{-2} d^{-1})[‡]		
Net	6.8	2.6
Gross	90	280
N nitrification (mg N m^{-2} d^{-1})[‡]		
Net	6.6	−0.4
Gross	67	45

*Data are for top 9 cm of mineral soil.
†Mean of 7 dates from November through September.
‡Mean of 3 dates from November through April.

In agreement with past studies, the researchers found higher inorganic-nitrogen concentrations and greater net nitrogen mineralization and net nitrification in soil from the young, recently disturbed stand. However, gross rates of inorganic-nitrogen production were 10 to 100 times greater than net rates, with high rates of nitrification (gross nitrate production) being nearly as high in the soil from the old-growth forest as from the young stand. Thus, significant nitrification occurred in soils of both stands but immobilization rates of ammonium and nitrate were relatively higher compared to net production rates in the older stand. Probably a more significant point drawn from these data was how rapidly both the ammonium and nitrate pools turned over, with mean residence times on the order of days or hours and with greater turnover in the old-growth stands. This work agrees with the dogma that as forests age, soil nitrogen cycling becomes more conservative, not because nitrification is lessened but because nitrate immobilization is greater and not because turnover slows down, because it may actually increase. Nitrogen seems to be conserved because of enhanced immobilization, which is likely fueled by greater carbon availability (Hart et al., 1994).

water contain < 10 mg $NO_3{}^-$–N L^{-1}. High concentrations of nitrate are associated with methemoglobinemia (blue-baby syndrome), which is now quite rare. A further environmental hazard may be the production of carcinogenic nitrosamines from reactions between nitrite and secondary amines.

Assimilatory Nitrate Reduction

Nitrate can be assimilated by plants and microorganisms. The process of assimilatory nitrate reduction requires energy for the conversion of nitrate to ammonium and subsequent incorporation of ammonium into amino acids. Consequently, this process is regulated by nitrogen availability, and nitrate utilization is expected when energy is in excess relative to the concentrations of ammonium or organic-nitrogen compounds. For this reason, soil scientists believed assimilation of nitrate (also called *nitrate immobilization,* a term which emphasizes that the nitrogen has been made unavailable to other organisms) by soil microorganisms to be minor. However, there is growing evidence that nitrate immobilization is an important process in some soils (Box 12–4).

Plants vary in their ability and preference for ammonium and nitrate uptake. When both ammonium and nitrate are equally available in soil solution, it is energetically more favorable for plants to use ammonium because nitrate must be reduced prior to use by the plant. In many cases, however, plants are not energy limited, so reducing power is available to convert nitrate to ammonium. This is particularly true for plants that reduce nitrate in leaf tissue, where this reduction is coupled to light energy and photosynthesis. The relative energy cost of ammonium versus nitrate metabolism is even more difficult to calculate in a heterogeneous medium like soil, because it may be more efficient for a plant to use nitrate than to put energy into growing a more extensive root system to access the less mobile ammonium.

Dissimilatory Nitrate Reduction

Nitrate can also be reduced by dissimilatory processes (Table 12–6). In acidic soils of pH 5 or less, nitrogen gases can be produced chemically, with NO formation from the dismutation of nitrite being the major reaction. Nitrite can also react with the amino groups of organic-nitrogen compounds to form dinitrogen by the van Slyke reaction. These *chemodenitrification* reactions are typically minor compared to biological dissimilatory processes.

In most soils, respiratory denitrification is usually the major dissimilatory process that reduces nitrate. In *nonrespiratory denitrification,* organisms produce nitrous oxide under aerobic conditions but do not gain energy from this reaction. Nonrespiratory denitrification is accomplished by a wide range of bacteria, fungi, and algae; it has even been associated with higher plants and animals, although in these latter cases nitrous oxide is probably produced by associated microorganisms. With the exception of a few genera (e.g., *Propionibacterium, Lactobacillus,* and *Fusarium*), the fraction of nitrate converted to nitrous oxide is generally less than 25%. The importance of nonrespiratory denitrification in converting nitrate into nitrous oxide in nature is currently unknown, primarily because of the difficulty of distinguishing this process from others that produce nitrous oxide.

Nitrate-respiring bacteria convert nitrate to nitrite under anaerobic conditions. In doing so, they gain energy via oxidative phosphorylation (161 kJ or 38 kcal mol^{-1} $NO_3{}^-$).

Table 12–6 Processes that reduce nitrate.

Process	Products	Energy conserved	Regulated by	Soil condition where expected
Assimilatory				
NO_3^- assimilation*	NH_4^+	no	NH_4^+, organic N	low NH_4^+ concentration
Dissimilatory				
Chemodenitrification	$NO >> N_2$, N_2O	no		acidic
Nonrespiratory denitrification	N_2O	no	?	aerobic
NO_3^- respiration†	NO_2^-	yes	O_2	anaerobic
Dissimilatory NO_3^- reduction to NH_4^+	$NH_4^+ >> N_2O$	a few strains	O_2	anaerobic
Respiratory denitrification	$N_2 > N_2O > NO$	yes	O_2	anaerobic

Based on Tiedje (1994).
*Also known as NO_3^- immobilization.
†All known organisms that dissimilate NO_3^- to NH_4^+ are also NO_3^- respirers, but most NO_3^- respirers accumulate NO_2^-.

The enteric bacteria, which are facultative anaerobes, are typical examples; however, many of these can also further reduce nitrite to ammonium. Complete reduction of nitrate to ammonium is known as **dissimilatory nitrate reduction to ammonium,** or DNRA. Under anaerobic conditions, several genera of bacteria are capable of DNRA (Table 12–7). The overall reaction for DNRA is:

$$NO_3^- + 4H_2 + 2H^+ \rightarrow NH_4^+ + 3H_2O$$

A total of $8e^-$ are transferred during this reduction, with an energy yield of 600 kJ (143 kcal) mol^{-1} NO_3^-, or 150 kJ (36 kcal) mol^{-1} $2e^-$ transferred. The first step in this reaction is the conversion of nitrate to nitrite, which is linked to energy production via oxidative phosphorylation as it is with the nitrate respirers. Most bacteria that carry out DNRA do not gain any additional energy from the subsequent reduction of nitrite to ammonium. Some species of *Campylobacter, Desulfovibrio,* and *Wolinella* are exceptions that do apparently generate ATP from this final reduction step. Because most DNRA bacteria gain only minimal energy from this reduction, some researchers have suggested that this process serves to either detoxify the nitrite intermediate or to regenerate reducing equivalents through the reoxidation of $NADH_2$. The latter process seems to be the most important because:

- Under conditions of energy (carbon) limitation, nitrite accumulates in the medium, which suggests that nitrite may not be particularly toxic and that energy is not produced by further reduction to ammonium.

- Under conditions of excess carbon, ammonium is the major product, presumably because of the need to regenerate NAD.

These observations in pure culture agree with ecological studies that find DNRA bacteria to predominate over respiratory denitrifiers in carbon-rich environments, such as sediments and sewage sludge, whereas denitrifiers predominate in more carbon-poor habitats, such as soils.

Table 12–7 Bacteria that can dissimilate nitrate to ammonium (DNRA).

Genus	Typical habitat
Obligate anaerobes	
Clostridium	Soil, sediment
Desulfovibrio	Sediment
Selenomonas	Rumen
Veillonella	Intestinal tract
Wolinella	Rumen
Facultative anaerobes	
Citrobacter	Soil, wastewater
Enterobacter	Soil, wastewater
Erwinia	Soil
Escherichia	Soil, wastewater
Klebsiella	Soil, wastewater
Photobacterium	Seawater
Salmonella	Sewage
Serratia	
Vibrio	Sediment
Microaerophile	
Campylobacter	Oral cavity
Aerobes	
Bacillus	Soil, food
Neisseria	Mucous membranes
Pseudomonas	Soil, water

Based on Tiedje (1988).

Denitrification

The major form of dissimilatory nitrate reduction in soil is respiratory denitrification, more commonly known simply as **denitrification**. This refers to the reduction of nitrate to gaseous nitrogen products, principally dinitrogen and nitrous oxide, coupled to energy production via oxidative phosphorylation. It is an example of anaerobic respiration, where an alternate electron acceptor other than oxygen is used. The overall stoichiometry of the reaction is:

$$2NO_3^- + 5H_2 + 2H^+ \rightarrow N_2 + 6H_2O$$

Denitrification gains slightly less energy per mole of NO_3^- than DNRA (560 kJ or 134 kcal); however, it gains more per mole of $2e^-$ transferred (224 kJ or 53 kcal). The higher thermodynamic yield per $2e^-$ is consistent with the observation that denitrification is likely to be the most important reductive process in soils where heterotrophic organisms are often limited by available carbon.

Denitrifying bacteria comprise 0.1 to 5% of the total bacterial population of soils and represent a wide range of taxonomic groups (Table 12–8). This taxonomic diversity spans phylogenetic groups. Despite this diversity, soil denitrifiers are dominated by members of the genus *Pseudomonas,* with species of *Alcaligenes, Flavobacterium,* and *Bacillus* also common. Thus aerobic heterotrophs predominate, although autotrophic denitrifiers are also known. Furthermore, bacteria normally associated with other nitrogen transformations (e.g., *Azospirillum, Nitrosomonas,* and *Rhizobium*) denitrify under certain conditions. Because denitrification is described as an anaerobic process carried

Table 12–8 Genera of denitrifying bacteria.

Genus	Interesting characteristics of some species
Organotrophs	
Alcaligenes	Commonly isolated from soils
Agrobacterium	Some species are plant pathogens
Aquaspirillum	Some are magnetotactic, oligotrophic
Azospirillum	Associative N_2 fixer, fermentative
Bacillus	Spore former, fermentative, some species thermophilic
Blastobacter	Budding bacterium, phylogenetically related to *Rhizobium*
Bradyrhizobium	Symbiotic N_2 fixer with legumes, e.g., soybean
Branhamella	Animal pathogen
Chromobacterium	Purple pigmentation
Cytophaga	Gliding bacterium; cellulose decomposer
Flavobacterium	Common soil bacterium
Flexibacter	Gliding bacterium
Halobacterium	Halophilic
Hyphomicrobium	Grows on one-C substrates, oligotrophic
Kingella	Animal pathogen
Neisseria	Animal pathogen
Paracoccus	Halophilic, also lithotrophic
Propionibacterium	Fermentative
Pseudomonas	Commonly isolated from soil, very diverse genus
Rhizobium	Symbiotic N_2 fixer with legumes, e.g., alfalfa, clover
Wolinella	Animal pathogen
Phototrophs	
Rhodopseudomonas	Anaerobic, reduce SO_4^{2-}
Lithotrophs	
Alcaligenes	Use H_2, also heterotrophic, commonly isolated from soil
Bradyrhizobium	Use H_2, also heterotrophic, symbiotic N_2 fixer with legumes
Nitrosomonas	NH_3 oxidizer
Paracoccus	Use H_2, also heterotrophic, halophilic
Pseudomonas	Use H_2, also heterotrophic, commonly isolated from soil
Thiobacillus	S oxidizer
Thiosmicrospira	S oxidizer
Thiosphaera	S oxidizer, heterotrophic nitrifier, aerobic denitrification

Based on Firestone (1982) and Tiedje (1988, 1994)

out by prokaryotes, it is interesting to note the recent isolation of an aerobic denitrifying bacterium (*Thiosphaera pantotropha*) and of fungi that appear to have the capability for respiratory denitrification (Robertson and Kuenen, 1984).

Denitrification Enzymes

The denitrification pathway involves four reductive steps and their corresponding enzymes (Fig. 12–7). Dissimilatory nitrate reductase (Nar) is a membrane-bound enzyme that contains molybdenum/iron, and labile sulfur groups. It catalyzes the reduction of nitrate to nitrite, with the generation of ATP. This step is common to all organisms that dissimilate nitrate. Synthesis of Nar is inhibited by oxygen, as is the activity of existing enzyme.

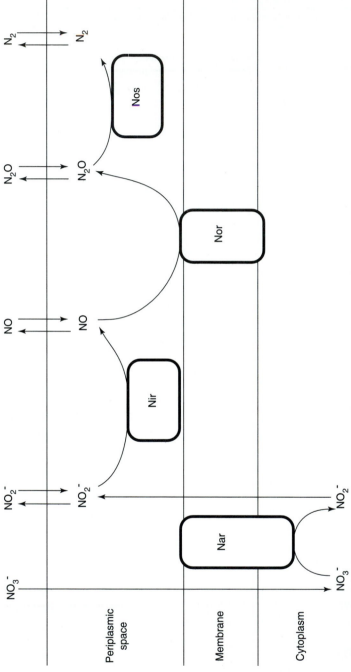

Figure 12–7 Denitrification pathway showing the location of the four denitrification enzymes: Nar—**ni**trate **r**eductase, Nir—**ni**trite **r**eductase, Nor—**ni**tric **o**xide **r**eductase, Nos—**ni**tro**us** oxide reductase. *Adapted from Ye et al. (1994). Used with permission.*

The reduction of nitrite to nitric oxide is a defining characteristic of denitrifiers. This step is catalyzed by nitrite reductase (Nir). Two forms of Nir are known: one contains copper (Cu-Nir), and the other contains cytochromes c and d_1 (heme-Nir). Heme d_1 is the active site of the heme-Nir. About two-thirds of the denitrifiers contain heme-Nir, including most of the *Pseudomonas* strains, *Alcaligenes, Paracoccus, Thiobacillus,* and *Azospirillum.* Copper is required for activity of Cu-Nir, which is less common than heme-Nir but more widespread taxonomically. It is found in some *Pseudomonas* and *Alcaligenes* strains, *Bacillus, Rhizobium, Thiosphaera, Nitrosomonas,* and others. Researchers have not determined whether the two types of Nir are related to performance of the denitrifiers in the environment. Nitrite reductase appears to be associated with the periplasmic (i.e., between the membrane and cell wall) side of the cell membrane. Like Nar, Nir synthesis is repressed by oxygen; however, it is less clear if its activity is directly inhibited by oxygen. The presence of nitrate induces the expression of Nir.

We now know that nitric oxide is an obligate intermediate in the denitrification pathway, and it is converted to nitrous oxide by the activity of nitric oxide reductase (Nor). Nitric oxide reductase is membrane bound and contains cytochromes b and c. Because it is membrane bound, the electron transport associated with Nor activity is presumably linked to ATP synthesis. An interesting point about nitric oxide reduction is that this is the step in which the N=N double bond is formed; however, the exact mechanism of this reaction is not yet known. Synthesis of Nor is repressed by oxygen and induced by nitrogen oxides.

The final enzyme of the denitrification pathway is nitrous oxide reductase (Nos), which reduces nitrous oxide to dinitrogen. Nitrous oxide reductase is a periplasmic protein that contains eight copper atoms. Although Nos is not associated with the cell membrane, energy production must be associated with this final reductive step because denitrifiers can grow with nitrous oxide as their sole electron acceptor. Like other denitrification enzymes, synthesis of Nos is regulated by oxygen and nitrogen oxides. Its activity is more strongly inhibited by oxygen, and is more sensitive to low pH than the other enzymes. Thus, under high oxygen or low pH conditions, relatively more nitrous oxide than dinitrogen is produced. Nitrous oxide reductase is also strongly inhibited by sulfide and acetylene. This latter characteristic is the basis for the so-called **acetylene block method,** which greatly spurred the study of denitrification in natural environments. In the presence of acetylene, nitrous oxide accumulates and can be measured with a gas chromatograph to quantify rates of denitrification.

Factors Affecting Denitrification in the Environment

As with nitrification, the regulation of denitrification can be thought of as a hierarchy from more to less important factors. The presence of denitrifiers is seldom a limitation. Denitrifiers make up a reasonably large fraction of the soil bacteria, probably because most of them normally exist as aerobic heterotrophs that switch to nitrate as an alternate electron acceptor when oxygen becomes unavailable. Therefore, in most soils the formation of anaerobic conditions is the most important controlling factor, followed in order by the availability of nitrate and carbon. Of course, there are always instances when other factors, such as temperature and soil pH, may be extreme enough to limit denitrification.

Soil aeration. Although examples of aerobic denitrification have been observed fairly recently, denitrification in soils and other natural habitats is predominantly an anaerobic process. Oxygen affects denitrification by regulating enzyme synthesis and by inhibiting enzyme activity. Enzyme synthesis is less sensitive to oxygen than is activity. Synthesis of Nar and Nir is derepressed when oxygen concentrations reach about one-tenth of atmospheric concentrations (2 kPa O_2 in the gas phase, which is in equilibrium with 29 μmol O_2 L^{-1} H_2O at 20°C). As oxygen concentrations decrease, inhibition of denitrifier enzyme activities is relieved sequentially, with Nar being the least oxygen sensitive and Nos being the most sensitive. The differential sensitivity of denitrifier enzyme activity explains why the ratio of nitrous oxide to dinitrogen increases as oxygen concentrations increase.

The oxygen concentration experienced by denitrifiers in soil is a complex function of many interacting factors that control soil aeration. Aeration in soils occurs predominantly by diffusion, although there may be some transport by convection or even of oxygen dissolved in percolating water. Diffusion of oxygen is directly proportional to the concentration gradient of oxygen, which is largely a function of heterotrophic respiratory activity consuming oxygen in the soil and inversely proportional to the path-length for diffusion. The proportionality constant is called the diffusion coefficient (D). It varies over several orders of magnitude depending on soil texture, water content, and tortuosity. For example, D varies from 0.208 cm^2 s^{-1} in air to 2.6×10^{-5} cm^2 s^{-1} in water (i.e., oxygen diffuses through water about 10,000 times more slowly than through air). A detailed example will be covered later as a case study, but the general principle is that rates of denitrification are generally greatest in wet soils when more than 80% of the pore space is filled with water and where there is reasonably high respiratory activity (Box 12–5).

Once anaerobic conditions are established, denitrification rates are most often limited by either nitrate or carbon availability. Which of these two is more limiting depends on their relative abundance, which is often related to soil type, plant community, or management practices.

Nitrate availability. In most natural soil systems, nitrate, the alternate electron acceptor, is more limiting than carbon even if heterotrophic microorganisms are carbon-limited under aerobic conditions. There are several reasons for this. First, the obligately aerobic heterotrophs, which make up the bulk of the microbial biomass, can no longer compete for carbon in the absence of oxygen. In effect, carbon is less limiting under anaerobic conditions. Second, in many wildland soils, such as forest soils, grasslands, and natural wetlands, net nitrogen mineralization and net production of nitrate are small. Habitats that are very anaerobic or experience long periods of anaerobiosis, such as some sediments and anaerobic digestors, are similarly limited in nitrate availability because nitrification is inhibited under anaerobic conditions. Finally, even in soils with relatively high rates of nitrification, there are many competing fates for the nitrates produced. Many of these fates, including leaching and DNRA, are enhanced under wet, anaerobic soil conditions.

Carbon availability. Researchers often find correlations between measures of carbon availability (e.g., respiration rates) and denitrification rates. This is evidence that denitrification rates are influenced by carbon availability; however, it is confounded because carbon utilization also influences oxygen supply. Nevertheless, some controlled studies have shown a positive response of denitrification to additions of carbon under anaerobic conditions. It is likely that carbon limitation is greatest in soils with high nitrification rates or large nitrate pools, such as fertilized agricultural soils.

Box 12–5

Anaerobic Microsites. An intriguing puzzle about denitrification is how this anaerobic process can occur in aerobic soils with nearly atmospheric concentrations of oxygen. Although at least one aerobic denitrifier is now known, most of the evidence points to the establishment of anaerobic zones within an otherwise aerobic soil profile. The most studied and best documented of these anaerobic microsites occur within large soil aggregates.

Using several simplifying assumptions and typical values for diffusion and consumption, it is possible to calculate how large a soil aggregate must be for an anaerobic microsite to develop (Smith, 1980). This calculation can be extended to estimate the anaerobic volume of soils using aggregate size distributions and to relate the anaerobic volume to denitrification rates.

The most direct demonstration of the existence of anaerobic microsites and their relationship to denitrification activity is the work of Sexstone et al. (1985). Using oxygen microelectrodes, they measured oxygen profiles within saturated soil aggregates, detected anaerobic zones in their centers, and mapped oxygen contours.

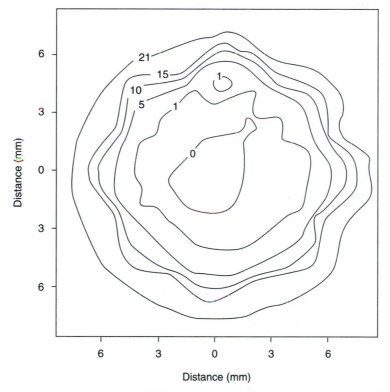

Oxygen contours in a saturated soil aggregate as measured with a microelectrode. Numbers on contour lines are volume percent of oxygen. *Adapted from Sexstone et al. (1985b).*

The measured anaerobic radii were highly correlated with those calculated with the oxygen consumption-diffusion model, and measurable denitrification rates were associated only with aggregates that had anaerobic zones. However, denitrification did not occur in all aggregates that had anaerobic zones, probably because factors other than aeration limited denitrification.

Miscellaneous soil and environmental factors affecting denitrification.
Denitrification responds to temperature as do most biological processes, increasing
as temperature increases until a maximum is reached, above which activity declines
rapidly. Mesophilic denitrifiers predominate in most soils, although activity has been
measured near freezing and also under thermophilic conditions. Temperature is
likely to have a more complex effect on denitrification than on some other soil
processes because it also affects oxygen and nitrous oxide solubility, gas diffusion
coefficients, and the oxygen consumption activity of other heterotrophs.

The response of denitrifiers to pH is similar to that of other soil heterotrophs,
which usually function best near neutrality. Biological denitrification has been measured in some acidic soils, but rates are usually low and the measurements can be
potentially confounded by chemodenitrification. Relatively little research has focused
on denitrification in soils of high pH.

Spatial scale and appropriate controlling factors. When considering the
regulation of denitrification, it is useful to integrate the concept of spatial scale with
the list of controlling factors (Fig. 12–8). Denitrifying bacteria are ultimately influenced by concentrations of oxygen, nitrate, and carbon compounds just external to
their cell surfaces. Consequently, measuring these concentrations is appropriate for
physiological studies in the laboratory. Other properties are likely to be more useful and insightful when studying denitrification in soils, particularly as the spatial
scale increases from the microbial cell through microcolonies, soil aggregates, soil
columns, and field plots up to the landscape scale (Box 12–6). We are just beginning to understand how to scale up such processes as denitrification.

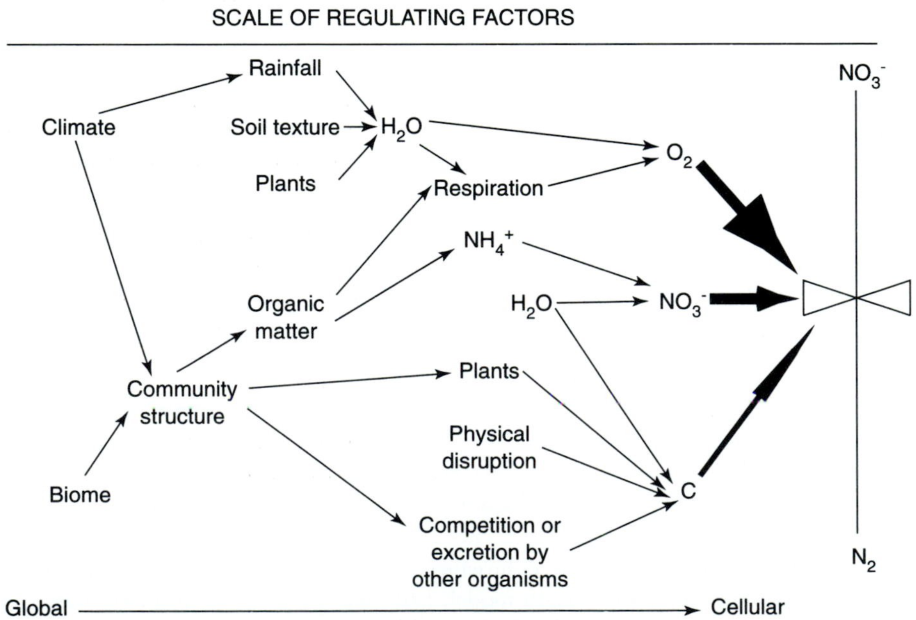

Figure 12–8 Regulation of denitrification at various spatial scales. The thickness of the
right most arrows reflects the relative importance of oxygen, nitrate, and carbon as regulators of denitrification. *Adapted from Tiedje (1988). Used with permission.*

Box 12–6

Spatial Variability of Denitrification. Denitrification is notorious for being highly variable in time and space. Denitrification rates can vary more than 100-fold from one day to the next.

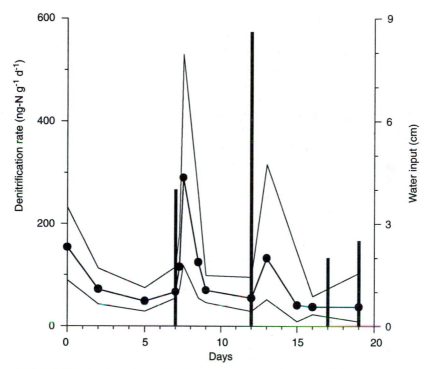

Temporal variation in daily denitrification rates of a clay loam in Michigan in autumn. The fine solid lines represent the upper and lower 95 percent confidence intervals and vertical bars represent water input. Note the response of denitrification to the addition of water. *Adapted from Sexstone et al. (1985a).*

Such short-term changes are often associated with precipitation, irrigation, or sometimes nutrient additions (e.g., inorganic fertilizer or manures). Seasonal responses, largely to soil temperature or precipitation patterns, are also observed. For example, denitrification rates in the Pacific Northwest of the U.S. are highest in the fall and spring when both soil temperature and water content are relatively high, whereas low rates are found in the winter, because of low soil temperatures, and in the summer, because of very dry soil conditions.

Variations in soil denitrification rates span several spatial scales, although the field plot level has probably been studied the most. The most common observation when sufficient numbers of denitrification measurements are made is that most rates are low with just a

Box 12–6 cont'd

few high or very high rates. This results in a skewed frequency distribution that is most often described as lognormal. This observation has often been attributed to the formation of "hot spots" of activity where optimal conditions of anaerobiosis, adequate nitrate, and available carbon concentrations coincide. The existence of such hot spots was perhaps best shown by the clever experiment of Parkin (1987). In this experiment, the denitrification activity of a soil core was monitored as it was subdivided into smaller and smaller units. Ultimately, over 85% of the activity of the core was found to be associated with one small piece of decaying vegetation.

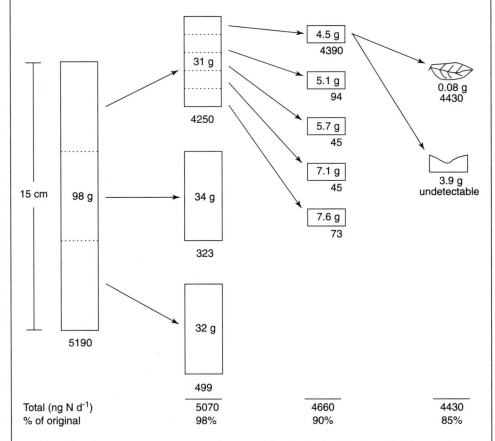

Evidence for "hot spots" as important locations for denitrification in soil. The numbers at the bottom of each soil section are denitrification rates (ng-N day^{-1}). Note that almost all the denitrification activity was associated with a single leaf. *Adapted from Parkin (1987).*

Summary

The nitrogen cycle is both fascinating and frustrating in its complexity. A thorough knowledge of the cycle is fundamental, though, if one wishes to understand the functioning of natural ecosystems, to manage agricultural ecosystems for productivity and sustainability, and to ameliorate environmental problems.

Most nitrogen in soil is in organic forms. Organic nitrogen serves as a reservoir of nitrogen, slowly supplying the more dynamic and much smaller inorganic nitrogen pools. The conversion of organic nitrogen to inorganic nitrogen is called mineralization. The first step of this process is the production of ammonium by ammonification, which is carried out by a wide variety of soil microorganisms and soil animals. Ammonification is always counterbalanced by the opposite pattern of immobilizing ammonium into the soil biomass through assimilation. A major controlling factor determining whether net mineralization or immobilization of nitrogen occurs is the C/N ratio of the decomposing organic matter.

The mineralization process is continued further by the conversion of ammonium to nitrate by the nitrifiers, which are a relatively restricted group of bacteria. Ammonia-oxidizing bacteria perform the first step of this two-step process, the transformation of ammonium to nitrite. Nitrite is further oxidized to nitrate by the nitrite-oxidizing bacteria. The ammonia and nitrite oxidizers are autotrophic bacteria that gain their energy from these inorganic oxidations.

Nitrate has many fates in the soil environment. It is readily taken up by plants and can also be immobilized by heterotrophic microorganisms. Because nitrate is relatively mobile, it can be readily leached, which represents not only a loss from the system, but also a potential environmental problem. A final fate of nitrate is to be lost to the atmosphere through denitrification.

Denitrification occurs under anaerobic conditions, which can exist as microsites even in well-aerated soils. The reduction of nitrate to gaseous nitrous oxide and dinitrogen is accomplished by a wide range of bacteria, most of which normally function as aerobic heterotrophs. Denitrification is a relatively benign loss of nitrogen when dinitrogen is the dominant product; however, nitrous oxide production can be an environmental concern because it acts as a greenhouse gas and has been implicated in the destruction of ozone in the stratosphere.

The nitrogen cycle is closed by the process of N_2 fixation, which is ultimately the source of all the nitrogen that is transformed within the soil ecosystem. That process is covered in detail in Chapters 13 and 14.

Cited References

Anderson, J.M., and T. Spencer. 1991. Carbon, nutrient and water balances of tropical rainforests subject to disturbance. MAB Digest No. 7. UNESCO, Paris, France.

Currie, J.A. 1961. Gaseous diffusion in the aeration of aggregated soils. Soil Sci. 92:40–45.

Davidson, E.A., S.C. Hart, and M.K. Firestone. 1992. Internal cycling of nitrate in soils of a mature coniferous forest. Ecology 73:1148–1156.

De Boer, W., and H.J. Laanbroek. 1989. Ureolytic nitrification at low pH by *Nitrosospira* species. Arch. Microbiol. 152:178–181.

Firestone, M.K. 1982. Biological denitrification. pp. 289–326. *In* F.J. Stevenson (ed.) Nitrogen in agricultural soils. American Society of Agronomy, Madison, Wis.

Freney, J.R., D.L. Chen, A.R. Mosier, I.J. Rochester, G.A. Constable, and P.M. Chalk. 1993. Use of nitrification inhibitors to increase fertilizer nitrogen recovery and lint yield in irrigated cotton. Fertil. Res. 34:37–44.

Hankinson, T.R., and E.L. Schmidt. 1988. An acidophilic and a neutrophilic *Nitrobacter* strain isolated from the numerically predominant nitrite-oxidizing population of an acid forest soil. Appl. Environ. Microbiol. 54:1536–1540.

Hart, S.C., G.E. Nason, D.D. Myrold, and D.A. Perry. 1994. Dynamics of gross nitrogen transformations in an old-growth forest: The carbon connection. Ecology 75:880–891.

Holt, J.G., N.R. Krieg, P.H.A. Sneath, J.T. Staley, and S.T. Williams (eds.). 1994. Bergey's manual of determinative bacteriology. 9th ed. Williams and Wilkins Co., Baltimore.

Jansson, S.L., and J. Persson. 1982. Mineralization and immobilization of soil nitrogen. pp. 229–252. *In* F.J. Stevenson (ed.), Nitrogen in agricultural soils. American Society of Agronomy, Madison, Wis.

Keerthisinghe, D.G., J.R. Freney, and A.R. Mosier. 1993. Effect of wax-coated calcium carbide and nitrapyrin on nitrogen loss and methane emission from dry-seeded flooded rice. Biol. Fertil. Soils 16:71–75.

Ladd, J.N., and R.B. Jackson. 1982. Biochemistry of ammonification. pp. 173–228. *In* F.J. Stevenson (ed.), Nitrogen in agricultural soils. American Society of Agronomy, Madison, Wis.

McCarty, G.W., and J.M. Bremner. 1986. Inhibition of nitrification in soil by acetylenic compounds. Soil Sci. Soc. Am. J. 50:1198–1201.

McCarty, G.W., and J.M. Bremner. 1990. Evaluation of 2-ethynylpyridine as a soil nitrification inhibitor. Soil Sci. Soc. Am. J. 54:1017–1021.

Olson, R.A., and L.T. Kurtz. 1982. Crop nitrogen requirements, utilization, and fertilization. pp. 567–604. *In* F.J. Stevenson (ed.), Nitrogen in agricultural soils. American Society of Agronomy, Madison, Wis.

Parkin, T.B. 1987. Soil microsites as a source of denitrification variability. Soil Sci. Soc. Am. J. 51:1194–1199.

Post, W.M., J. Pastor, P.J. Zinke, and A.G. Strangenberger. 1985. Global patterns of soil nitrogen storage. Nature 317:613–616.

Prosser, J.I. 1989. Autotrophic nitrification in bacteria. Adv. Microb. Physiol. 30:125–180.

Robertson, L.A., and J.G. Kuenen. 1984. Aerobic denitrification: A controversy reviewed. Arch. Microbiol. 139:351–354.

Sexstone, A.J., T.B. Parkin, and J.M. Tiedje. 1985a. Temporal response of soil denitrification to rainfall and irrigation. Soil Sci. Soc. Am. J. 49:99–103.

Sexstone, A.J., N.P. Revsbech, T.B. Parkin, and J.M. Tiedje. 1985b. Direct measurement of oxygen profiles and denitrification rates in soil aggregates. Soil Sci. Soc. Am. J. 49:645–651.

Smith, K.A. 1980. A model of the extent of anaerobic zones in aggregated soils, and its potential application to estimates of denitrification. J. Soil Sci. 31:263–277.

Stevenson, F.J. 1982. Origin and distribution of nitrogen in soil. pp. 1–42. *In* F.J. Stevenson (ed.), Nitrogen in agricultural soils. American Society of Agronomy, Madison, Wis.

Tiedje, J.M. 1988. Ecology of denitrification and dissimilatory nitrate reduction to ammonium. pp. 179–244. *In* A.J.B. Zehnder (ed.), Biology of anaerobic microorganisms. John Wiley and Sons, New York.

Tiedje, J.M. 1994. Denitrifiers. pp. 245–267. *In* R.W. Weaver, J.S. Angle, and P.J. Bottomley (eds.), Methods of soil analysis, Part 2. Microbiological and biochemical properties. Soil Science Society of America Book Series No. 5, Madison, Wis.

Waring, R.H., and W.H. Schlesinger. 1985. Forest ecosystems—Concepts and management. Academic Press, Inc., Orlando, Fla.

Ye, R.W., B.A. Averill, and J.M. Tiedje. 1994. Denitrification: Production and consumption of nitric oxide. Appl. Environ. Microbiol. 60:1053–1058.

General References

Knowles, R., and T.H. Blackburn. 1993. Nitrogen isotope techniques. Academic Press, New York.

Prosser, J.I. 1986. Nitrification. IRL, Oxford, England.

Revsbech, N.P., and J. Sørensen. 1990. Denitrification in soil and sediment. Plenum Press, New York.

Sprent, J.I. 1987. The ecology of the nitrogen cycle. Cambridge University Press, Cambridge, England.

Stevenson, F.J. 1982. Nitrogen in Agricultural Soils. American Society of Agronomy, Madison, Wis.

Stevenson, F.J. 1986. Cycles of soil. John Wiley and Sons, Inc., New York.

Study Questions

(Note, starred questions are more difficult.)

1. Draw your own diagram of the terrestrial nitrogen cycle including all major pools and transformations.

2. Describe the composition of soil organic nitrogen and how this relates to its biological availability.

3. Describe the steps involved in releasing ammonium from chitin.

4. A crop residue (45% C, 1.5% N) is chopped and mixed into soil. Assume that the soil microbial biomass has an average C/N ratio of 8 and an efficiency (microbial yield coefficient) of 0.6 in mineralizing this residue. Will net mineralization or immobilization of nitrogen occur? Show your calculations and explain your reasoning.

5. Describe how nitrate is formed and list its possible fates in the soil ecosystem.

6. Explain why the pH of agricultural soils declines when they are fertilized with ammo-niacal nitrogen. Would this also happen if nitrate fertilizers were used?

7. Give an example of why net rates of nitrogen-cycle processes may be misleading.

8. Contrast nitrification and denitrification with respect to the bacterial genera capable of carrying out each process and discuss how the respective substrates are utilized bio-chemically.

9. Denitrification rates are known to be highly variable in time and space. Provide a ra-tionale for this variability based on what you know about the factors that regulate den-itrification activity.

*10. The following data are based on an original paper that examined the effects of differ-ent grass species on nitrogen cycling (Oecologia 84:433–441). Use the data given in the table to fill in the missing values, calculate how much nitrogen is required for mi-crobial growth in each grass system, and estimate the annual net nitrogen mineraliza-tion rate for the Grass B system. Assume that roots are the only plant input, that roots are 45% carbon, that all the grass nitrogen is available, that lignin-carbon is unavail-able, and that 80% of the nonlignin-carbon is metabolized in one year ($k=1.6$ yr^{-1}). The microbial growth yield is 0.5, and the microbial C/N ratio is 8. Clearly state any other assumptions that you think are necessary. Show your work.

Root/soil property	Grass A	Grass B
Root biomass (g m^{-2})	50	1,200
Root C (g m^{-2})		
Root N (%)	2.0	0.5
Root N (g m^{-2})		
Root C/N	22.5	90
Root lignin-C (%)	5	15
Net N mineralization (g-N m^{-2} yr^{-1})	14	

*11. Production of allelochemicals that specifically inhibit autotrophic nitrification has been proposed as a mechanism through which plants conserve nitrogen within an ecosys-tem. Develop an experiment that would conclusively prove or disprove this theory.

*12. Discuss whether denitrification is likely to be enhanced in the rhizosphere compared to bulk soil. Develop a logical argument to support your view and describe an experi-ment that could be done to test your hypothesis.

Chapter 13

♦

Biological Dinitrogen Fixation: Introduction and Nonsymbiotic

David A. Zuberer

♦

In nature there is an enormous reserve of organic matter poor in nitrogen. One should ask how all this organic carbon could be circulated in nature without the existence of organisms capable of fixing free nitrogen. These organisms . . . could be nothing else but microbes.

Sergei Winogradsky

After photosynthesis, biological dinitrogen fixation, the reduction of atmospheric dinitrogen (N_2) to two molecules of ammonia, is the second most important biological process on earth. In the absence of modern fertilizers or animal wastes, natural ecosystems rely on the biological conversion of atmospheric dinitrogen to forms available for plant and microbial growth by a variety of prokaryotic microbes. Dinitrogen fixation is mediated exclusively by prokaryotes, including many genera of bacteria, cyanobacteria, and the actinomycete *Frankia*. The N_2-fixing microbes can exist as independent, free-living organisms or in associations of differing degrees of complexity with other microbes, plants, and animals. These range from loose associations, such as **associative symbioses,** to complex **symbiotic** associations (Fig. 13–1) in which the bacterium and host plant communicate on an exquisite molecular level and share physiological functions (Chapter 14 details information on symbiotic N_2-fixing plant/microbe associations). Central to all of these systems is an N_2-fixing prokaryote containing the enzyme complex **nitrogenase,** which is responsible for the conversion of dinitrogen to ammonia. Only certain prokaryotes fix dinitrogen. Thus, when one sees statements about "N_2-fixing" plants, keep in mind that they do so by virtue of their prokaryotic partner, not on their own. Organisms that can use atmospheric dinitrogen as their sole source of nitrogen for growth are called **diazotrophs** ("diazo" for dinitrogen).

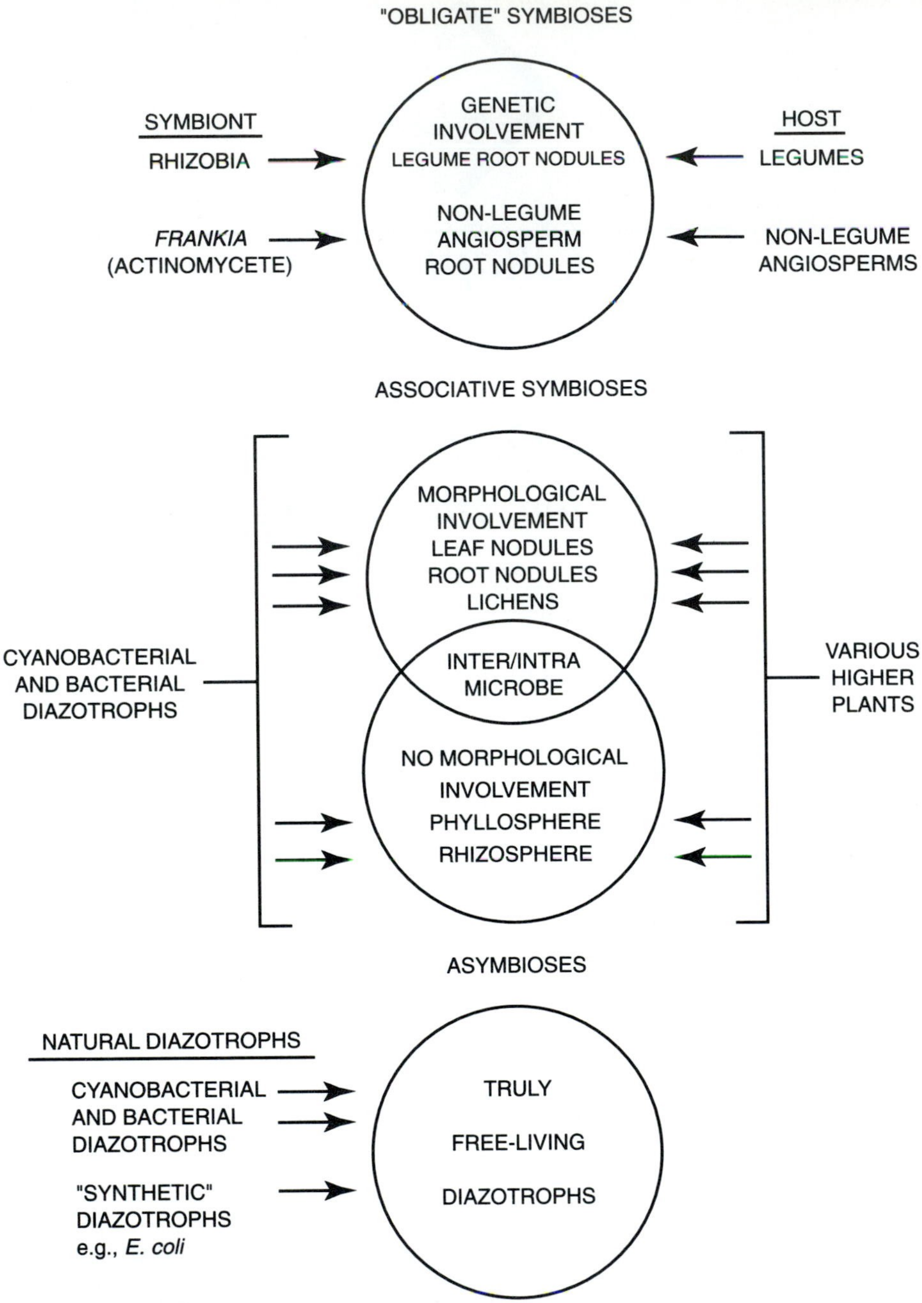

Figure 13–1 Biological N_2 fixation is mediated exclusively by free-living diazotrophs and a wide variety of plant-microbe associations of varying complexities. The most elaborate of these are the root-nodule symbioses between rhizobia and legumes and between the actinomycete *Frankia* and a variety of nonleguminous trees and shrubs. Between the free-living diazotrophs and the root-nodule symbioses lie a broad array of associations between plants and diazotrophs. These are the so-called "associative symbioses" characterized by a loose association between the plant and diazotroph. *Modified from Burns and Hardy (1975). Used with permission.*

Historical Background

The agricultural importance of legumes was recognized very early in the history of humans, reaching a milestone in 1888 with the isolation and description of the first organism, the root-nodule bacterium *Rhizobium,* determined (though much later) to have the ability to fix dinitrogen. *Rhizobium* and its relatives fix only small amounts of dinitrogen in pure culture, and significant rates of N_2 fixation only occur when the bacteria are in root nodules. In 1893, Winogradsky isolated the first free-living bacterium, an anaerobe, capable of significant N_2 fixation. He named this organism *Clostridium pasteurianum* in recognition of Louis Pasteur; Winogradsky, though a Russian, did much of his later scientific work at the Pasteur Institute in Paris. A few years later, in 1901, the Dutch microbiologist Martinus Beijerinck isolated an aerobic free-living, N_2-fixing bacterium, which he named *Azotobacter chroococcum* (derived from the French word "azote," meaning "lifeless" in recognition of the inertness of nitrogen).

Thus by the late nineteenth century, the foundation was laid for the extensive studies of N_2 fixation that would follow. These early discoveries began what has become more than 100 years of research on the fascinating and important subject of biological N_2 fixation.

The Significance of Biological Dinitrogen Fixation

A fundamental principle of agriculture and plant physiology is that plants require relatively high levels of nitrogen to produce abundant biomass, or yield. All forms of life require nitrogen to synthesize proteins and other important biochemicals, and nitrogen is often the limiting nutrient for plant and microbial growth in soils. In natural systems, nitrogen for plant growth comes from the soil, from rainfall or other atmospheric deposition, or through biological N_2 fixation.

Industrial N_2 fixation amounts to about 85 million metric tons per year (Hauck, 1985; Waggoner, 1994) and requires substantial inputs of energy, usually in the form of natural gas (Box 13–1). However, this output is less than the contribution of fixed nitrogen acquired through biological N_2 fixation. Accurate estimates of the magnitude of biological N_2 fixation are hard to derive, but reported values range from 100 to 180 million metric tons per year. Biological processes contribute 65% of the nitrogen used in agriculture (Burris and Roberts, 1993). While much of this is through symbiotic N_2 fixation, nonsymbiotic and associative fixation are of some significance in crops, such as sugar cane and sorghum, which have a C_4 photosynthetic pathway, and in specific ecosystems where nitrogen for plant growth is a limiting factor. Because of its culture in flooded soils, rice derives significant benefit from the activities of free-living diazotrophic bacteria and cyanobacteria.

Biological N_2 fixation offers an alternative to the use of expensive ammonium-based fertilizer nitrogen. However, the high-yielding agricultural systems of the United States and elsewhere are difficult to sustain solely on biological N_2 fixation. Keeping up with an expanding population will probably always require the judicious use of fertilizers. Yet, legumes and perhaps other N_2-fixing systems (e.g., the *Azolla-Anabaena* symbiosis) have an important place in sustainable agricultural production. Dinitrogen-fixing plants are also of high value in restoring disturbed or impoverished

| Box 13–1 |

The Energy Costs of N_2 Fixation. Dinitrogen is described as the most stable diatomic molecule known. The two atoms of nitrogen in the diatomic molecule are joined by a very stable triple bond. This triple bond requires a lot of energy (945 kJ, or 226 kcal, per mole) to break and therein lies one of the major challenges of fixing dinitrogen chemically or biologically. Dinitrogen fixation is "energy expensive" because it requires much energy to break the triple bond of dinitrogen and also to provide the hydrogen necessary to reduce dinitrogen to two ammonia molecules. The chemical fixation of dinitrogen is accomplished most widely through the high-pressure catalytic method called the Haber-Bosch process, named after the German scientists who developed the process in 1914. The process served as a source of ammonium and nitrates for manufacture of explosives during World War I. The reaction for chemical fixation of dinitrogen can be written as follows:

$$N_2 + 3H_2 \longrightarrow 2\ NH_3\ (aq.)\quad \Delta G = -53kJ\ (12.7kcal)$$

$$K_2O,\ Al_2O_3$$

To obtain high yields from this reaction in reasonable time, high pressures (approximately 200 atm. or 20 MPa) are applied to the reaction vessel and the temperature is raised to 400–500°C. Natural gas (methane) is usually employed as the feedstock for the hydrogen needed to reduce the dinitrogen and is also required to heat the reaction vessel. To put these energy requirements in terms more easily understood, it takes about 875 cubic meters (31,000 cubic feet) of natural gas, 5.5 barrels of oil, or 2 metric tons of coal to fix 1 metric ton (2,200 lb) of ammonia (Dixon and Wheeler, 1986). About 40% of this fuel is used to provide the necessary heat, and the remainder provides the hydrogen needed for the reaction. Because we rely on fossil fuel (e.g., natural gas) in the production of ammonium-based fertilizers, costs are high; at present, those costs are somewhat inextricably linked to fossil fuel prices and the whims of that market. For example, an oil embargo against the United States in the early 1970s brought about long lines at gas pumps and quadrupled fertilizer prices. The high fertilizer prices also were due, in part, to short supplies at a time of peak demand.

Just as the chemical fixation of dinitrogen is energy intensive, so too is its fixation in biological systems. The principal differences lie in the sources of reductant and energy and the fact that biological N_2 fixation takes place at ambient pressures and temperatures. That is quite a feat when we consider the rigors of the industrial process. Energy for biological N_2 fixation comes from the oxidation of organic carbon sources, such as glucose, or from light in the case of photosynthetic diazotrophs.

---

soils. They serve as excellent cover crops, green manures, and forage crops for livestock production. Use of N_2-fixing crops also has the potential to reduce the contamination of groundwater with nitrate. Further aspects of the utility of symbiotic N_2 fixation in agriculture and other ecosystems are discussed in Chapter 14.

The Nitrogenase Enzyme Complex

Biological N_2 fixation is mediated by the enzyme complex nitrogenase. It is most appropriately called the "nitrogenase complex" because it consists of two protein components each composed of multiple subunits. The nomenclature of the nitrogenase complex is as follows (Evans and Burris, 1992):

- The overall complex is known as **nitrogenase.**

- The molybdenum-iron (MoFe) protein is *dinitrogenase* (the "type species" substrate is dinitrogen; enzymes conventionally are named relative to their substrate).

- The iron (Fe) protein is designated *dinitrogenase reductase;* the general consensus is that its function is the reduction of dinitrogenase.

Characteristics of the two proteins are summarized in Table 13–1. A diagrammatic representation of the nitrogenase complex is shown in Figure 13–2. What sets the nitrogenase complex apart is that it:

- consists of two proteins, the MoFe protein (dinitrogenase) and the iron protein (dinitrogenase reductase),

- is destroyed by oxygen,

- contains iron and molybdenum or vanadium,

- needs Mg^{2+} ions to be active,

- converts ATP to ADP when functioning,

- is inhibited by ADP,

- reduces dinitrogen and several other small triply bonded molecules, and

- reduces H^+ to H_2 even when dinitrogen is present.

The overall reaction for biological N_2 fixation using nitrogenase is shown in Figure 13–2. Two MgATP are required for each electron transferred from dinitrogenase reductase to dinitrogenase; thus the reaction shows a requirement of 16 molecules of ATP (112 kcal). Under natural conditions, however, probably 20 to 30

Table 13–1 **Characteristics of dinitrogenase (the MoFe protein) and dinitrogenase reductase (the Fe protein).**

Dinitrogenase
  Mol. wt.: 220,000–270,000
  Has 4 subunits: 2 approximately 50,000 mol. wt.
                  2 approximately 59,000 mol. wt.
  2 molybdenum atoms per molecule
  22–24 iron atoms per molecule
  Half-life in air: up to 10 min.

Dinitrogenase reductase
  Mol. wt.: 55,000–66,000
  Has 2 subunits: 27,500–34,000
  No molybdenum
  4 iron atoms per molecule
  4 labile sulfur atoms per molecule
  Half-life in air: 0.5–0.75 sec.

Adapted from Postgate (1987). Used with permission.

MgATP are needed as the process is less efficient than when observed under optimum laboratory conditions (Burris and Roberts, 1993). A consensus on the general model for the mechanism of nitrogenase has evolved over quite a few years. The mechanism can be summarized as follows (Evans and Burris, 1992):

- Dinitrogenase reductase (the Fe protein) accepts electrons from a low-redox donor, such as reduced ferredoxin (Fd_{red}) or flavodoxin, and binds two MgATP.

- It transfers electrons, one at a time, to dinitrogenase (the MoFe protein).

- Dinitrogenase reductase and dinitrogenase form a complex, the electron is transferred, and two MgATP are hydrolyzed to two MgADP + Pi (phosphate).

- Dinitrogenase reductase and dinitrogenase dissociate, and the process is then repeated.

- When dinitrogenase has collected enough electrons, it binds a molecule of dinitrogen, reduces it, and releases ammonium.

- Dinitrogenase then accepts additional electrons from dinitrogenase reductase to repeat the cycle.

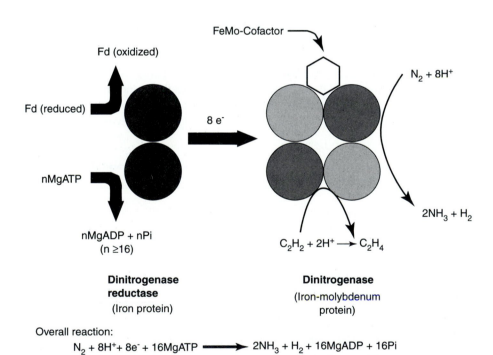

Overall reaction:

$$N_2 + 8H^+ + 8e^- + 16MgATP \longrightarrow 2NH_3 + H_2 + 16MgADP + 16Pi$$

Figure 13–2 The nitrogenase enzyme complex consists of two protein components, dinitrogenase reductase (the Fe protein) and dinitrogenase (the MoFe protein). The dinitrogenase reductase gathers electrons from low-redox carriers such as ferredoxin and flavodoxin and transfers them to dinitrogenase. Nitrogen is bound to the dinitrogenase protein where it is reduced.

In each cycle of N_2 fixation, dinitrogenase and dinitrogenase reductase bind together, MgATP is hydrolyzed, and an electron is transferred. The dissociation of the MoFe protein–Fe protein complex is the rate-limiting step of the process. In fact, the nitrogenase complex is "remarkably slow—it takes 1.25 sec for a molecule of enzyme to form two of NH_3. The two proteins have to come together and separate 8 times to reduce one N_2 molecule" (Postgate, 1987). A consequence of the slowness of nitrogenase is that N_2-fixing bacteria must synthesize a lot of the protein. Nitrogenase can commonly account for 10% of the cell's proteins, and levels up to 40% have been recorded (Postgate, 1994).

Returning to the reaction in Figure 13–2 of biological N_2 fixation, we observe that for every $8e^-$ transferred via the nitrogenase complex, $2e^-$ are consumed in the formation of H_2. The production of H_2 that accompanies the fixation of N_2 is obligatory. One H_2 (requiring 4 MgATP) is released for each N_2 reduced to $2NH_3$ (requiring 12 MgATP). Thus, 25% of the energy from MgATP is "lost" in the production of H_2. Interestingly, some diazotrophs contain an uptake hydrogenase that allows them to oxidize some of the H_2 and to generate a reduced electron carrier or MgATP. This can then be used in the N_2 fixation reaction, thereby recapturing some of the energy lost.

The physiological requirements for a free-living diazotroph to fix dinitrogen are summarized in Table 13–2. These requirements stem in large part from the unique properties and requirements of the nitrogenase complex, including its exceptional sensitivity to molecular oxygen, the metal content of the complex (iron and molybdenum or vanadium), and the need for adequate supplies of reducing power and MgATP.

Substrates for Nitrogenase

The principal substrate for nitrogenase is dinitrogen (N≡N). Note that the two atoms of nitrogen are joined by a triple bond. In addition to reducing protons to H_2,

Table 13–2 **Physiological needs for N_2 fixation by free-living diazotrophic microbes.**

Trace elements
    Molybdenum or vanadium, iron—for nitrogenase
    Magnesium—for production of MgATP
ATP—a minimum of 16 ATP per N_2 fixed. Probably 20–30 under natural conditions.
    Needed for nitrogenase activity and for nitrogenase synthesis. The high ATP requirement means an abundant supply of energy-yielding substrates must be readily available for vigorous N_2 fixation.
Acceptable temperature
    Most diazotrophs are mesophiles.
    Many will not grow on media at 37°C.
    Nitrogenase activity falls off rapidly at about 40°C.
Oxygen excluded from the enzyme complex
    Nitrogenase is destroyed by O_2.
Source of low-redox reductant
    Restricted to the naturally occurring ferredoxins and flavodoxins.
Reduce hydrogen evolution—up to 35% of the ATP diverted to nitrogenase may be consumed in H_2 evolution.

Adapted from Postgate (1987). Used with permission.

Table 13–3 **Substrates for nitrogenase: The triple bond connection.**

Name	Formula	Major products
Proton	H^+	H_2
Dinitrogen	$N\equiv N$	$NH_3 + H_2$
Nitrous oxide	$N\equiv N^+ - O^-$	$N_2 + H_2O$
Azide	$[N\equiv N^+ - N]^-$	$N_2 + NH_3 + N_2H_4$
Acetylene	$HC\equiv CH$	$H_2C=CH_2$
Cyanide	$[C\equiv N]^-$	$CH_4 + NH_3$
Carbon monoxide	$C\equiv O$	$C=O$ apparently binds to the N-binding site but is not reduced. It blocks reaction with other substrates.

nitrogenase also reduces several other small, triply bonded molecules (Table 13–3). Of particular interest on this list of substrates is the gas acetylene. In the late 1960s researchers discovered that nitrogenase can reduce acetylene ($HC\equiv CH$) to ethylene ($H_2C=CH_2$) (Burris, 1974). This finding is significant because acetylene and ethylene can be very easily measured using a gas chromatograph to separate the two gases. Thus was born the simple, sensitive, and rapid **acetylene reduction assay** for nitrogenase activity. The development of this assay led to an explosion in research on N_2 fixation because it removed the need to have an expensive mass spectrometer or to deal with the tedious and insensitive Kjeldahl analyses to measure nitrogen in soil or biological materials. However, the technique is not without limitations, which will be discussed later in the chapter.

The Alternative Nitrogenases

Prior to the early 1980s, researchers thought that only one type of nitrogenase existed and that molybdenum was essential for N_2 fixation. However, in the early 1980s, Bishop and colleagues isolated a second nitrogenase (nitrogenase 2) from *Azotobacter chroococcum,* which was produced only under conditions of molybdenum starvation (Bishop and Premakumar, 1992). With this discovery, it became clear that molybdenum was not required for N_2 fixation in all bacteria and could be supplanted with vanadium. Nitrogenase 2 is similar to nitrogenase 1 because it consists of two proteins (a VFe protein and an Fe protein), produces hydrogen, and is sensitive to oxygen. Nitrogen binds to the VFe protein, suggesting that metals other than molybdenum can participate in binding the nitrogen to the protein for reduction. Subsequently, a third nitrogenase, nitrogenase 3, has been described. This complex does not appear to contain either molybdenum or vanadium and is synthesized under starvation for both of the metals.

All the nitrogenases are complexes of a dinitrogenase reductase and a dinitrogenase component. Nitrogenases 2 and 3 appear to evolve more hydrogen than nitrogenase 1, and nitrogenases 2 and 3 reduce acetylene to ethane (C_2H_6) rather than ethylene.

The finding that dinitrogen could be fixed in the absence of molybdenum is interesting because prior to the discovery of the alternative nitrogenases, microbiologists and biochemists believed that the element was an obligate requirement for N_2 fixation. It now seems that the predominance of nitrogenase 1 (the molybdenum-containing enzyme) may simply be because researchers have routinely used molybdenum-containing media to isolate N_2-fixing bacteria. The other nitrogenases may be quite common in nature, though this possibility has not yet been fully explored. The study of alternative nitrogenase systems is quite recent, and many fundamental questions remain unanswered.

The Genetics of Nitrogenase

As one might surmise from the complexities of the N_2-fixing system, the genetics of the process are equally complex. Much of our knowledge of the genetics of the nitrogenase system has come from the intensive study of the bacterium *Klebsiella pneumoniae,* a member of the family Enterobacteriaceae and a close relative of *Escherichia coli.* Twenty-five years of research on this bacterium and, more recently, other diazotrophic bacteria have shown that the nitrogenase complex and supporting systems are under the control of no less than 21 genes. Because of this complexity, the system does not lend itself to easy genetic manipulation and transfer to higher organisms. The prospect of genetically engineering crops like corn to fix their own nitrogen remains an elusive goal for researchers.

The Free-Living Dinitrogen–Fixing Bacteria

Biological N_2 fixation is restricted to prokaryotes This includes typical bacteria and specialized bacteria like cyanobacteria and the actinomycetes. Of the 10,000 or so bacterial genera named, only about 100 contain bonafide diazotrophic species. Although this may appear to be a somewhat limited representation, species representing all sorts of physiological types and occupying all sorts of ecological niches have been described. Some of the genera of free-living diazotrophs are listed in Table 13–4. These organisms encompass such groups as heterotrophic and chemoautotrophic bacteria, photoautotrophs (bacteria and cyanobacteria), and photoheterotrophs with respect to carbon metabolism. Note also that they are well represented by aerobes, **microaerophiles** that grow best at low oxygen tension, facultative anaerobes, and obligate anaerobes. Such metabolic diversity enables some type of diazotroph to colonize almost any imaginable sort of habitat. Indeed, diazotrophs are widespread in nature as free-living microbes and in a large number of associations with plants and animals. This great metabolic diversity means that in all sorts of environments, diazotrophs can make contributions to the supply of fixed nitrogen for growth of nonfixing microbes and higher forms of life. For example, diazotrophs colonize the roots and rhizospheres of many plant species and make small quantities of fixed nitrogen available to the plants.

Factors Affecting Dinitrogen Fixation by Free-Living Diazotrophs

Thus far, we have discussed the complexities and uniqueness of the nitrogenase complex. In this section the factors that must be successfully integrated for a nonsymbiotic diazotroph to fix dinitrogen will be discussed, followed by a description of the integration of these factors with respect to the functioning of nonsymbiotic diazotrophs in association with higher plants, the so-called **associative symbioses.**

Table 13–4 List of genera of microbes which include free-living, N_2-fixing species or strains. This list is not intended to be all-inclusive.

	Genus or type	Species (examples only)
Aerobes	*Azotobacter*	*A. chroococcum*, A. vinelandii**
	Azotococcus	*A. agilis**
	Azomonas	*A. macrocytogenes**
	Beijerinckia	*B. indica*, B. fluminis**
	Derxia	*D. gummosa**
	Pseudomonas	*P. stutzeri, P. saccharophila*
	Azoarcus	*A. communis, A. indigens*
	Acetobacter	*A. diazotrophicus*
Facultative (aerobic when not fixing N_2)	*Klebsiella*	*K. pneumoniae, K. oxytoca*
	Bacillus	*B. polymyxa, B. macerans*
	Enterobacter	*E. agglomerans (Erwinia herbicola)*
	Citrobacter	*C. freundii*
	Escherichia	*E. intermedia*
	Propionibacterium	*P. shermanii, P. petersonii*
Microaerophiles (normal aerobes when not fixing N_2)	*Xanthobacter*	*X. flavus*, X. autotrophicus*
	Thiobacillus	*T. ferro-oxidans*
	Azospirillum	*A. lipoferum*, A. brasilense**
	Aquaspirillum	*A. perigrinum*, A. fascicilus**
	Methylosinus	*M. trichosporum*
Strict anaerobes	*Clostridium*	*C. pasteurianum*, C. butyricum*
	Desulfovibrio	*D. vulgaris, D. desulficuricans*
	Methanosarcina	*M. barkeri*
Phototrophs (aerobic) cyanobacteria	*Anabaena*	*A. cylindrica, A. inaequalis*
	Nostoc	*N. muscorum*
	Calothrix	
	(7 other genera of heterocystous cyanobacteria)	
	Gloeothece	*G. alpicola*
Phototrophs (microaerophiles) cyanobacteria	*Plectonema*	*P. boryanum*
	Lyngbya	*L. aestuarii*
	Oscillatoria	
	Spirulina	
Phototrophs (facultative) bacteria	*Rhodospirillum*	*R. rubrum*
	Rhodopseudomonas	*R. palustris*
Phototrophs (anaerobes) bacteria	*Chromatium*	*C. vinosum*
	Chlorobium	*C. limicola*
	Thiopedia	
	Ectothiospira	*E. shapovnikovii*

Adapted from Postgate (1987). Used with permission. See also Young (1992).
*Signifies that all reported strains of the species fix dinitrogen.

Sources of Energy for Diazotrophs

With the exception of the phototrophic bacteria and cyanobacteria, all diazotrophs require an organic or inorganic, in the case of chemoautotrophs, energy source. A requirement for abundant energy sources is dictated by the high energy demands of nitrogenase. Remember, a minimum of 16 ATPs are required to make two molecules of ammonia from dinitrogen. A wide variety of carbon sources, ranging from methane (CH_4) to complex carbohydrates, can be used by one diazotroph or another. It is not usually the type of carbon source that limits dinitrogen fixation, but the lack of an ample supply in many habitats that limits fixation by the nonsymbiotic bacteria. The soil is not an "organic soup" abundantly supplied with readily available carbon sources, and diazotrophs must compete with all the other soil microbes for the same carbon.

In terms of carbon sources for energy, it is interesting to consider the efficiency of N_2 fixation according to the amount of carbon consumed per nitrogen fixed (Table 13–5). Note that the efficiencies of N_2 fixation in terms of mg of N fixed g^{-1} of carbon source are low, averaging about 15 mg N g^{-1} of carbon source. Also, the assimilation of ammonium is about twice as efficient as N_2 fixation, and that fixation by anaerobes is generally much less efficient than by aerobes. An exception to this

Table 13–5 Carbon and energy source requirements for heterotrophs grown in chemostats limited by the carbon and energy source.

Organism	Carbon and energy source	Efficiency of nitrogen incorporation during growth (mg N g^{-1} C and energy source used)		Carbon and energy expenditure for N_2 fixation	
		N_2	NH_4^+	(g C g^{-1} N_2)	lb C used per 100 lb of N_2 fixed
Anaerobic growth by fermentation					
Clostridium pasteurianum	Sucrose	11	22	46	4,600 (2,054)*
Klebsiella pneumoniae	Glucose	8	19	72	7,200 (3,214)
Aerobic growth					
Klebsiella pneumoniae (O_2-limited)	Glucose	15	35	38	3,800 (1,696)
Azospirillum brasilense (9 $\mu M\ O_2$)	Malate	26	48	19	1,900 (848)
Azotobacter vinelandii (2–10 $\mu M\ O_2$)	Sucrose	16	63	47	4,700 (2,098)
Azotobacter vinelandii (180 $\mu M\ O_2$)	Sucrose	7	38	117	11,700 (5,223)

Adapted from Hill (1992). Used with permission. See also Giller and Day (1985).
*The 100-lb figure is chosen as representative of a typical amount of fertilizer nitrogen that might be applied to a crop. Numbers in ( ) are values in kg.

generalization is the reduced efficiency of fixation by *Azotobacter* at higher levels of oxygen. An efficiency of 15 mg g^{-1} of carbon source is equivalent to about 13 kg (30 lb) of nitrogen fixed per ton of carbon source. The need for large amounts of available energy sources is obvious.

In nature, the carbon sources for N_2 fixation by heterotrophic bacteria normally come from the remains of crop residues and roots decomposing in the soil and carbon released from active roots. Addition of readily available carbon sources to soil generally stimulates N_2 fixation for short periods of time.

Effects of Combined Nitrogen on Dinitrogen Fixation

Since N_2 fixation is so costly to a cell, it is not surprising that the nitrogenase complex is under strong regulation by the level of combined nitrogen (i.e., ammonium, nitrate, and organic nitrogen) in the medium or the environment. The formation of the nitrogenase complex is repressed by ammonium at fairly low levels. For example, Alexander and Zuberer (1989a) showed that as little as 4.2 μg ml^{-1} of ammonium nitrogen in nutrient solution eliminated acetylene reduction associated with corn roots at an oxygen concentration of 2% (0.02 atm; 2 kPa) around the roots. Nitrate and organic nitrogen also prevent the synthesis of nitrogenase. By repressing the formation of nitrogenase at very low levels of combined nitrogen the organisms avoid the high expense of synthesizing and operating an enzyme system that is not needed under conditions of nitrogen sufficiency.

Effects of Oxygen on Nitrogenase Activity

An interesting characteristic of nitrogenase, and one that poses considerable problems for most free-living diazotrophs, is its extreme sensitivity to molecular oxygen. In most bacteria the enzyme complex is irreversibly "poisoned" by exposure to O_2. This oxygen sensitivity has led to the development of some unique strategies among the diazotrophs to protect the enzyme system. Most unusual among these is the ability of some *Azotobacter* species to protect their enzymes through respiratory or conformational protection. Box 13–2 illustrates the many adaptations that allow N_2 fixation to occur in environments with widely divergent aeration conditions.

Other Environmental Factors

The supply of available carbon, presence or absence of sufficient combined nitrogen, and the abundance of molecular oxygen exert primary control over the synthesis and level of nitrogenase activity expressed by bacteria in a given environment. Other environmental factors are of lesser significance but cannot be ignored. Nitrogenase is active over a fairly narrow temperature range. At the lower limits of 5 to 10°C, nitrogenase activity is low, whereas at the upper limits, 37 to 40°C, nitrogenase activity falls off rapidly because of the sensitivity of the enzyme to heat. Most diazotrophs are mesophiles, but there are a few exceptions. For example, some cyanobacteria grow in hot springs.

Numerous other factors can affect the growth and survival of diazotrophs and thus directly or indirectly influence N_2 fixation. Among these are adequate supplies of phosphorus (N_2 fixation requires high levels of phosphorus), other inorganic nutrients, especially trace metals, and the acidity or alkalinity of the environment.

Box 13–2

How Microbes Solve the Oxygen Problems for Nitrogenase.

- **Avoidance**—Anaerobes and facultative anaerobes fix dinitrogen only in the absence of oxygen with the exception of *Klebsiella pneumoniae,* which can tolerate very low levels of oxygen. In fact, oxygen is one of the factors that regulates the synthesis of nitrogenase in this bacterium.

- **Microaerophily**—Most aerobic diazotrophic bacteria fix dinitrogen maximally at low partial pressures of oxygen, thereby lessening the exposure of nitrogenase to oxygen. For example, Okon et al. (1977) reported that *Azospirillum* fixes dinitrogen most rapidly at 0.7 kPa (0.007 atm) oxygen.

- **Respiratory protection**—Respiration functions in all aerobes to divert oxygen away from nitrogenase to some extent. In certain *Azotobacter* species, an exaggerated form of respiratory protection is observed. In fact, this bacterium exhibits one of the highest respiration rates of all life forms. The high respiration rate serves to scavenge oxygen and to keep it away from nitrogenase. As a consequence, the bacterium must consume large amounts of substrate to scavenge oxygen and growth is very inefficient in terms of carbon consumed under these conditions.

- **Production of specialized cells**—Many diazotrophic cyanobacteria produce specialized, thick-walled cells, called **heterocysts,** in which the nitrogenase is compartmentalized. These cells do not evolve oxygen in photosynthesis, and the thick wall excludes external oxygen. The diazotrophic actinomycete, *Frankia,* produces vesicles to protect nitrogenase from oxygen. The vesicle wall becomes thicker as the oxygen concentration increases in the medium.

- **Slime**—Aerobic diazotrophs grown on nitrogen-free agar media frequently produce large, gummy colonies caused by the production of extracellular polysaccharides. The gum serves as a diffusion barrier to the free flow of oxygen into the colony so that cells in the interior of the colony are not as exposed to oxygen. The efficacy of this mechanism is questionable.

- **Conformational protection**—Some *Azotobacter* species produce a protein that binds to the nitrogenase and protects it from damage by oxygen. In the presence of oxygen and when respiration cannot keep up with the incoming oxygen supply, the protein binds to the nitrogenase complex and alters its conformation (shape) to protect it from oxygen. When exposed to oxygen, the organism stops N_2 fixation abruptly, but when conditions of oxygenation return to a more favorable state, the organism resumes N_2 fixation. The enzyme is not destroyed.

- **Temporal or spatial separation of N_2 fixation and oxygen-evolving processes**—Nonheterocystous cyanobacteria solve the oxygen problem by fixing dinitrogen primarily during the dark phase of growth (i.e., temporal separation) when oxygen evolution is not occurring, and respiration also serves to scavenge oxygen away from nitrogenase. Others form aggregates of cells, and some of these cells then function in a microaerophilic environment with spatial separation.

After Postgate (1987). Used with permission.

The Associative Dinitrogen–Fixing Bacteria

The Nature of the Association

Examples abound of associations in which an otherwise free-living diazotroph associates with a plant. These more or less casual associations generally exhibit no morphological modification of the host or overt genetic interaction between the plant and the bacterium, such as those observed in the root-nodule symbioses described in Chapter 14. Instead, they tend to result from the bacterium establishing itself in an environment where a carbon source is made available, albeit somewhat inadvertently, by the plant. Diazotrophic bacteria have been found on leaf surfaces, the so-called phyllosphere, and in and around the roots, or the rhizosphere, of a wide range of plant species.

Diazotrophic bacteria have been observed in the intercellular spaces of the outer cell layers of roots as well as in the moribund cells of the outer layers and in vascular elements of some plants. The bacteria residing in the rhizosphere and at the root surface (Fig. 13–3 and Fig. 13–4) use root exudates, secretions, lysates, and sloughed cells as a carbon source for growth and N_2 fixation. Some bacteria, such as *Acetobacter diazotrophicus,* may occupy internal root tissues of sugarcane, including the vascular elements. As a consequence of their activities, some dinitrogen is fixed and is eventually made available for plant growth. Bacteria that take up residence in the vascular elements gain the obvious advantage of being "first in line" for substrates and thereby reduce the problem of competition from nondiazotrophs. It is widely accepted that diazotrophic bacteria can be active colonizers of plant roots.

Knowledge that diazotrophic bacteria are part of the root and rhizosphere bacterial population has prompted interest in using the bacteria as an alternate source of nitrogen for crop growth. In the mid to late 1970s considerable research was launched as a result of reports of diazotrophic bacteria fixing large amounts of nitrogen (up to 90 kg ha^{-1} yr^{-1}) in association with the roots of tropical grasses, such as *Paspalum notatum* cv. batatais and *Digitaria* species. The association between *Paspalum notatum* and *Azotobacter paspali* is interesting because of the apparent specificity involved. The bacterium associates only with the tetraploid cultivar "batatais" of the grass. These and other studies stimulated much more research on the use of free-living diazotrophs as inoculants for crop plants. Further aspects of the use of these bacteria as inoculants are discussed in Box 13–3.

Today, the widely accepted rates of N_2 fixation by root-associated bacteria are probably in the range of 5 to 25 kg N ha^{-1} yr^{-1} or growing season. Compare these rates with those reported for a range of N_2-fixing plant-microbe associations in Table 13–6. These low rates of N_2 fixation by associative bacteria are probably insignificant for production of the major cereal crops; however, they are significant for range and prairie grasses where the only other nitrogen inputs are from atmospheric deposition and possibly from the excretion products of animals. They may also prove yet to be significant in the culture of low-maintenance grasses for amenity planting and roadside vegetation, and they may be important for growth of tropical plants other than grasses.

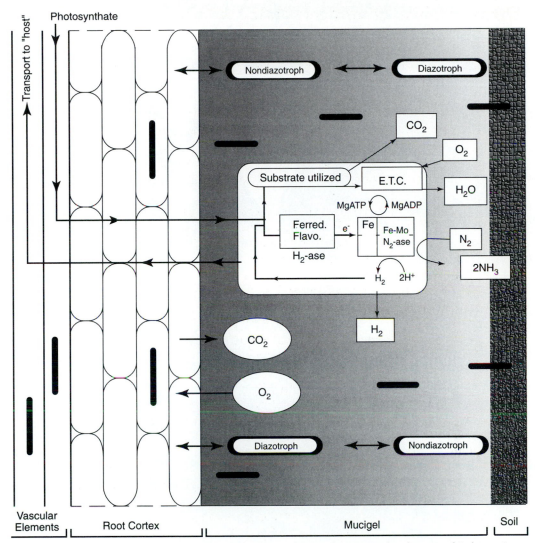

Figure 13–3 A model for root and rhizoplane-rhizosphere colonization by diazotrophic bacteria. The central reactions of the diazotroph are in the enlarged "cell" at center. Diazotrophs and nondiazotrophs (indicated by black bars) colonize the rhizosphere soil, the mucigel layer, the rhizoplane (root surface), and the outer cell layers of the root. In some instances, certain bacteria can colonize the vascular elements. The bacteria grow at the expense of carbon sources (photosynthate, exudates, secretions, lysates, and sloughed cells) provided by the plant and forming a gradient of substrates external to the root (shaded area). Diazotrophs must compete for substrates with nondiazotrophs. Dinitrogen is fixed by bacteria in all of the locations shown and eventually becomes available to the plant or other microbes when the cells die and become mineralized. Interactions between diazotrophs and nondiazotrophs may be beneficial (e.g., a nondiazotroph lowering the oxygen concentration or immobilizing ammonium or nitrate) or detrimental (e.g., nondiazotrophs consuming carbon that otherwise might have been available to support N_2 fixation). Abbreviations: Ferred. = ferredoxin, Flavo. = flavodoxin, E.T.C. = electron-transport chain.

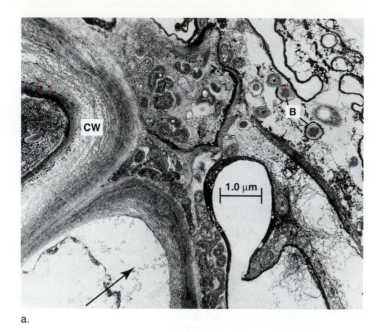

a.

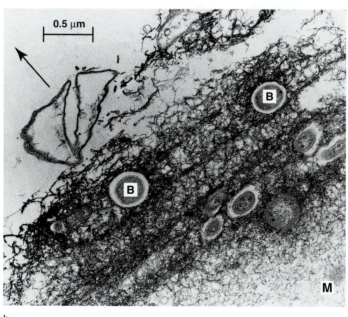

b.

Figure 13–4 Photomicrographs of ultra-thin sections of mangrove roots viewed with a transmission electron microscope. (a) Junction between two root cells (note the multilayered cell wall (cw) material at left) and bacterial cells (B) in the mucilage and sloughed cell debris at the root surface. Numerous bacteria are in this material. The arrow points away from the central axis of the root. (b) Bacteria embedded in this mucilage. In this micrograph the fibrillar mucilage appears in layers probably because the plant root had periods of increased exudation of carbon. For reference orientation, these photos represent the immediate exterior of the intact root which could be encompassed in a zone such as that marked with the box in Figure 13–5 and enlarged graphically in Figure 13–3.

Box 13–3

Does Inoculation with Free-Living Diazotrophs Benefit Crop Production? Among the most widely studied of the associative symbioses is the association between *Azospirillum* and roots of numerous grass species, including most of the important cereal crops. The subject has been thoroughly reviewed by Okon (1994). We should point out that the idea of exploiting free-living diazotrophs for crop production is not new. Throughout the 1950s and 1960s, the Russians treated vast acreages of wheat with *Azotobacter* in a formulation called "azotobacterin" in hopes of increasing yields without applications of fertilizer nitrogen. This practice met with little success. A small percentage of the experiments showed increased growth due to inoculation and when increases were observed, they were on the order of 10% or just barely detectable.

Okon and Labandera (1994) summarized the results of 20 years of field inoculation experiments with the bacterium *Azospirillum*. They concluded that the majority of the experiments (60 to 70%) led to statistically significant yield responses on the order of 5 to 30%. They further concluded that successful experiments "appeared to be those in which researchers paid special attention to the optimal numbers of cells of *Azospirillum* in the inoculant, using inoculation methods where the optimal number of cells remained viable and available to colonize the roots."

Although these results are encouraging for the use of bacterial inoculants to increase or maintain crop yields, we must point out that the scientific literature tends to be skewed by the traditional bias against publishing papers reporting negative results. Thus, many experiments where no inoculation responses were observed have never been published. Another important point to be learned from the large body of work dealing with inoculation of plants with *Azospirillum* is that it is now widely accepted that increases in plant growth due to inoculation with the bacterium are in all likelihood not due to increased N_2 fixation. Rather, they may be attributed to the production of plant growth-promoting hormones (e.g., auxins, giberellins, and cytokinins) or to a variety of plant physiological responses. For example, inoculated plants are reported to be more effective in taking up inorganic nutrients, and it has been reported that the bacteria applied as inoculants elicit changes in the permeability of the membranes of root cells. *Azospirillum,* then, is not only an N_2-fixing associate of the roots of many plant species, it also has the attributes of other so-called **plant-growth-promoting rhizobacteria** (Chapter 17 and 19).

Why are the rates of N_2 fixation by root- and rhizosphere-colonizing bacteria so low? The answer lies in the requirements of the nitrogenase system. The enzyme complex is irreversibly harmed by exposure to oxygen and is stringently regulated by the level of combined nitrogen in the vicinity of the cell. Also large amounts of energy—and therefore large amounts of substrates—are needed to reduce dinitrogen. Next, consider the conditions that might be encountered around the roots of a plant growing in a normally well-aerated (i.e., not waterlogged) soil. In the associative grass-bacteria systems the diazotrophic bacteria live on or in the root or in the rhizosphere soil immediately surrounding the root. They are encouraged there by the liberation of substrates from the "host" root (Fig. 13–3).

The oxygen concentration around roots can vary considerably, but in general it is high enough that some limitation on N_2 fixation would result from the effect of

Table 13–6 **Estimated average rates of biological N$_2$ fixation for specific organisms and associations.**

Organism or system	Dinitrogen fixed (kg ha^{-1} yr^{-1})
Free-living microorganisms	
Cyanobacteria ("blue-green algae")	25
Azotobacter	0.3
Clostridium pasteurianum	0.1–0.5
Grass-bacteria associative symbioses	5–25
Plant-cyanobacterial associations	
Gunnera	12–21
Azolla	313
Lichens	39–84
Legumes	
Soybeans (*Glycine* max L. Merr.)	57–94
Cowpeas (*Vigna, Lespedeza, Phaseolus,* and others)	84
Clover (*Trifolium hybridum* L.)	104–160
Alfalfa (*Medicago sativa* L.)	128–600
Lupines (*Lupinus* sp.)	150–169
Nodulated nonlegumes	
Alnus (alders, e.g. red and black alders)	40–300
Hippophae (sea buckthorn)	2–179
Ceanothus (snow brush, New Jersey tea, California lilac)	60
Coriaria ("tutu" in New Zealand)	60–150
Casuarina (Australian pine)	58

Adapted from Stevenson (1982). Used with permission.

molecular oxygen on nitrogenase. A secondary effect of oxygen is that the diazotrophs would respire more carbon to protect the nitrogenase and reduce the efficiency of N$_2$ fixation. In contrast, the formation of microsites with low pO$_2$ (low molecular oxygen concentrations) result from the consumption of oxygen by root and microbial respiration so that some diazotrophs might actually benefit from the reduced pO$_2$. Alexander and Zuberer (1989a) showed that nitrogenase activity associated with roots of corn and sorghum was greatest at pO$_2$ ranging from 1.3 to 2.1% (1.3 to 2.1 kPa) and markedly lowered as the pO$_2$ approached 5%. They also demonstrated that plants whose roots were exposed to ^{15}N$_2$ at the low oxygen concentration contained more than 200 times as much ^{15}N as plants whose roots were exposed to air (Alexander and Zuberer, 1989b). Thus, the inhibition of nitrogenase activity by oxygen is reduced as the oxygen content of the environment decreases, but even at values approaching 5% the activity is nearly shut down.

Since nitrogenase activity is also closely regulated by combined nitrogen, the bacteria only fix dinitrogen when it is absolutely necessary. Thus, nitrogen levels in the soil solution around roots of fertilized crop plants may be sufficient to cause limitations on root-associated N$_2$ fixation. Dinitrogen fixation would be expected to occur only in areas where plant and microbial uptake by nondiazotrophs depleted the available supply of soil nitrogen to levels sufficient to derepress nitrogenase synthesis. This may be likely in grasslands where dense rooting and microbial colonization of roots consistently keep the supply of available nitrogen at very low levels, but in highly fertilized soils the prospects for high rates of N_2 fixation are limited.

Perhaps the most significant constraint on root- and rhizosphere-associated diazotrophs is lack of an adequate supply of readily available carbon sources coupled with the extremely high competition for any carbon liberated from the root. The rhizosphere is not populated exclusively with diazotrophic bacteria. In fact, they comprise only 1 to 10% of the recoverable rhizosphere population. Competition among these organisms for the limited supply of available carbon can place severe restrictions on N_2 fixation in the root zone. However, if the diazotroph can become established inside the root, the competition for substrates is lessened, particularly for cells in the vascular elements of the plant (see Box 13–4 on N_2 fixation in sugarcane). Many studies have shown that N_2 fixation in the rhizosphere is carbon-limited; marked stimulation of N_2 fixation usually occurs upon addition of readily available carbon sources to soil. The addition of the ready supply of carbon and the resulting immobilization of combined soil nitrogen make conditions favorable for high rates of N_2 fixation. Even incorporation of straw into soil can support some N_2 fixation.

To put the limitation of carbon supply on N_2 fixation into perspective, consider a statement from Giller and Day (1985) about estimating maximal possible rates of N_2 fixation for a given amount of root carbon. They calculated a maximum N_2 fixation rate by wheat of 15 kg N ha^{-1}, based on four assumptions:

1. The efficiency of conversion of root carbon is 10 g C g^{-1} N fixed.

2. Dinitrogen-fixing bacteria comprise 10% of the total rhizosphere population (a high estimate) and acquire organic carbon in proportion to their numbers.

3. All of the carbon translocated below ground is equally available for use by all bacteria.

4. About 1,500 kg C ha^{-1} are translocated below ground, as estimated for a wheat crop.

Recall that most estimates of N_2 fixation in grasslands range from 5 to 25 kg N ha^{-1}. It is clear that conditions are conducive for N_2 fixation in the rhizosphere, at least part of the time during the growing season. It is also likely that the integration of biotic and abiotic factors provides favorable microsites for N_2 fixation to occur at relatively high rates for short periods or at low rates for longer periods.

The factors (biotic and abiotic) affecting N_2 fixation by bacteria associated with below-ground plant parts are summarized in Figure 13–5. Remember that the bacteria are dependent on the "host" plant for provision of carbon sources; thus anything that affects the photosynthesis of the plant indirectly affects the bacteria associated with the roots or other below-ground structures such as nodules. Also important are the combined effects of the soil variables (moisture, aeration, pH, temperature, and nitrogen content) on plant and microbial growth. By studying the simple models presented in Figures 13–3 and 13–5, we can develop an appreciation of how all factors must come together at appropriate levels to allow for expression of N_2 fixation by bacteria in the roots or rhizospheres of nonleguminous

Box 13–4

Dinitrogen-Fixing Rice and Sugar Cane? Within the last three decades, a great deal of effort has been expended in looking for N_2-fixing plant-bacteria associations capable of providing agriculturally significant amounts of fixed nitrogen for crop production. The optimistic reports of the 1970s suggesting high rates of N_2 fixation associated with roots of grasses were soon tempered by numerous reports that failed to corroborate the earlier findings. However, from this work, scientists have identified some systems where free-living diazotrophs can enhance their "host" plants.

Two crops which appear to derive substantial benefit from their diazotrophic associates are rice grown under flooded conditions and sugarcane. Interestingly, both of these crops have been observed to produce appreciable yields in the absence of added fertilizers. The ability of plants to grow well without added fertilizers raised the suspicion that these plants were somehow benefiting from N_2-fixing bacteria.

Consider the case for N_2 fixation in wetland rice culture. Rice is the major staple crop for two-thirds of the world's people. For centuries in Asian countries, rice has been cultured without the use of fertilizers other than small inputs of human and animal manures, yet the crop bears good yields. Because of the tolerance of rice plants for growth in flooded, often anoxic soils, they derive benefit from a variety of N_2-fixing microbes, including free-living heterotrophic bacteria, N_2-fixing cyanobacteria, phototrophic bacteria, and a symbiotic partnership between the floating aquatic fern *Azolla* and the heterocystous cyanobacterium *Anabaena* (discussed further in Chapter 14). The flooded conditions of the rice paddy overcome some of the environmental limitations on N_2-fixing bacteria and cyanobacteria. First, the floodwater is a much better environment for growth of algae and cyanobacteria than is an upland "dry" soil. Second, bacteria in the mud and associated with roots of rice plants find the low-oxygen environment protective of their nitrogenase enzyme. Thus, the oxygen poisoning of nitrogenase is at least partially alleviated by the naturally reduced oxygen status of the mud and floodwater. Third, before the rice plants grow tall enough to shade the paddy surface, considerable light is available for growth of diazotrophic photosynthetic bacteria and cyanobacteria. Rice may gain 30 to 50 kg N ha^{-1} per growing season through the combined activities of these microbes.

While sugarcane is not grown in flooded soils, it too has been observed to yield well with seemingly little input of fertilizer nitrogen. During the 1980s, Döbereiner and colleagues in Brazil discovered that some cultivars of sugarcane contained a diazotrophic bacterium that they named *Acetobacter diazotrophicus*. The organism can be isolated from the internal tissues of roots, stems, and leaves. It is unique in that it will grow in a 30% sucrose solution and fix dinitrogen optimally at pH 5.5 (and less well at values approaching pH 3); in addition, its N_2-fixing capability seems to be more tolerant of oxygen than other free-living bacteria. The bacterium has been observed in the intercellular spaces and in the xylem elements of sugarcane stems, and it may achieve numbers on the order of 10^6 to 10^7 cells g^{-1} of these tissues and 10^3 to 10^4 cells ml^{-1} of fluid in the intercellular spaces (James et al., 1994; Dong et al., 1994). By occupying such a unique location, the organism is poised to intercept a large supply of sugar to drive N_2 fixation, and it is free of the competition from other bacteria that organisms in the rhizosphere must deal with. Some sugarcane cultivars may derive as much as 100 to 150 kg N ha^{-1} from this unique association. Time will tell what the real significance of this bacterium is in the commercial production of sugarcane.

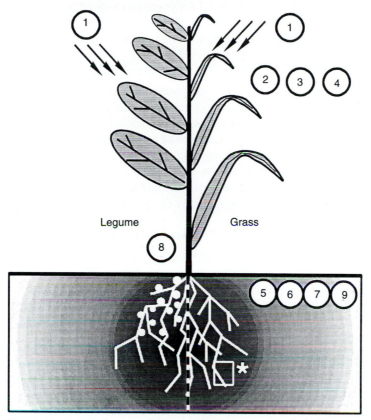

Figure 13–5 Main factors affecting nitrogenase activity below ground: (1) light intensity, (2) plant photosynthetic potential (e.g., C_3 vs. C_4 plants), (3) air temperature, (4) humidity, (5) soil temperature, (6) soil redox and water content, (7) other soil factors—pH, O_2, and N content, (8) plant genetic factors, and (9) nature and numbers of the diazotrophs. The boxed area next to the asterisk is represented in Figure 13–3.

plants like the grasses. We can also gain an appreciation of the many factors that are "taken care of" by the host plants in the exquisite root nodule symbioses discussed in Chapter 14.

Methods for Measuring Dinitrogen Fixation

The agronomic and ecological importance of biological N_2 fixation has driven intensive study of the process for the past century. Early studies of biological N_2 fixation were limited by the lack of techniques sensitive enough to measure nitrogen gains in plants or to measure the activity of the enzyme complex itself. As developments in electronics have evolved rapidly so too has the development of sophisticated instrumentation for measurement of biological processes, including N_2

fixation. Here the methods routinely used for measurement of N_2 fixation or nitro-genase activity will be introduced. Detailed reviews of the subject are available else-where (Bergerson, 1980; Weaver and Danso, 1994; Sprent and Sprent, 1990).

Measurement of Dinitrogen Fixation by the Nitrogen Difference Method

It is possible to estimate N_2 fixation by comparing the yields and nitrogen content of plants grown with and without N_2-fixing bacteria. One example is comparing the yields of nodulated and nonnodulated legumes grown under similar conditions of soil nitrogen. The nitrogen in the nonfixing plant is a measure of the nitrogen acquired from soil and is subtracted from the nitrogen of the fixing plant; the difference is attributed to N_2 fixa-tion. Weaver and Danso (1994) indicate that the sensitivity of the method is such that an increase of about 20 kg N ha^{-1} would be required for detection of significant differences. The method is limited to the study of systems where N_2 fixation rates are high.

Measurement of Dinitrogen Fixation by the Acetylene Reduction Assay

The ability of the nitrogenase complex to reduce acetylene (C_2H_2) to ethylene (C_2H_4) forms the basis for the acetylene reduction assay (see Burris, 1974, for a description of development of the method). The sensitivity of this method, based on gas chromatog-raphy, allowed detection of very low levels of nitrogenase activity. In fact, the sensitiv-ity of the acetylene reduction assay facilitated attempts to measure N_2 fixation by grasses.

The reaction for the reduction of dinitrogen to ammonia is:

$$N_2 + 8H^+ + 8e^- \longrightarrow 2NH_3 + H_2$$

The conversion of acetylene to ethylene occurs by the following reaction:

$$C_2H_2 + 2H^+ + 2e^- \longrightarrow C_2H_4$$

The reduction of dinitrogen (including reductant used in H_2 evolution) requires eight electrons, whereas the reduction of acetylene requires two electrons. Thus, reduc-tion of four molecules of acetylene is equivalent to reduction of one molecule of N_2, that is, the ratio of acetylene reduced to nitrogen fixed is 4/1. In practice, this ratio has been shown to vary anywhere from 1.5/1 to 25/1, depending on the system be-ing measured. Most investigators assume a ratio of 3/1 to 4/1.

In the acetylene reduction assay, the system to be measured (whole plants, iso-lated roots, soil cores, or bacterial cultures) is exposed to an atmosphere containing 11% acetylene and incubated under appropriate conditions. The concentration of acetylene is such that the nitrogenase complex is saturated with this substrate and dinitrogen, the normal substrate (Fig. 13–2), is no longer reduced. Samples of the gas phase are periodically removed and injected into the gas chromatograph for quantification of ethylene production from acetylene.

The acetylene reduction assay is not without limitations—careful attention must be paid to these limitations to avoid gross over-estimations of N_2 fixation—and the method is still a point of some contention among researchers. Pitfalls of the acety-lene reduction assay for measurement of associative N_2 fixation include (Giller, 1987):

• production of ethylene which is not from acetylene,

• oxidation of ethylene by other soil bacteria,

- nitrogen starvation during prolonged incubations due to inhibition of nitrogenase by acetylene, and

- failure to calibrate the acetylene reduction vs. dinitrogen fixation ratio appropriately.

Measurement of Dinitrogen Fixation by Stable Isotope (^{15}N) Methods

The most definitive measurements of biological N_2 fixation make use of the stable, heavy isotope, ^{15}N, and require access to a mass spectrometer. The availability of these instruments is not as major a limitation as it once was but ^{15}N methods are still expensive, both because of the high costs of the ^{15}N-labeled nitrogen sources and the cost of analysis.

The principal methods involved in studies using ^{15}N to measure N_2 fixation are:

- measurement of the incorporation of $^{15}N_2$ (labeled dinitrogen) into plant or microbial cells,

- isotope dilution methods in which the content of ^{15}N in plant tissue is measured and the ratio of ^{15}N to ^{14}N is calculated, and

- natural abundance of ^{15}N in soils or plants.

The first method is straight-forward. Samples are exposed to an atmosphere of about 10% $^{15}N_2$, usually in a balance of argon or helium (with sufficient oxygen and carbon dioxide) to eliminate competition from $^{14}N_2$. Control plants not exposed to $^{15}N_2$ are included to measure background ^{15}N. Following incubation, the samples (cells or ground plant materials) are digested and the ^{15}N content of the materials is determined using a mass spectrometer. Detection of ^{15}N in tissues or cells provides definitive proof of N_2 fixation and allows very accurate quantification of the amount of dinitrogen fixed.

The *isotope dilution* and *natural abundance* methods rely on the fact that in nature the ratio of ^{15}N/^{14}N is remarkably constant. ^{15}N and ^{14}N constitute about 0.3663 and 99.6337% of the nitrogen in the atmosphere, respectively. A substance containing more ^{15}N is said to be ^{15}N-enriched. Such materials can be exploited to measure N_2 fixation in the field. In practice, the soil is fertilized to a low level with the ^{15}N-enriched fertilizer, and the plants are grown to maturity or a suitable stage for harvest. The ^{15}N content in the tissues is analyzed and the nitrogen derived from the atmosphere can be calculated. The fixing plant will derive part of its nitrogen from the atmosphere and will contain more ^{14}N than the soil nitrogen thus the ^{15}N in the fixing plant is "diluted" with the ^{14}N from atmospheric dinitrogen.

The ^{15}N natural abundance method is based on the same principle as the isotope dilution method. Natural materials contain nitrogen that is naturally enriched with ^{15}N because of isotope discrimination during biological transformations of nitrogen in soil. Therefore, plants containing less ^{15}N than the natural abundance can be suspected of supporting N_2 fixation. The assumptions and formulae for calculating the amount of nitrogen fixed and the rationale for the calculations are given by Weaver and Danso (1994).

Summary

In this chapter we introduced the fundamentals of the process of biological N_2 fixation and the nature of the enzyme complex nitrogenase, which mediates the conversion of dinitrogen to ammonia. The process is unique in that only prokaryotic microbes, as far as we know, possess this enzyme system. Just as it is expensive to fix dinitrogen industrially, it is also expensive biologically for the microbes that fix dinitrogen. Therefore, high rates of biological N_2 fixation occur only when a ready supply of energy—light for phototrophs and organic carbon for chemoheterotrophs—is available. For this reason, the chemoheterotrophic N_2-fixing microbes are usually found in association with plants or in soils receiving organic carbon additions.

Despite the cumbersome nature of the nitrogenase enzyme complex, about two-thirds of the global supply of fixed nitrogen is produced by biological N_2 fixation, mostly through the legume-rhizobia symbiosis discussed in Chapter 14. While the contributions of fixed nitrogen from the nonsymbiotic diazotrophs are generally low (5 to 25 kg N ha^{-1} yr^{-1}), they are significant in areas where no other source of nitrogen is available. The associative symbiosis between grasses and free-living diazotrophs continues to receive attention not only for its interesting biology but with an eye toward enhancing such associations for practical use in sustaining agricultural and natural ecosystems.

It is difficult to foresee what breakthroughs continued investigations of nitrogenase and the microbes that possess it will bring. Imagine if chemists, biochemists, and microbiologists could someday mimic the action of nitrogenase in huge tanks to fix dinitrogen without the high costs of the fossil fuel currently needed. Imagine if molecular biologists could someday get N_2-fixing root nodules on nonleguminous plants, such as wheat, rice, and corn. Then a whole new set of questions would arise. For example, can these plants provide the carbon necessary to support N_2-fixing nodules and still produce abundant grain? These are pipe dreams perhaps, but it is reasonably certain that N_2 fixation has a strong future in agricultural research and production and in the maintenance of our relatively unmanaged ecosystems such as forests and grasslands.

Cited References

Alexander, D.B., and D.A. Zuberer. 1989a. Impact of soil environmental factors on rates of N_2 fixation associated with intact maize and sorghum plants. pp. 273–285. *In* F.A. Skinner, R.M. Boddey, and I. Fendrik (eds.), Nitrogen fixation with non-legumes. Kluwer Academic Press, Dordrecht, The Netherlands.

Alexander, D.B., and D.A. Zuberer. 1989b. $^{15}N_2$ fixation by bacteria associated with maize roots at a low partial O_2 pressure. Appl. Environ. Microbiol. 55:1748–1753.

Bishop, P.E, and R. Premakumar. 1992. Alternative nitrogen fixation systems. pp. 736–762. *In* G. Stacey, R.H. Burris, and H.J. Evans (eds), Biological nitrogen fixation. Chapman and Hall, New York.

Burns, R.C., and R.W.F. Hardy. 1975. Nitrogen fixation in bacteria and higher plants. Springer-Verlag, New York.

Burris, R.H. 1974. Methodology. pp. 10–33. *In* A. Quispel (ed.), The biology of nitrogen fixation. American Elsevier Publishing Co., New York.

Burris, R.H., and G.P. Roberts. 1993. Biological nitrogen fixation. Annu. Rev. Nutr. 13:317–335.

Dong, Z., M.J. Canny, M.E. McCully, M.R. Roberedo, C.F. Cabadilla, E. Ortega and R. Rodés. 1994. A nitrogen-fixing endophyte of sugarcane stems. A new role for the apoplast. Plant Physiol. 105:1139–1147.

Dixon, R.O.D., and C.T. Wheeler. 1986. Nitrogen fixation in plants. Chapman and Hall, New York.

Evans, H.J., and R.H. Burris. 1992. Highlights in biological nitrogen fixation during the last 50 years. pp. 1–42. *In* G. Stacey, R.H. Burris, and H.J. Evans (eds.). Biological nitrogen fixation. Chapman and Hall, New York.

Giller, K.E. 1987. Use and abuse of the acetylene reduction assay for measurement of "associative" nitrogen fixation. Soil Biol. Biochem. 19:783–784.

Giller, K.E., and J.M. Day. 1985. Nitrogen fixation in the rhizosphere: significance in natural and agricultural systems. pp. 127–147. *In* A.H. Fitter (ed.), Ecological interactions in soil. Blackwell Scientific Publications, Oxford, England.

Hauck, R.D. 1985. Agronomic and technological approaches to improving the efficiency of nitrogen use by crop plants. pp. 317–326. *In* K.A. Malik, S.H.M. Naqvi,and M.I.H. Aleem (eds.), Nitrogen and the environment. Nuclear Institute for Agriculture and Biology, Faisalabad, Pakistan.

Hill, S. 1992. Physiology of nitrogen fixation in free-living heterotrophs. pp. 87–134. *In* G. Stacey, R.H. Burris, and H.J. Evans (eds.), Biological nitrogen fixation. Chapman and Hall, New York.

James, E.K., V.M. Reis, F.L. Olivares, J.I. Baldani, and J. Döbereiner. 1994. Infection of sugar cane by the nitrogen-fixing bacterium *Acetobacter diazotrophicus*. J. Expt. Bot. 45:756–766.

Okon, Y., J.P. Houchins, S.L. Albrecht, and R.H. Burris. 1977. Growth of *Spirillum lipoferum* at constant partial pressures of oxygen and the properties of its nitrogenase in cell-free extracts. J. Gen. Microbiol. 98:87–93.

Okon, Y., and C. Labandera. 1994. Agronomic application of *Azospirillum:* An evaluation of 20 years worldwide field inoculation. Soil Biol. Biochem. 26:1591–1601.

Okon, Y. 1994. *Azospirillum*/plant associations. CRC Press, Boca Raton, Fla.

Postgate, J.R. 1987. Nitrogen fixation. 2nd ed. Edward Arnold, London.

Postgate, J. 1994. The outer reaches of life. Cambridge University Press, Cambridge, England.

Stevenson, F.J. 1982. Origin and distribution of nitrogen in soil. pp. 1–42. *In* F.J. Stevenson (ed.), Nitrogen in agricultural soils. Agronomy No. 22, American Society of Agronomy, Madison, Wis.

Sumner, M.E. 1990. Crop responses to *Azospirillum* inoculation. pp. 53–123. *In* B.A. Stewart (ed.), Advances in soil science. Vol. 12. Springer-Verlag, New York.

Young, J.P.W. 1992. Phylogenetic classification of nitrogen-fixing organisms. pp. 43–86. *In* G. Stacey, R.H. Burris and H.J. Evans (eds.), Biological nitrogen fixation. Chapman and Hall, New York.

Waggoner, P.E. 1994. How much land can ten billion people spare for nature? Council on Agricultural Science and Technology, Report 121, Ames, Ia.

Weaver, R.W., and S.K.A. Danso. 1994. Dinitrogen fixation. pp. 1019–1045. *In* R.W. Weaver, S. Angle, P. Bottomley, D. Bedzdicek, S. Smith, A. Tabatabai, and A. Wollum (eds.), Methods of soil analysis, Part 2. Microbiological and biochemical properties. Soil Science Society of America, Book Series, No. 5. Madison, Wis.

General References

Döbereiner, J., and F.O. Pedrosa. 1987. Nitrogen-fixing bacteria in nonleguminous crop plants. Science Tech Publishers, Madison, Wis.

Elmerich, C., W. Zimmer, and C. Vielle. 1992. Associative nitrogen-fixing bacteria. pp. 212–258. *In* G. Stacey, R. H. Burris, and H. J. Evans (eds.), Biological nitrogen fixation. Chapman and Hall, New York.

Postgate, J.R. 1982. The fundamentals of nitrogen fixation. Cambridge University Press, Cambridge, England.

Schlesinger, W.H. 1991. Biogeochemistry. An analysis of global change. Academic Press, New York.

Sprent, J.I., and P. Sprent. 1990. Nitrogen fixing organisms. Pure and applied aspects. Chapman and Hall, New York.

Zuberer, D.A. 1990. Soil and rhizosphere aspects of N_2-fixing plant-microbe associations. pp. 317–353. *In* J.M. Lynch (ed.), The rhizosphere. John Wiley and Sons, New York.

Study Questions

1. Discuss the significance of biological N_2 fixation in terms of its relative contribution to the world's supply of fixed nitrogen.

2. Write the biochemical reaction for the process of biological N_2 fixation.

3. What kinds of microorganisms are capable of carrying out biological N_2 fixation? Why are these microbes called "diazotrophs"?

4. Describe, in general terms, the nitrogenase enzyme complex. What are some unusual properties of the nitrogenase enzyme complex?

5. What are some problems, inherent in the nitrogenase system, that limit our abilities to use nonsymbiotic N_2-fixing bacteria to improve the yields of agriculturally important crop plants?

6. Compare and contrast the processes of nonsymbiotic and symbiotic N_2 fixation.

7. What is an "associative symbiosis"? How do rates of N_2 fixation compare in associative symbioses and root-nodule symbioses between legumes and *Rhizobium?*

8. Why is it not a simple matter to genetically engineer higher plants to fix their own nitrogen for growth?

9. Why do rice and sugarcane benefit from the activities of nonsymbiotic N_2-fixing bacteria more so than other nonleguminous plants?

10. If a free-living diazotroph consumes about 35 g of carbon to fix 1 g of N, how much carbon would have to be supplied to fix 45 kg (about 100 pounds) of nitrogen? Give your answer in kilograms and pounds. Show your calculations.

Chapter 14

♦

Biological Dinitrogen Fixation: Symbiotic

Peter H. Graham

♦

An' I am blest, becos me feet 'ave trod
A land 'oo's fields reflect the smile o' God
C.J. Dennis

A mutualistic symbiosis is an association between two organisms from which each derives benefit. It is usually a long-term relationship and, in the case of symbiotic dinitrogen (N_2) fixation, often involves development of a special structure to house the microbial partner. Each N_2-fixing symbiotic association involves an N_2-fixing prokaryotic organism, the microsymbiont (e.g., *Rhizobium, Klebsiella, Nostoc,* or *Frankia*) and a eukaryotic, usually photosynthetic, host (e.g., leguminous or non-leguminous plant, water fern, or liverwort). These symbioses contribute more than 100 million metric tons of combined nitrogen per year to the global nitrogen economy and account for more than 65% of the nitrogen used in agriculture. Rates of N_2 fixation vary with host, microsymbiont, and environment but in temperate clover pastures may reach 600 kg N fixed ha^{-1}. Grain legumes fix 100 to 200 kg N ha^{-1} growth $cycle^{-1}$ and supply 40 to 85% of the plant's nitrogen needs through symbiosis. The direct availability of the fixed nitrogen permits the host to grow in soils that are nitrogen deficient and, at the same time, reduces losses by denitrification, volatilization, and leaching, thus improving the sustainability of an agricultural system. Dinitrogen fixation is likely to become even more important in the future as population increases in many developing countries necessitate sharply increased crop production, while pollution, energy, and cost concerns limit significant increases in the use of fertilizer nitrogen.

Box 14–1

The Importance of Symbiotic Dinitrogen Fixation. Symbiotic N_2 fixation is the single greatest contributor to the global nitrogen economy. Principal contributors include:

- leguminous plants and their associated rhizobia,

- actinorhizal plants in symbiosis with *Frankia*,

- the water fern *Azolla* and its microsymbiont *Anabaena*, and

- lichen symbioses involving cyanobacteria.

The Symbiosis Between Legumes and Rhizobia

Legumes have been used in crop rotations since the time of the Romans. Theophrastus (370–285 B.C.) stated " . . . beans are not a burdensome crop to the ground, they even seem to manure it. . . . wherefore the people of Macedonia and Thessaly turn over the ground when it is in flower." However, it was not until detailed nitrogen balance studies became possible that leguminous plants were shown to accumulate nitrogen from sources other than soil and fertilizer. In 1886 Hellriegel and Wilfarth demonstrated that the ability of legumes to convert dinitrogen from the atmosphere into compounds that could be used by the plant was associated with the presence of swellings or **nodules** on the legume root. They related this association to the presence of particular bacteria within the nodule. Later in this chapter we discuss the different types of root-nodule bacteria, but for the moment can refer to them collectively as **rhizobia.** It was then a series of short, but important, steps to the isolation of rhizobia from nodules by Beijerinck in 1888 and to the completion of Koch's postulates by the demonstration of their ability to reinfect the legume and to fix dinitrogen in symbiosis (Chapter 1).

Groupings of Rhizobia and Their Separation into Species

Early studies showed that each rhizobial strain or isolate had a finite host range, nodulating certain legumes but not others. This led to the concept of cross-inoculation, with legumes grouped according to the different rhizobia with which they formed nodules. Thus rhizobia isolated from species of *Medicago* (e.g., alfalfa) would also nodulate *Melilotus*, and vice versa, though rhizobia isolated from these hosts would not nodulate *Trifolium* spp. (clovers). More than 20 different cross-inoculation groups were identified, with the bacteria from the clover, medic, bean, lupine, pea, and soybean groups named as separate species within the single genus *Rhizobium* (e.g., *R. trifolii* for clover). Host specificity is still important in the identification of rhizobia, but more recently other traits have assumed greater significance in their classification. There were a number of reasons for this:

- Initial studies involved mainly legumes of agricultural importance. Study of less traditional legumes blurred cross-inoculation boundaries. For example, the bacterial strain NGR234, originally isolated from *Lablab purpureus*, the hyacinth bean, nodulates with 34 different species of legume and with a nonlegume, *Parasponia andersonii* (Stanley and Cervantes, 1991). Some nonlegumes, such as rice and wheat, can be induced to form nodule-like structures with rhizobia if pretreated with plant enzymes or hormones. Even now, though, less than 15% of the roughly 19,700 species of legumes have been evaluated for nodulation.

- Many anomalous results have been reported. One study gave more than 500 examples where strains were either promiscuous (i.e., they nodulated legumes from other cross-inoculation groups) or failed to nodulate legumes from their own group.

- Nodulation genes of some rhizobia are plasmid-borne. Strains losing the plasmid as the result of exposure to high temperature also lose the ability to form nodules and for many years could not be identified. In one soil, noninfective rhizobia lacking the symbiotic plasmid outnumbered those capable of nodule formation by 40 to 1.

- Taxonomic methods were developed to compare strains on the basis of many different traits. Computer-based numerical classification, along with taxonomic methods based on differences in cell DNA or RNA, often resulted in groupings at odds with those based on host range.

Some of the traits now used in the classification of rhizobia are listed in Table 14–1. The original genus *Rhizobium* is now divided into 4 genera and 16 species, as shown in Table 14–2. Most of these changes have occurred since 1985.

Similarities between the rhizobia and other organisms have been identified. Thus the fast-growing species *Rhizobium tropici* shows a close affinity to species of *Agrobacterium*, a plant-pathogenic bacterium causing crown gall and hairy root diseases, while some strains of the slower-growing *Bradyrhizobium* produce bacteriochlorophyll and are more closely related to the photoautotroph, *Rhodopseudomonas palustris*.

Table 14–1 Characteristics for the phenotypic and phylogenetic characterization of rhizobia.

Phenotypic traits*

    Range of substrates used as sources of energy (e.g., sugars, sugar alcohols, and complex carbohydrates)

    Range of substrates used as sources of nitrogen (e.g., amino acids, urea, and nitrate)

    Resistance to specific antibiotics

    Electrophoretic mobility of different cell enzymes

    Tolerance to different stresses (e.g., salt, temperature, and pH)

Phylogenetic traits[†] (refer to Chapters 8)

    Pattern of banding of DNA restriction fragments (RFLPs)

    Degree of hybridization with specific DNA probes

    16S rRNA sequence analysis

*Phenotypic traits can be observed in culture.

†Phylogenetic traits are related to cell DNA or RNA composition.

Table 14–2 **Genera and species of the root-nodule bacteria of legumes. Genera in the square brackets refer to host legumes nodulated by each species of root-nodule bacteria. Common names are included for well-known legume genera. In several examples in this list, different species of root-nodule bacteria nodulate the same legume.***

Rhizobium[†]

 R. leguminosarum (with three biovars: trifolii [*Trifolium*, clovers], viciae [*Pisum*, peas; *Vicia*, field beans; *Lathyrus;* and *Lens*, lentil], and phaseoli [*Phaseolus*, bean]
 R. loti [*Lotus*, trefoil]
 R. tropici [*Phaseolus*, bean; *Leucaena*, Ipil-Ipil, and *Macroptilium*]
 R. etli [*Phaseolus*]
 R. galegae [*Galega, Leucaena*]
 R. huakuii [*Astragalus*, milkvetch]
 R. ciceri [*Cicer*, chickpea]
 R. mediterraneum [*Cicer*, chickpea]

Sinorhizobium

 S. meliloti [*Melilotus*, sweetclover; *Medicago*, alfalfa; and *Trigonella*, fenugreek]
 S. fredii [*Glycine*, soybean]
 S. saheli [*Sesbania*]
 S. teranga [*Sesbania, Acacia*, wattle]

Bradyrhizobium

 B. japonicum [*Glycine*, soybean]
 B. elkanii [*Glycine*]
 B. liaoningense [*Glycine*]

Azorhizobium

 A. caulinodans [*Sesbania*]

*Other genus and species names exist in the literature. Some predate the present names. Others (e.g., Photobacterium) have not been accepted as valid.

[†]Strains of *Rhizobium* and *Bradyrhizobium* that do not belong in any named species are usually identified by the host from which they were isolated, e.g., *Rhizobium* spp. (*Acacia*) or *Bradyrhizobium* spp. (*Lupinus*).

The Infection Process

Nodule Initiation and Development

The process of nodule formation is outlined here. Greater detail can be obtained by reading Hirsch (1992).

Infection. Mechanisms by which rhizobia infect their hosts and induce root- or stem-nodule formation include:

- penetration of root hairs and formation of infection threads as found in plants such as clovers and beans,

- entry via wounds or sites of lateral root emergence, as found in peanut and the pasture legume *Stylosanthes*, and

- penetration of root primordia, as found on the stem of plants such as *Sesbania*.

Root-hair infection has been studied for many years, using small-seeded legumes that were inoculated with rhizobia, embedded in agar, and grown between glass slides. Such

Fåhreus slides permitted observation of the different steps of the infection process under the microscope. Infection begins with rhizobial attachment to immature, emerging root hairs of a compatible host. Deformation and curling of the root hair follows (Fig. 14–1), with the root hair surface at the point of infection hydrolyzed to permit penetration of the rhizobia. Rhizobia then move down the root hair toward the root cortex.

Rhizobia never gain free intracellular access to their host. During infection, and as they move down the root hair, they become enclosed within a plant-derived

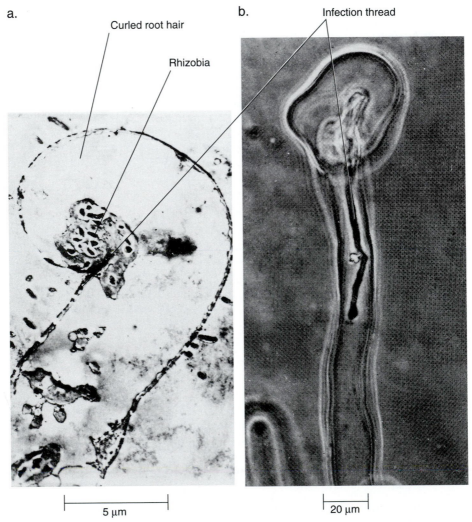

a.

Curled root hair

Rhizobia

b.

Infection thread

5 µm

20 µm

Figure 14–1 Root hair deformation, curling, and infection: an early stage in the nodulation of clover by rhizobia. (a) The initiation of infection thread formation showing the plant-derived gelatinous matrix surrounding 10 to 20 cells of rhizobia. *From Sahlman and Fåhreus (1963). Used with permission.* (b) Movement of the infection thread down the root hair toward the root cortex. *From Fåhreus (1957). Used with permission.*

infection thread. They remain surrounded by this material until released into modified cells of the root cortex, where again they are enclosed within a plant-derived **peribacteroid membrane.** These membranes protect the bacteria from the defense responses of the host.

The demonstration that the nodulation genes of many rhizobia are carried on plasmids led to more detailed molecular studies of infection. Two different groups of genes are required for infection:

- *Common* nodulation genes (*nod* A,B,C) are found in all rhizobia. A fourth gene (*nod* D) is sometimes included in this group but is unique in being the only *nod* gene expressed in the absence of a suitable host.

- *Host-specific* nodulation genes (*nod* E, F, G, H, I, J, L, M, P, Q in the case of *S. meliloti*) differ with type of rhizobia and define the host range.

Because *nod* D is the only *nod* gene expressed in the absence of the host, studies to determine the host factor(s) needed to trigger expression of other *nod*-genes soon followed. *Flavonoids,* complex phenolic compounds exuded from the legume root, were implicated and hypothesized to interact with the product of *nod* D prior to the expression of the other genes. Considerable specificity has been shown in this interaction. For example, luteolin is the principal flavonoid involved in *nod*-gene expression in *S. meliloti,* whereas naringenin and genistein are required for *B. japonicum.*

Characterization of the different nodulation genes led to the detection and characterization of a series of substances, termed lipo-oligosaccharides or *nod factors,* which are responsible for root hair deformation and curling and the division of cortical cells in the root at concentrations lower than $10^{-9}M$. Composition of these nod factors varies with microsymbiont but in each case includes a backbone of β 1, 4-linked, N-acetylglucosamine units specified by the common *nod* genes. The chemical composition of the major lipo-oligosaccharide produced by *S. meliloti* is shown in Figure 14–2. The *nod* genes in *Bradyrhizobium* are not located on plasmids, but are otherwise analogous to those found in *Rhizobium.* Even in *Rhizobium,* not all of the genes contributing to nodule formation and function are found on plasmids. Figure 14–3 shows the effect of mutation in the chromosomal genes required for bacterial **lipopolysaccharide** production on nodule morphology and nitrogenase activity.

Nodule development and function. As the infection thread penetrates the root cortex and the rhizobia it contains are released into host cells, cell division and enlargement of these cells results in the formation of a visible nodule. Root nodules differ in appearance and structure, a trait determined by the host legume. Determinate nodules, such as those which occur on soybean and *Phaseolus,* are round and have no pronounced meristematic region. In contrast, the indeterminate nodules of peas, medics, and clovers are elongated with a pronounced meristematic region, and increase in length over the growing season.

Examination of an alfalfa nodule under the microscope reveals four distinct zones, as shown in Figure 14–4:

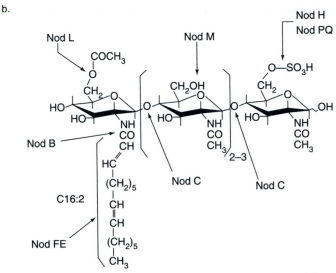

Figure 14–2 Structure of the major lipo-oligosaccharide nod factor produced by *S. meliloti* and the role of different *nod* genes in its biosynthesis. (a) A simplified genetic map of the *nod* gene region showing structural and regulatory genes and their organization. (b) The lipo-oligosaccharide molecule in this species has a backbone of β 1,4-linked glucosamine residues, and carries N-acylated, N-acetylated, and O-sulfated side chains. Site of action of the nod gene products (enzymes) are shown. As an example of *nod* gene function, the products of *nod* PQ exhibit homology with the enzymes ATP sulfurylase and APS kinase. *From Denarie and Cullimore (1993). Used with permission.*

- Meristematic region in which host cells undergo active division but show little infection by rhizobia (M).

- A region in which many plant cells are infected but in which the bacteria have not undergone changes in size and shape (TI). Dinitrogen fixation is limited.

- Region of active N$_2$ fixation (ES), often red or pink in color due to the presence of **leghemoglobin.** Host cells will contain many rhizobia, which may be misshapen. Such bacteria are referred to as **bacteroids.**

- Region of nodule senescence in which the symbiosis is breaking down (LS). Bacteroids may undergo lysis, and the degradation of leghemoglobin results in a green or brown coloration.

Nodules with a large pink or red region usually are active in N$_2$ fixation and are said to be *effective*. If the nodule is white or greenish brown, either the symbiosis is ineffective or the nodule is undergoing breakdown and is said to be senescing.

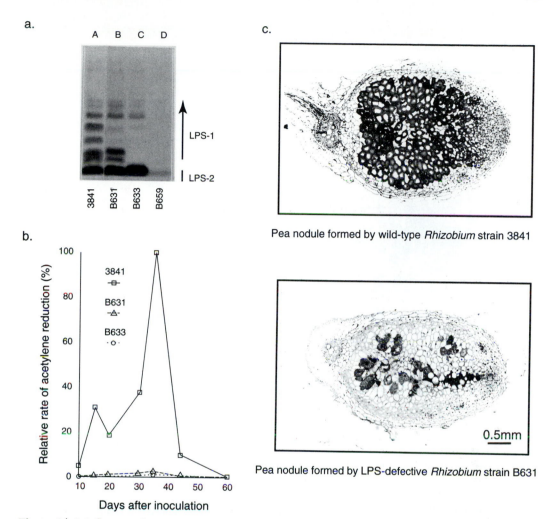

Figure 14–3 Influence of mutations in lipopolysaccharide (LPS) production on nodulation and nitrogenase activity (an index of N_2 fixation potential) in pea seedlings. (a) Periodate silver stained polyacrylamide gels showing differences in the lipopolysaccharide of strain 3841 (wild type) and of three mutants derived from it. (b) Seasonal profiles of nitrogenase activity achieved by the wild type strain and by two LPS mutants. (c) Differences in the morphology of pea nodules produced by strain 3841 and B631. Nodules produced by such mutant strains contain few infected cells. *From Perotto et al. (1994). Used with permission.*

In indeterminate nodules like those in alfalfa, bacteroids produce ammonia (see Fig. 13–2), which is exported to the host cell and there converted via glutamine, glutamate, and aspartate to asparagine. Asparagine is then exported to the shoot. The overall reaction for this series of changes is:

2 ammonia + 3ATP + oxaloacetate →

$$\text{L-asparagine} + \text{AMP} + 2\text{ADP} + 2\text{P}_i + \text{NAD} + \text{PP}_i$$

Determinate nodules export a very different end product. Glutamate and aspartate are produced but are then used to synthesize purines such as xanthine. These

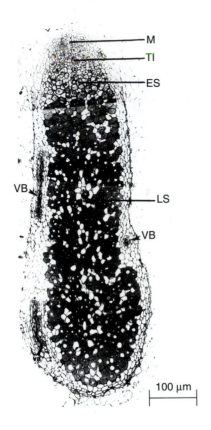

Figure 14-4 The internal organization of an inde-terminate root nodule: M, meristematic tissue; TI, region with infection thread penetration; ES, region of early symbiotic activity; LS, region of late symbiotic activity; VB, vascular bundle. The nodule is attached to the root at the bottom of the photo. *Photo courtesy C.P. Vance.*

are converted in neighboring noninfected cells to the ureides, allantoin and allantoic acid. Because the level of these substances in nodules and xylem sap is usually correlated with recent N_2 fixation, ureide analysis of bleeding sap is often used to estimate N_2 fixation.

Relatively few infections result in root-nodule formation. Successful infections may be visible as early as five to six days after inoculation with rhizobia, with active N_2 fixation beginning eight to fifteen days thereafter. During this period a number of proteins are produced in the root hair or nodule that are not found in host or bacteria alone. Expression of these substances, called **nodulins,** may be both time-and-tissue dependent. "Early" nodulins have been recovered from infected root hairs less than six hours after inoculation. "Late" nodulins, which are more often related to nodule function and N_2 fixation, include leghemoglobin and the nodule enzymes nitrogenase, uricase, and glutamine synthetase.

Host-Rhizobium Specificity

Specificity can occur at each stage of the nodulation process. Thus, only 25 to 30% of legumes in the subfamily Caesalpinioideae ever form nodules, while infectiveness subgroups have been identified within several cross-inoculation groups. One such subgroup includes *Trifolium tembense,* a pasture species com-

Box 14–2

Events Leading to Nodulation and N_2 Fixation in Legumes.

- Attachment of rhizobia to the root begins within 1 minute of inoculation.

- Number of attached rhizobia increases with time up to several hours.

- Root hair curling begins within 5 hours.

- Infection threads become visible within 3 days of inoculation.

- Nodules become visible within 5 to 12 days.

- N_2 fixation is often evident in 15-day-old plants.

Table 14–3 **Levels of host-*Rhizobium* specificity affecting nodule formation in legumes.**

Legume species that are never nodulated (e.g., *Cassia bicapsularis*)
Cross inoculation group specificity
Infectiveness subgroups
Nonnodulating plant genotypes, e.g.,
   the rj_1 gene of soybean
   the nod 125 mutant of *Phaseolus vulgaris*
Nodulation preference

---

mon in Kenya and Tanzania. This species nodulates with rhizobia from *T. rueppellianum* and *T. usambarense* but either fails to nodulate or is ineffective with rhizobia from other African and European clovers. Similarly, the pea variety Afghanistan will nodulate with pea rhizobia from the center of origin of this crop in the Middle East, but not with most of the rhizobia from pea varieties in Europe. These and additional examples of specificity listed in Table 14–3 have often been a problem in the introduction of new plant germplasm.

Host-rhizobial interactions also influence levels of N_2 fixation. Thus, peanuts and cowpeas are nodulated by and fix dinitrogen with many different soil bradyrhizobia, whereas *Centrosema* and *Desmodium* species often nodulate with these strains but fix little dinitrogen. A consequence is that when cowpeas or peanuts are introduced into a new area, they are often well nodulated and grow vigorously, even in the absence of inoculation, whereas *Centrosema* and *Desmodium* species may have many nodules but grow poorly.

Box 14–3

Infectiveness and Effectiveness.

- *Infectiveness* is the ability of a rhizobial strain to form nodules with a particular legume.

- *Effectiveness* is the ability of those nodules to fix dinitrogen.

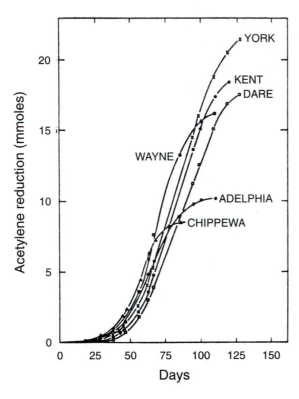

Figure 14–5 Influence of soybean maturity group on potential N_2 fixation, as estimated with the acetylene reduction method. The varieties shown are of increasing vegetative period, with Chippewa early to flower and York and Kent having much longer vegetative periods. Once podfill begins, available energy for N_2 fixation can be insufficient for continued nodule activity. *From Hardy et al. (1973). Used with permission.*

Even different varieties of the same species can vary in their ability to fix dinitrogen with the same rhizobial strain. This often relates to differences in maturity, with early-flowering varieties more likely to be limited in their N_2-fixing ability (Fig. 14–5). Genetic differences in earliness of nodulation, nodule mass, nodule senescence, and enzyme function have also been reported, and attempts to improve current levels of N_2 fixation through plant breeding are under way in a number of laboratories. Perhaps the simplest approach to improve N_2 fixation of new varieties is to inoculate plants with rhizobia, and to grow and select the varieties under conditions of nitrogen deficiency. The growth and yield of the plant genotypes being tested will then depend heavily on their ability to fix dinitrogen.

Influence of Environmental Factors

Environmental factors influence all aspects of nodulation and symbiotic N_2 fixation, in some cases reducing rhizobial survival in soil and in others affecting nodulation or even growth of the host. Critical factors include acidity, temperature, mineral nutrition, and salinity and alkalinity.

Acidity

In Latin America alone there are more than 800 million ha of Oxisols and Ultisols, the pH of which is usually less than 5.0. In this region, as in other regions of the world

Box 14–4

The Symbiotic Environment. Environmental factors influence all aspects of the symbiosis between legumes and rhizobia, including:

- survival of the rhizobia on the seed and in soil,

- infection and nodule formation, and

- N_2 fixation.

---

where soils are acid (Fig. 14–6), N_2 fixation may be markedly reduced. This may be because of the direct effect of H^+ concentration, the presence of toxic levels of aluminum and manganese, or deficiencies of calcium, phosphorus, and molybdenum.

Soil acidity per se can limit rhizobial growth and persistence in soil. Fast-growing rhizobia are generally considered more sensitive to acidity than bradyrhizobia, but strains of *R. tropici* and *R. loti* may also be acid tolerant. In contrast, most isolates of *S. meliloti* are particularly acid sensitive. In one study an average of 89,000 *S. meliloti* cells g^{-1} were reported in soils of near neutral pH, but only 37 cells g^{-1} in soils of pH less than 6.0. Surprisingly, not all of the strains recovered from acid soils are acid tolerant, suggesting that microsites of more favorable pH can occur.

Failure to nodulate is also common in acid soils, in part because of lowered numbers of rhizobia, but also because pH limits rhizobial attachment to infectible root hairs. Although it is common in the United States to lime acid soils, the large areas involved and the cost and availability of limestone preclude this approach in many other countries. Alternative practices include the use of acid-tolerant inoculant strains and host cultivars and the pelleting (coating) of inoculated seed with a layer of ground rock phosphate or limestone. In Australia, the use of relatively acid-tolerant *S. meliloti* strains such as WSM419, together with *Medicago* species collected from acid soils in Sardinia, has permitted extension of the area sown to annual medics by some 350,000 ha since 1985 (Ewing and Howieson, 1989). Similarly, an acid-tolerant *R. tropici* strain, CIAT899, is now the strain recommended for inoculating *Phaseolus vulgaris* in the acid soils of Brazil.

Plant species vary in tolerance to aluminum and manganese, but are generally more affected by these ions than are the rhizobia. Thus, some rhizobia tolerate 100 μM aluminum and 300 μM manganese, but reduced root growth of alfalfa (*Medicago sativa*) occurs at only 8 μM aluminum, and nodulation in cowpea is inhibited at 25 μM aluminum.

Temperature

Rhizobia are mesophiles and most do not grow below 10°C or above 37°C. Exceptions are rhizobia associated with certain Arctic legumes, and bradyrhizobia collected from the hot, dry Sahel savannah of Africa. Maintaining favorable temperatures during the shipment and storage of inoculant, and seed inoculation and planting, is particularly critical. Exposure to high temperatures at these times can lead to the loss of the symbiotic plasmid in *Rhizobium* or reduce cell numbers below the levels needed for good nodulation.

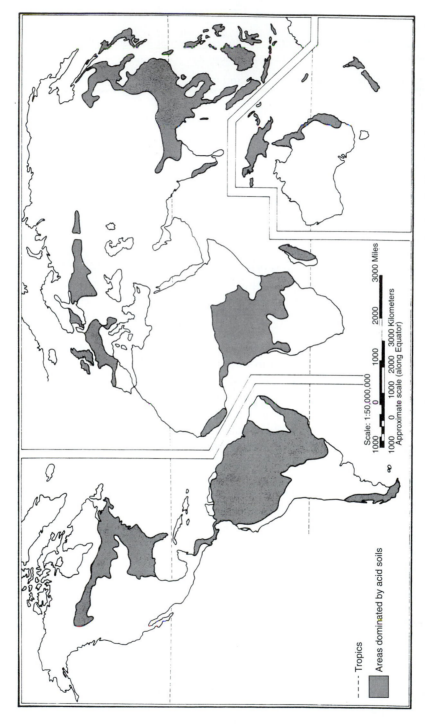

Figure 14–6 Regions of the world where the dominant soils are sufficiently acid to constrain the growth and nodulation of important leguminous crops. *From van Wambeke (1976).*

Temperature also influences nodule growth, N_2 fixation, and the time period for which nodules are active. The optimum temperature for many legumes is around 25°C; exposure to temperatures of greater than 40°C, even for short periods, can cause irreparable loss of nodule function.

Mineral Nutrition

Although well-nourished plants generally nodulate and fix dinitrogen better than those that are nutrient limited, several elements have specific functions in nodulation and symbiotic N_2 fixation (Table 14–4). Adequate levels of these elements are essential for effective N_2 fixation (see also Table 13–2), and failure to supply them results in the generalized yellow chlorosis typical of nitrogen deficiency. Several of these elements warrant specific mention.

Phosphorus. Leguminous plants dependent on N_2 fixation commonly require more phosphorus than similar plants supplied fertilizer nitrogen. Nodules are an important phosphorus sink and commonly have the highest concentration of that element in the plant. The high energy cost of N_2 fixation with its need for large amounts of ATP leads to the elevated phosphorus requirement. In the case of the legume nodule, this is compounded by the energy cost of building and maintaining functioning nodules. Bacterial strains and host cultivars differ in their phosphorus-use efficiencies, but cultivar variation has been studied mainly in host plants fertilized with nitrogen. Field-grown legumes form tripartite symbioses with both *Rhizobium* or *Bradyrhizobium* and arbuscular mycorrhizal fungi. This has an added energy cost to the host, but the benefit from additional phosphorus uptake for N_2 fixation can be considerable.

Molybdenum. The principal function of molybdenum in the legume is as a component of the nitrogenase enzyme complex. As such, the requirement for molybdenum is satisfied by supplying as little as 100 to 500 g ha^{-1} of this element. How to supply this small amount can be a problem, especially under acid soil conditions where adsorption reduces the availability of soil molybdenum to the plant. Molybdenum salts have sometimes been incorporated with the inoculant. While the effect varies with the molybdenum salt added, this can drastically reduce survival of rhizobia in the inoculant and its addition is not recommended.

Iron. Iron is a component of leghemoglobin, which functions in the regulation of oxygen supply to bacteroids. It is also a component of both the Fe and FeMo proteins

Table 14–4 **Elements having specific functions in the nodulation or N_2 fixation of legumes.**

Molybdenum	FeMoCo protein of nitrogenase
Phosphorus	Energy transformations in the nodule
Iron	Fe and FeMo proteins of nitrogenase
	Leghemoglobin
	FeS centers of nitrogenase
	Unspecified function in nodule development
Calcium	Attachment of rhizobia to root hairs
	Cell wall integrity in *Rhizobium*
Sulfur	FeS centers of nitrogenase
Cobalt	Nodule coenzyme function
Nickel	Hydrogenase function

of the nitrogenase complex and is essential for early nodule development. Plants which are iron-deficient develop many nodule initials but few functioning nodules. Both host and bacterial strain can differ in efficiency of iron utilization. In the case of the bacteria, some strains can produce iron-sequestering siderophores and so compete more effectively for iron in the rhizosphere.

Salinity and Alkalinity

The effects of saline or alkaline conditions are likely to be greater on the host or symbiosis than on the rhizobia. Alkaline soil conditions limit the availability of iron, zinc, manganese, and boron in the soil, thereby reducing plant growth and N_2 fixation. Foliar fertilization with micronutrients is often an effective remedy. Legumes as a group are also markedly sensitive to salt, with some species affected by concentrations as low as 80 mM. In contrast, strains of rhizobia from *Medicago* and *Acacia* often tolerate 500 mM salt. Cells of *Rhizobium* exposed to high salt concentrations often accumulate osmoregulants such as glutamic acid, trehalose, glycine betaine, and proline, which help to maintain turgor in the cell and limit the damage caused by salts.

Legume Inoculation

When a new legume species is introduced into a region, soils are unlikely to contain appropriate rhizobia, and inoculation is usually needed for adequate nodulation and N_2 fixation. Yield increases following this initial inoculation are often of the order of 50%, with clear differences evident between inoculated and noninoculated plants, as seen in Figure 14–7. Inoculation in subsequent years is usually not necessary. In fact, where a legume has an extensive history of cultivation in a region, most soils contain abundant rhizobia, and even noninoculated plants are heavily nodu-

Figure 14–7 Response of soybeans to inoculation in Florida. The right-hand plot in the foreground received inoculation, while the adjacent plot did not. *Photo courtesy of W. Scudder and D.H. Hubbell. Used with permission.*

Box 14–5

Inoculation in the American Midwest. In the American Midwest, agricultural soils usually contain 1,000 to 10,000 soybean rhizobia g^{-1}. These indigenous rhizobia limit nodule formation by inoculant strains, which often produce less than 20% of the nodules formed. Because the indigenous or naturalized populations often fix less dinitrogen than the inoculant strain, the full benefits of the symbiosis may not be realized. Soybeans grown in this region derive only 30 to 40% of their nitrogen needs from N_2 fixation.

---

lated. If inoculation is practiced in such an area, the inoculant strain usually produces only a small fraction of the nodules formed, and a yield response is unlikely.

Need for Inoculation

A simple, three-treatment experiment will establish the need for inoculation. The treatments are:

- noninoculated control plots,

- plots inoculated with a strain of *Rhizobium* or *Bradyrhizobium* effective on the host legume, and

- plots inoculated with the same strain, but also supplied fertilizer nitrogen.

Extensive nodulation of the noninoculated plants means that the soil already contains indigenous rhizobia able to nodulate this host. The contrast between the noninoculated plants and those supplied with nitrogen will then be a measure of the effectiveness of the native rhizobia. If the noninoculated plants are green and vigorous, inoculation is probably not necessary. Absence of nodulation in the noninoculated plants, coupled with heavy nodulation of plants receiving inoculation, indicates that inoculation is needed. The differences in plant growth among the three treatments are indicators of the efficiency of N_2 fixation by the inoculant strain. If excellent plant growth is achieved in all three treatments then either the native rhizobia are highly effective and inoculation is not necessary or the site was higher in available nitrogen than was anticipated. Poor growth in all treatments would imply that a factor other than nitrogen was limiting plant growth.

Strain Selection and Testing

If inoculation is required, the strain or strains employed must meet the following criteria:

- form highly effective nodules with all commonly used varieties of the legume species for which it is recommended,

- be competitive in nodule formation and persistent in the soil,

- tolerate extremes of acidity, temperature, and other soil conditions,

- grow well in simple, inexpensive culture media,

- be genetically stable, and

- survive well on the seed prior to seed germination.

Inoculant-quality rhizobia should be selected after screening at several levels. The initial step is usually a growth chamber or greenhouse study of numerous strains obtained from other collections or from the field. Marked variation in nodulation and N_2 fixation usually is evident. Poorer strains are discarded, and the remaining strains may be tested with different varieties of the legume to eliminate any possibility of host-strain interaction or may be further tested under field conditions. Ideally such field trials are conducted at sites that vary in numbers of indigenous rhizobia; they should be followed up in subsequent growing seasons to ensure that the inoculant strains persist in the soil. Finally, environmental and cultural factors that could influence strain performance in the field need to be considered.

Inoculants and Inoculation

The number of rhizobia per seed necessary to ensure good nodulation varies with seed size and environmental conditions. In countries where inoculant quality is regulated by law, the usual standard is from 1,000 rhizobia per seed for small-seeded legumes such as clover to 100,000 rhizobia per seed for bean and soybean. In the early 1900s, soil from previously planted fields was the principal, but far from ideal, inoculant. The supply was limited, moving soil was cumbersome, and the possibility of transferring root pathogens or nematodes was a major concern. The inoculant industry now manufactures pure cultures of rhizobia for seed and soil inoculation. Inoculants range from simple tube cultures sufficient for small quantities of exotic seed to large-scale, fermenter-grown cultures mixed with peat or other carrier material and used in the commercial inoculation of large areas of crops such as soybean, bean, peanut, and clover.

The large-scale production of inoculants is a simple process designed to supply a minimum of 10^8 to 10^9 highly effective rhizobia per gram of product. Inoculants from many countries do not meet this standard. Factors contributing to this failure include:

- inadequate testing of the inoculant strain,

- mutation in the inoculant strain after repeated subculture or storage at high temperature,

- inappropriate culture media,

- contamination of the rhizobial culture,

- carrier materials that will not support suitable populations of rhizobia, and

- poor storage and transport conditions.

Characteristics of a good inoculant carrier are shown in Table 14–5. The most common carrier is peat, but compost, sterile bagasse (derived from the milling of sugarcane), coal, polyacrylamide, vegetable oils, and soil have all been used successfully. No listing of physical or chemical properties can fully explain why some peats make suitable inoculant carriers and others do not.

Four procedures are commonly used in legume inoculation. For additional detail, refer to Somasegaran and Hoben (1994).

Table 14–5 **Qualities of a good inoculant carrier material.**

High water-holding capacity
Nontoxic to rhizobia
Readily available, inexpensive, and easily processed
Sterilizable by autoclaving (pressurized steam) or radiation
Good adherence to seed
Good buffering capacity

- *Seed inoculation:* The inoculant is mixed with milk or some other slightly adhesive material (called the *sticker*), and the seed is uniformly covered with this suspension. The seed is dried in the shade and sown the same day.

- *Seed pelleting:* The sticker is a stronger adhesive, such as gum arabic or methyl cellulose, and the seed, once inoculated as above, is rolled in finely ground limestone or rock phosphate. Pelleting combats unfavorable soil conditions such as low pH and allows aerial sowing. Preinoculation of seed for subsequent sale is not recommended because rhizobial numbers on the seed can decline dramatically during storage.

- *Soil inoculation with a granular peat or liquid:* The inoculant is banded below the seed and makes contact with the emerging radicle. Soil inoculation is most useful for seed that has been treated with fungicide or for conditions where higher than normal inoculation rates are desirable.

- *Inoculation in the planter seed box:* The inoculant is mixed directly with seeds in a planter box attached to a tractor. Inoculant and seed tend to separate, providing uneven coverage. This simple method is usually an "insurance measure" when soils are already likely to contain rhizobia.

Strain Competitiveness and Persistence

Even without inoculation, it is common for a newly introduced legume to have a few nodules. These arise from seed- or dust-borne rhizobia or from native legumes having compatible rhizobia. When these nodules rot, they can release more than 10^{10} rhizobia g^{-1}, ensuring a buildup in the soil to levels of 10^3 to 10^4 rhizobia g^{-1} of soil. Unfortunately, many of these organisms are not particularly effective, and do little to benefit subsequent plantings. Worse, they can limit the ability of inoculant rhizobia to form nodules and become established in the soil.

Attempts to overcome this problem using strains selected for superior competitive ability, heavier than normal inoculation rates, and improved carrier and delivery systems have all had limited success. Accentuating this problem, rhizobia are not very mobile in soil, and as the root elongates, they may be left behind. One solution is host cultivars that nodulate preferentially with the inoculant strain or exclude indigenous rhizobia. Several soybean cultivars that can restrict nodulation by the indigenous strains but nodulate normally with specific inoculant strains have been identified (Table 14–6). Differences in the response of the wild *Phaseolus vulgaris* accession G21117 to inoculation with strain CIAT632 and CIAT 899 are shown in Figure 14–8.

Table 14–6 **Differences in the nodule occupancy of three genotypes of soybean differing in ability to restrict nodulation by strain USDA123. Data are from a 2-year field evaluation.**

Genotype	% of nodules occupied by strain		
	USDA123	USDA122 or USDA138	Other
Williams (nonrestrictive)	76	20	4
PI371607 (restrictive)	3	89	8
PI377578 (restrictive)	5	92	3

From Keyser and Li (1992). Used with permission.

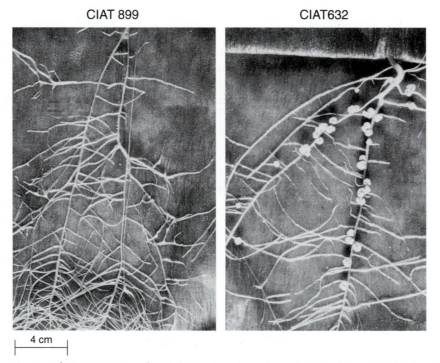

Figure 14–8 Restriction of nodulation by *Phaseolus vulgaris* cultivar G21117. Plants on the left were inoculated with *R. etli* strain CIAT899 and those on the right with *R. tropici* strain CIAT632. Both strains produce normal nodules and are effective on most cultivars of *P. vulgaris*. However, restricted nodulation occurs when CIAT899, but not CIAT632, is used to inoculate G21117. *From Kipe-Nolt et al. (1992). Used with permission.*

Other Important Symbiotic Dinitrogen-Fixing Associations

Frankia and the Actinorhizal Symbiosis

Frankia is an actinomycete that forms actinorhizal, N_2-fixing nodules with a range of angiosperms (Table 14–7). The host species are not typical crop plants, but several are important in agroforestry, the ecology and nitrogen economy of marginal soils, mine spoil reclamation, or the stabilization of sand dunes. Rates of N_2 fixation are highly variable but can be equivalent to those achieved by leguminous symbioses.

Table 14-7 Families and genera of actinorhizal plants

Family	Genera with nodules
Betulaceae	*Alnus* (alder)*
Casuarinaceae	*Allocasuarina, Casuarina* (Australian pine), *Ceuthostoma*, and *Gymnostoma*
Coriariaceae	*Coriaria*
Datiscaceae	*Datisca*
Elaeagnaceae	*Elaeagnus, Hippophae* (sea buckthorn), and *Shepherdia*
Myricaceae	*Comptonia* and *Myrica* (myrtle)
Rhamnaceae	*Ceanothus* (snowbrush), *Colletia, Discaria, Kentrothamnus, Retanilla, Talguenea*, and *Trevoa*
Rosaceae	*Cercocarpus* (mountain mahogany), *Chamaebatia, Cowania, Dryas*, and *Purshia*

*Common names for some important examples are included in parentheses.

Isolation of *Frankia* from nodules was not achieved until 1978, and still requires very specific methodologies. However, numerous isolates are now available, and information on morphological, genetic, and specificity differences is beginning to accumulate. Particular points of interest include:

- *Frankia* is a Gram-positive, filamentous organism characterized by multilocular sporangia and N_2-fixing vesicles *in vitro*.

- Nodule formation results from root hair infection or intercellular invasion. Nodules are perennial, modified lateral roots with lobes up to 5 cm in length.

- Host specificity exists, but needs further definition. Three to four host-specificity groups have been suggested, though species of *Myrica* and *Gymnostoma* appear to be promiscuous and nodulate with strains from all groups.

- Few *Frankia* isolates produce spores within the nodule. Although this trait is regulated by the organism, nodules in which spore formation occurs seem to have a higher energy cost for N_2 fixation and so contribute less to the host.

- Vesicle production occurs under conditions of nitrogen limitation, with the type of vesicle produced dependent on the host plant. Vesicles are borne as terminal swellings or on short hyphal branches; at maturity, they show a pronounced lipid envelope that protects the nitrogenase from oxygen.

Additional information on the biology and symbiotic specificity of *Frankia* is provided by Benson and Silvester (1993).

Azolla/Anabaena Symbiosis

The aquatic fern *Azolla* is a common green manure used in Vietnam and China for rice production. *Azolla* maintained in slow-flowing creeks or overwintered in protected beds is introduced into paddies between plantings of rice and is then either incorporated before rice seedlings are transplanted or left to be shaded out as the rice canopy develops. The low C/N ratio of the fern facilitates rapid nitrogen mineralization after incorporation, with yields in the subsequent rice crop increased by as much as 1,000 kg ha^{-1}. In this case N_2 fixation is because of the heterocystous cyanobacterium

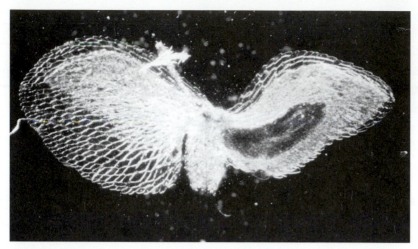

Figure 14–9 Location of the symbiotic cyanobacterium *Anabaena azollae* within the leaf (frond) of the water fern *Azolla*. The frond has been cleared, with the *Anabaena* filaments visible as the darkened region to the center-right of the frond. *Photo courtesy T.A. Lumpkin. Used with permission.*

Anabaena azollae growing within cavities in the dorsal leaf lobe (Fig. 14–9). Under favorable conditions, rates of N_2 fixation can reach 2 kg ha^{-1} day^{-1}.

Four aspects of this symbiosis warrant particular comment:

- *Azolla* can be cured of its microsymbiont, but no one has succeeded in reintro-ducing *Anabaena*.

- Dinitrogen fixation in *Anabaena* occurs predominantly in specialized cells called **heterocysts.** Under free-living conditions, only 6 to 10% of the cells in the filament are heterocysts, but in the mature *Azolla* frond, this frequency rises to 20 to 30%.

- Gram-positive bacteria that do not fix dinitrogen, identified as species of *Arthrobacter,* have also been found in the leaf cavity and are thought to play a role in this symbiosis. However, no definite function is known for these organ-isms.

- During heterocyst formation by free-living *Anabaena,* a small piece of DNA is deleted, leaving the *Nif* HDK genes all regulated by a single promoter. In con-trast, all cells of the *Azolla* microsymbiont have this arrangement.

Although the yield benefits from the use of *Azolla* can be appreciable, they are offset to some extent by labor costs for storage, propagation, and field distribution of the fern, by the need in some areas to tie up scarce land while the inoculum is multiplied, and by insect and disease problems. Nonetheless, the *Azolla/Anabaena* symbiosis continues to be important in Asia, with some estimates of the area sown to *Azolla* in China being as high as 1.5 million ha annually. Additional information on this symbiosis is provided by Lumpkin and Plucknett (1980) and Peters and Meeks (1989).

Summary

This chapter highlights the potential of N_2 fixation in legumes, and problems associated with utilizing this symbiosis. It also introduces other N_2-fixing symbioses, some of which have potential for use in agriculture. As with the legumes, their greater exploitation will require a multidisciplinary and ecological approach.

Symbiotic N_2 fixation currently accounts for more than 65% of the nitrogen used in agriculture. As the world's population increases, this contribution must increase. Problems in the availability of fertilizer nitrogen and groundwater pollution resulting from excessive fertilization will limit the degree to which fertilizer nitrogen usage can be increased. Although inoculant production in many regions of the world leaves much to be desired, the inoculant technology reviewed in this chapter is within the reach of most countries and needs only to be properly and consistently applied. However, other bottlenecks need to be remedied if symbiotic N_2 fixation is to assume a more important role in the agriculture of the twenty-first century. Areas of research that need to be resolved include:

- improving the ability of different varieties to fix dinitrogen,

- overcoming the problem of low nodule occupancy by inoculant strains,

- improving the persistence of inoculant strains in soil,

- enhancing the tolerance of both host and microsymbiont to environmental stresses, and

- understanding better the contribution of N_2 fixation to the nitrogen economy of both modern and traditional farming systems.

Cited References

Benson, D.R., and W.B. Silvester. 1993. Biology of *Frankia* strains, actinomycete symbionts of actinorhizal plants. Microbiol. Rev. 57:293–319.

Denarie, J., and J. Cullimore. 1993. Lipo-oligosaccharide nodulation factors: A minireview. New class of signalling molecules mediating recognition and morphogenesis. Cell 74:951–954.

Ewing, M.A., and J.G. Howieson. 1989. The development of *Medicago polymorpha* L. as an important pasture species for southern Australia. pp. 197–198. *In* Proc.16th. International Grasslands Congress, Nice, France.

Fåhreus, G. 1957. The infection of clover root hairs by nodule bacteria studied by a simple glass slide technique. J. Gen. Microbiol. 16:374–381.

Hardy, R.W.F., R.C. Burns, and R.D. Holsten. 1973. Applications of the acetylene-ethylene assay for measurement of nitrogen fixation. Soil Biol. Biochem. 5:47–81.

Hirsch, A.M. 1992. Developmental biology of legume nodulation. New Phytol. 122:211–237.

Keyser, H.H., and F. Li. 1992. Potential for increasing biological nitrogen fixation in soybean. Plant Soil 141:119–135.

Kipe-Nolt, J.A., C. Montealegre, and J. Tohme. 1992. Restriction of nodulation by a broad host range *Rhizobium tropici* strain CIAT899 in wild accessions of *Phaseolus vulgaris* L. New Phytol. 120:489–494.

Lumpkin, T.A., and D.L. Plunkett. 1980. *Azolla:* Botany, physiology and use as a green manure. Econ. Bot. 34:111–153.

Perotto, S., N.J. Brewin, and E.L. Kannenberg. 1994. Cytological evidence for a host defense response that reduces cell and tissue invasion in pea nodules by lipopolysaccharide-defective mutants of *Rhizobium leguminosarum* strain 3841. Mol. Plant Microbe Interact. 7: 99–112.

Peters, G.A, and J.C. Meeks. 1989. The *Azolla-Anabaena* symbiosis: Basic biology. Rev. Plant Physiol. Plant Mol. Biol. 40:193–210.

Sahlman, K., and G. Fåhreus. 1963. An electron microscope study of root hair infection by *Rhizobium*. J. Gen. Microbiol. 33:425–427.

Stanley, J., and E. Cervantes. 1991. Biology and genetics of the broad host range *Rhizobium* sp. NGR234. J. Appl. Bacteriol. 70:9–19.

van Wambeke, A. 1976. Formation, distribution and consequences of acid soils in agricultural development. pp. 15–24. *In* M.J. Wright (ed.), Plant adaptation to mineral stress in problem soils. Technical Assistance Bureau, Agency for International Development, Washington, DC.

General References

Ladha, J.K., T. George, and B.B. Bohlool (eds.). 1992. Biological nitrogen fixation for sustainable agriculture. Kluwer Academic Press, Dordrecht, The Netherlands.

Roughley, R.J. 1970. The preparation and use of legume seed inoculants. Plant Soil 32:675–701.

Somasegaran, P., and H. Hoben. 1994. Handbook for rhizobia. Springer-Verlag, New York.

Weaver, R.W., and P.H. Graham. 1994. Legume nodule symbionts. pp. 199–222. *In* R.W. Weaver, S. Angle, P. Bottomley, D. Bezdicek, S. Smith, A. Tabatabai, and A. Wollum (eds.). Methods of soil analysis, Part 2. Microbiological and biochemical properties. Soil Science Society of America, Book series, No. 5, Madison, Wis.

Study Questions

1. Competition for energy between developing pods and nodules illustrates the influence of energy supply on nodule function. What experimental treatments can you suggest to study the importance of energy supply in nodulation and N_2 fixation?

2. Dinitrogen-fixing symbioses adopt different strategies to protect nitrogenase from the inhibitory effects of oxygen. Give three examples discussed in this chapter.

3. You join the Peace Corps in Nepal and are assigned to the development of an inoculant industry for that country. Describe the steps you would need to take to achieve such a development.

4. In Latin America and Africa, N_2-fixing crops such as bean and cowpea are often grown in association with corn. How might this affect N_2 fixation?

5. Develop a method to screen strains of *Rhizobium* for differences in pH tolerance.

6. Legumes abound in most situations. Find one such plant and describe its nodulation. How many nodules does it have, how are they distributed, and what type of nodule are they? Describe the internal appearance of the nodule. How would you determine whether the nodules were formed by a *Rhizobium* or *Bradyrhizobium* strain?

Chapter 15

◆

Transformations of Sulfur

James J. Germida

◆

. . . sulfur is a devilish substance. . . . discharged from the bowels of the earth, by volcanoes or evil-smelling hot springs . . . surely the effluent of Hell itself. . . .

J.R.Postgate

Sulfur (S) is an essential element for the growth and activity of all living organisms. It is one of the ten major bio-elements required by organisms in relatively high concentrations (i.e., $> 10^{-4}M$). Sulfur is required for the synthesis of the amino acids cysteine, cystine, and methionine. It plays an active role in plants, animals, and microorganisms as an important constituent of vitamins, hormones, and structural components, and for other metabolically important molecules such as coenzyme A. For example, the **disulfide bond** formed between cysteine residues helps stabilize the tertiary structure of proteins (Box 15–1).

Sulfur is an important source of metabolic energy for many bacteria. For example, certain chemoautotrophic bacteria obtain energy for cell growth and division by oxidizing reduced sulfur compounds. In fact, some exotic ecosystems, such as hot sulfur springs and hydrothermal vent communities, are driven by energy generated in the oxidation of sulfur, which they use to fix carbon (Box 15–2).

This chapter introduces the basic principles of the sulfur cycle and the microorganisms that drive it. This information will give one an appreciation of how important it is to understand the biogeochemistry of sulfur and other elements (Box 15–3).

| **Box 15–1** |

Disulfide Bonds Stabilize Protein Structure. A disulfide bond is a covalent linkage between two sulfur atoms. This type of linkage is commonly found in polypeptides and protein molecules and may occur as an interchain or intrachain bond between sulfur-containing amino acids. This type of bond is not common to all proteins, but is usually critical to those that do possess it.

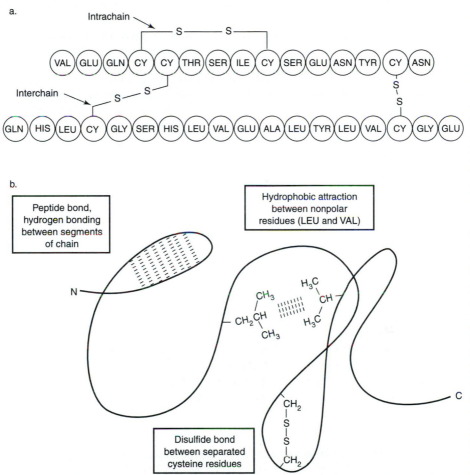

(a) Amino acid sequence of two polypeptide chains illustrating how intrachain and interchain disulfide covalent bonds form between cysteine residues. (b) Diagrammatic representation of how disulfide bonds and other stabilizing forces confer a unique spatial conformation on proteins. Not all types of forces or their frequency of occurrence are shown.

---| **Box 15–2** |---

Life Without Light: The Importance of Chemosynthesis. All life requires energy. Furthermore, it is clear that the sun's energy, trapped through photosynthesis, is the driving force for **primary productivity** in all ecosystems. Almost all, that is! About 20 years ago scientists made one of the most startling discoveries of twentieth-century biology. They found bizarre, exotic biological communities living in the deep ocean where hydrothermal fluids rise through the earth's crust. The communities at these deep sea thermal vents consist of giant tubeworms and masses of large clams. Life here is sustained by primary productivity based on microbial sulfur oxidation (Jannasch and Mottl, 1985).

Hydrogen sulfide is the most abundant compound in the deep sea thermal vent fluid. This compound is extremely toxic to higher animals, but some sulfur-oxidizing bacteria can use it as an energy substrate. Considerable energy is released when hydrogen sulfide is oxidized. Through a process called **chemosynthesis,** analogous to photosynthesis, vent bacteria employ chemical rather than light energy to fix inorganic carbon to make organic compounds. More surprisingly, the giant tubeworms and clams living at the vent are in a symbiotic relationship with these sulfur-oxidizing bacteria that live in their tissue. This is truly a remarkable example of how life adapts to extreme environments and is found where least expected. In fact, it now appears that microbial life exists in all environments with an oxidizable energy source and favorable conditions for microbial life.

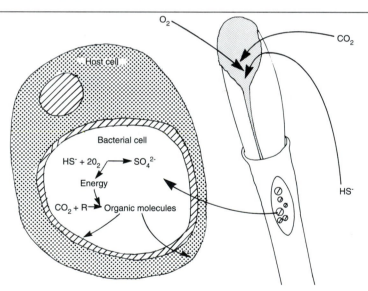

Bacteria found in symbiosis with tube worms at hydrothermal vents use chemosynthesis to fix carbon to make organic compounds. Shown is the breathing organ attached to the body of the worm, *Riftia pachyptila,* which contains the symbiotic association. The bacteria use the hydrosulfide ion (HS^-) and oxygen (O_2) to generate chemical energy (chemosynthesis). The hydrosulfide ion serves as the electron donor; the energy is released when HS^- and O_2 are combined, driving a series of reactions called the Calvin cycle. In the Calvin cycle, organic compounds are formed as the carbon dioxide (CO_2) present in both vent water and sea water is fixed. In this illustration, R stands for the auxiliary compounds, including CO_2-fixing enzymes, involved in the cycle. Note that sulfuric acid is produced as a by-product. The process is analogous to photosynthesis, in which green plants use light energy to fix carbon dioxide and form organic compounds. *Redrawn from Tunnicliffe (1992). Used with permission.*

| Box 15–3 |

Sulfur Can Be Good or Bad for the Environment. Although sulfur is considered a macronutrient, its availability to the biota can be limiting in some ecosystems; then it must be applied as a supplement or fertilizer. However, in other instances, too much sulfur is a pollutant. For example, burning high-sulfur-content (i.e., bituminous and lignite) coals produces gaseous sulfur emissions that can become components of acid precipitation. Microbial oxidation of reduced sulfur minerals associated with coal seams and the subsequent leaching of the resulting oxyanions (e.g., SO_4^{2-}) gives rise to acid mine drainage that can pollute streams or acidify reclaimed soils. Finally, sulfur is important for many industrial processes; it can be used, for example, in the production of chemicals, concrete, and asphalt. As one reads this chapter, it will become evident that sulfur is an essential, versatile, and economically and ecologically important element.

The Sulfur Cycle in Agroecosystems and Terrestrial Environments

The sulfur cycle bears many similarities to the nitrogen cycle. Both of these elements exist in a number of oxidation states and undergo similar types of chemical reactions and biological transformations, including volatilization. The majority of sulfur is found in the **lithosphere** (Table 15–1). Most nitrogen on earth is also in the lithosphere; however, dinitrogen in the atmosphere is the major pool of biologically available nitrogen. Only a small portion of the sulfur pool is found in the atmosphere, and most sulfur that cycles through the atmosphere is because of human activities. In fact, since the Industrial Revolution, increased burning of fossil fuels has almost doubled the rate of sulfur entering the atmosphere to approximately 1.5×10^{11} kg S yr^{-1}. The volatilization

Table 15–1 **Estimated quantities of sulfur in major sulfur pools.**

Pool	Mass (kg S)
Atmosphere	3.6×10^9
Hydrosphere	1.3×10^{18}
Oceans	1.3×10^{18}
Marine organisms	2.4×10^{10}
Fresh waters	3.0×10^{12}
Ice	6.0×10^{12}
Lithosphere	24.1×10^{18}
Igneous rocks	5.0×10^{18}
Metamorphic rocks	11.4×10^{18}
Sedimentary rocks	7.7×10^{18}
Evaporites	5.1×10^{18}
Shales	2.0×10^{18}
Limestones	0.1×10^{18}
Sandstones	0.3×10^{18}
Soils	2.6×10^{16}
Soil organic matter	1.0×10^{13}
Biosphere	7.6×10^{12}

From Pierzynski et al. (1994). Used with permission.

Table 15–2 **Amounts and distribution of sulfur in some world soils.**

Location	Type of soil	Total sulfur ($\mu g\ g^{-1}$)
Saskatchewan, Canada	Agricultural	88–760
British Columbia, Canada	Grassland	286–928
	Forest	162–2,328
	Organic	1,122–30,430
	Agricultural	214–438
Iowa, U.S.	Agricultural	57–618
Carolinas, U.S.	Tidal marsh	3,000–35,000
Hawaii, U.S.	Volcanic	180–2,200
Eastern Australia	Agricultural	38–545
Nigeria	Agricultural	25–177
Brazil	Agricultural	43–398

Modified from Paul and Clark (1989).

of sulfur as hydrogen sulfide, carbonyl sulfide, and dimethyl sulfide, for example, from marine algae, marsh lands, mud flats, plants, and soils also contributes to the global circulation of sulfur through the atmosphere.

The nature and quantities of the various sulfur pools in surface soils are the basis for sulfur cycling in terrestrial environments. These sulfur pools are influenced by pedogenic factors such as climate, regional vegetation, and local topography. For example, the total sulfur content of soils ranges from 0.002 to 10%, with the highest levels in tidal flats, and in saline, acid sulfate, and organic soils. The impact of pedogenic factors on the total sulfur concentrations in the surface is clear when comparing values for soils from diverse geographic areas, as shown in Table 15–2.

Nature and Forms of Organic and Inorganic Sulfur in Soil

Organic sulfur constitutes more than 90% of the total sulfur present in most surface soils. However, the precise nature of the organic sulfur compounds in soil cannot be clearly identified. Thus, organic sulfur is grouped into two broad categories, organic sulfates and carbon-bonded sulfur. Examples of these organic compounds are given in Figure 15–1. Organic sulfates (R–O–S) include sulfate esters (C–O–S), sulfamates (C–N–S), and sulfated thioglycosides (N–O–S). Organic sulfates constitute 30 to 75% of total organic sulfur in soil. Carbon-bonded S (C–S) includes the sulfur present in amino acids, proteins, polypeptides, heterocyclic compounds (e.g., biotin and thiamin), sulfinates, sulfones, sulfonates, and sulfoxides. A large portion of carbon-bonded sulfur present in soil has yet to be identified; however, in some cases the carbon-bonded sulfur of amino acids may constitute up to 30% of the organic sulfur in soil.

Inorganic forms of sulfur account for less than 25% of the total sulfur in most agricultural soils. Sulfur exists in a number of forms with a wide range of oxidation states (Table 15–3). Sulfide, elemental sulfur, sulfite, thiosulfate, tetrathionate, and sulfate are the main forms of inorganic sulfur in agricultural soils. Sulfate is the most common form of inorganic sulfur found in well-aerated agricultural soils, whereas sulfides account for less than 1% of total sulfur and measurable quantities of thiosulfate and tetrathionate are usually detected only in soils treated with sulfur fertilizer or those receiving pollutants.

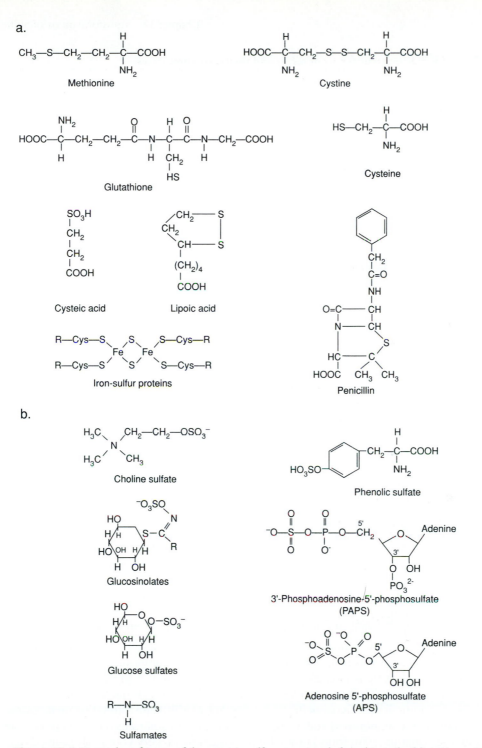

Figure 15–1 Examples of some of the organic sulfur compounds found in soils. (a) Amino acids and other compounds containing carbon-bonded sulfur. Note that lipoic acid also has a disulfide bond. (b) Compounds that possess ester sulfate bonds (e.g., C–O–S and C–N–S). This class of compounds is sometimes used by microorganisms to store sulfur, although some bacteria are able to store elemental sulfur. *From Paul and Clark, 1989. Used with permission of Academic Press, Inc.*

Table 15–3 Important forms of sulfur and their oxidation states.

Compound	Formula	Oxidation state(s) of sulfur
Sulfide	S^{2-}	-2
Polysulfide	S_n^{2-}	$-2, 0$
Sulfur*	S_8°	0
Hyposulfite (dithionite)	$S_2O_4^{2-}$	$+2$
Sulfite	SO_3^{2-}	$+4$
Thiosulfate[†]	$S_2O_3^{2-}$	$-1, +5$
Dithionate	$S_2O_6^{2-}$	$+6$
Trithionate	$S_3O_6^{2-}$	$-2, +6$
Tetrathionate	$S_4O_6^{2-}$	$-2, +6$
Pentathionate	$S_5O_6^{2-}$	$-2, +6$
Sulfate	SO_4^{2-}	$+6$

From Vairavamurthy et al. (1993).
*Occurs in an octagonal ring in crystalline form.
[†]Outer S has a valence of -1; inner S has a valance of $+5$.

Microbial Transformations of Sulfur in Soil

The sulfur cycle—emphasizing soil and plant sulfur transformations in agro-ecosystems—is illustrated in Figure 15–2. The major forms of sulfur in soil include elemental sulfur (S°), sulfides (S^{2-}), sulfates (S^{6+}), and organic sulfur compounds. Most of this sulfur enters the soil as soluble inorganic forms produced during the weathering of minerals, from fertilizers and atmospheric deposition, or as soluble organic and inorganic forms from the decomposition of organic matter. Losses of sulfur occur through leaching, surface runoff, volatilization, and crop removal.

Soil microorganisms drive the sulfur cycle. Hence, sulfur undergoes many microbially mediated transformations in soil, including:

• oxidation and reduction reactions,

• mineralization and immobilization reactions, and

• volatilization reactions.

The soil microbial biomass is the key driving force behind all sulfur transformations. This biomass acts as both a source and sink for inorganic sulfate, whereas microbial activity regulates both the fluxes of sulfur between different pools (inorganic sulfate, labile organic sulfur, and resistant organic sulfur) and the losses of sulfur from these pools (e.g., conversion of complex organic sulfur compounds into mobile forms that may be lost by leaching). Most of the sulfur in soil (75 to 90%) is found in organic complexes. These complexes are either stable, passive fractions that turn over very slowly or active, dynamic fractions that are readily transformed or metabolized. Actually, the continuum of sulfur organic complexes in soil ranges from very old, stable (e.g., organic sulfur found in soil humus) fractions to very young, short-lived (e.g., organic sulfur found in the amino acid cysteine) fractions. The microbial biomass is the engine for the conversion of passive fractions into active fractions, and vice versa. This is illustrated in the following diagram, where the relative flux of sulfur between pools is reflected in the size of the arrows depicting microbial conversion:

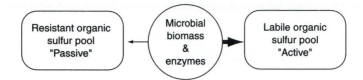

Understanding the sources, sinks, and transformations of sulfur in soil is crucial to ensure adequate supplies of sulfur for the biota of various ecosystems and to protect the environment from the detrimental effects of too much sulfur. Because microorganisms are so important for the conversions between the active and passive organic matter pools, any factor that disturbs or otherwise has an impact on

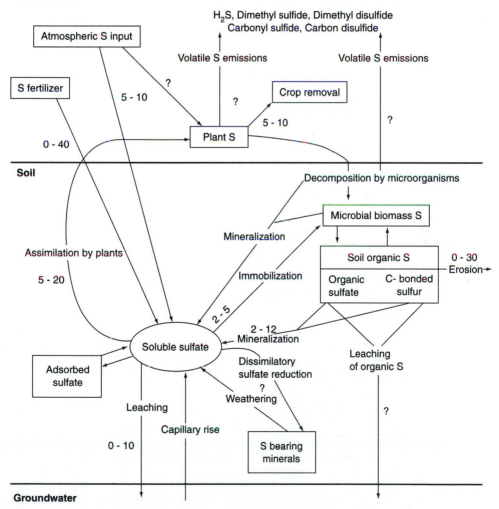

Figure 15–2 Conceptual sulfur cycle in agroecosystems. Numbers represent flux estimates (kg ha^{-1} yr^{-1}) for sulfur transformations in western Canadian soils. *Adapted from Schoenau and Germida (1992). Used with permission.*

Table 15–4 **Total sulfur and ester sulfate content of selected microorganisms grown in culture with varying sulfur concentrations.**

Organism	Total sulfur (μg g^{-1} cells)			Ester sulfur (%)		
	S1*	S2	S3	S1	S2	S3
Arthrobacter globiformis	1,626	1,706	1,850	23	10	14
Bacillus licheniformis	ND	1,667	1,700	ND	6	8
Bacillus sp., soil isolate	1,142	1,054	ND	19	10	ND
Micrococcus flavus	2,398	1,500	1,950	7	8	9
Pseudomonas cepacia	2,477	ND	ND	16	ND	ND
Fusarium solani	ND	4,750	4,900	ND	13	21
Penicillium nalgiovensis	ND	1,815	2,450	ND	45	45
Soil isolate J-20	2,800	3,764	4,017	45	14	27
Soil isolate P-44	ND	5,527	6,400	ND	25	32
Streptomyces isolate 34L	ND	3,043	3,072	ND	14	16

From Gupta (1989).
*S$_1$ = Complex organic medium; S$_2$, S$_3$ = mineral salts medium containing 16 and 32 μg S ml^{-1}, respectively.

the microbial biomass influences sulfur cycling. For example, crop rotations and soil cultivation typically increase sulfur cycling in soil. When a soil is cultivated, it is mixed and churned and broken up into smaller pieces. This increases soil aeration and exposes soil particles and organic matter that were previously "hidden" from the soil microflora. Fresh organic matter, containing different ratios of carbon, nitrogen, phosphorus, and sulfur, is mixed into the soil. Some microbial biomass is activated due to the flush of available nutrients, and some of the biomass might be killed. For example, fungal hyphae help to hold soil aggregates together; as cultivation breaks up these aggregates, the hyphae are broken, resulting in dead biomass. This dead biomass is now available to be mineralized. As new microbial biomass is formed during decomposition of the newly exposed or added "active" organic material, nutrient elements (carbon, nitrogen, phosphorus, sulfur, and micronutrients) are cycled back and forth between active and passive states as organic and inorganic forms of the element. The overall consequence is increased nutrient cycling.

Similarly, interactions among different microbial groups, such as predation and parasitism, tend to increase the turnover rate of microbial biomass sulfur and hence affect sulfur fluxes. One can think of this in the following way. Bacteria and fungi store sulfur in the organic and inorganic sulfur compounds that comprise their cells (Table 15–4). When predators, such as soil amoebae, eat bacterial cells or pieces of fungal hyphae, that biomass material (labile or active sulfur pool) is broken down into smaller, nonmetabolized sulfur containing organic fragments resistant to further metabolism by the amoebae, which are excreted into the soil, along with any inorganic sulfur or metabolizable organic sulfur not needed by the amoebae for growth. The excess inorganic sulfur or metabolizable organic sulfur is now available for plants and other organisms to use. This is an example of the processes of **mineralization** and **immobilization** of nutrients.

$$\text{Organic S} \xrightleftharpoons[\text{Immobilization}]{\text{Mineralization}} \text{Inorganic S}$$

Biological mineralization and immobilization are processes that occur concurrently and exhibit a strong relationship with the soluble sulfate pool in soil.

$$\text{sulfate (outside cell)} \xrightarrow{\text{active transport}} \text{sulfate (inside cell)} \quad [1]$$

$$\text{ATP + sulfate} \xrightarrow{\text{ATP sulfurylase}} \text{APS + PPi} \quad [2]$$

$$\text{ATP + APS} \xrightarrow{\text{APS phosphokinase}} \text{PAPS} \quad [3]$$

$$\text{2 RSH + PAPS} \xrightarrow{\text{PAPS reductase}} \text{sulfite + AMP-3-phosphate + RSSP} \quad [4]$$

$$\text{sulfite + 3 NADPH} \xrightarrow{\text{sulfite reductase}} \text{H}_2\text{S + 3 NADP} \quad [5]$$

$$\text{O-acetyl-L-serine + H}_2\text{S} \xrightarrow[\text{sulfhydrylase}]{\text{O-acetylserine}} \text{L-cysteine + acetate + H}_2\text{O} \quad [6]$$

Figure 15–3 Assimilatory reduction of sulfate and formation of cysteine. RSH stands for thioredoxin; its reduced form is regenerated from the oxidized form through reduction by NADPH. *Redrawn from Gottschalk (1979). Used with permission.*

Immobilization occurs as a result of the microbial assimilation of nutrients that are then rendered unavailable for further plant or microbial uptake until the cell dies and is remineralized. Immobilization of sulfur may also involve precipitation as metal sulfide, especially pyrite, as in salt marshes. Because these transformations are mediated by microorganisms, soil factors that influence the growth and activity of microorganisms (e.g., pH, temperature, and moisture) also affect the rate of sulfur transformations. To estimate or predict the available sulfur status of soils, it is necessary to understand the factors that influence these processes.

Immobilization (Assimilation)

Microbial assimilation and conversion of inorganic sulfate into organic sulfur through the *assimilatory sulfate reduction pathway* leads to temporary immobilization of sulfur from plant or microbial availability. This process involves ATP sulfurylase and two energy-rich sulfate nucleotides, APS (adenosine 5′-phosphosulfate) and PAPS (3′-phosphoadenosine-5′-phosphosulfate). The overall reaction of SO_4^{2-}–S incorporation into amino acids is shown in Figure 15–3.

Most of the sulfur accumulated by microorganisms is in the form of amino acids in proteins; however, microorganisms also accumulate sulfate esters, sulfonates, vitamins, and cofactors. Some microorganisms, such as fungi, accumulate especially large amounts of sulfate esters (Table 15–4; Fig. 15–1). This is important because organic sulfates (e.g., sulfate esters and thioglucosides) are considered to be the most labile form of organic sulfur in soil and may comprise up to 30 to 70% of the organic sulfur in surface soils. The relative proportion of fungal biomass to bacterial biomass in soil (approximately 2:1) underscores the potential importance of microorganisms accumulating ester-sulfur compounds.

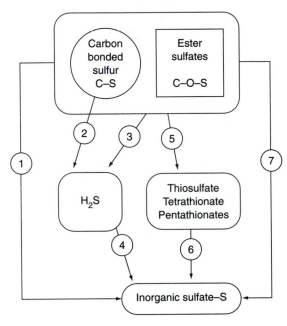

Figure 15–4 Known pathways for mineralization of organic sulfur compounds in soil: (1) biological (direct) mineralization during the oxidation of carbon as an energy source; (2) hydrolysis of cysteine by cysteine desulfhydrolase; (3) anaerobic mineralization (desulfurization) of organic matter; (4) biological oxidation of hydrogen sulfide to sulfate through elemental sulfur and sulfite; (5) incomplete oxidation of organic sulfur into inorganic sulfur compounds; (6) biological oxidation of tetrathionate to sulfate through sulfide; and (7) biochemical (indirect) mineralization when sulfate esters are hydrolyzed by sulfatases. *Adapted from Lawrence (1987).*

Typically, the addition of inorganic SO_4^{2-}–S to soil leads to its quick incorporation into the organic sulfur fractions via microbial assimilation. The rate and magnitude of this immobilization is increased in the presence of an energy source, such as metabolizable organic matter or addition of easily degradable carbon sources like glucose. Later, much of this accumulated sulfur is found in the fulvic acid fractions, especially as organic sulfates.

Mineralization

Mineralization of organic sulfur in soil is largely mediated by microbial activity. The various known pathways of sulfur mineralization are summarized in Figure 15–4. Carbon-bonded sulfur is mineralized either through oxidative (aerobic) decomposition or desulfurization (anaerobic) processes, whereas various sulfatases are involved in the mineralization of sulfate esters (Box 15–4). The mineralization process may be direct (i.e., cell mediated), involving viable microbial cells, or indirect (i.e., cell-free, enzyme mediated), involving enzymes such as sulfatases. In the case of direct mineralization, elements such as nitrogen and sulfur in direct association with carbon are mineralized as microorganisms oxidize the organic carbon compounds to obtain energy. Heterotrophic soil microorganisms decompose organic sulfur compounds to grow; as the carbon-sulfur bond is broken, the sulfur is released, usually as sulfide. Because this process involves actively growing microorganisms, their requirement for sulfur may meet or even exceed the sulfur supplied by the substrate. Thus net mineralization of sulfur by this process may not be reflected by increases in the sulfate-sulfur pool in soil. In the case of indirect mineralization, those elements that exist as sulfate esters are hydrolyzed by intracellular or extracellular enzymes. This process, also known as *enzymatic mineralization,* occurs mainly outside the cell and may be regulated by end product inhibition, or the sulfate level. Direct mineralization is controlled by the microbial need for carbon and energy sources, whereas indirect mineralization is controlled by factors influencing enzyme synthesis, activity, and kinetics.

Box 15–4

Enzymes Drive Mineralization of Sulfate Esters in Soil. Sulfatases (technically referred to as sulfohydrolases EC 3.1.5) are enzymes that hydrolyze sulfuric acid esters, where the linkage with sulfate is in the form of R–O–S and R represents a diverse group of organic moieties. This reaction can be shown as follows:

$$R\text{–}O\text{–}SO_3^- + H_2O \longrightarrow R\text{–}OH + SO_4^{2-} + H^+$$

Arylsulfatase is by far the most studied enzyme involved with sulfur cycling in soil. This is because organic sulfates are abundant in soil, and hence arylsulfatase may play an important role in the mineralization of organic soil sulfur. The enzyme may be extracellular or associated with cell debris, and its activity is easily determined based on colorimetric assays.

Factors Affecting the Mineralization of Sulfur in Soil

Mineralization is generally measured as net mineralization, either the amount of SO_4^{2-}–S accumulated during the period of study or the difference between gross mineralization and assimilation. Thus, for higher net mineralization to occur, the mineralization—assimilation balance has to be driven toward mineralization. A break-even point for mineralization and immobilization can be calculated based on the carbon-to-sulfur ratio of the substrate, the decomposing organisms, and the yield coefficient. For example, if we consider the decomposition of crop residues, net mineralization will generally occur with a carbon-to-sulfur ratio of 200 or less, whereas net sulfur immobilization will occur when the ratio is greater than 400/1. Because microbial activity is the driving force for mineralization and immobilization, these processes are significantly influenced by all factors affecting microbial metabolism, such as:

- energy and nutrient supply,

- carbon-to-sulfur ratio,

- abundance of organic sulfur,

- water availability,

- pH,

- temperature, and

- redox potential.

For example, actively growing plants may significantly increase sulfur mineralization in soils. Plants supply energy sources to the rhizosphere in the form of root exudates that increase microbial growth and activity, thus increasing sulfur mineralization. However, the re-assimilation of inorganic sulfates released by the growing microorganisms may result in no increase in the sulfur pool available to plants and may even result in reduction of that pool when the microbial demand exceeds the rate of sulfur mineralization. Different plants excrete different types and amounts of root exudates and require different amounts of sulfur for growth. Hence, crop rotations can have a significant impact on sulfur cycling in soil.

Microbial Oxidation of Inorganic Sulfur Compounds

Chemoautotrophic and Chemoheterotrophic Sulfur Oxidation

The abiotic oxidation of reduced sulfur compounds can occur to a limited extent in soils, but microbial reactions clearly dominate the process. For example, the biological oxidation of elemental sulfur in soils apparently takes place primarily via the following sequence of reactions (i.e., those most common to heterotrophs), although some of the products may result from abiotic side reactions:

$$S^\circ \rightarrow \qquad S_2O_3^{2-} \rightarrow \qquad S_4O_6^{2-} \rightarrow \qquad SO_4^{2-}$$

Elemental sulfur   Thiosulfate   Tetrathionate   Sulfate

Many different microorganisms are important for the oxidation, reduction, and cycling of sulfur in soil and other ecosystems (Table 15–5). In the case of sulfur oxidation, the microorganisms can be divided into:

- chemoautotrophs (lithotrophs), including species of the genus *Thiobacillus,*

- photoautotrophs, including species of purple and green sulfur bacteria, and

- chemoheterotrophs (organotrophs), including a wide range of bacteria and fungi.

The chemoautotrophs and chemoheterotrophs are largely responsible for oxidizing sulfur in most aerobic, agricultural soils.

Many chemoautotrophic bacteria (e.g., thiobacilli) are capable of oxidizing reduced inorganic sulfur compounds. The biochemistry of sulfur oxidation by thiobacilli growing *in vitro* has been extensively reviewed (Postgate and Kelly, 1982; Pronk et al., 1990). For acidophilic thiobacilli, the most common sequence of reactions involved in sulfur oxidation is:

$$S^\circ \rightarrow \qquad SO_3^{2-} \rightarrow \qquad SO_4^{2-}$$

Elemental sulfur   Sulfite   Sulfate

A great variety of thiobacilli can be isolated from natural habitats. They include obligate acidophilic chemoautotrophs, facultative chemoautotrophs (thiobacilli that grow autotrophically with reduced inorganic sulfur compounds as energy sources, but are also capable of heterotrophic growth), and **mixotrophs,** which can use mixtures of inorganic and organic compounds simultaneously. The thiobacilli differ in

Table 15–5 **Sulfur-using bacteria occurring in soil and aquatic habitats.**

Group	Sulfur conversion	Habitat requirements	Habitat example	Examples of Genera
Heterotrophs that use oxidized S species as electron acceptors	$SO_4^{2-} \rightarrow HS^-$ $S_2O_3^{2-} \rightarrow HS^-$ or S^o $S^o \rightarrow HS^-$ $SO_3^- \rightarrow HS^-$	anaerobic; organic substrates available; light not required	anoxic sediments and soils	*Desulfomonas* *Desulfovibrio* *Desulfotomaculum* *Desulfurmonas* *Campylobacter*
Obligate and faculative autotrophs that use reduced S as an energy source	$HS^- \rightarrow S^o$ $S^o \rightarrow SO_4^{2-}$ $S_2O_3^{2-} \rightarrow SO_4^{2-}$	$H_2S - O_2$ interface; light not required	mud; hot springs; mine drainage; soils	*Thiobacillus* *Thiomicrospira* *Achromatium* *Beggiatoa*
Phototrophs that use reduced S as an electron donor	$HS^- \rightarrow S^o$ $S^o \rightarrow SO_4^{2-}$	anoxic; H_2S; light	shallow water; anoxic sediments; metalimnion or hypolimnion; anoxic water	*Chlorobium* *Chromatium* *Ectothiorhodospira* *Thiopedia* *Rhodopseudomonas*
Heterotrophs that use organic S compounds as energy sources or that hydrolyze esters	org S $\rightarrow HS^-$ org S $\rightarrow$ volatile org S ester $SO_4 \rightarrow SO_4^{2-}$	source of organic S compounds	sediments; soils; water column	Many
Microorganisms that use SO_4^{2-} or H_2S in biosynthesis	$SO_4^{2-} \rightarrow$ protein $HS^- \rightarrow$ protein $SO_4^{2-} \rightarrow$ DMSP*	nonspecific	sediments; soils; water column	Many

From Cook and Kelly (1992). Used with permission.
*dimethylsulfoniumpropionate

their physiological characteristics and in the reduced sulfur compounds used as energy sources (Table 15–6). The majority of these thiobacilli are obligate aerobes, although some, like *Thiobacillus denitrificans,* can grow anaerobically by using nitrate as a terminal electron acceptor. Other species of thiobacilli use electron donors such as ferrous iron (*T. ferrooxidans*) and thiosulfate (*T. thioparus*) in addition to sulfur.

Although thiobacilli can oxidize sulfur to plant-available sulfate in some soils, this process is also evidently mediated by many different heterotrophic soil microorganisms. Bacteria, such as *Arthrobacter, Bacillus, Micrococcus,* and *Pseudomonas,* some actinomycetes, and a wide range of fungi are also capable of oxidizing elemental and reduced forms of sulfur. Many of these sulfur-oxidizing heterotrophs have been isolated from soil and may:

• oxidize sulfur, producing mainly thiosulfate,

• oxidize sulfur, producing sulfate, and

• oxidize thiosulfate to sulfate.

The pathway by which heterotrophic microorganisms produce these sulfur oxyanions has not been established, although several studies suggest that it is enzymatic in fungi. Apparently, no energy is derived by the organisms through these oxidations, and the transformations are incidental to the major metabolic pathways.

Table 15–6 Characteristics of species of the genus *Thiobacillus*.

Species	Electron donor	Electron acceptor	Facultative heterotroph	Facultative anaerobe	pH optimum
T. thiooxidans	H_2S, $S°$, $S_2O_3^{2-}$	O_2	−	−	2.2
T. ferrooxidans	$S°$, $S_2O_3^{2-}$, Fe^{2+}	O_2	+	−	3.0
T. neapolitanus	$S°$, $S_2O_3^{2-}$	O_2	−	−	6.6
T. kabobis	$S°$	O_2	−	−	ND
T. tepidarius	$S_2O_3^{2-}$,	O_2	−	−	7.0
T. thioparus	$S°$, $S_2O_3^{2-}$, NCS^-	O_2, NO_2^-	−	+	6.9
T. denitrificans	$S°$, $S_2O_3^{2-}$, $S_4O_6^{2-}$	O_2, NO_3^-	−	+	7.0
T. intermedius	$S_2O_3^{2-}$	O_2	+	−	ND
T. novellus	$S_2O_3^{2-}$,	O_2	+	−	8.4
T. acidophilus	ND	O_2	+	−	3.0
T. organoparus	ND	O_2	+	−	ND
T. versutus	$S_2O_3^{2-}$,	O_2, organic C	+	+	8.2
T. perometabolis	$S_2O_3^{2-}$, $S°$	O_2	+	−	ND

Adapted from Germida and Janzen (1993), Konopka et al. (1986), and Kuenen and Beudeker (1982).
ND = no data

Because the heterotrophic organisms are generally more numerous in soils than chemoautotrophs, mixed populations of heterotrophs probably play the dominant role in sulfur oxidation in many aerobic, neutral, and alkaline agricultural soils.

The opposing view that thiobacilli play the dominant role in sulfur oxidation in soils is largely based on the observation that these bacteria achieve rates of sulfur oxidation in culture far in excess of those achieved by heterotrophs growing under the same conditions. Most thiobacilli are facultative or obligate chemoautotrophs, which means that they can oxidize sulfur independently of the supply of available organic carbon. Marked increases in numbers of thiobacilli may follow the addition of reduced forms of sulfur to some soils, supporting the concept that populations of thiobacilli are important oxidizers of the added sulfur. However, no consistent correlation has been found between sulfur-oxidation rates and the incidence of thiobacilli, except that rates of sulfur-oxidation are generally low in soils that lack these organisms and are accelerated in soil inoculated with thiobacilli. It is probable that in many soils the initial oxidizers of reduced sulfur compounds are heterotrophic organisms until the pH is reduced sufficiently to permit oxidation by chemolithotrophs. In addition, there is good evidence that **consortia** of heterotrophs and autotrophs working together to bring about the oxidation of sulfur in agricultural soils.

Other Sulfur Bacteria

Other bacteria may also oxidize sulfur compounds. The gliding sulfur oxidizers include those bacteria that have a gliding motion on the substrate; their cells are arranged in **trichomes.** The most important members of this group in relation to sulfur-oxidation in soils are species of *Beggiatoa,* bacteria that participate in sulfide oxidation in the root zone of rice. All strains of *Beggiatoa* deposit sulfur intracellularly in the presence of hydrogen sulfide. Phototrophic bacteria, such as *Chromatium* and *Chlorobium,* also play an important role in sulfide oxidation in rice paddy soil, but not in aerobic agricultural soils. A number of nonfilamentous, chemolithotrophic sulfur-oxidizing bacteria, such as

Sulfolobus, Thiospira, or *Thiomicrospira,* have also been isolated from special habitats, but the importance of these bacteria in sulfur oxidation in soils has yet to be determined. The activity of different groups of sulfur-oxidizing bacteria may be predicted based on the relative turnover rates of inorganic sulfur compounds and organic substrates during energy-limiting growth conditions (Kuenen and Beudeker, 1982; Box 15–5).

Box 15–5

Predicting the Occurrence of Sulfur-Oxidizing Microorganisms in Different Habitats. Consider a hot sulfur spring like those found in Yellowstone National Park U.S.. The molar ratio of inorganic sulfur compounds to organic substrates in the water is very large (i.e., lots of dissolved, reduced inorganic sulfur compounds in the water and very little organic matter); hence, we predict that obligate autotrophic (i.e., chemolithotrophic) sulfur-oxidizing bacteria would be abundant. We also know these bacteria are thermophiles because they are living in water at a temperature greater than 60°C. Alternatively, in a soil that contains 1% organic matter and 0.01% thiosulfate or elemental sulfur, the ratio of inorganic sulfur compounds to organic substrates would be very small, and thus heterotrophic sulfur-oxidizing organisms would be abundant.

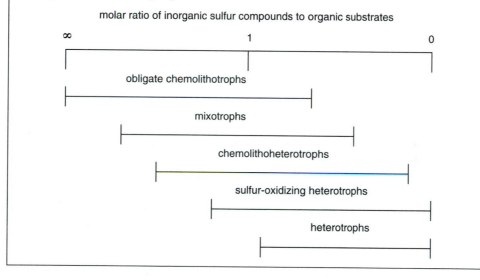

Biogenesis and Oxidation of Metal Sulfides

Metal sulfides may be formed through biotic or abiotic reactions. In both cases the metal sulfide results from the interaction between a metal ion and a sulfide ion:

$$M^{2+} + S^{2-} \rightarrow MS$$

Many sulfate-reducing bacteria, such as *Desulfovibrio* spp. or *Desulfotomaculum* spp., are involved in the biogenesis of sulfides of antimony, cobalt, cadmium, iron, lead, nickel, and zinc. The extent of metal-sulfide genesis depends on many factors, the most important of which is the relative toxicity of the metal ion. In nature, this toxicity is probably reduced when the metal ions are adsorbed on clays or complexed with organic matter. The formation of metal sulfides during the mineralization of organic sulfur compounds is also possible, although little is known about this phenomenon.

The oxidation of metal sulfides in soil involves both chemical and microbial processes and, as a result, is a more complex process than is sulfur oxidation. Chalcocite (Cu_2S), chalcopyrite ($CuFeS_2$), galena (PbS), pyrite (FeS_2), and nickel sulfide (NiS) are just a few examples of metal sulfides that are subject to microbial transformations. For example, the biological oxidation of pyrite follows a series of oxidation steps described in the following equations. These biotic oxidations are responsible for the formation of acid mine drainage and acid soil formation in surface mine spoils. First, ferrous sulfate is formed as the result of an abiotic oxidation step:

$$2FeS_2 + 2H_2O + 7O_2 \rightarrow 2FeSO_4 + 2H_2SO_4$$

This reaction is then followed by the bacterial oxidation of ferrous sulfate, generally by *T. ferrooxidans*:

$$4FeSO_4 + O_2 + 2H_2SO_4 \rightarrow 2Fe_2(SO_4)_3 + 2H_2O$$

This reaction occurs chemically but can be accelerated 10^6-to 10^8-times by thiobacilli. Subsequently, ferric sulfate is reduced and pyrite oxidized by a strictly chemical reaction.

$$Fe_2(SO_4)_3 + FeS_2 \rightarrow 3FeSO_4 + 2S^\circ$$

$$2S + 6Fe_2(SO_4)_3 + 8H_2O \rightarrow 12FeSO_4 + 8H_2SO_4$$

The elemental sulfur produced is finally oxidized by *T. thiooxidans* and *T. ferrooxidans,* and the acidity produced helps the whole process to continue.

$$2S^\circ + 3O_2 + 2H_2O \rightarrow 2H_2SO_4$$

Note the net production of 10 molecules of H_2SO_4 during the process.

Although several sulfur-oxidizing thiobacilli and heterotrophs can be isolated from acid sulfate soils in which pyrite is being oxidized, only *T. ferrooxidans* appears to play an important role in the process. The biological oxidation of sulfides and other reduced sulfur compounds can have severe consequences for the environment (Box 15–6). For example, acid mine drainage contaminates several thousand kilometers of streams in the Appalachian coal mining region of the United States.

Box 15–6

Problems Associated with Sulfur Oxidation. Oxidation of reduced sulfur compounds leads to the formation of acidic products. As a consequence, sulfur oxidation can have detrimental effects on the environment. This is especially true in the case of sulfide minerals such as pyrite (FeS_2). Some of the more serious problems associated with sulfur oxidation include:

- formation of acid mine drainage,

- formation of acid sulfate soils,

- corrosion of concrete structures, and

- corrosion of metals.

Microbial Reduction of Inorganic Sulfur Compounds

Bacterial Sulfate Reduction

The reduction of sulfate to hydrogen sulfide is mediated mainly by anaerobic, sulfate-reducing bacteria. This process may be significant in anaerobic, waterlogged soils, but is usually not important in well-aerated agricultural soils, except in anaerobic microsites. Nevertheless, sulfate reduction is a major component of the sulfur-cycle in soils exposed to waterlogging or periodic flooding, especially where readily decomposable plant residues are present.

Microorganisms reduce oxidized sulfur compounds by either an assimilatory or dissimilatory process. Some use *assimilatory sulfate reduction* to meet their sulfur requirements. In *dissimilatory sulfate reduction,* bacteria use sulfate as a terminal electron acceptor, and large quantities of hydrogen sulfide (H_2S) are released. This process is analogous to the denitrification process discussed in Chapter 12. Like most denitrification, dissimilatory sulfate reduction is a strictly anaerobic process. In this case it is carried out by bacteria such as *Desulfovibrio* spp., *Desulfomonas* spp., and *Desulfotomaculum* spp. (Table 15–7). These bacteria use end products of other fermentations such as lactate, malate, and ethanol as electron donors.

Factors Influencing Sulfate Reduction

When a soil is flooded, electron acceptors become reduced in an ordered sequence: first oxygen, followed by nitrate, nitrite, manganic, and ferric compounds, and finally sulfate and carbon dioxide. Although the reduction of one compound does not have to be completed before another is reduced, oxygen and nitrate must be removed before the reduction of ferric and sulfate ions can occur. Because of this reaction sequence, sufficient ferrous ions generally are available to react with any hydrogen sulfide produced, and as a result, free hydrogen sulfide is rarely liberated from soils. Sulfate reduction increases with the period of soil submergence and following the addition of organic matter. Sufficient organic substrates to stimulate the process are also

Table 15–7 **Dissimilatory sulfate-reducing bacteria.**

Genera	*Desulfobacter, Desulfobulbus, Desulfococcus, Desulfonema, Desulfosarcina, Desulfotomaculum, Desulfovibrio*
General characteristics	Strict anaerobes Grow at mildly acid to mildly alkaline pH Generally mesophilic, but some species thermophilic
Substrates	Most sulfate reducers will also reduce sulfite and thiosulfate Some species reduce elemental sulfur Organic matter utilization varies with genus and species As a group, capable of completely oxidizing fatty acids from C1 to C18, lactate, pyruvate, low-molecular-weight alcohols, and some aromatic compounds
Habitats	Anaerobic sediments of freshwater, brackish water, and marine environments, thermal regions, water-logged soils, and animal intestines.

From Trudinger (1986). Used with permission.

liberated from seeds and from roots into the rhizosphere, with the result that in paddy soils, blackening caused by ferrous sulfide deposits often occurs in the root region. There is evidence, however, that rice roots can aerate the soil sufficiently that ferric iron is observed on the root surface. In general, the rate of sulfate reduction increases with decreasing redox potential, with the optimum being a function of soil pH, around -300 mV at pH 7. Sulfate-reducing bacteria are active in soil, sediments, polluted water, oil-bearing strata, and shales. Their activity may be beneficial or detrimental to the surrounding environment and have serious economic consequences (Box 15–7).

| Box 15–7 |

Consequences of Sulfate Reduction. The activity of sulfate-reducing bacteria and the problems they pose can be seen in many examples from our daily lives:

- Sulfate-reducing bacteria are a major cause of corrosion of underground iron pipes, costing between $1.6 billion and $5.0 billion in the United States in 1990.

- Turf managers find that sulfate-reducing bacteria can produce a black layer under golf course greens by using the organic matter in root exudates to reduce soil sulfates to ferrous sulfides.

- The water in the canals of Venice is polluted with hydrogen sulfide and, as a consequence, the gondolas of Venice turn black regardless of their original color.

Volatilization of Inorganic and Organic Sulfur Compounds from Soil

A number of sulfur gases are released from soils, marshes, peats, and sediments or from anthropogenic sources. These gases may be inorganic or organic and play an important role in the cycling of sulfur through the atmosphere (Chapter 23). Many different fungi and heterotrophic bacteria are responsible for the formation of these volatile compounds during the metabolism of organic sulfur compounds (Table 15–8).

Table 15–8 **Biochemical origin of volatile sulfides produced in soils by microbial degradation of organic matter under aerobic and anaerobic conditions.**

Volatile Name	Formula	Biochemical precursors
Hydrogen sulfide	H_2S	Proteins, polypeptides, cystine, cysteine, glutathione
Methyl mercaptan	CH_3SH	Methionine, methionine sulphoxide, methionine sulphone, S-methylcysteine
Dimethyl sulfide	CH_3SCH_3	Methionine, methionine sulphoxide, methionine sulphone, S-methylcysteine, homocysteine
Dimethyl disulfide	CH_3SSCH_3	Methionine, methionine sulphoxide, methionine sulphone, S-methylcysteine
Carbon disulfide	CS_2	Cysteine, cystine, homocysteine, lanthionine, djenkolic acid
Carbonyl sulfide	COS	Lanthionine, djenkolic acid

From Andreae and Jaeschke (1992). Used with permission.

Environmental Aspects of Sulfur Pollutants

Acid Sulfate Soils

Acid sulfate soils contain sulfides, mainly in the form of pyrites, which may be oxidized to yield free and adsorbed sulfates. They are characterized by yellow mottling due to the formation of jarosites [$AFe_3(SO_4)_2(OH)_6$, where $A = K^+$, NH_4^+, Na^+ or H_3O^+], and have a pH typically below 4. Although these soils cover large areas of the tropics, they tend to be of only local importance in temperate regions. The acidification of these soils results from the abiotic and microbial oxidation of pyrite. Problems in producing crops on these soils occur because of aluminum and manganese toxicity rather than to the direct effects of acidity. Acid sulfate soils can be reclaimed by:

- controlling the water table through adequate drainage,

- adding lime,

- planting crops tolerant of aluminum, manganese, and iron, and

- improving soil fertility generally.

Deposition of Atmospheric Sulfur in Soils

Soils subject to atmospheric pollution receive sulfur from the atmosphere largely in the form of dilute sulfuric acid. Thus, sulfate is the main sulfur ion entering soils from the atmosphere; smaller quantities of sulfite and bisulfite may also contaminate these soils. Atmospheric pollution deposits, consisting largely of soot, may also be locally important sources of reduced sulfur compounds, particularly in areas adjacent to industrial plants, such as coking and steel works. Because sulfate is the major sulfur ion entering soil from atmospheric pollution, we expect that the major sulfur transformations that occur involve sulfur assimilation and sulfur reduction rather than sulfur oxidation. However, when reduced sulfur compounds in the atmosphere are deposited on soils, they are rapidly oxidized.

Summary

Sulfur is an essential element for all living organisms and is the basis for primary productivity in some exotic communities. It exists in a number of oxidation states as inorganic and organic compounds that undergo a number of biotic and abiotic transformations. These transformations can be beneficial or detrimental to ecosystems depending on the forms and fluxes of sulfur. The cycling of sulfur through aquatic, terrestrial, and atmospheric ecosystems is similar to that of other elements, such as carbon and nitrogen, and is influenced by natural and anthropogenic processes.

We currently have a basic understanding of the forms and amounts of sulfur in terrestrial ecosystems and the processes controlling the supply of sulfur to plants. The exchange of sulfur gases between the soil-plant system and the atmosphere is less well documented. Understanding how key processes in the sulfur cycle respond to environmental factors (e.g., construction of models of mineralization and volatilization processes that include temperature, moisture, substrate, and microbial composition response functions) will help us predict accurately the impact of human-induced or natural changes on sulfur fluxes in all components of the biosphere.

Cited References

Andreae, M.O., and W.A. Jaeschke. 1992. Exchange of sulphur between biosphere and atmosphere over temperate and tropical regions. pp. 27–61. *In* R.W. Howarth, J.W.B. Stewart, and M.V. Ivanov (eds.), Sulphur cycling on the continents: Wetlands, terrestrial ecosystems, and associated water bodies. SCOPE 48. John Wiley and Sons, Chichester, England.

Cook, R.B., and C.A. Kelly. 1992. Sulphur cycling and fluxes in temperate dimictic lakes. pp. 145–188. *In* R.W. Howarth, J.W.B. Stewart, and M.V. Ivanov (eds.), Sulphur cycling on the continents: Wetlands, terrestrial ecosystems, and associated water bodies. SCOPE 48. John Wiley and Sons, Chichester, England.

Germida, J.J., and H.H. Janzen. 1993. Factors affecting the oxidation of elemental sulfur in soils. Fertil. Res. 35:101–114.

Gottschalk, G. 1979. Bacterial metabolism. Springer-Verlag, New York.

Gupta, V.V.S.R. 1989. Microbial biomass sulfur and biochemical mineralization of sulfur in soils, Ph.D. dissertation. University of Saskatchewan, Saskatoon, Sask., Canada.

Jannasch, H.W., and M.J. Mottl. 1985. Geomicrobiology of deep-sea hydrothermal vents. Science 229:717–725.

Konopka, A.E., R.H. Miller, and L.E. Sommers. 1986. Microbiology of the sulfur cycle. pp. 23–56. *In* M.A. Tabatabai (ed.), Sulfur in agriculture. Agron. Monogr. 27, American Society of Agronomy, Madison, Wis.

Kuenen, J.G., and R.F. Beudeker. 1982. Microbiology of thiobacilli and other sulphur-oxidising autotrophs, mixotrophs and heterotrophs. Phil. Trans. Royal Soc. Lond. Ser. B. 298:473–497.

Lawrence, J.R. 1987. Microbial oxidation of elemental sulfur in agricultural soils, Ph.D. dissertation. University of Saskatchewan, Saskatoon, Sask., Canada.

Paul, E.A., and F.E. Clark. 1989. Soil microbiology and biochemistry. Academic Press, San Diego, Calif.

Pierzynski, G.M., J.T. Sims, and G.F. Vance. 1994. Soils and environmental quality, CRC Press, Boca Raton, Fla.

Postgate, J.R., and D.P. Kelly (eds.). 1982. Sulphur bacteria. The Royal Society, Cambridge, England.

Pronk, J.T., R. Meulenberg, W. Hazeu, P. Bos, and J.G. Kuenen. 1990. Oxidation of reduced inorganic sulphur compounds by acidophilic thiobacilli. FEMS Microbiol. Rev. 75: 293–306.

Schoenau, J.J., and J.J. Germida. 1992. Sulphur cycling in upland agriculture systems. pp. 261–277. *In* R.W. Howarth, J.W.B. Stewart, and M.V. Ivanov (eds.), Sulphur cycling on the continents: Wetlands, terrestrial ecosystems, and associated water bodies. SCOPE 48. John Wiley and Sons, Chichester, England.

Trudinger, P.A. 1986. Chemistry of the sulfur cycle. pp. 1–22. *In* M.A. Tabatabai (ed.), Sulfur in agriculture. Agron. Monogr. 27, American Society of Agronomy, Madison, Wis.

Tunnicliffe, V. 1992. Hydrothermal-vent communities of the deep sea. Am. Scientist 80:336–349.

Vairavamurthy, A., B. Manowitz, G.W. Luther III, and Y. Jeon. 1993. Oxidation state of sulfur in thiosulfate and implications for anaerobic energy metabolism. Geochim. Cosmochimica Acta. 57:1619–1623.

General References

Brady, N.C., and R.R. Weil. 1996. The nature and properties of soils. 11th ed. Prentice Hall, Upper Saddle River, N.J.

Ehrlich, H.L. 1996. Geomicrobiology. Marcel Dekker, New York.

Freney, J.R. 1986. Forms and reactions of organic sulfur compounds in soils. pp. 207–232. *In* M.A. Tabatabai (ed.), Sulfur in agriculture. Agron. Monogr. 27, American Society of Agronomy, Madison, Wis.

Germida, J.J., M. Wainwright, and V.V.S.R. Gupta. 1992. Biochemistry of sulfur cycling in soil. pp. 1–53. *In* J.-M. Bollag and G. Stotzky (eds.), Soil biochemistry. Vol. 7. Marcel Dekker, New York.

Howarth, R.W., J.W.B. Stewart, and M.V. Ivanov (eds.). 1992. Sulphur cycling on the continents: Wetlands, terrestrial ecosystems, and associated water bodies. SCOPE 48. John Wiley and Sons, Chichester, England.

Postgate, J.R. 1992. Microbes and man. 3rd ed. Cambridge University Press. Cambridge, England.

Schlegel, H.G., and B. Bowien (eds.). 1989. Autotrophic bacteria. Science Tech Publishers, Madison, Wis.

Schlesinger, W.H. 1991. Biogeochemistry: An analysis of global change. Academic Press, San Diego, Calif.

Tabatabai, M.A. (ed.). 1986. Sulfur in agriculture. Agron. Monogr. 27, American Society of Agronomy, Madison, Wis.

Study Questions

1. What is meant by the term *chemosynthesis,* and how does this relate to the sulfur cycle?

2. Discuss how the oxidation of reduced sulfur compounds drives primary productivity at deep sea thermal vents.

3. Why are the gondolas on the canals of Venice painted black?

4. Sketch a model that helps predict the types of sulfur-oxidizing microorganisms that will be found in certain habitats.

5. What is a sulfur mixotroph?

6. Sketch the sulfur cycle, listing examples of organisms involved in each phase of the cycle, and then discuss how soil microorganisms drive this cycle.

7. List and discuss some of the similarities between the sulfur and nitrogen cycles.

8. What is a disulfide bond and why is it important?

9. Why might crop rotations influence sulfur cycling in soil?

10. Compare and contrast biological (direct) mineralization and enzymatic (indirect) mineralization processes.

11. What types of microorganisms could be used to mine metals?

12. Discuss how and why a group of microorganisms might act together as a consortium to oxidize sulfur in soil.

13. In what type of soil would you expect to find active sulfate-reducing bacteria?

14. Why is the enzyme arylsufatase important in the sulfur cycle in soil?

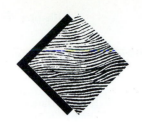

Chapter 16

♦

Transformations of Other Elements

Michael D. Mullen

♦

The land belongs to itself. If anything, we belong to it . . . as much as earthworms
or corn plants. We rise up a while and sink back in. We borrow our lives from it.

Nancy Paddock

We have seen that microorganisms are important in the cycling of carbon, nitrogen, and sulfur. Soil microorganisms are also actively involved in the cycling of many other elements. In fact, most of the nutrients required by living organisms and many other elements, including several potentially toxic materials, are biologically transformed in the environment (examples are shown in Table 16–1). The mechanisms by which soil microorganisms transform these elements include:

- **Mineralization** and **immobilization** reactions that mediate the transformation of an element from organic to inorganic and inorganic to organic forms, respectively.

- **Redox** reactions that involve the transfer of electrons from (oxidation) or to (reduction) the element in question.

- **Solubilization** by which some soil minerals, such as the relatively insoluble phosphate minerals, are made available to plants and other organisms.

- **Methylation** of elements, such as mercury, that results in increased mobility of the element in the environment.

In this chapter we discuss microbial transformations of several important elements. Phosphorus (P), iron (Fe), and manganese (Mn) are important nutrient

Table 16–1 **Microbially mediated transformations of several essential and nonessential elements in soil.**

Element	Microbial transformation		
	Mineralization-immobilization	Oxidation-reduction	Methylation
Phosphorus, potassium, calcium, magnesium, copper, zinc	yes	no	no
Iron, manganese	yes	yes	no
Arsenic, mercury, selenium	yes*	yes	yes

*Mercury is transformed between inorganic and organic forms; arsenic and selenium can act like phosphorus and sulfur in biological systems.

elements, while mercury (Hg) and selenium (Se) are elements that are associated with environmental problems. Our discussion of these elements is not meant to minimize the importance of the microbial community in the transformations of other elements. Many other elements are also subject to one or more of these processes. The elements discussed here serve only to illustrate the range of transformations that microorganisms mediate.

Phosphorus

Forms of Phosphorus in Soil

Phosphorus is critical to all life forms because of the role it plays in many important biomolecules such as DNA (deoxyribonucleic acid), phospholipids, and ATP (adenosine triphosphate). The amount of total phosphorus found in a surface soil can vary greatly, ranging from < 100 μg P g^{-1} ($\cong$ 200 kg ha^{-1}) in very sandy soils to > 1,000 μg g^{-1} ($\cong$ 2,000 kg ha^{-1}) in soils derived from basic rocks (Stevenson, 1986). The primary mineral form of phosphorus is rock phosphate, or *apatite* (Table 16–2). Highly weathered soils, such as those of the southeastern United States, may have very little remaining apatite, while relatively unweathered and alkaline soils formed from basic parent materials, such as those of the northwestern United States, may have high contents. The chemical weathering of apatite results in the release of orthophosphate ($H_2PO_4^-$ is the dominant species at pH values below 7.2; HPO_4^{2-} dominates above pH 7.2). Very little orthophosphate is present in the soil solution at any one time, usually < 1% of the total phosphorus.

Solution phosphorus concentrations of 0.1 to 1 mg L^{-1} are common in soil. Of this, more than half may be in the form of soluble organic compounds released by dead cells or in colloidal organic compounds. Addition of soluble phosphate fertilizers or mineralization of organic phosphorus results in the release of orthophosphate into the soil solution, followed by precipitation as iron and aluminum phosphates in acid soils or calcium phosphates in alkaline soils (stable-inorganic phosphorus) or by adsorption to iron and aluminum oxides (labile-inorganic phosphorus). These reactions result in low orthophosphate concentrations in the soil solution. The optimum availability of orthophosphate occurs at a soil pH of about 6.5 because precipitation as both aluminum and calcium phosphates is minimized. One

Table 16–2 **Examples of inorganic phosphorus minerals in soil and their solubilities with respect to dissolution of the cation and PO_4^{3-}.**

Name	Formula	Solubility product (log)
Fluoroapatite	$Ca_5(PO_4)_3F$	-59
Hydroxyapatite	$Ca_5(PO_4)_3OH$	-57
Tricalcium phosphate	$Ca_3(PO_4)_2$	-29
Variscite	$AlPO_4 \cdot 2H_2O$	-21
Strengite	$FePO_4 \cdot 2H_2O$	-26

Calculated from equilibrium data in Lindsay (1979).

practical aspect of liming acid soils or acidifying alkaline soils is the improved availability of orthophosphate.

Many organic forms of phosphorus are also found in soils. As plant and animal remains or waste products are returned to the soil, readily mineralized organic phosphorus compounds are introduced. Microorganisms also produce organic phosphorus compounds as organic materials which are transformed in soil. Organic phosphorus in most soils may account for as little as 3 and as much as 90% or more of the total soil phosphorus but usually represents 30 to 50% of total phosphorus in most soils. The total amount of organic phosphorus in a soil is usually strongly correlated with total organic carbon; organic phosphorus decreases with depth in the soil profile, as does organic carbon.

The chemical nature of the soil organic phosphorus fraction is not well known. There are many different forms of organic phosphorus in soil; some common compounds are illustrated in Figure 16–1. Of these compounds, inositol phosphates, often called phytins or phytic acids, are typically present in the greatest quantity, comprising 10 to 50% of the total organic phosphorus. These inositol phosphates are often in a polymeric state in the soil and are relatively resistant to decomposition. The inositol hexaphosphates are found in the greatest amounts, followed by inositol compounds with one to five phosphate groups. Most of the inositol phosphates, and other polymeric forms of organic phosphorus in soil, are thought to be of microbial origin (Cosgrove, 1977). Phospholipids and nucleic acids may account for 1 to 5% of the organic phosphorus. Other identifiable organic phosphorus compounds are typically present in trace amounts only.

Another important pool of organic phosphorus is the soil microbial biomass. This fraction represents the actively cycling pool of organic phosphorus in the soil environment and is part of the labile, or readily available, organic phosphorus. Through this pool, the active mineralization and immobilization of phosphorus in soil occurs. Concentrations of 5 to 75 µg biomass P g^{-1} soil are common. Brookes et al. (1984) observed that the biomass phosphorus fraction ranged from 2 to 5% of the total organic phosphorus in arable soils to upwards of 20% in some grassland and forest soils in Great Britain. Cultivation of soil results in a reduction of soil organic matter and total organic phosphorus and reduces the proportion of biomass phosphorus to total organic phosphorus. Therefore, tillage depletes the labile organic phosphorus pool more rapidly than the stable organic phosphorus fraction.

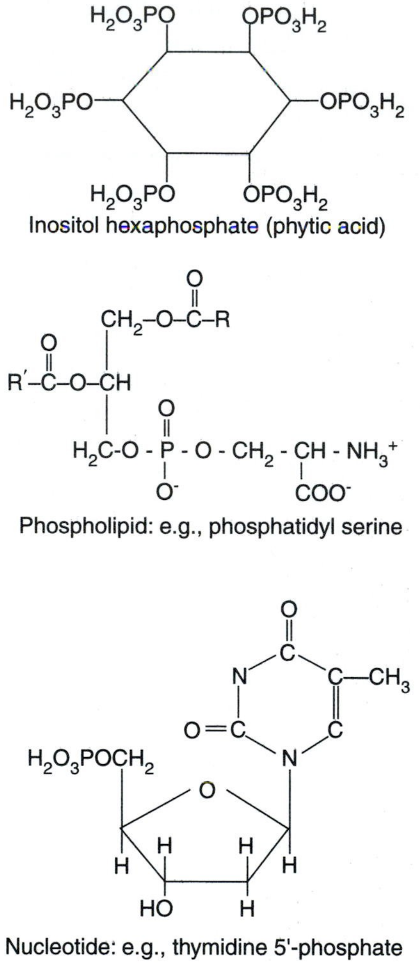

Inositol hexaphosphate (phytic acid)

Phospholipid: e.g., phosphatidyl serine

Nucleotide: e.g., thymidine 5'-phosphate

Figure 16–1 Representative forms of organic phosphorus inputs to the soil environment.

The Phosphorus Cycle

A model of the phosphorus cycle, illustrated in Figure 16–2, shows the various compartments of phosphorus in the terrestrial environment. Phosphorus is affected by both biological and chemical reactions. This model divides the phosphorus cycle into a geochemical subcycle and a biological subcycle, with the solution phosphorus pool serving as the central point in the overall cycle. Solution phosphorus is the source of orthophosphate for plants and soil microorganisms.

In the biological phosphorus subcycle, orthophosphate can be taken up by plants or immobilized into microbial biomass. As plant residues and animal remains and wastes are returned to soil, the organic phosphorus may be directly incorporated into stable humus, mineralized to orthophosphate, or immobilized into the microbial biomass. Biomass phosphorus is subject to incorporation into humic substances and mineralization and immobilization reactions. The turnover or cycling of the biomass contributes significantly to the labile organic phosphorus pool. Crop removal and erosion are two mechanisms for loss of organic phosphorus.

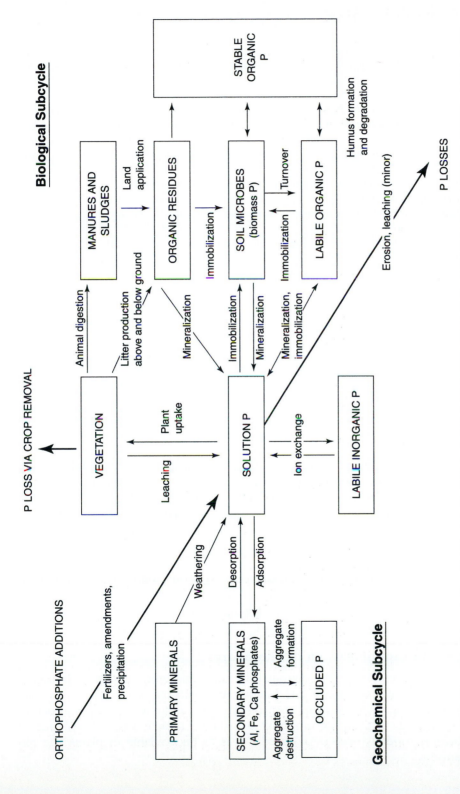

Figure 16–2 The phosphorus cycle, showing inputs, losses, and major transformations in the soil environment. The phosphorus cycle consists of two subcycles: one geochemical (lower left diagonal) and one biological (upper right diagonal). The boxes represent pools of phosphorus, while arrows indicate phosphorus inputs, transfers, and losses. *Adapted from Walbridge (1991). Used with permission.*

| Box 16–1 |

Measurement of Microbial Biomass Phosphorus. The method involves fumigating the soil with chloroform ($CHCl_3$) for 24 hours and extracting the soil with 0.5 M $NaHCO_3$. The chloroform extraction lyses microbial cells, releasing the contents of the cells to the soil. The inorganic orthophosphate (Pi) extracted from a nonfumigated control is subtracted from the Pi released in the fumigated soil. This Pi underestimates the total microbial phosphorus, and a correction factor, k_p, is used to account for this. A k_p of 0.4 is often used; this extraction efficiency was determined by evaluating the recovery of known amounts of phosphorus from microorganisms added to soil. However, this factor may actually be higher or lower, depending on soil properties. Because orthophosphate is readily adsorbed in soil, a correction for phosphorus fixation is also necessary. A known amount of $H_2PO_4^-$, typically 25 $\mu g\ g^{-1}$ soil, is added to a third soil sample with the $NaHCO_3$ extracting solution. The difference between the Pi in this extract and the nonfumigated control represents the amount of added Pi recovered. Brookes et al. (1982) provide further details on this method. Microbial biomass phosphorus is calculated as follows:

$$\frac{\mu g\ \text{biomass P}}{g\ \text{soil}} = \left(\frac{25}{c-a}\right) \times \left(\frac{b-a}{0.4}\right)$$

where: a = μg Pi extracted from the nonfumigated control

   b = μg Pi extracted from the fumigated soil

   c = μg Pi extracted from a nonfumigated soil extracted with 0.5 M $NaHCO_3$ providing 25 μg Pi g^{-1} soil

   25 = known concentration of Pi added to nonfumigated soil, in μg Pi g^{-1} soil

   0.4 = k_p

In the geochemical phosphorus subcycle, orthophosphate is solubilized from primary and secondary minerals by chemical and biochemical weathering processes. Dissolution of these compounds in the soil solution or solubilization through microbially produced organic acids releases orthophosphate to the soil solution for plant and microbial uptake. Orthophosphate can adsorb to aluminum and iron oxides (labile inorganic phosphorus) or precipitate as aluminum, iron, or calcium phosphates (secondary minerals).

There is limited evidence for microbial reduction of orthophosphate to phosphine gas (PH_3) and oxidation of orthophosphite ($H_2PO_3^-$) to orthophosphate. However, the thermodynamic state of soils and sediments would preclude the occurrence of these phosphorus species in such systems.

Mineralization and Immobilization of Phosphorus

Phosphorus availability is controlled by mineralization and immobilization through the organic fraction and the solubilization and precipitation of phosphate in inorganic forms. As the remains of plants, animals, and microbes are returned to the soil, they are actively decomposed by soil microorganisms. Phosphorus in these organic residues must be released if it is to be available to plants and microorganisms.

Phosphorus mineralization is an enzymatic process. As a group, the enzymes involved, called *phosphatases*, catalyze a variety of reactions that release phosphate from organic phosphorus compounds to the soil solution. Phosphatases are released by microorganisms extracellularly into the soil solution to catalyze these hydrolytic mineralization reactions:

- *Phosphomonoesterases* hydrolyze the phosphate from monoester forms of phosphorus, such as those in nucleotides or phospholipids (Box 16–2).

- *Phosphodiesterases* hydrolyze phosphate from diester forms of phosphorus, such as in nucleic acids.

- *Phytases* hydrolyze phosphate from inositol phosphates.

Once phosphorus is mineralized, it can be taken up by plants or immobilized back into microbial cells, or it can form insoluble inorganic complexes. The microbial biomass can affect phosphorus availability through immobilization, the incorporation of orthophosphate ions into organically bound forms in the organism. For example, orthophosphate reacts with ADP (adenosine diphosphate) and a suitable input of energy to form ATP. The extent of immobilization is affected by the carbon to phosphorus ratio of the organic materials being decomposed and the amount of available phosphorus in the soil solution.

The carbon to phosphorus ratio of an added residue can determine the extent to which inorganic phosphorus is mineralized or immobilized. If insufficient phosphorus is available in the residue for assimilation of the added carbon, then inorganic phosphorus from the soil solution must be used and net immobilization occurs. Conversely, if more phosphorus is present in the residue than is needed for carbon assimilation, net mineralization of orthophosphate occurs. Generally, a C/P ratio $< 200/1$ results in mineralization, while a C/P ratio $> 300/1$ results in immobilization. Ratios between 200 to 300 result in little net change in solution phosphorus concentrations. These processes are similar to those for nitrogen and sulfur mineralization and immobilization. In addition to the phosphorus content of a residue, other soil and environmental variables (e.g., pH, temperature, aeration, and soil moisture) affect microbial activity and phosphorus mineralization. The element that is most limiting controls the overall mineralization rate of a residue. If rapid carbon mineralization of a phosphorus-limited residue is occurring, then immobilization of phosphorus from the soil results. As mineralizable organic carbon disappears, a portion of the phosphorus-rich microbial biomass will be mineralized as well, resulting in the eventual release of the previously immobilized phosphorus.

Solubilization of Inorganic Phosphorus

Inorganic phosphorus minerals are generally found as aluminum and iron phosphates in acidic soils, while calcium phosphates dominate in alkaline soil (Table 16–2). These slightly soluble compounds provide orthophosphate to soil solution to the extent allowed by the solubility of the compound. Solubility products (Table 16–2) provide a relative measure of solubility in pure water; however, these compounds are also affected by soil pH (Lindsay, 1979). Orthophosphate is supplied to the roots primarily by diffusion. Thus chemical equilibria between orthophosphate, adsorbed orthophosphate, and inorganic phosphate minerals are important in supplying phosphorus to plants and microorganisms. As phosphorus is taken up, it is replenished from these sources.

Box 16–2

The Phosphomonoesterase Assay. This assay measures the potential of a soil to mineralize orthophosphate by hydrolysis of phosphomonoester bonds in organic phosphorus sources. The assay includes an organic phosphate analog, p-nitrophenylphosphate, as a substrate (Tabatabai, 1994). The soil is treated with toluene to inhibit microbial activity and a buffer solution to maintain the reaction pH. As the phosphate-ester bond is hydrolyzed, p-nitrophenol is formed. After incubation, NaOH is added to the soil to stop the reaction and adjust the soil to an alkaline pH where p-nitrophenol forms a yellow color. Colorimetric determination of the p-nitrophenol concentration permits calculation of the rate of enzyme activity. The reaction is:

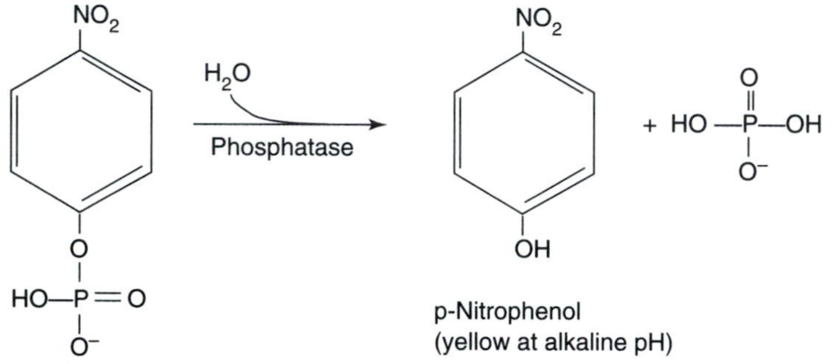

$$\text{and phosphatase activity} = \frac{\mu g\ p\!-\!nitrophenol\ released}{g\ soil\ hour}$$

p-Nitrophenylphosphate → (Phosphatase, H_2O) → p-Nitrophenol (yellow at alkaline pH) + $HO-\overset{O}{\underset{O^-}{\overset{\|}{P}}}-OH$

Plant roots and soil microorganisms can enhance the dissolution of phosphate compounds by the release of carbon dioxide and organic acids to the soil solution. Carbonic acid can promote the acid dissolution of calcium and magnesium phosphate compounds. Similarly, the acidity produced by the nitrifying bacteria and sulfur-oxidizing bacteria promotes solubilization of insoluble phosphate salts. A wide range of organic acids is produced by microorganisms and plants, and many act as **chelating agents** to solubilize aluminum, iron, calcium, and magnesium phosphates, resulting in the release of orthophosphate into the soil solution (Fig. 16–3; Stevenson, 1967). One group of organisms that may be important in this regard is the mycorrhizal fungi, which form symbioses with plant roots and enhance the uptake of phosphorus and other nutrients. Under water-logged conditions, hydrogen sulfide, produced by sulfate-reducing bacteria or other processes, can also displace metal cations from insoluble phosphates, with the release of phosphate.

Environmental Aspects of Phosphorus Cycling

Phosphorus can cause environmental damage if applied in excess to soils. Phosphorus is the most limiting nutrient for primary productivity in many ecosystems, particularly aquatic systems. If a large amount of phosphorus is applied to

Figure 16–3 Example of Ca^{2+} chelation by citric acid.

Box 16–3

Phosphorus-Solubilizing Bacteria. Some bacteria are very effective in solubilizing phosphorus from rock phosphate. One example is *Bacillus megaterium* var. *phosphaticum*. This bacterium has been formulated into inocula referred to as *phosphobacterin* and applied to soils to enhance the solubilization of phosphorus minerals. This practice was reported to be successful in the former Soviet Union; however, it has had only limited success in the United States. Nonetheless, the incorporation of a readily mineralizable carbon source, such as manure, can enhance phosphorus solubilization through increased biological activity. The increased organic carbon also can serve to complex aluminum in acid soils, thereby reducing the extent to which aluminum phosphates precipitate.

---

soil, it may move to surface waters with runoff water and eroded soil particles that bind phosphorus. The major consequence is a process called **eutrophication.** Eutrophication can occur in surface waters when excessive phosphorus or nitrogen accumulates. The process occurs most noticeably in relatively still bodies of water such as ponds and lakes. If phosphorus is the limiting nutrient in the pond, excess phosphorus additions can result in rapid growth of algae in the water. This algal bloom is often noticeable as a green "scum" or film on the water surface. As the algae die, they settle to the bottom, where bacterial decomposition of the nutrient-rich algae occurs. The decomposition of the algae results in oxygen depletion. If the process continues, the lake slowly becomes anaerobic, resulting in a poor environment for many forms of aquatic life.

Nonpoint sources of phosphorus, such as agricultural activities, are often associated with increased phosphorus in surface water. For example, runoff from cropland and soils receiving animal manures can lead to significant phosphorus inputs into surface waters. In industrialized countries, nonpoint sources are the major source of phosphorus. In nonindustrialized countries with limited sewage treatment capacity, raw sewage discharge into surface waters (a point source) is a significant source of phosphorus pollution.

The natural ability of soils to adsorb phosphorus from solution removes phosphorus from treated wastewater in the United States. Natural and constructed wetlands can serve as a sink for the removal of phosphorus from these wastewaters before their release to a river or lake.

Iron, Manganese, Mercury, and Selenium

Many metallic and metalloid elements, some of them micronutrients and others that are environmental pollutants, are transformed by microorganisms during metabolic activities. These transformations may provide effective means for remediation of contaminated soils and sediments (Frankenberger and Losi, 1995). The transformations of iron, manganese, mercury, and selenium can affect their solubility and availability to plants and other organisms and their mobility in the environment.

Oxidation and Reduction Processes: Iron and Manganese

Microorganisms are involved in redox reactions of many elements. Biologically mediated redox reactions are often linked to energy production in an organism. Chemoautotrophic bacteria oxidize reduced inorganic compounds to extract electrons for use in ATP production. Alternatively, the reduction of many elements occurs during energy production under anaerobic conditions when the element is an alternative to oxygen as a terminal electron acceptor.

Elements involved in redox reactions, in addition to carbon, nitrogen, and sulfur, are listed in Table 16–3. Of those listed, iron and manganese are the most commonly transformed elements in the soil environment. Oxidation and reduction of iron and manganese can affect availability of these elements to plants, the corrosion of iron and steel structures, and soil properties indicative of imperfect drainage (i.e., mottling and manganese concretions). Because these elements are also subject to oxidation and reduction by chemical means, it is difficult to assess the extent to which these transfor-

Table 16–3 **Examples of common elements (excluding carbon, nitrogen, and sulfur) subject to microbial oxidation-reduction reactions in soils and sediments and examples of bacterial genera involved with each reaction.**

Element	Common oxidation states	Reaction, significance, and redox couple		Some bacterial genera reported to be involved
Cr	Cr^{6+}, Cr^{3+}	Oxidation—NR*		
		Reduction—AR, D	$Cr^{6+} + 3e^- \rightarrow Cr^{3+}$	*Aeromonas, Bacillus, Chlorella, Pseudomonas*
Fe	Fe^{3+}, Fe^{2+}	Oxidation—E	$2Fe^{2+} \rightarrow 2Fe^{3+} + 2e^-$	*Thiobacillus*
		Reduction—AR	$2Fe^{3+} + 2e^- \rightarrow 2Fe^{2+}$	*Geobacter, Desulfovibrio, Pseudomonas, Thiobacillus*
Hg	Hg^{2+}, Hg^0	Oxidation—NE	$Hg^0 \rightarrow Hg^{2+} + 2e^-$	*Bacillus, Pseudomonas*
		Reduction—D	$Hg^{2+} + 2e^- \rightarrow Hg^0$	*Chlorella, Pseudomonas, Streptomyces*
Mn	Mn^{4+}, Mn^{2+}	Oxidation—E, D	$Mn^{2+} \rightarrow Mn^{4+} + 2e^-$	*Arthrobacter, Pseudomonas*
		Reduction—AR	$Mn^{4+} + 2e^- \rightarrow Mn^{2+}$	*Bacillus, Geobacter, Pseudomonas*
Se	Se^{6+}, Se^{4+}, Se^0, Se^{2-}	Oxidation—E	$Se^{2-} \rightarrow Se^0 + 2e^-$	*Bacillus, Thiobacillus*
		Reduction—AR	$SeO_4^{2-} + 8e^- \rightarrow Se^{2-}$	*Clostridium, Desulfovibrio, Micrococcus*

From Ehrlich (1990), Frankenberger and Losi (1995) and Lovely (1993).
AR, element used as a terminal-electron acceptor in anaerobic respiration; D, detoxification mechanism; E, energy source; NE, nonenzymatic reaction, microorganism alters the physicochemical environment.
*NR, not reported to be biologically mediated.

mations are biotic or abiotic. As Ghiorse (1994) points out, the presence of organisms capable of iron and manganese oxidation and reduction in soils is often evidence for potential activity in the soil. However, the contribution of microorganisms to iron and manganese oxidation and reduction cannot be easily measured by direct means.

Iron is among the most abundant elements in the soil environment, but concentrations of soluble iron are typically very low in aerobic soil environments. Iron contents range from less than 0.05% in coarse-textured soils to more than 10% in highly weathered Oxisols found in the tropics. The iron cycle (Fig. 16–4) is characterized by the oxidation and reduction of iron compounds in soils and sediments. However, mineralization from organically bound iron and the solubilization of iron from inorganic compounds by microorganisms are also important processes. Manganese cycling is similar to that of iron. Total manganese concentrations in soil may be less than 0.01 to 0.3%. The form that is most available for plant growth is the reduced Mn^{2+} ion. Chemical and microbial oxidation results in the formation of relatively insoluble manganese oxides.

Chemical oxidation of Fe^{2+} occurs very rapidly under aerobic conditions at pH > 3 and is the major pathway of iron oxidation in most soil environments. Under acidic conditions, ferrous iron (Fe^{2+}) can be oxidized to ferric iron (Fe^{3+}) by chemoautotrophic bacteria such as *Thiobacillus ferrooxidans*. A representative reaction for ferrous iron oxidation by *T. ferrooxidans* is the oxidation of ferrous sulfate ($FeSO_4$) to ferric sulfate and ferric hydroxide:

$$12FeSO_4 \quad + \quad 3O_2 \quad + \quad 6H_2O \quad \rightarrow \quad 4Fe_2(SO_4)_3 \quad + \quad 4Fe(OH)_3$$

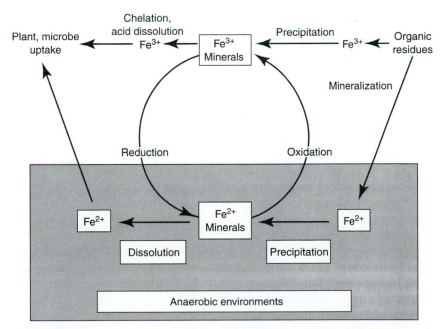

Figure 16–4 The iron cycle. Transformations include oxidation and reduction of iron-minerals, precipitation of Fe^{2+} and Fe^{3+} minerals, dissolution of iron-minerals to plant-available forms, and mineralization of organically bound iron.

Manganese oxidation is also microbially mediated in soil at pH > 5. The oxidation of Mn^{2+} occurs as follows, and the rate increases with increasing alkalinity up to about pH 8:

$$Mn^{2+} + 2OH^- \rightarrow MnO_2 + H_2O$$

The MnO_2 formed is very insoluble. Manganese concretions, or nodules, are often found in soils with cyclical oxidation and reduction processes, such as those with fluctuating water tables. Several bacteria enzymatically oxidize Mn^{2+}, although few are true chemoautotrophs. Some of the bacteria are **mixotrophs**; they derive some energy from the oxidation, but obtain carbon from organic carbon. Other organisms oxidize Mn^{2+} to MnO_2 with H_2O_2 via the *catalase* enzyme:

$$Mn^{2+} + H_2O_2 \rightarrow MnO_2 + 2H^+$$

This reaction does not provide energy for the organism, but serves to remove potentially harmful H_2O_2. Manganese can also be oxidized nonenzymatically when microorganisms alter the pH of the microenvironment to 8 or above, where chemical oxidation occurs readily, or by producing compounds that can chemically oxidize Mn^{2+}. These reactions may serve to protect the cell from high concentrations of Mn^{2+}.

Microorganisms also carry out dissimilatory iron and manganese reduction (Lovley, 1993). Most organisms that reduce Mn^{4+} also reduce Fe^{3+}. Here we will discuss only iron reduction. Dissimilatory iron reduction is often coupled to the oxidation of fermentation products from other organisms. Complex organic matter, sugars, and amino acids can be partially oxidized by fermentation, resulting in the production of organic acids, alcohols, hydrogen, and methane. Bacteria from several genera can oxidize these fermentation products with Fe^{3+} serving as the terminal electron acceptor. The reduction of Fe^{3+} by *Geobacter metallireducens* with the oxidation of acetate occurs as follows:

$$CH_3COO^- + 8Fe^{3+} + 4H_2O \rightarrow 2HCO_3 + 8Fe^{2+} + 9H^+$$

Some organisms can couple the complete oxidation of monoaromatic compounds with Fe^{3+} reduction. *Geobacter metallireducens* oxidizes environmental contaminants, such as toluene and phenol, to carbon dioxide in the presence of Fe^{3+}. These organisms may play important roles in the remediation of aquifers contaminated by polyaromatic hydrocarbons. There are also reports of bacteria that couple the oxidation of H_2 to H^+ with the reduction of Fe^{3+}.

Dissimilatory Fe^{3+} and Mn^{4+} reductions are important in the environment. For example:

- Fe^{3+} is an important electron acceptor in organic matter decomposition in aquatic sediments and saturated soils. This can lead to gleying of the soil, or the characteristic gray to gray-blue colors associated with reduced iron compounds that indicate drainage problems.

- Fe^{3+} reduction in phosphate minerals can lead to the release of phosphate for subsequent uptake by plants and microbes.

- Fe^{3+} reduction can lead to steel corrosion, a problem of great concern to many industries.

- Mn^{4+} reduction can lead to the release of soluble Mn^{2+} into soil solution. Reoxidation of this manganese can result in the formation of manganese concretions in the soil, an indicator of fluctuating water tables in the soil profile.

Microbial Enhancement of Iron Activity

In well-aerated soils, Fe^{3+} is the dominant form of iron. The activity of Fe^{3+} in soil solution under aerobic conditions is low. For example, assume that $Fe(OH)_3$ is the solid phase controlling Fe^{3+} activities in soil solution. At pH 7, the calculated activity of Fe^{3+} in soil solution would be approximately 10^{-17} M (Lindsay, 1979). The activity continues to decrease as pH increases. The low solubility of Fe^{3+} in alkaline soils results in iron-deficient conditions for plants and microorganisms.

Many microorganisms produce Fe^{3+}-complexing compounds known as **siderophores.** These are low-molecular-weight, iron-transporting compounds with high Fe^{3+} affinity (Leong, 1986). The purpose of these compounds is to solubilize Fe^{3+} for uptake by the organism. Fungi and some bacteria produce siderophores with hydroxymate active sites, while other bacterial siderophores are catechol derivatives (Fig.16–5). The organism has siderophore receptor proteins on its outer membrane that bind the siderophore-Fe^{3+} complex. The iron is released from the siderophore via reduction of Fe^{3+} to Fe^{2+}. The Fe^{2+} is then transported into the cell and the siderophore is released to complex more Fe^{3+}. Interestingly, the chelation

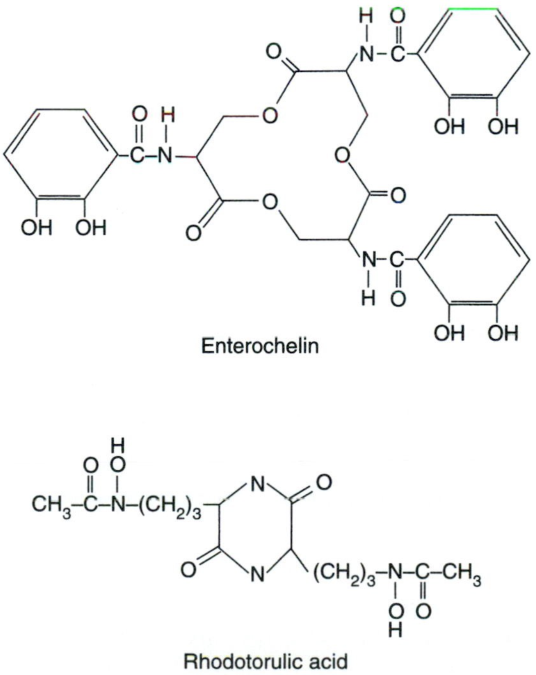

Enterochelin

Rhodotorulic acid

Figure 16–5 Examples of microbial siderophores. Enterochelin is a catechol derivative produced by the bacterium *Salmonella*. Rhodotorulic acid is a hydroxymate siderophore produced by the yeast *Rhodotorula*. The Fe^{3+} is complexed by the hydroxyl and keto groups on the molecules.

of Fe^{3+} from iron phosphate minerals also results in the liberation of soluble orthophosphate.

Some pseudomonads produce a yellow-green fluorescent siderophore, called
pseudobactin, that binds iron tightly, prohibiting its use by other organisms. By
this mechanism the pseudomonad can lower the availability of Fe^{3+} to other microbes and thereby control some fungal root pathogens in iron-limited environments.

Microbial Transformations of Mercury and Selenium

Mercury and selenium are environmentally important elements that are affected
by another interesting microbial transformation, **methylation.** Methylation is a
detoxification mechanism for the organisms involved. Mercury can be microbially
reduced, while selenium can be microbially oxidized and reduced.

Mercury is used or produced in a variety of industries. It can enter the environment as
a result of ore smelting, production of chlorine and caustic soda, agricultural practices,
and other human activities. At one time, mercury compounds were common pesticides.

Mercury can cause defects in the central nervous system. In the 1950s, industrial
discharges of methylmercury into Minamata Bay in Japan resulted in the contamination of fish with methylmercury, the poisoning of thousands of people, and several
hundred deaths.

Mercury undergoes several transformations by microorganisms in the environment (Fig. 16–6). It can be present in many forms, including elemental mercury
(Hg^0), inorganic mercuric ion (Hg^{2+}), methylmercury (CH_3Hg^+), or dimethylmer-
cury [$(CH_3)_2Hg$]. The methylation of mercury occurs under anaerobic conditions,
such as those found in wetland soils and lake sediments (Zillioux et al., 1993).
This reaction increases the solubility and volatility of mercury, resulting in its increased movement into the food chain. Methylmercury is lipophilic and can be
concentrated in fish in contaminated waters.

Recent evidence indicates that the methylation is mediated by sulfate-reducing
bacteria. The methylation reaction occurs via the transfer of methyl groups from
methylcobalamine, or methylated vitamin B_{12}, to Hg^{2+}.

$$Hg^{2+} + B_{12}{-}CH_3 \rightarrow CH_3Hg^+ + \text{reduced - } B_{12}$$

The methylcobalamine is produced during fermentation by sulfate reducers, and
the methylation generally stops when fermentation ceases, although sulfate reduction may continue (Choi and Bartha, 1993). As methylmercury migrates into
more aerobic zones, demethylation of the methylmercury can occur, followed by
reduction of the Hg^{2+}, resulting in the formation of elemental mercury. Elemental
mercury is volatile and can escape to the atmosphere.

Many organisms reduce mercury to volatile Hg^0 as a detoxification mechanism.
Reduction can occur during aerobic growth and is not linked to energy produc-
tion. Several aerobic and facultative bacteria (e.g., *Bacillus, Pseudomonas,
Corynebacterium, Micrococcus,* and *Vibrio*) catalyze the reduction of Hg^{2+} to Hg^0
(Nakamura et al., 1990). In mercury-contaminated environments, Hg^{2+}-resistant
communities of bacteria are generally present, and Hg^{2+} reduction is fairly rapid.

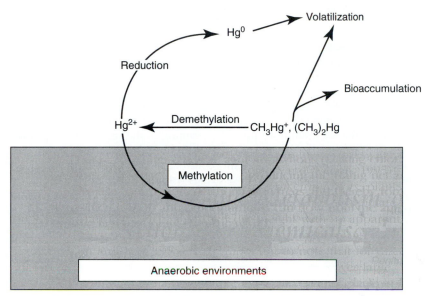

Figure 16–6 The mercury cycle. Transformations of mercury include reducing of ionic Hg^{2+} to elemental Hg^0, methylating of Hg^{2+} under anoxic conditions, and demethylating under aerobic conditions. Methylmercury and elemental mercury are volatile and can be lost to the atmosphere. Methylmercury is toxic and can accumulate in fatty tissues of animals, which is increasing concentrated as methylmercury moves up the food chain.

Bacterial reduction of Hg^{2+} to Hg^0 may be a useful remediation strategy for contaminated sediments.

Selenium is essential for animals in trace quantities, but is toxic at high concentrations. The biochemistry of selenium is similar to that of sulfur (Masscheleyn and Patrick, 1993). Common species of selenium in aerobic environments are selenate (SeO_4^{2-}) and selenite (SeO_3^{2-}). Selenide (Se^{2-}) is present as H_2Se and insoluble metal precipitates in reduced environments. Both selenate and selenite are soluble. Several bacterial genera, including *Pseudomonas, Clostridium,* and *Flavobacterium,* use oxidized selenium, such as SeO_4^{2-}, SeO_3^{2-} and Se^0, as terminal electron acceptors. The reduction of oxidized selenium to insoluble elemental selenium is a principle means for immobilizing soluble selenate and selenite from contaminated environments (Doran, 1982).

Selenium can also be methylated to dimethylselenide [$(CH_3)_2Se$] by a mechanism similar to that described for mercury. Volatilization of dimethylselenide from contaminated waters and soils then results in the removal of the selenium from the environment. In contrast to mercury methylation, selenium methylation reduces the toxicity of selenium and is apparently a detoxification mechanism that is catalyzed by bacteria, fungi, and some plants.

Box 16–4

An Example of Selenium Contamination. Kesterson Reservoir is a series of 12 ponds covering approximately 500 ha in the southern part of the Kesterson National Wildlife Refuge in California, U.S. (Ohlendorf, 1989). The ponds were designed as collection and evaporation basins for agricultural drainage water from the San Joaquin Valley. Between 1981 and 1986, the drainage waters were found to be high in selenium and salinity. Selenium concentrations approached 300 μg L^{-1} (the EPA drinking water standard for selenium is 10 μg L^{-1}), resulting in the transport of approximately 9 Mg of selenium into the reservoir during this time. **Bioaccumulation** and **biomagnification** of selenium in the food chain occurred, resulting in abnormal rates of mortality and deformity in waterfowl. These events resulted in the closure of the reservoir to drainage waters. Since closure, many of the low lying areas have been filled with soil, and management plans have been implemented to dissipate selenium through plant bioaccumulation and volatilization by soil microorganisms. Biologically mediated removal of selenium from contaminated sites may prove more cost effective than soil removal and disposal.

Summary

The importance of microbial cycling of elements in soil extends beyond carbon, nitrogen, and sulfur. Most nutrients and other elements are actively transformed by a variety of mechanisms. Microbially mediated transformations of a number of elements are important to plant and microbial nutrition and environmental concerns. Some of the nutrient elements include phosphorus, iron, and manganese, while elements of environmental significance include, but are not limited to, mercury and selenium.

The phosphorus cycle includes mineralization and immobilization reactions through the microbial biomass pool, and the solubilization of inorganic phosphorus minerals by organic and inorganic acids and chelating agents produced by soil organisms. The mineralization of organic phosphorus is an important source of phosphorus for plant growth, particularly in nonagricultural ecosystems.

Two other nutrient elements, iron and manganese, are also transformed in soils. Reduced forms (Fe^{2+} and Mn^{2+} minerals) can be oxidized by chemoautotrophic bacteria, providing a source of electrons for energy production. Oxidized forms (Fe^{3+} and Mn^{4+} minerals) can serve as terminal electron acceptors during dissimilatory reduction. Reduction results in increased solubility of these elements. Reduced iron minerals give rise to the characteristic gray colors present in soils with poor drainage.

Mercury is a contaminant in many environments, particularly aquatic sediments. Ionic Hg^{2+} can be microbially methylated in sediments, resulting in lipophilic methylmercury. Sulfate-reducing bacteria are thought to be primarily responsible for the methylation reaction. Methylmercury is highly toxic and can accumulate in the food chain. The ionic form can also be reduced to volatile elemental mercury, resulting in removal of mercury from a given environment to the atmosphere.

Selenium is an essential element for animals, but is toxic at high concentrations. The microbiology of selenium is very similar to that of sulfur; it can be oxidized and reduced and can also undergo methylation reactions similar to those of mercury. However, methylation of selenium results in decreased toxicity.

Cited References

Brookes, P.C., D.S. Powlson, and D.S. Jenkinson. 1982. Measurement of microbial biomass phosphorus in soil. Soil Biol. Biochem. 14:319–329.

Brookes, P.C., D.S. Powlson, and D.S. Jenkinson. 1984. Phosphorus in soil microbial biomass. Soil Biol. Biochem. 16:169–175.

Choi, S.-C., and R. Bartha. 1993. Cobalamin-mediated mercury methylation by *Desulfovibrio desulfuricans* LS. Appl. Environ. Microbiol. 59:290–295.

Cosgrove, D.J. 1977. Microbial transformations in the phosphorus cycle. Adv. Microbial Ecol. 1:95–134.

Doran, J.W. 1982. Microorganisms and the biological cycling of selenium. pp. 1–32. *In* K.C. Marshall (ed.), Advances in microbial ecology. Plenum Publishing, New York.

Ehrlich, H.L. 1990. Geomicrobiology. 2nd ed. Marcel Dekker, New York.

Frankenberger, W.T., and M.E. Losi. 1995. Applications of bioremediation in the cleanup of heavy elements and metalloids. pp. 173–210. *In* H.D. Skipper and R.F. Turco (eds.), Bioremediation: Science and applications. SSSA Special Publication 43. Soil Science Society of America, Madison, Wis.

Ghiorse, W.C. 1994. Iron and manganese oxidation and reduction. pp. 1079–1096. *In* R.W. Weaver et al. (eds.), Methods of soil analysis, Part 2. Microbiological and biochemical properties. Soil Science Society of America, Madison, Wis.

Leong, J. 1986. Siderophores: Their biochemistry and possible role in the biocontrol of plant pathogens. Ann. Rev. Phytopathol. 24:187–209.

Lindsay, W.L. 1979. Chemical equilibria in soils. John Wiley and Sons, New York.

Lovley, D.R. 1993. Dissimilatory metal reduction. Ann. Rev. Microbiol. 47:263–290.

Masscheleyn, P.H., and W.H. Patrick, Jr. 1993. Biogeochemical processes affecting selenium cycling in wetlands. Environ. Toxicol. Chem. 12:2235–2243.

Nakamura, K., M. Sakamoto, F. Uchiyama, and O. Yagi. 1990. Organomercurial-volatilizing bacteria in the mercury-polluted sediment of Minamata Bay, Japan. Appl. Environ. Microbiol. 56:304–305.

Ohlendorf, H.M. 1989. Bioaccumulation and effects of selenium in wildlife. pp. 133–177. *In* L.W. Jacobs (ed.), Selenium in agriculture and the environment. SSSA Special Publication 23. Soil Science Society of America, Madison, Wis.

Stevenson, F.J. 1967. Organic acids in soil. pp. 119–146. *In* A.D. McLaren and G.H. Peterson (eds.), Soil biochemistry. Vol. 1. Marcel Dekker, New York.

Stevenson, F.J. 1986. Cycles of soil: Carbon, nitrogen, phosphorus, sulfur, micronutrients. John Wiley and Sons, New York.

Tabatabai, M.A. 1994. Soil enzymes. pp. 775–833. *In* R.W. Weaver, S. Angle, P. Bottomley, D. Bezdicek, S. Smith, A. Tabatabai, and A. Wollum (eds.), Methods of soil analysis, Part 2. Microbiological and biochemical properties. Soil Science Society of America, Madison, Wis.

Walbridge, M.R. 1991. Phosphorus availability in acid organic soils of the lower North Carolina coastal plain. Ecology 72:2083–2100.

Zillioux, E.J., D.B. Porcella, and J.M. Benoit. 1993. Mercury cycling and effects in freshwater wetland ecosystems. Environ. Toxicol. Chem. 12:2245–2264.

Study Questions

1. Why would tillage of soil result in a more rapid decrease in microbial biomass phosphorus relative to stable organic phosphorus?

2. Describe the importance of phosphorus mineralization to plant nutrition in undisturbed soil ecosystems. Why would this process be of primary importance to plant growth?

3. If a waste material were added to soil that had a C/N/P ratio of 500/20/1, would phosphorus likely be mineralized from the waste or immobilized from the soil solution? Why?

4. Why is a soil pH of approximately 6.5 important for plant and microbial growth?

5. You have measured total organic phosphorus and biomass phosphorus for two soils, one tilled and the other a long-term, no-till cropping system. For biomass determinations, you measured inorganic phosphorus (Pi) in a nonfumigated sample (a), a fumigated sample (b), and a nonfumigated sample extracted with 0.5 M $NaHCO_3$ containing KH_2PO_4 to provide $25\mu g$ Pi g^{-1} soil(c).

The data are as follows:

Sample	Soil 1	Soil 2
	(μg Pi g^{-1} soil)	
a	5.8	4.3
b	20.9	7.9
c	25.2	23.0
Total organic P	262.4	212.6

   a. Determine the microbial biomass phosphorus for each soil.

   b. Indicate which soil has been in cultivation and which has been in continuous no-till for several years. Justify your answer.

6. Under what conditions might microbial reduction of Fe^{3+} compounds occur? If NO_3^- is present, will Fe^{3+} reduction be important? Explain.

7. How might a microorganism facilitate the solubilization of both iron and phosphorus simultaneously?

8. Both mercury and selenium can be methylated by microorganisms. Why is the methylation of mercury a potential problem while selenium methylation is not?

Part 3

♦

Applied and Environmental Topics

Chapter 17

♦

The Rhizosphere and Spermosphere

Ann C. Kennedy

♦

To own a bit of ground, to scratch with a hoe, to plant seed,
and watch the renewal of life—this is the most commonest delight.
Charles Dudley Warner

To a soil microorganism, the rhizosphere is like a lush oasis in the desert. In comparison to the near-starvation conditions of the bulk soil, the rhizosphere is the place where nutrients are plentiful, life is good, and microorganisms flourish. The rhizosphere is the zone of altered microbial diversity, increased activity and number of organisms, and complex interactions of microorganisms and the root. The significance of the rhizosphere arises from the release of organic material from the root and the subsequent effect of increased microbial activity on nutrient cycling and plant growth. The microbial community in the rhizosphere can influence plant growth in beneficial, neutral, variable, or harmful ways (Fig. 17–1).

The term **rhizosphere** was first used by Hiltner in 1904 to describe the zone of soil under the influence of roots (Hiltner, 1904; Box 17–1). The rhizosphere can extend more than 5 mm from the root and, more importantly, is the area of increased microbial activity (Figure 17–2). The area of increased microbial activity around the seed is called the **spermosphere** (Slykhuis, 1947; Figure 17–3). This term arose because the seed was commonly referred to as the sperm; hence the word "spermosphere" paralleled the areas of influence of the root, or rhizosphere. The spermosphere can extend 1 to 10 mm from the seed, but distances up to 20 mm have been reported. Seed colonization is the first step in root colonization in the soil. Microorganisms established on the germinating seed can multiply and colonize the length of the root as it emerges and grows through the soil. Thus colonization of the imbibing seed may predispose future colonization of the root.

Box 17–1

Terminology. Hiltner coined the term "rhizosphere" based on observations of the legume root in relation to symbiotic N_2-fixing bacteria. The use of this terminology has expanded to represent, in more general terms, the volume of soil surrounding and under the influence of the roots of all plants. It is important to recognize that definitions may change as our understanding changes.

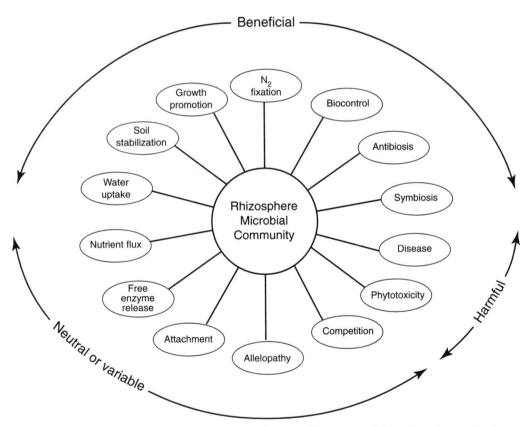

Figure 17–1 The beneficial, harmful, and neutral or variable effects of the rhizosphere microbial community on plant growth.

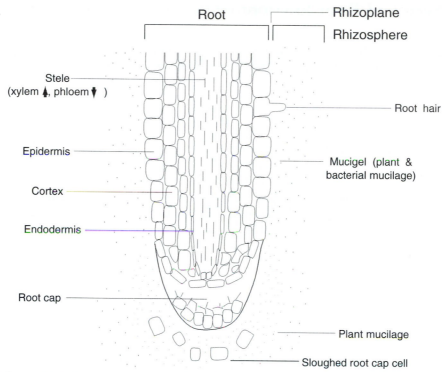

Figure 17–2 A root and corresponding rhizosphere and rhizoplane.

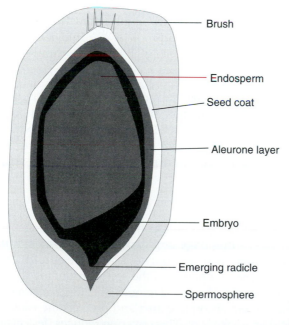

Figure 17–3 A representative seed and corresponding spermosphere.

The Seed and Root Environment

The volume of soil influenced by seed and root is neither uniform nor well understood. The rhizosphere and spermosphere consist of several different gradients, which change as the distance from the root or seed increases. The heterogeneity of this zone and the many different processes occurring within it make it even more difficult to measure and study.

Spermosphere

The area around the germinating seed is not as well studied as that area surrounding the root, yet the potential for altering microbial effects to change the plant is immense. Those events that occur in the spermosphere can change plant establishment and affect physiological maturity. Many of these effects are seen at harvest. Compounds from the seed plug are the first source of nutrients for microbial populations. Once the seed coat is damaged or broken and moisture is present, the seed imbibes water and, in so doing, releases nutrients for nearby microorganisms. This release of nutrient-rich material begins the spermosphere and rhizosphere effect. The greatest amount of exudate originates from the embryo end of the seed; therefore, the corresponding initial microbial colonization is greater at this end.

Root Growth Through Soil

The elongating tip of a root is sterile, and as the root grows into the soil, contact is made with colonies of microorganisms distributed throughout the soil. Roots normally grow into air-filled pores larger than 10 μm in diameter. Properties of the soil affect root diameter, root hairs, and the branching pattern of lateral roots. The distribution of roots is influenced by species and cultivar, soil texture, physical and chemical properties, and temporal changes. Root growth rates often are more influenced by soil type and physical conditions than by plant variety.

Root Hairs

Root hairs are extensions of the epidermal cells of the plant root. The functions of root hairs are still debated among researchers, but they are important in ion uptake from soil, especially with immobile ions like $H_2PO_4^-$ and Zn^{2+}. More root hairs are found in coarsely aggregated soils (aggregate size of 3 to 4 mm) than in finely aggregated soils (less than 1 mm). Root hairs may penetrate adjacent soil aggregates and take up nutrients there.

Root Cap

The primary function of the root cap is to protect the meristem as the root moves through the soil. The cap produces **mucilage** and may release substances that enhance plant cell growth. This nutrient-rich area provides a lush region for microbial growth. Root cap cells may elongate and persist in the soil surrounding the root for many centimeters behind the root tip. The mucilaginous sheath (sometimes called mucigel layer) functions as a nutrient absorptive area and protects the root tip from chemicals in the soil that could injure the root. This covering reduces desiccation and abrasion damage to the root tip or apical region of the root.

Box 17–2

Overview of Scale. Roots are much larger than bacteria and fungi. For example, a soil bacterium may be 0.5 μm wide and 5 μm long; a fungal strand may be up to 20 μm wide and more than 1,000 μm long (it is difficult to estimate the length of individual fungal hyphae in soil since most sampling techniques break these filamentous structures). In contrast, plant roots are about 1,000 μm wide and extend 1,000,000 μm down the soil profile. An average root hair is 10 μm wide and 500 μm long. Compared to an average soil bacterium and fungus, root hairs and roots may be 1 to 1,000 times wider and 1 to 100,000 times longer.

Rhizosphere Boundaries

The boundaries of the various sections of the rhizosphere and spermosphere are difficult to demarcate and are the source of much discussion (Box 17–3). A simple solution is to divide this area into the rhizosphere, rhizoplane, and root. As previously defined, the rhizosphere is the zone of soil influenced by roots and their exudation of substances that affect microbial activity. The **rhizoplane** is the surface of a plant root and any strongly adhering soil particles. The area within the root has been defined in many ways, but it is best considered as the root itself rather than a portion of the rhizosphere. The microbial populations that colonize the interior of the root and form intimate associations with the root are considered **endophytes.**

Box 17–3

Other Definitions of the Rhizosphere. The division or compartmentalization of the rhizosphere has led to much controversy. The rhizosphere had been further divided into the ectorhizosphere and endorhizosphere. The *ectorhizosphere* was considered to be the zone outside the root. The *endorhizosphere* included the root epidermis and cortex delimited by the Casparian strip, a thickening in endodermal cells that made them impermeable to water and ions. This region did not include those organisms found within the stele, which also may be important in plant-microbe interactions.

The term *endorhizosphere* was useful to identify and illustrate the importance of those microorganisms that exist in the internal portions of the root; however, if the rhizosphere is soil-based, then the term endorhizosphere may be inappropriate because it identifies the region within the root and not the rhizosphere. The three terms—rhizosphere, rhizoplane, and root—reduce questions or problems that may arise. To avoid confusion, these terms always should be defined when they are used.

The rhizosphere, rhizoplane, and root are easier to conceptualize than to study in practice (Box 17–4). For example, if the rhizosphere can extend from 1 to 4 mm or more from the root surface and the rhizoplane with adhering soil can be up to 2 mm in width, then the rhizoplane and the rhizosphere are difficult to differentiate. Furthermore, inhabitants in the rhizosphere may be difficult to distinguish from those of the rhizoplane because of mucilages secreted by the root. The soil-root interface is the area of greatest number and variety of plant-microbial associations.

Box 17–4

Practical Rhizosphere Analysis. How does one determine what organisms are inhabiting the various areas of the rhizosphere when these regions are indistinct?
Rhizosphere analyses are often limited by the procedures to identify the microorganisms of these zones. Rhizosphere microorganisms are considered to be those microorganisms removed when the roots and adhering soil are shaken gently in water.
Rhizoplane microorganisms are considered to be those left when the washed roots
are transferred to fresh diluent and shaken vigorously. Finally, root organisms are
those microorganisms that are recovered from the root after it is surface-sterilized
and macerated.

The Rhizosphere Environment

The physical and chemical properties of soil are discussed in Chapter 2. Here we focus on how changes in some of these soil properties in the vicinity of a plant root
affect microbial activity. The nutritional aspect of the rhizosphere will be covered in
the next section.

Rhizosphere moisture will affect microbial colonization directly or indirectly as it
affects plant growth. Cells at the surface of the root may experience severe
changes in water potential (Fig. 17–4). On one hand, soils above or near field capacity (approximately -0.03 MPa or less negative potential) have little change in
soil water potential even though plant transpiration can move a large amount of
the soil water. On the other hand, in soils of -0.2 MPa or more negative potential,
the soil water potential near the root can stress microbial cells in the rhizosphere.
At low soil water potential, motility and diffusion of nutrients can be reduced, dramatically influencing microbial growth. This does not mean that high soil water potentials are necessarily beneficial, because a large percentage of pore space is water-filled and oxygen may be limiting.

Soil texture can affect transport and root colonization by microorganisms. In
general, bacteria move greater distances down the root in coarse-textured soils
than in fine-textured soils. The extent of the rhizosphere effect also may be greatest in sandy soils and least in heavy clay and organic soils.

Temperature can play a role in colonization rates of the rhizoplane. Often microorganisms colonize best at temperatures below their optimum growth temperature. This difference may be because of lower competition from the indigenous microflora at the lower temperatures or a reduction in root growth rate, thus allowing
greater root colonization or infection.

Soil pH can alter the survival and competitive abilities of microorganisms. The
root-induced production of H^+, HCO_3^-, or organic compounds and their subsequent release into the rhizosphere affects ion uptake and thus pH (Box 17–5).
This in turn alters the environment of the rhizosphere. Fluctuations in rhizosphere
pH can be as great as 1 pH unit and can either be higher or lower than the bulk
soil. The magnitude and direction of the changes in rhizosphere pH depend
mainly on nitrogen nutrition, but symbiotic relationships and microbial activity
also influence pH changes.

Box 17–5

Effect of Nitrogen Source on Rhizosphere pH. The source of nitrogen influences ion uptake by the root and thus affects rhizosphere pH. When nitrate is supplied to the plant, the rhizosphere pH increases because a higher rate of HCO_3^- release occurs than H^+ excretion. When ammonium is supplied, the reverse is true. Ammonium-based fertilizers such as urea, anhydrous ammonia, and aqua ammonia can reduce rhizosphere pH, vastly altering the rhizosphere populations and nutrient release from the root and soil. In acid soils this rhizosphere acidification may reduce bacterial numbers.

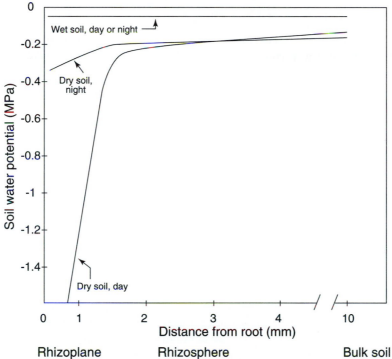

Figure 17–4 Effect of distance from root on soil water potential for day versus night, and wet versus dry soil conditions. *From Papendick and Campbell (1975). Used with permission.*

Plant-Derived Compounds

Plant roots excrete various organic substances into the rhizosphere, and these substances provide a rich source of nutrients for the microbial community. Because most soils are low in readily available carbon for microbial growth, carbon-containing substances leaked by seeds and roots are of major importance to microbial growth. This is not too surprising because a typical microbial cell is approximately 50% carbon. Too often, the term *exudate* is used to define all compounds that come from the root. However, the organic fractions from the root are of various forms and composition and are better defined as exudates, secretions, mucilages, mucigel, and lysates (Table 17–1).

Table 17–1 **Definitions of compounds originating from plant roots.**

Exudates	Compounds of low molecular weight that leak from all cells either into the intercellular spaces and then to the soil via the cell junctions or directly through the epidermal cell walls into the soil. The release of these compounds is not metabolically mediated.
Secretions	Mucilages of low and high molecular weight released from roots as a result of metabolic processes.
Mucilages	Four sources of plant mucilages contribute to the organic materials in the rhizosphere: • mucilage originating in the root cap and secreted by the Golgi apparatus • hydrolysates of the polysaccharide of the primary cell wall between the epidermal cells and sloughed root cap cells • mucilage secreted by the epidermal cells which still only have primary walls • mucilage produced by bacterial degradation of the outer multilamellate primary cell walls of old, dead epidermal cells
Mucigel	Gelatinous material at the surface of roots grown in normal nonsterile soils. It includes natural and modified plant mucilages, bacterial cells and their metabolic products, colloidal mineral and organic matter from the soil. The mucigel is important in maintaining contact between the root and the soil as the root shrinks during daytime water stress, permitting continuous uptake of nutrients and water.
Lysates	Compounds released by lysis of epidermal cells.

From Rovira et al. (1979). Used with permission.

Exudate Composition

Exudate composition varies with the soil environment and the growth stage of the seed and root. It is difficult to determine which compounds are true exudates and which compounds are released from lysed root cells. Compounds in exudates include sugars, amino acids, vitamins, tannins, alkaloids, phosphatides, and other unidentified substances (growth factors, fluorescent substances, nematode cyst or egg-hatching factors, and fungal growth stimulants and inhibitors).

Sugars provide readily available sources of carbon for microbial growth. Amino acids, whose percentages vary with plant species, are a readily available source of nitrogen for microbial growth. These compounds are the most-studied root exudates and are thought to be associated with susceptibility or resistance of plants to root-infecting fungi. Organic acids can chelate metals, affecting pH and the absorption and translocation of nutrient elements. The range in quantity and variety of vitamins in exudates can account to some extent for differences in bacterial populations. Although most research has focused on water-soluble and nonvolatile exudates, water insoluble and volatile exudates may affect the growth of soil microorganisms. Little is known about these later exudates, although soluble and insoluble compounds are thought to be present in equivalent proportions in the exudate.

Exudation Rate

The actual rate of exudation from roots is difficult to calculate. Various gradients of release probably exist, with the greatest amount of exudate arising from

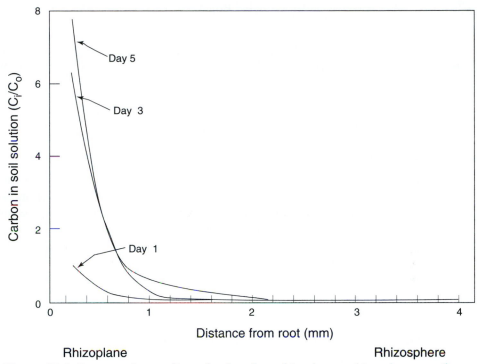

Figure 17–5 Concentration gradient of carbon from rhizoplane to rhizosphere to soil. Soluble ^{14}C–based carbon concentration in the soil solution (C_i) relative to the source of carbon (C_o). *From Yeates and Darrah (1991). Used with permission.*

those cells involved in cell elongation and lateral root formation. The concentration of carbon in the rhizosphere declines exponentially as the distance from the root increases (Fig. 17–5). Increased exudation can result from extremes in temperature, water stress, phosphorus deficiency, increased light intensity, and increased populations of microorganisms. Also influencing exudation rate can be herbicides, pathogens, foliar treatments, and symbiotic associations. Decreased exudation can result from nitrogen deficiency, decreased light intensity, and decreased microbial populations.

Plant exudates are the substrates for microbial growth. As microbial activity increases, the growth of the root and plant is affected. Greater populations of microorganisms in the rhizosphere competing for nutrients may create a nutrient deficiency for the plant. In contrast, microorganisms may make nutrients from insoluble sources, such as phosphates and trace metals, available to plants. In addition, symbiotic associations of roots with rhizobia or mycorrhizal fungi increase root exudates. Mycorrhizal roots have enhanced nutrient-absorbing capacity, which subsequently increases plant growth. Plant pathogenic microorganisms (both foliage-infecting and root-infecting) can alter both the quality and amount of exudate and result in greater populations of microorganisms in the rhizosphere.

| Box 17–6 |

Effect of Seed Scarification. Exudation rates of carbon from nonblemished, moistened seeds of soybean with intact seed coats is estimated at 5.3 µg hour $^{-1}$ seed $^{-1}$ over a 24-hour period. Scarification, which breaks or injures the seed coat, increased exudation to 217 µg C hour^{-1} seed $^{-1}$! Damage to the seed coat can vastly change the exudation pattern and thus the microbial community around that seed.

Mucilage and Mucigel

The mucilage exuded by roots forms a sheath of slime on the external surface of the root (Table 17–1). The principal site of mucilage release is the root tip or apical zone of a root, particularly the meristematic zone immediately behind the root cap. By definition, mucigel contains plant mucilages and plant products as well as bacterial cells and their products. The mucigel creates an intimate contact between roots or root hairs and soil particles. Contained in this sheath are organic materials excreted from living root cells or released from cortical cells or senescent root hairs sloughed off from the root. This mucigel is more likely found on the root and root hairs than at the root tip (see also figures 13–3 and 13–4).

Microbial Populations of the Rhizosphere

A vast number of species of microorganisms are present in the rhizosphere, and their numbers generally decrease as the distance from the root increases (Figure 17–6). To measure the effect of the rhizosphere on a particular population, the number of microorganisms in the rhizosphere (R) and the number of microorganisms in the bulk soil (S, soil not influenced by the root) are compared. This R/S ratio provides an estimate of how strongly the rhizosphere affects a particular organism (Table 17–2). This relationship can also differ with plant species (Table 17–3). The R/S ratio is especially helpful in determining the **rhizosphere competence,** the ability of an organism to colonize the rhizosphere, of various organism-plant combinations. A microorganism with good rhizosphere competence is a good candidate as a microbial inoculant.

Bacteria

Bacteria, including actinomycetes, are the most numerous inhabitants of the rhizosphere—typically numbering 10^6 to 10^9 organisms g^{-1} rhizosphere soil—although they account for only a small portion of the total biomass because of their small size. Many bacteria have a large R/S ratio, indicating marked stimulation in the rhizosphere. In general, nonsporulating rods are found in great abundance in the rhizosphere. Pseudomonads and other Gram-negative bacteria are especially competitive in the rhizosphere and occupy a large portion of the total bacterial population on the root. Actinomycetes account for approximately 10 to 30% of total microflora in the rhizosphere, depending on the season or timing of nutrient additions. Actinomycetes have smaller R/S ratios than other bacteria.

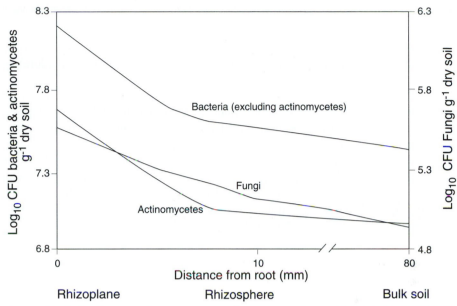

Figure 17–6 Distribution of naturally occurring groups of organisms on the rhizoplane and in the rhizosphere of spring wheat and bulk soil (CFU = colony forming units). *Modified from Papavizas and Davey (1975). Used with permission.*

Table 17–2 **Numbers of microorganisms in the rhizosphere (R) of wheat (*Triticum aestivum* L.) and nonrhizosphere soil (S) and their resultant R/S ratio.**

Microorganisms	Rhizosphere	Nonrhizosphere	R/S ratio
	CFU* g^{-1} soil		
Bacteria	120×10^7	5×10^7	24.0
Fungi	12×10^5	1×10^5	12.0
Protozoa	24×10^2	10×10^2	2.4
Ammonifiers	500×10^6	4×10^6	125.0
Denitrifiers	$1,260 \times 10^5$	1×10^5	1,260.0

From Rouatt et al. (1960). Used with permission.
*CFU = colony forming units.

Table 17–3 **Colony counts of bacteria in the rhizoplane and rhizosphere of different crop plants and in nonrhizosphere soil.**

Plant	Rhizoplane	Rhizosphere	Nonrhizosphere	R/S ratio
	CFU* $\times 10^6 \ g^{-1}$ root dry mass			
Red clover (*Trifolium pratense*)	3,844	3,255	134	24
Oats (*Avena sativa*)	3,588	1,090	184	6
Flax (*Linum usitatissum*)	2,450	1,015	184	5
Wheat (*Triticum aestivum*)	4,119	710	120	6
Maize (*Zea mays*)	4,500	614	184	3
Barley (*Hordeum vulgare*)	3,216	505	140	3

From Rouatt and Katznelson (1961). Used with permission.
* CFU = colony forming units.

Fungi

Although plate counts of rhizosphere fungi generally are less than bacteria and actinomycetes, these counts are based on spores and mycelial propagules and may be misrepresentative. Direct microscopy indicates considerable fungal growth along root surfaces; there may be as much as 12-14 mm of hyphae mm^{-2} of root surface. Population numbers may best be reported as ranges, most frequently 10 to 20% of the total microflora or 10^5 to 10^6 g^{-1} rhizosphere soil. The rhizosphere effect, as indicated by R/S ratios ranging from 10 to 20, can be substantial for fungi. The rhizosphere and spermosphere favor initial colonization of fungal species, such as Zygomycetes and asexual forms of many Ascomycetes, able to grow on simple sugars. Several rhizosphere-inhabiting fungi are pathogenic, while others form symbiotic (mycorrhizal) associations with roots. Mycorrhizal fungi may be a major component of the microbial biomass in the rhizosphere but, because many of these fungi do not grow on artificial media in the laboratory, they are often overlooked.

Movement of Microflora in the Rhizosphere

Dispersal

Microorganisms can move through the soil along the root extensions and colonize the rhizosphere to its full extent. Microorganisms are thought to move down the root by three different mechanisms:

- motility of the organism (active transport),

- water movement on the surface of the root (passive transport), and

- movement of microorganisms on the root apex as the root cells elongate (passive transport).

Although bacteria inoculated on the seed will colonize roots within 3 cm of the seed in the absence of percolating water, their further dispersal along the roots is increased by water movement through the soil.

Root Colonization

A number of factors affect root colonization by soil microorganisms. These include characteristics of the microorganisms themselves, the plant, and the environment (Table 17–4). Although most scientists report colonization on a per root or root dry mass basis, electron micrographs show that only 7 to 15% of the root surface is actually occupied by microorganisms. The rhizoplane bacteria are scattered across the root in small colonies associated with only a small portion of the root surface (Fig. 17–7). These colonies are usually associated with holes, tears, crevices, and junctions of epidermal cells in roots. Recognition signals may play an important role in root colonization. For example, bacteria that agglutinate root exudates can better follow the downward growth of the root in the soil and colonize the rhizosphere than those with no agglutination activity. The entire rhizosphere may be colonized by an introduced organism after

Table 17–4 Factors influencing rhizosphere colonization.

Microbial characteristics
    Nutrient requirements
    Early growth rate
    Cellulase production
    Antibiotic tolerance or production
    Siderophore production
    Unique physiological attributes
    Tolerance to fungicides or other chemicals
Plant characteristics
    Plant species
    Plant age
    Altered plant genetics
    Foliar treatments
Environment
    Soil type
    Soil texture
    Soil moisture
    Soil atmosphere
    Soil temperature
    Soil fertility
    Soil-applied pesticides

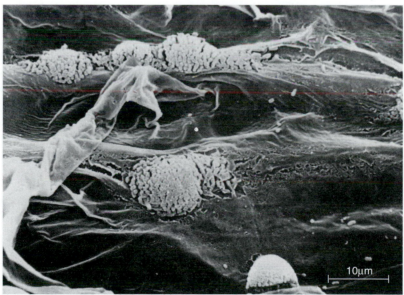

Figure 17–7 Colonization of the root by bacteria. Note the clumping of colonies at intersection of cells. *Photo courtesy of Begonia et al. (1990). Used with permission.*

seeds are treated with that organism. In such a case, transport (active or passive) obviously is not a problem. There are, however, many cases where neither bacteria nor fungi applied as an inoculant to the seed are transported along the root surface. Thus, transport of microorganisms in the rhizosphere can be highly variable and depends on the microorganism, the plant, and the environment.

Faunal Populations

As with the microfloral populations, microfaunal and macrofaunal populations are affected by the rhizosphere. Protozoa (i.e., amoebae, flagellates, and ciliates), nematodes, mites (Acarina), and springtails (Collembola) exhibit the greatest rhizosphere effect. The highest populations of these species are found at food sites and are not uniformly distributed throughout the soil. Protozoa feed selectively and therefore can influence the composition of the bacterial population in the rhizosphere.

Root exudates influence abundance and behavior of nematodes directly or indirectly in a variety of crops, and the populations are much greater in the rhizosphere than in root-free soils. Egg-hatching of nematodes can be induced by root exudates. Factors responsible for stimulating hatching are produced by some plants, but are not limited to host plants. Nematode larvae are usually found in the highest concentration at the region of greatest cell elongation behind the root tip. This enhanced population is not only a response to an increased food supply, but also in response to substances released that attract nematodes. Conversely, nematodes may also be repelled by some plant exudates.

Because many microarthropods are mycophagous (fungal-eating), they are more abundant near root surfaces and on organic matter where fungi are abundant. Mites and springtails are predominantly found in the rhizosphere and feed on woody tissue, leaf tissue, fungi, bacteria, or algae. They follow plant roots to the available food supply. Like protozoa, mites and springtails can be plant- and bacteria-specific. The high populations of bacteria and fungi in the rhizosphere encourage the buildup of many of these faunal species that graze on microorganisms. The faunal component is often overlooked in rhizosphere studies and yet it is a dynamic, integral factor in rhizosphere functioning.

Microbe–Plant Interactions in the Rhizosphere

The growth of a root through the soil increases the population of many microorganisms. This phenomenon is exciting; however, of even greater interest and possible use are the microbial interactions that benefit plant growth (Table 17–5). The most intensely studied of these interactions are N_2 fixation, mycorrhizal associations, plant growth promotion, decomposition, and nutrient cycling.

Dinitrogen-fixing associations can add otherwise unavailable nitrogen to the plant. Rhizobia and related bacteria form nodules on the roots of plants, take dinitrogen from the air, and transform it to plant-available nitrogen. The plant provides nutrients and a competition-free home for the bacteria, while the bacteria fix valuable dinitrogen for the plant.

Mycorrhizal fungi form symbiotic associations with plant roots. They are involved in the nutrient cycling process, especially in stressed environments (e.g.,

***Table 17–5* Beneficial activities of microorganisms in the rhizosphere.**

Decomposing plant residue and organic material
  Humus synthesis
  Mineralization of organic nitrogen, sulfur, and phosphorus
Increasing plant nutrient availability of phosphorus, manganese, iron, zinc, and copper
  Symbiotic mycorrhizal associations
  Production of organic chelating agents
  Oxidation-reduction reactions
  Phosphorus solubilization
Helping biological N_2 fixation
  Free-living bacteria and cyanobacteria
  Associative microorganisms
  Symbiotic legume and nonlegume
Promoting plant growth
  Production of plant growth hormones
  Protection against root pathogens and pseudopathogens
  Enhanced nutrient use efficiency
Controlling deleterious microorganisms and plants
  Plant diseases
  Soil nematodes and insects
  Weeds
Biodegrading synthetic pesticides or contaminants
Enhancing drought tolerance of plants
Improving soil aggregation

phosphorus- and water-deficient soils). Mycorrhizal associations enhance nutrient solubilization and nutrient uptake in the rhizosphere and expand the volume of soil the root can explore. Benefits include increased rate of nutrient absorption, selective ion uptake, and protection from extremes in the environment. This relationship is beneficial under moisture-limiting conditions and may also impart some protection to plant roots from pathogens. Mycorrhizal associations are enhanced by crop rotation and management practices favoring minimum disturbance.

Plant growth-promoting rhizobacteria (PGPR) are specific strains of bacteria in the rhizosphere that enhance seed germination and plant growth. Many different mechanisms are responsible for plant growth promotion. Soil scientists initially believed the plant growth that resulted from inoculation with *Azotobacter* and *Azospirillum* was due to N_2 fixation; however, other factors may impart growth stimulation. These and other bacteria (e.g., *Bacillus*) produce plant growth-stimulating hormones such as gibberellic and indoleacetic acid. Plant growth promotion may also be indirect by reducing pathogen colonization of the seed and root. An example of this effect is the microbial production of the iron-chelating siderophores. These compounds can make iron more available to a PGPR and less available to a plant pathogen. Because of the roles iron play in cellular metabolism, this may affect microbial growth.

Microorganisms play a large part in the decomposition and subsequent mineralization of organic matter in the soil. Their activity yields various compounds, including nitrates, phosphates, sulfates, carbon dioxide, and water. These processes, as measured by microbial respiration, may be four times greater in the rhizosphere than in the nonrhizosphere soil, though it is difficult to separate root and microbial respiration.

The principal interface for many plant-microbe interactions is the rhizosphere. The dynamics of these interactions will determine the health of the plant. Microorganisms inhabiting the rhizosphere, such as pathogens or microorganisms that produce phytotoxins, may be detrimental to plant health. The challenge is to enhance beneficial relationships and minimize harmful interactions.

Inoculants

Recognizing the impact microorganisms have on plant growth has led to the development of inoculants for enhancing plant growth. Exposing specific microorganisms to a root increases the probability that they will flourish. The introduction of a microorganism with the seed and exposure to exuding nutrients enhances microbial growth and establishment of the organism on the root. Dinitrogen-fixing bacteria, PGPR, mycorrhizal fungi, and biological control agents that reduce unwanted insects, pathogens, and weeds all have potential as inoculants (Table 17–6). However, inoculation will only be successful if the microorganisms flourish when they are intro-

Table 17–6 **Examples of microorganisms that have shown potential as inocula for soil or rhizosphere amendment.**

Microorganism	Effect
Agrobacterium tumefaciens	Controlled crown gall in grapes; some avirulent strains controlled crown gall in other plant species
Alcaligenes spp.	Reduced *Fusarium* wilt of carnation
Bacillus subtilis	Increased germination and growth of cabbage
Azospirillum brasilense	Enhanced uptake of NO_3^-, K^+, and $H_2PO_4^-$. Improved growth and dry mass of corn and sorghum
Pseudomonas fluorescens	Increased growth of carnation, sunflower, *Vinca,* and zinnia from seeds or cuttings Reduced damping-off of cotton seedlings Increased yield of potato and sugar beet Increased yield of radish Changed serogroup distribution of *Bradyrhizobium* strains on soybean roots Suppressed "take-all" of wheat Reduced stand, growth, and seed production of the weed downy brome
Pseudomonas putida	Increased yield of sugar beet
Pseudomonas spp.	Reduced incidence of *Fusarium* wilt of flax, cucumber, and radish
Pseudomonas syringae pv. *tabaci*	Enhanced alfalfa growth, plant nitrogen nodulation, and N_2 fixation
Trichoderma harzianum	Faster seed germination and increased seed emergence Earlier flowering and increased number of flowers Increased growth, height, and dry mass of plants Faster rooting of cuttings

duced into the soil or on the seed. We can improve inoculation success by better understanding the processes of microbial competition and survival. The challenge for scientists is to determine which factors increase rhizosphere competence and colonization ability of a microorganism in the rhizosphere and spermosphere.

Modeling Rhizosphere Function

Modeling the processes within the rhizosphere and spermosphere should allow us to more fully understand the biology of these regions. In order to form a *conceptual model* of the rhizosphere, the separate components operating within the system must be brought together to form a network representing the interactions that occur. A *mathematical model* can then be developed to quantify the various interactions represented by the conceptual model. Although these models are often simplistic—perhaps a cylindrical porous membrane to represent the root, root segments, and various types of roots—the models do serve to indicate the outcome of events within the more complex rhizosphere or spermosphere regions. Any mathematical model is based on fundamental assumptions, and the validity of those assumptions in individual situations will determine the overall accuracy of the modeling results.

Models have been developed to:

- predict the effect of environmental changes on microbial populations, substrate concentrations, soil water, root density, and root exudation rate,

- determine the effect of root development and environment on the release and deposition of carbon, and possibly root turnover, in the rhizosphere,

- predict the microbial biomass in the rhizosphere as influenced by distance from the root and the amounts of soluble root exudate present, and

- predict root water potential or carbon dynamics within the rhizosphere or spermosphere to examine spatial distribution of nutrients, microbes, and microbial biomass in microsites.

Inoculants, plant infection and disease, and symbiotic colonization also lend themselves well to mathematical modeling. Investigations on the complexity and overlapping boundaries of the various regions of the rhizosphere may be assisted by modeling, which may lead to further understanding in delineating these controversial regions. Studies of rhizosphere and spermosphere ecology can use these models as tools for testing subsequent hypotheses about the zone of soil around the root and seed.

Future Research Needs

Our understanding of the rhizosphere and the myriad of interactions within this zone has increased greatly since the term was first introduced in 1904. Yet the rhizosphere and spermosphere still hold much unknown territory. The challenge is to understand these regions of soil-root and soil-seed interfaces to manage microorganisms, increase plant growth, and reduce the impact of crop cultivation on the environment.

To successfully manage the rhizosphere, scientists need to work toward a number of goals, including:

- identifying more completely the factors in microorganisms and plants that control root colonization,

- understanding microbial ecology of the rhizosphere in greater detail,

- expanding the view of the rhizosphere from a single root to overlapping rhizospheres, and

- developing better methods to measure microbial populations and their competitive ability in the rhizosphere.

The challenge is great because the rhizosphere is complex. Despite this complexity, manipulations of microorganisms in the rhizosphere have already demonstrated some success with increased crop yields, suppression of pests, and improved soil quality. With further study, many more benefits are likely.

Summary

The rhizosphere and spermosphere are areas where organic materials are released from the root or seed. This results in altered microbial diversity and increased numbers of organisms, microbial activity, and interactions among microorganisms and the seed, roots, and adjacent soil. The microbial community in the rhizosphere can influence plant growth in ways that are beneficial, neutral, or detrimental. These interactions influence the health, vigor, and productivity of the plant.

Cited References

Begonia, M.F.T, R.J. Kremer, L. Stanley, and A. Jamshedi. 1990. Association of bacteria with velvetleaf roots. Trans. Missouri Acad. Sci. 24:17–26.

Hiltner, L. 1904. Über neuere Erfahrungen und Probleme auf dem Gebiet der Bodenbakteriologie und unter besonderer Berücksichtigung der Gründüngung und Brache. Arb. Dtsch. Landwirtsch. Ges. Berlin 98:59–78.

Papavizas, G.C., and C.B. Davey. 1961. Extent and nature of the rhizosphere of *Lupinus*. Plant Soil 14:215–236.

Papendick, R.I., and G.S. Campbell. 1975. Water potential in the rhizosphere and plant and methods of measurement and experimental control. pp. 39–49. *In* G.W. Bruehl (ed.), Biology and control of soil-borne pathogens. The American Phytopathological Society, St. Paul, Minn.

Rouatt, J.W., and H. Katznelson. 1961. A study of the bacteria on the root surface and in the rhizosphere soil of crop plants. J. Appl. Bacteriol. 24:164–171.

Rouatt, J.W., H. Katznelson, and T.M.B. Payne. 1960. Statistical evaluation of the rhizosphere effect. Soil Sci. Soc. Amer. Proc. 24:271–273.

Rovira, A.D., R.C. Foster, and J.K. Martin. 1979. Note on terminology: Origin, nature and nomenclature of the organic materials in the rhizosphere. pp. 1–4. *In* J.L. Harley and R.S. Russell (eds.), The soil-root interface. Academic Press, London.

Slykhuis, J.T. 1947. Studies on *Fusarium culmorum* blight of crested wheat and brome grass seedlings. Can. J. Res. Sect. C. 25:155–180.

Yeates, G., and P.R. Darrah. 1991. Microbial changes in a model rhizosphere. Soil Biol. Biochem. 23:963–971.

General References

Curl, E.A., and B. Truelove. 1986. The rhizosphere. Springer-Verlag, Berlin.

Keister, D.L., and P.B. Cregan. 1991. The rhizosphere and plant growth. Kluwer Academic Publishers, Dordrecht, The Netherlands.

Lynch, J.M. 1990. The rhizosphere. John Wiley and Sons, Chichester, U.K.

Lynch, J.M. 1983. Soil biotechnology: Microbiological factors in crop productivity. Blackwell Scientific Publications, Boston.

Study Questions

1. Why is the rhizosphere a zone of increased microbial activity?

2. Why is the spermosphere an important area of study?

3. What role do exudates play in the rhizosphere?

4. Distinguish among rhizosphere, rhizoplane, and root.

5. How does microbial activity compare among the rhizosphere, rhizoplane, and soil?

6. By what mechanisms can bacterial populations be beneficial, detrimental, or neutral to plant growth?

7. What factors affect root colonization?

Chapter 18

♦

Mycorrhizal Symbioses

David M. Sylvia

♦

To survive and flourish in the long run, we usually need to develop mutualistic relationships. Nonmutualistic members will eventually disappear or lose influence in the community.

J.M. Trappe

Mycorrhiza (plural, -zae or -zas) refers to an association, or symbiosis, between plants and fungi that colonize the cortical tissue of roots during periods of active plant growth. The association is characterized by the movement of:

* plant-produced carbon to the fungus and

* fungal-acquired nutrients to the plant.

The term **mycorrhiza,** which literally means *fungus-root,* was first applied to fungus-tree associations described in 1885 by the German forest pathologist A.B. Frank. Since then we have learned that the vast majority of land plants form symbiotic associations with fungi: an estimated 95% of all plant species belong to genera that characteristically form mycorrhizae. The mycorrhizal condition is the rule among plants, not the exception.

The benefits afforded plants from mycorrhizal symbioses can be characterized either agronomically by increased growth and yield or ecologically by improved fitness (i.e., reproductive ability). In either case, the benefit accrues primarily because mycorrhizal fungi form a critical linkage between plant roots and the soil. Mycorrhizal fungi usually proliferate both in the root and in the soil. The soilborne or *extramatrical hyphae* take up nutrients from the soil solution and transport them to the root. By this mechanism, mycorrhizae increase the effective absorptive surface area of the plant. In nutrient-poor or moisture-deficient soils, nutrients taken up by the extramatrical hyphae can lead to improved plant growth and reproduction. As a result, mycorrhizal plants are often more competitive and better able to tolerate environmental stresses than are nonmycorrhizal plants.

Global Perspective

Mycorrhizal associations vary widely in structure and function. Despite the many exceptions, it is possible to state broad generalizations about latitude (or altitude), soil properties, and structure and function of the different mycorrhizal types that colonize the dominant vegetation in a gradient of climatic zones (Fig. 18–1). The major types of mycorrhizae are summarized in Table 18–1. Ericaceous plants (any of a large genus of the heath family of low, evergreen shrubs), which dominate the acidic, high-organic heathland soils of subarctic and subalpine regions, are colonized by a group of ascomycetous fungi, giving rise to the **ericoid** type of mycorrhiza. This mycorrhizal type is characterized by extensive growth within (i.e., intracellular) cortical cells, but little extension into the soil. The fungi produce extracellular enzymes that break down organic matter, enabling the plant to assimilate nutrients mineralized from organic compounds present in the colloidal material surrounding roots. Moving along the environmental gradient, coniferous trees replace ericaceous shrubs as the dominant vegetation. These trees are colonized by a wide range of mostly basidiomycetous fungi that grow between (i.e., intercellular) root cortical cells, forming the **ectomycorrhizal** type of mycorrhiza. Ectomycorrhizal fungi may produce large quantities of hyphae on the root and in soil. These hyphae function in the absorption and translocation of inorganic nutrients and water, but also release nutrients from litter layers by-production of enzymes involved in mineralization of organic matter.

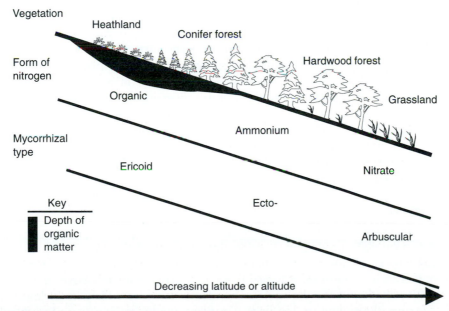

Figure 18–1 Changes in dominant vegetation, nitrogen form, and mycorrhizal types with decreasing latitude or altitude. *Adapted from Read (1984). Used with permission.*

Table 18–1 **Plant host, fungi and important characteristics of major types of mycorrhiza.**

Mycorrhizal type	Hosts involved	Fungi involved	Characteristic structures	Characteristic functions
Ectomycorrhizae	Mostly gymnosperms Some angiosperms Restricted to woody plants	Mostly basidiomycetes Some ascomycetes Few zygomycetes	Hartig net Mantle Rhizomorphs	Nutrient uptake Mineralization of organic matter Soil aggregation
Arbuscular	Bryophytes Pteridophytes Some gymnosperms Many angiosperms	Zygomycetes (Glomales)	Arbuscules Vesicles Auxiliary cells	Nutrient uptake Soil aggregation
Ericaceous	Ericales Monotropaceae	Ascomycetes Basidiomycetes	Some with hyphae in cell, some with mantle and net	Mineralization of organic matter Transfer between plants
Orchidaceous	Orchidaceae	Basidiomycetes	Hyphal coils	Supply carbon and vitamins to embryo
Ectendomycorrhizae	Mostly gymnosperms	Ascomycetes	Hartig net with some cell penetration Thin mantle	Nutrient uptake Mineralization of organic matter

At the warmer and drier end of the environmental gradient, grasslands often form the dominant vegetation. In these ecosystems nutrient use is high and phosphorus is frequently a limiting element for growth. Grasses and a wide variety of other plants are colonized by fungi belonging to the order Glomales. These fungi form arbuscules, or highly branched structures (the term literally means *little trees*) within (intracellular) root cortical cells, giving rise to the **arbuscular** type of mycorrhiza. The Glomalean fungi may produce extensive extramatrical hyphae (i.e., hyphae outside of the root) and can significantly increase phosphorus inflow rates of the plants they colonize.

The diversity of these root-fungal associations provides plants with a range of strategies for efficient functioning in an array of plant-soil systems. The objective of this chapter is to provide an overview of this diversity and to evaluate the roles and potential for management of the mycorrhizal symbioses in native and managed ecosystems. Because ectomycorrhizae and arbuscular mycorrhizae are the most widespread, we will emphasize these types of associations.

Types of Mycorrhizae

Ectomycorrhizae

The diagnostic feature of ectomycorrhizae (EM) is the presence of hyphae between root cortical cells, producing a netlike structure called the *Hartig net,* after Robert Hartig who is considered the father of forest biology (Fig. 18–2 a, b). Many EM also have a sheath, or *mantle,* of fungal tissue that may completely cover the absorbing root (usually the fine feeder roots). The mantle can vary widely in thickness, color, and texture depending on the particular plant-fungus combination. The mantle increases the surface area of absorbing roots, and often affects fine-root morphology,

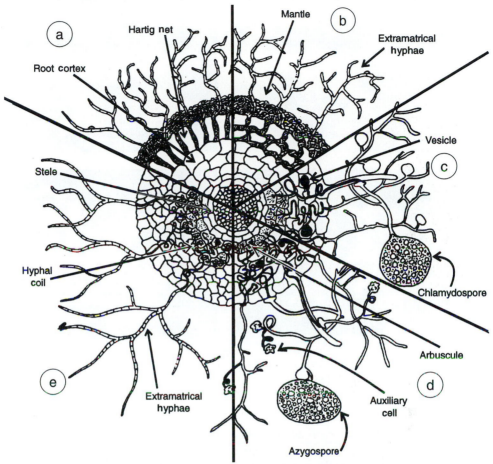

Figure 18–2 Diagrammatic cross-section of a fine root illustrating key features of several types of mycorrhizae. Ectomycorrhizae typically form on (a) Fagaceae and (b) Pinaceae hosts. Arbuscular mycorrhizae typically form by fungi classified in the (c) Glomineae and (d) Gigasporineae. (e) Orchidaceous type of mycorrhiza. *Modified from original drawing by V. Furlan. Used with permission.*

resulting in root bifurcation and clustering (Fig. 18–3). Contiguous with the mantle are hyphal strands that extend into the soil. Often the hyphal strands will aggregate to form **rhizomorphs** that may be visible to the unaided eye. The internal portion of rhizomorphs can differentiate into tubelike structures specialized for long-distance transport of nutrients and water.

Ectomycorrhizae are found on woody plants ranging from shrubs to forest trees. Many of the host plants belong to the families Pinaceae, Fagaceae, Betulaceae, and Myrtaceae. Over 4,000 fungal species, belonging primarily to the *Basidiomycotina* and fewer to the *Ascomycotina,* are known to form ectomycorrhizae. Many of these fungi produce mushrooms and puffballs on the forest floor. Some fungi have a

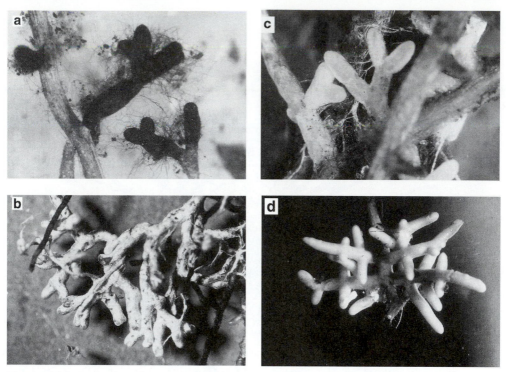

Figure 18–3 Examples of ectomycorrhizal short roots of conifers formed by various fungi. (a) *Cenococcum geophilum,* (b) *Pisolithus arhizus,* (c) *Thelephora terrestris,* and (d) *Amanita* sp. Roots tips range from 1 to 2 mm in diameter. *Figure (d) is courtesy of the Department of Plant Pathology, Cornell University. Used with permission.*

narrow host range, such as *Boletus betulicola* on *Betula* spp., while others have very broad host ranges, such as *Pisolithus arhizus = P. tinctorius,* which forms ectomycorrhiza with more than 46 tree species belonging to at least eight genera.

Arbuscular Mycorrhizae

The diagnostic feature of arbuscular mycorrhizae (AM) is the development of a highly branched arbuscule within root cortical cells (Fig. 18–4). The fungus initially grows between cortical cells but soon penetrates the host cell wall and grows within the cell. The general term for all mycorrhizal types where the fungus grows within cortical cells is **endomycorrhiza.** In this association neither the fungal cell wall nor the host cell membrane are breached. As the fungus grows, the host cell membrane invaginates and envelops the fungus, creating a new compartment where material of high molecular complexity is deposited. This apoplastic space prevents direct contact between the plant and fungus cytoplasm and allows for efficient transfer of nutrients between the symbionts. The arbuscules are relatively short-lived, less than 15 days, and are often difficult to see in field-collected samples.

Other structures produced by some AM fungi include vesicles, auxiliary cells, and asexual spores (Fig. 18–2 c, d). *Vesicles* are thin-walled, lipid-filled structures that usually form in intercellular spaces. Their primary function is thought to be for stor-

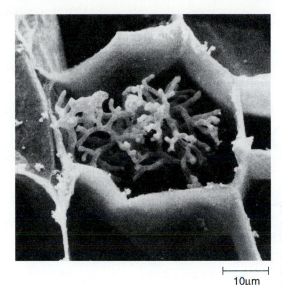

Figure 18–4 Scanning electron photomicrograph of an arbuscule formed by *Glomus mosseae* within a cortical cell of *Liriodendron tulipifera* (yellow poplar). *From Kinden and Brown (1975). Used with permission.*

10μm

age; however, vesicles can also serve as reproductive propagules for the fungus. *Auxiliary cells* are formed in the soil and can be coiled or knobby. The function of these structures is unknown. Reproductive spores can be formed either in the root or more commonly in the soil. Spores produced by fungi forming AM associations are asexual, forming by the differentiation of vegetative hyphae. For some fungi (e.g., *Glomus intraradices*), vesicles in the root undergo secondary thickening, and a septum (cross wall) is laid down across the hyphal attachment leading to spore formation, but more often spores develop in the soil from hyphal swellings.

The fungi that form AM are currently all classified in the order Glomales (Morton, 1988; Fig. 18–5). The taxonomy is further divided into suborders based on the presence of:

- vesicles in the root and formation of *chlamydospores* (thick wall, asexual spore) borne from subtending hyphae for the suborder Glomineae or

- absence of vesicles in the root and formation of auxiliary cells and *azygospores* (spores resembling a zygospore but developing asexually from a subtending hypha resulting in a distinct bulbous attachment) in the soil for the suborder Gigasporineae.

The term **vesicular-arbuscular mycorrhiza** (VAM) was originally applied to symbiotic associations formed by all fungi in the Glomales, but because a major suborder lacks the ability to form vesicles in roots, AM is now the preferred acronym. The order Glomales is further divided into families and genera according to the method of spore formation. The spores of AM fungi are very distinctive. They range in diameter from 10 μm for *Glomus tenue* to more than 1,000 μm for some *Scutellospora* spp. The spores can vary in color from hyaline (clear) to black and in surface texture from smooth to highly ornamented. *Glomus* forms spores on the ends of hyphae, *Acaulospora* forms spores laterally from the neck of a swollen hyphal terminus, and *Entrophospora* forms spores within the neck of the hyphal terminus. The Gigasporineae are divided into two genera based upon the presence of inner

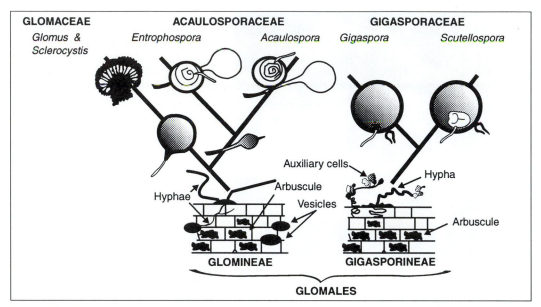

Figure 18–5 Taxonomic relationships among arbuscular mycorrhizal fungi. This figure arranges all taxonomic groups hierarchically according to their phylogenetic relationships. It is based on structure and development of the whole organism. All species are grouped together in the Glomales because they form mycorrhizae with arbuscules in root cortical cells. Members of each suborder (Glomineae and Gigasporinae) differ in kinds of mycorrhizal structures and the sequence in which they form. Families (Glomaceae, Acaulosporaceae, Gigasporaceae) and genera (italicized names) differ in mode of spore formation, organization, and germination processes. Species are distinguished by differences in subcellular properties of the spore wall or internal flexible walls usually associated with germination (not shown). *Original drawing by J.B. Morton. Used with permission.*

membranous walls and a germination shield (wall structure from which the germ tube can arise) for *Scutellospora* or the absence of these structures for *Gigaspora*.

The AM type of symbiosis is very common as the fungi involved can colonize a vast taxonomic range of both herbaceous and woody plants, indicating a general lack of host specificity among this type. However, it is important to distinguish between *specificity* (innate ability to colonize), *infectiveness* (amount of colonization), and *effectiveness* (plant response to colonization). AM fungi differ widely in the level of colonization they produce in a root system and in their impact on nutrient uptake and plant growth.

Ericaceous Mycorrhizae

The term *ericaceous* is applied to mycorrhizal associations found on plants in the order Ericales. The hyphae in the root can penetrate cortical cells (endomycorrhizal habit); however, no arbuscules are formed. Three major forms of ericaceous mycorrhiza have been described:

- Ericoid—cells of the inner cortex become packed with fungal hyphae. A loose welt of hyphae grows over the root surface, but a true mantle is not formed. The ericoid mycorrhizae are found on plants such as *Calluna* (heather), *Rhododendron* (azaleas and rhododendrons), and *Vaccinium* (blueberries) that have very fine root systems and typically grow in acid, peaty soils. The fungi involved are ascomycetes of the genus *Hymenoscyphus*.

- Arbutoid—characteristics of both EM and endomycorrhizae are found. Intracellular penetration can occur, a mantle forms, and a Hartig net is present. These associations are found on *Arbutus* (e.g., Pacific madrone), *Arctostaphylos* (e.g., bearberry), and several species of the Pyrolaceae. The fungi involved in the association are basidiomycetes and may be identical to the fungi that colonize EM tree hosts in the same region.

- Monotropoid—the fungi colonize achlorophyllous (lacking chlorophyll) plants in Monotropaceae (e.g., Indian pipe), producing the Hartig net and mantle. The same fungi also form EM associations with trees and thereby form a link through which carbon and other nutrients can flow from the autotrophic host plant to the heterotrophic, parasitic plant.

Orchidaceous Mycorrhizae

Mycorrhizal fungi have a unique role in the life cycle of plants in the Orchidaceae. Orchids typically have very small seeds with little nutrient reserve. The plant becomes colonized shortly after germination, and the mycorrhizal fungus supplies carbon and vitamins to the developing embryo. For achlorophyllous species, the plant depends on the fungal partner to supply carbon throughout its life. The fungus grows into the plant cell, invaginating the cell membrane and forming hyphal coils within the cell (Fig. 18–2e). These coils are active for only a few days, after which they lose turgor and degenerate and the nutrient contents are absorbed by the developing orchid. The fungi participating in the symbiosis are basidiomycetes similar to those involved in decaying wood (e.g., *Coriolus, Marasmius*) and pathogenesis (e.g., *Armillaria* and *Rhizoctonia*). In mature orchids, mycorrhizae also have roles in nutrient uptake and translocation.

Mixed Infections

Several fungi can colonize the roots of a single plant, but the type of mycorrhiza formed is usually uniform for a host. In some cases, however, a host can support more than one type of mycorrhizal association. *Alnus* (alders), *Salix* (willows), *Populus* (poplars), and *Eucalyuptus* can have both AM and EM associations on the same plant. Some ericoid plants have occasional EM and AM colonization.

An intermediate mycorrhizal type can be found on coniferous and deciduous hosts in nurseries and burned forest sites. The *ectendomycorrhiza* type forms a typical EM structure, except the mantle is thin or lacking and hyphae in the Hartig net may penetrate root cortical cells. The ectendomycorrhiza is replaced by EM as the seedling matures. The fungi involved in the association were initially designated "E-strain" but were later shown to be ascomycetes and placed in the genus *Wilcoxina*.

Uptake and Transfer of Soil Nutrients

When a nutrient is deficient in soil solution, the critical root parameter controlling its uptake is surface area. Hyphae of mycorrhizal fungi have the potential to greatly increase the absorbing surface area of the root. For example, Rousseau et al. (1994)

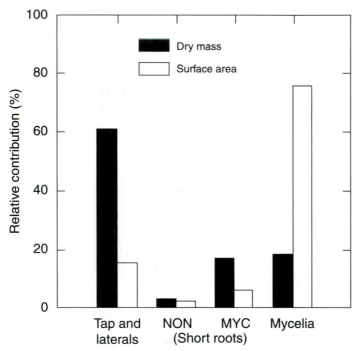

Figure 18–6 Relative distribution of dry mass and surface area among root system components of mycorrhizal seedlings colonized by *Pisolithus arhizus*. NON is nonmycorrhizal short roots, MYC is mycorrhizal short roots. *From Rousseau et al. (1994). Used with permission.*

found that while mycelia (aggregates of hyphae) in soil accounted for less than 20% of the total nutrient absorbing surface mass, they contributed nearly 80% of the absorbing surface area of pine seedlings (Fig. 18–6). It is also important to consider the distribution and function of the extramatrical hyphae. If the mycorrhiza is to be effective in nutrient uptake, the hyphae must be distributed beyond the *nutrient depletion zone* that develops around the root. A nutrient depletion zone develops when nutrients are removed from the soil solution more rapidly than they can be replaced by diffusion. For a poorly mobile ion such as phosphate, a sharp and narrow depletion zone develops close to the root (Fig. 18–7). Hyphae can readily bridge this depletion zone and grow into soil with an adequate supply of phosphorus. Uptake of micronutrients such as zinc and copper is also improved by mycorrhizae because these elements are also diffusion-limited in many soils. For more mobile nutrients such as nitrate, the depletion zone is wide and it is less likely that hyphae grow extensively into the zone that is not influenced by the root alone. Another factor contributing to the effective absorption of nutrients by mycorrhizae is their narrow diameter relative to roots. The steepness of the diffusion gradient for a nutrient is inversely related to the radius of the absorbing unit; therefore, the soil solution should be less depleted at the surface of a narrow absorbing unit such as a hypha. Furthermore, narrow hyphae can grow into small soil pores inaccessible to roots or even root hairs.

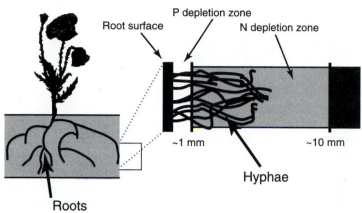

Figure 18–7 Diagram of how of a depletion zone develops next to the root surface. A narrow depletion zone (e.g., 1 mm) usually develops for phosphate, while a wide depletion zone (e.g., 10 mm) develops for nitrate. Mycorrhizal hyphae can generally grow beyond the phosphorus depletion zone, but not the nitrogen depletion zone.

Another advantage attributed to mycorrhizal fungi is access to pools of phosphorus not readily available to the plant. One mechanism for this access is the physiochemical release of inorganic and organic phosphorus by organic acids through the action of low-molecular-weight organic anions such as oxalate which can (Fox et al., 1990):

- replace phosphorus sorbed at metal hydroxide surfaces through ligand-exchange reactions,

- dissolve metal oxide surfaces that sorb phosphorus, and

- complex metals in solution and thus prevent precipitation of metal phosphates.

Some EM fungi produce large quantities of oxalic acid, and this may partially explain enhanced nutrient uptake by EM roots. Another mechanism by which mycorrhizal fungi release inorganic phosphorus is through mineralization of organic matter. This occurs by phosphatase-mediated hydrolysis of organic phosphate (C–O–P) ester bonds. Significant phosphatase activity has been documented for mycorrhizal fungi grown in pure cultures and for excised and intact EM short roots. In the field, a positive correlation has been reported between phosphatase activity and the length of fungal hyphae associated with EM mantles (Haussling and Marschner, 1989). Care must be exercised in interpreting these data because plant roots and the associated microflora also produce organic acids and phosphatases; however, mycorrhizal fungi certainly intensify this activity.

Ericoid and EM have a special role in the mineralization of nitrogen (Read et al. 1989). Most plant litter entering the soil has a high C/N ratio and is rich in lignin and tannins. Only a few mycorrhizal fungi can mobilize nutrients from these primary sources. However, a wide range of ericoid and EM fungi can obtain nitrogen and other nutrients from secondary sources of organic matter such as dead microbial biomass. A wide range of hydrolytic and oxidative enzymes capable of depolymerizing organic nitrogen has been demonstrated. These types of mycorrhizae may have an important role in nitrogen cycling in the acidic and highly organic soils where they predominate.

Carbon Fluxes in Mycorrhizal Plants

Mycorrhizal fungi range from **obligate symbionts,** which can only obtain carbon from the plant host as in the case of AM fungi, to **facultative symbionts,** which can also mineralize organic carbon from nonliving sources as in the case of some EM species. In nature the heterotrophic mycorrhizal fungi obtain all or most of their carbon from the autotrophic host plant. Ectomycorrhizae and ericoid mycorrhizae transform host carbohydrates into fungal-specific storage carbohydrates, such as mannitol and trehalose, which may produce a sink for photosynthate that favors transport of carbohydrate to the fungal partner. In AM, lipids accumulate in vesicles and other fungal structures and provide an analogous sink for host photosynthate.

As much as 20% of the total carbon assimilated by plants may be transferred to the fungal partner. This transfer of carbon to the fungus has sometimes been considered a drain on the host. However, the host plant may increase photosynthetic activity following mycorrhizal colonization, thereby compensating for carbon "lost" to the soil. Occasionally plant growth suppression has been attributed to mycorrhizal colonization, but usually this occurs only under low-light (photosynthate limiting) or high-phosphorus conditions.

In an ecosystem, the flow of carbon to the soil mediated by mycorrhizae serves several important functions. For some mycorrhizae, the extramatrical hyphae produce hydrolytic enzymes, such as proteases and phosphatases, that can have an important impact on organic matter mineralization and nutrient availability. Extramatrical hyphae of mycorrhizae also bind soil particles together and thereby improve soil aggregation (Fig. 18–8). Typically there are between 1 to 20 m of AM hyphae g^{-1} of soil (Sylvia, 1990). Another important consequence of carbon flow to the fungal partner is the development of a unique rhizosphere microbial community called the **mycorrhizosphere,** which we will discuss shortly. Soil scientists now realize that carbon flow to the soil is critical for the development of soil aggregation and the maintenance of a healthy plant-soil system. Enhanced carbon flow to the soil should be considered an important benefit of mycorrhizal colonization.

Interactions with Other Soil Organisms

Mycorrhizal fungi interact with a wide assortment of organisms in the rhizosphere. The result can be either positive, neutral, or negative on the mycorrhizal association or a particular component of the rhizosphere. For example, specific bacteria stimulate EM formation in conifer nurseries and are called *mycorrhization helper bacteria*. In certain cases these bacteria eliminate the need for soil fumigation (Garbaye, 1994).

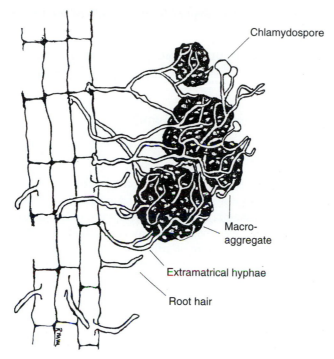

Chlamydospore

Macro-
aggregate

Extramatrical hyphae

Root hair

Figure 18–8 Diagram showing extramatrical mycorrhizal hyphae enmeshing soil microaggregates, leading to development of macroaggregates and soil stabilization within the rhizosphere. *From Miller and Jastrow (1992). Used with permission.*

The interaction between rhizobia and AM fungi has received considerable attention because of the relatively high phosphorus demand of N_2 fixation. The two symbioses typically act synergistically, resulting in greater nitrogen and phosphorus content in combination than when each is inoculated onto the legume alone. Legumes are typically coarse-rooted and therefore inefficient in extracting phosphorus from the soil. The AM fungi associated with legumes are an essential link for adequate phosphorus nutrition, leading to enhanced nitrogenase activity that in turn promotes root and mycorrhizal growth.

Mycorrhizal fungi colonize feeder roots and thereby interact with root pathogens that parasitize this same tissue. In a natural ecosystem where the uptake of phosphorus is low, a major role of mycorrhizal fungi may be protection of the root system from endemic pathogens such as *Fusarium* spp. Mycorrhizae may stimulate root colonization by selected biocontrol agents, but our understanding of these interactions is meager. Much more research has been conducted on the potential effects of mycorrhizal colonization on root pathogens. Mycorrhizal fungi may reduce the incidence and severity of root diseases. The mechanisms proposed to explain this protective effect include:

- development of a mechanical barrier—especially the mantle of the EM—to infection by pathogens,

- production of antibiotic compounds that suppress the pathogen,

- competition for nutrients with the pathogen, including production of siderophores, and

- induction of generalized host defense mechanisms.

Management of Mycorrhizae

The dramatic plant growth response achieved in pot studies following inoculation with mycorrhizal fungi in low fertility soils led to a flurry of activity in the 1980s aimed at using these organisms as biofertilizers. Field responses were often disappointing, especially in high-input agricultural systems, and many concluded that mycorrhizae had little practical importance in agriculture. Further studies, however, have confirmed that most agricultural plants are colonized by mycorrhizal fungi and that they can have a substantial impact, both positive and negative, on crop productivity (Johnson, 1993). Certainly, agriculturists should appreciate the distribution of mycorrhizae within their systems and understand the impact of their management decisions on mycorrhizal functioning.

Factors that should be considered when assessing the potential role of mycorrhizae in an agroecosystem include:

- *Mycorrhizal dependency (MD) of the host crop.* This is usually defined as the growth response of mycorrhizal (M) versus nonmycorrhizal (NM) plants at a given phosphorus level.

$$MD = ((M - NM) / NM) \times 100$$

Although most agricultural crops have mycorrhizae, not all benefit equally from the symbiosis. Generally, coarse-rooted plants benefit more than fine-rooted plants.

- *Nutrient status of the soil.* Assuming that the major benefit of the mycorrhizal symbiosis is improved phosphorus uptake, the management of mycorrhizal fungi will be most critical when soil phosphorus is limiting. Many tropical soils fix phosphorus, and proper mycorrhization of plants is essential to obtain adequate phosphorus. In temperate zones, phosphorus is sometimes applied in excess of crop demand. However, with increased concerns about environmental quality, phosphorus use in developed countries may be reduced, resulting in increased dependence on native mycorrhizae for nutrient uptake. Another factor to consider is the interaction of water stress with nutrient availability. As soils dry, phosphorus may become limiting even in soils that test high in available phosphorus.

- *Inoculum potential of the indigenous mycorrhizal fungi.* Inoculum potential is a product of the *abundance* and *vigor* of the propagules in the soil and can be quantified by determining the rate of colonization of a susceptible host under a standard set of conditions. Inoculum potential can be adversely affected by management practices such as fertilizer and lime application, pesticide use (especially fungicides), crop rotation, fallowing, tillage, and topsoil removal.

| Box 18–1 |

Examples of How Management Practices Affect Mycorrhizal Populations in Soil and Subsequent Growth of the Host Crop. Soil disturbance such as tillage can dramatically affect the function of mycorrhizae in an agricultural system. M.H. Miller and coworkers from the University of Guelph, Canada, documented an interesting case where disturbance of an arable, no-till soil resulted in reduced AM development and subsequently less absorption of phosphorus by seedlings of maize in the field (Miller et al., 1995). They hypothesized that soil disturbance reduced the effectiveness of the mycorrhizal symbiosis. To confirm this, they conducted a series of growth chamber studies with nondisturbed and disturbed soil cores collected from long-term field plots. Disturbance reduced both mycorrhizal colonization and phosphorus absorption by maize and wheat roots, but did not reduce phosphorus absorption by two nonmycorrhizal crops, spinach and canola. The authors concluded that under nutrient-limited conditions, the ability of mycorrhizal seedlings to associate with intact hyphal networks in soil may be highly advantageous for crop establishment.

Crop rotation and fallow systems can affect the diversity and function of mycorrhizal fungi. J.P. Thompson described the role of AM fungi in a long-fallow (more than 12 months) disorder of field crops in the state of Queensland, Australia (Thompson, 1987). In semiarid cropping systems, clean fallows conserve soil moisture and nitrate for the subsequent crop. Since the 1940s, some crops sown immediately after a long fallow grew poorly and had phosphorus and zinc deficiencies. The Australian researchers found that the fallow resulted in a decline in propagules of AM fungi in the soil and reduced colonization of the crop plants in the field. Furthermore, they conducted inoculation trials and found that increasing inoculum abundance in the soil overcame the deleterious impact of fallow. They recommended that farmers avoid planting mycorrhizal-dependent crops, such as linseed, sunflower, and soybean, following periods of fallow or after a nonhost plant such as canola that reduce AM propagules.

Problems and Potential for Inoculum Production and Use

In situations where native mycorrhizal inoculum potential is low or ineffective, providing the appropriate fungi for the plant is worth considering. With the current state of technology, inoculation is best for transplanted crops and in areas where soil disturbance has reduced native inoculum potential.

The first step in any inoculation program is to obtain an isolate that is both *infective,* or able to penetrate and spread in the root, and *effective,* or able to enhance growth or stress tolerance of the host. Individual isolates of mycorrhizal fungi vary widely in these properties, so screening trials are important to select isolates that will perform successfully. Screening under actual cropping conditions is best because indigenous mycorrhizal fungi, pathogens, and soil chemical and physical properties will influence the result.

Isolation and inoculum production of EM and AM fungi present very different problems. Many EM fungi can be cultured on artificial media. Therefore, isolates of EM fungi can be obtained by placing surface-disinfested portions of sporocarps or mycorrhizal short roots on an agar growth medium. The resulting fungal biomass can be used directly as inoculum but, for ease of use, inoculum often consists of the fungal material mixed with a carrier or bulking material such as peat. Obtaining isolates of AM fungi is more difficult because they will not grow apart from their hosts. Spores can be sieved from soil, surface disinfested, and used to initiate "pot cultures" on a susceptible host

| Box 18–2 |

Examples Where Inoculating with Either EM or AM Fungi Is Beneficial When Planting a Mycorrhizal-Dependent Crop In an Area Where Native Inoculum Potential Is Low. Pines were not native to Puerto Rico, and their fungal symbionts were absent from the soil (Vozzo and Hacskaylo, 1971). As far back as the 1930s, attempts to establish pine on the island were unsuccessful. Typically, the pines germinated well and grew to heights of 8 to 10 cm in a relatively short time, but then rapidly declined. Phosphorus fertilizers did not substantially improve plant vigor. In 1955, soil from under pine stands in North Carolina was transported to Puerto Rico where it was incorporated as inoculum into soil around 1-year-old "scrawny" pine seedlings growing at Maricao in the western mountains. Thirty-two seedlings were inoculated, and an equal number were monitored as noninoculated control plants. Within one year, inoculated plants had abundant mycorrhizal colonization and had achieved heights of up to 1.5 m, while most of the noninoculated plants had died. Further trials with mixtures of surface soil containing mycorrhizal fungi and with pure inocula, consisting of fungi growing in a peat-based medium, confirmed that inoculated seedlings were consistently more vigorous and larger than nonmycorrhizal ones. Subsequent surveys more than 15 years after inoculation indicated that the inoculated fungi continued to grow and sporulate in the pine plantations.

   Beach erosion is a problem in many coastal areas and replenishing the beaches with sand dredged from offshore is often the method of choice for restoring them. Native grasses are planted in the back beach to reduce further erosion and to initiate the dune-building process. In native dunes, beach grasses are colonized by a wide array of AM fungi. However, when these grasses are propagated in nurseries, they do not have mycorrhizae. Furthermore, the replenishment sand is typically devoid of AM propagules. In a series of studies (Sylvia, 1989), AM fungi were isolated from grasses growing in native dunes. The fungi were screened for effectiveness with the given host/soil combination and for compatibility with the nursery production system, and the effect of inoculation was documented on transplants placed on newly restored beaches. In the nursery, moderate amounts of colonization were achieved, even with high levels of pesticide and fertilizer use. After transfer of these plants to a low-nutrient beach environment, AM colonization spread rapidly and enhanced plant growth significantly compared to noninoculated control plants even though the plants were equal size when they left the intensively managed nursery. Compared to noninoculated plants after 20 months on the beach, AM-colonized plants had 219, 81, 64 and 53% more shoot dry mass, root length, plant height, and number of tillers, respectively. In most cases the objective of nursery inoculation is not to achieve a growth response, but rather to establish the symbiosis with the plant so that it can be effectively transferred to the field.

plant in sterile soil or an artificial plant-growth medium. Inoculum is typically produced in scaled-up pot cultures. Alternatively, hydroponic or aeroponic culture systems are possible; a benefit of these systems is that plants can be grown without a supporting substratum, allowing colonized roots to be sheared into an inoculum of high propagule number. Sylvia (1994) summarized the methods for working with AM inoculum.

   The goal of inoculation is to introduce propagules of selected mycorrhizal fungi into the rhizosphere of the target plant. The most common method is to place inoculum below the seed or seedling prior to planting. Alternative strategies include coating seed or encasing somatic embryos with inoculum. Such technologies should allow for more widespread use of inoculum in the future.

Aeroponic/Sheared-Root AM Inoculum. Aeroponic cultures consist of chambers where AM-colonized plants are suspended in a nutrient mist (a). The mist can be generated by an atomizing disk or pressurized spray through nozzles. The highly aerobic and low-nutrient environment promotes rapid and profuse colonization and sporulation by fungi in the Glomineae suborder. Following a culture period of 10 to 12 weeks, well-colonized roots (b) are removed from the chamber and cut into 1-cm lengths (c). Spores are separated from roots by washing over a coarse sieve (d). Cutting the roots with a food processor produces an inoculum of small and uniform size (e). The processed material is collected on a fine screen for use as an inoculum (f). Because roots are free of any adhering rooting medium, the food processor will cut roots cleanly and produce very high inoculum densities.

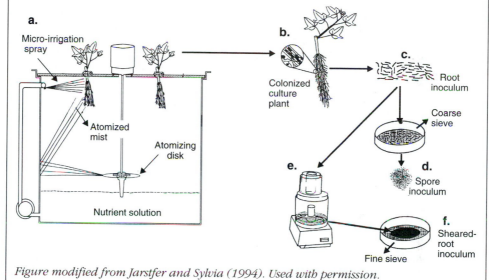

Figure modified from Jarstfer and Sylvia (1994). Used with permission.

Summary

Mycorrhizae are symbiotic associations that form between the roots of most plant species and fungi. These symbioses are characterized by bidirectional movement of nutrients where carbon flows to the fungus and inorganic nutrients move to the plant, thereby providing a critical linkage between the plant root and soil. In infertile soils, nutrients taken up by the mycorrhizal fungi can lead to improved plant growth and reproduction. As a result, mycorrhizal plants are often more competitive and better able to tolerate environmental stresses than are nonmycorrhizal plants.

Mycorrhizal associations vary widely in form and function. Ectomycorrhizal fungi are mostly basidiomycetes that grow between root cortical cells of many tree species, forming a Hartig net. Arbuscular mycorrhizal fungi belong to the order Glomales and form highly branched structures called arbuscules, within root cortical cells of many herbaceous and woody plant species.

Plant responses to colonization by mycorrhizal fungi can range from dramatic growth promotion to growth depression. Factors affecting this response include the mycorrhizal dependency of the host crop, the nutrient status of the soil, and the inoculum potential of the mycorrhizal fungi. Management practices such as tillage, crop rotation, and fallowing may adversely affect populations of mycorrhizal fungi in the field. Where native inoculum potential is low or ineffective, inoculation strategies may be helpful. With the current state of technology, inoculation is most feasible for transplanted crops and in areas where soil disturbance has greatly reduced the native inoculum potential.

Cited References

Fox, T.R., N.B. Comerford, and W.W. McFee. 1990. Kinetics of phosphorus release from spodosols: Effects of oxalate and formate. Soil Sci. Soc. Am. J. 54:1441–1447.

Garbaye, J. 1994. Helper bacteria: A new dimension to the mycorrhizal symbiosis. New Phytol. 128:197–210.

Haussling, M., and H. Marschner. 1989. Organic and inorganic soil phosphates and acid phosphatase activity in the rhizosphere of 80-year-old Norway spruce [*Picea abies* (L.) Karst.] trees. Biol. Fertil. Soils 8:128–133.

Jarstfer, A.G., and D.M. Sylvia. 1994. Aeroponic culture of VAM fungi. pp. 427–441. *In* A.K. Varma and B. Hock (eds.), Mycorrhiza: Structure, function, molecular biology and biotechnology. Springer-Verlag, Berlin.

Johnson, N.C. 1993. Can fertilization of soil select less mutalistic mycorrhizae? Ecol. Appl. 3:749–757.

Kinden, D.A., and M.F. Brown. 1975. Electron microscopy of vesicular-arbuscular mycorrhiza of yellow poplar. III. Host-endophyte interactions during arbuscular development. Can. J. Microbiol. 21:1930–1939.

Miller, M.H., T.P. McGonigle, and H.D. Addy. 1995. Functional ecology of vesicular arbuscular mycorrhizas as influenced by phosphate fertilization and tillage in an agricultural ecosystem. Crit. Rev. Biotech. 15:241–255.

Miller, R.M., and J.D. Jastrow. 1992. The role of mycorrhizal fungi in soil conservation. pp. 29–44. *In* G.J. Bethlenfalvay and R.G. Linderman (eds.), Mycorrhizae in sustainable agriculture. ASA Special Publ. No. 54, American Society of Agronomy, Madison, Wis.

Morton, J.B. 1988. Taxonomy of VA mycorrhizal fungi: Classification, nomenclature, and identification. Mycotaxon 32:267–324.

Read, D.J. 1984. The structure and function of vegetative mycelium of mycorrhizal roots. pp. 215–240. *In* D.H. Jennings and A.D.M. Rayner (eds.), The ecology and physiology of the fungal mycelium. Cambridge University Press, New York.

Read, D.J., J.R. Leake, and A.R. Langdale. 1989. The nitrogen nutrition of mycorrhizal fungi and their host plants. pp. 181–204. *In* L. Boddy, R. Marchant, and D.J. Read (eds.), Nitrogen, phosphorus and sulfur utilization by fungi. Cambridge University Press, New York

Rousseau, J.V.D., D.M. Sylvia, and A.J. Fox. 1994. Contribution of ectomycorrhiza to the potential nutrient-absorbing surface of pine. New Phytol. 128:639–644.

Sylvia, D.M. 1989. Nursery inoculation of sea oats with vesicular-arbuscular mycorrhizal fungi and outplanting performance of Florida beaches. J. Coastal Res. 5:747–754.

Sylvia, D.M. 1990. Distribution, structure, and function of external hyphae of vesicular-arbuscular mycorrhizal fungi. pp. 144–167. *In* J.E. Box and L.H. Hammond (eds.), Rhizosphere dynamics. Westview Press, Boulder, Colo.

Sylvia, D.M. 1994. Vesicular-arbuscular mycorrhizal fungi. pp. 351–378. *In* R.W. Weaver, S. Angle, P. Bottomley, D. Bezdicek, S. Smith, A. Tabatabai, and A. Wollum. (eds.), Methods of soil analysis, Part 2. Microbiological and biochemical properties. Soil Science Society of America, Madison, Wis.

Thompson, J.P. 1987. Decline of vesicular-arbuscular mycorrhizae in long fallow disorder of field crops and its expression in phosphorus deficiency of sunflower. Aust. J. Agric. Res. 38:847-867.

Vozzo, J.A., and E. Hacskaylo. 1971. Inoculation of *Pinus caribaea* with ectomycorrhizal fungi in Puerto Rico. For. Sci. 17:239–245.

General References

Allen, M.F. 1992. Mycorrhizal functioning: An integrative plant-fungal process. Chapman and Hall, New York.

Bethlenfalvay, G.J., and R.G. Linderman. 1992. Mycorrhizae in sustainable agriculture. ASA Special Publication No. 54. Agronomy Society of America, Madison, Wis.

Harley, J.L., and S.E. Smith. 1983. Mycorrhizal symbiosis. Academic Press, New York.

Norris, J.R., D.J. Read, and A.K. Varma. 1991. Methods in microbiology: Techniques for the study of mycorrhiza. Vol. 23. Academic Press, London.

Norris, J.R., D.J. Read, and A.K. Varma. 1992. Methods in microbiology: Techniques for the study of mycorrhiza. Vol. 24. Academic Press, London.

Read, D.J., D.H. Lewis, A.H. Fitter, and I.J. Alexander. 1992. Mycorrhizas in ecosystems. CAB International, Wallingford, England.

Robson, A.D., L.K. Abbott, and N. Malajczuk. 1994. Management of mycorrhizas in agriculture, horticulture and forestry. Kluwer Academic Publishers, Boston.

Study Questions

1. The term symbiosis literally means "living together" and usually implies a mutually beneficial association. Describe how each partner in a mycorrhizal symbiosis gains benefit from the association?

2. Describe the differences in structure and function between hyphae of mycorrhizal fungi that grow in the root (intramatrical) and the hyphae that grow in the soil (extramatrical).

3. When grown in phosphorus-limited soil, citrus is a mycorrhizal-dependent plant. However, when citrus is grown under high-phosphorus conditions (such as found in a nursery), inoculation with mycorrhizal fungi may lead to growth depression. Explain this phenomenon.

4. What are the major types of mycorrhizae and how does the structure and function of the symbiosis differ among them?

5. What is an arbuscule? Describe the unique features that make it an ideal structure for nutrient exchange between the fungus and the host plant.

6. Compare and contrast the fungi involved in arbuscular mycorrhizal versus ectomycorrhizal symbioses.

7. What is a nutrient depletion zone and why does it develop? How do mycorrhizae overcome the limitations to nutrient uptake caused by the development of depletion zones in the rhizosphere?

8. What are some of the consequences of host carbon that is transferred to the soil via mycorrhizal structures?

9. A tripartite association often forms between legumes, nitrogen-fixing rhizobia, and arbuscular mycorrhizal fungi. Why does the legume especially benefit from colonization by these two symbiotic microorganisms?

10. Under what circumstances would you consider inoculating plants with mycorrhizal fungi? What are the major limitations to inoculum production and use?

Chapter 19

♦

Biological Control of Soilborne Plant Pathogens and Nematodes

James H. Graham and David J. Mitchell

♦

The research worker in his laboratory and the farmer on his fields can do nothing to the soil that does not induce some change in the activity and balance of its microbial population. There must therefore be some element of biological control, or of its reverse, in almost all disease control practices.

S. D. Garrett

Microorganisms and plants are in dynamic equilibrium in the soil. As soil microorganisms interact with plants through space and time, their interactions can be beneficial, neutral, or detrimental. In this chapter we discuss how to promote beneficial interactions in order to limit plant disease caused by soilborne microorganisms.

If soil microorganisms cause continued irritation of the plant, the injurious alteration of a process(es) of energy utilization results in plant disease. The disease-causing agent usually weakens or destroys plant cells and tissues by altering normal physiological functions. The **pathogen** is the biotic agent that causes disease. The pathogen lives on or in the host plant and obtains nutrition from the host. The susceptibility of the plant determines whether the pathogen infects and obtains nutrition from the host. The environment affects host susceptibility and pathogen activity.

Soil is the source of many microorganisms, including bacteria, fungi, nematodes, protozoa, viruses, insects, and mites, that are pathogenic to plants. Many species of protozoa and microarthropods also can modify the rhizosphere microflora and act as plant-pathogen deterrents (Curl et al., 1988). Because relatively little is known about the contributions of viruses and microbe-feeding protozoans or fauna to the suppression of pathogens, this chapter will focus on the first three groups of soil microorganisms as plant pathogens and biological control agents.

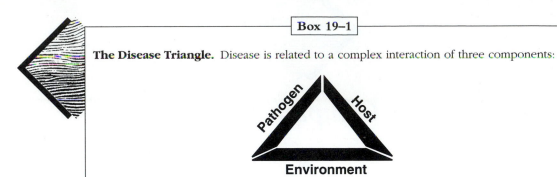

Box 19–1

The Disease Triangle. Disease is related to a complex interaction of three components:

Pathogen

Host

Environment

Fundamental Biological Control Concepts

Biological control as classically defined is the use of organisms for the control of insect pests and weeds. This involves the importation of natural enemies of a pest from its native range and its subsequent introduction to the pathogen's habitat. Biological control of soilborne pathogens and nematodes is concerned with methods other than the classical approaches used for weeds and insects (Cook and Baker, 1983).

Biological control of plant disease is the reduction in the numbers and activity of a plant pathogen, using one or more organisms. Biological control in the belowground environment can occur within plant roots, the rhizosphere, or bulk soil. Plant roots are the site where the pathogen initially infects host tissue. The **rhizosphere** is the narrow zone of soil surrounding plant roots that contains root exudates, sloughed root remains, and large populations of soil microorganisms of various nutritional groupings. Bulk soil, outside of the rhizosphere, has relatively smaller populations of microorganisms. Biological control is usually operative in the rhizosphere but can occur in other habitats as well.

Antagonists are biological agents that reduce the numbers or disease-producing activities of the pathogen through one or more of the following mechanisms.

- **Antibiosis** occurs when the pathogen is inhibited or killed by metabolic products of the antagonist; these products include lytic agents, enzymes, volatile compounds, and other toxic substances.

- **Competition** occurs when the antagonist reduces the pathogen's activities by gaining a measure of the supply of limited resources, such as organic and inorganic nutrients, growth factors, oxygen, or space.

- **Hyperparasitism** occurs when the antagonist invades the pathogen by secreting enzymes such as chitinases, cellulases, glucanases, and other lytic enzymes.

The Plant–Soil–Microbial Equilibrium

Because soil changes less rapidly than the atmosphere above it, soil microorganisms have become adapted to a relatively stable habitat. Populations of antagonists, pathogens, and host plants fluctuate within certain limits in response to abiotic and biotic interactions. Slight changes in soil environmental conditions may have profound effects on populations of the pathogen and antagonist alike. Changes produced by agricultural practices tend to be varied, rapid, and ongoing, which permits little time for reestablishment of the biological equilibrium. Intensive use of fertilizers, pesticides, tillage, irrigation, and other crop management practices for selected higher-yielding cultivars has resulted in increased severity or incidences of plant disease. The replacement of a diverse plant community with a single genotype, or monoculture, shifts the biological equilibrium toward development of plant disease epidemics that otherwise rarely occur in undisturbed ecosystems.

The occurrence of a plant disease epidemic indicates that one or more of the following conditions exist:

- The pathogen is highly virulent or is present in high numbers.

- The abiotic environment is more favorable for the pathogen than for the host or antagonist.

- The host plant is genetically homogeneous, highly susceptible, and continuously or extensively grown.

- The antagonist is absent or in low numbers because of a lack of a favorable environment or inhibition by other microorganisms.

Biological equilibrium in crop disease situations can be manipulated through the integrated use of chemical controls, such as fungicides and nematicides; cultural controls, such as sanitation or crop rotation; and biological control methods, such as plant resistance or introduction of antagonists. Usually, biological controls are more subtle and operate more slowly, but are generally more stable and longer lasting than chemical and cultural control measures. However, biological controls should be integrated with other control measures because different methods are effective at different times and locations under varying conditions.

Approaches to Biological Control

Pathogen-suppressive soils are defined by Cook and Baker (1983) as soils in which:

- the pathogen does not establish or persist,

- the pathogen establishes but causes no damage, or

- the pathogen causes some disease damage, but the disease becomes progressively less severe even though the pathogen persists in soil.

Most soils suppress soilborne pathogens to some degree. Soils may suppress pathogens in a general or specific sense. *General suppression* of a pathogen is directly related to the total amount of microbial activity in the soil or plant at a critical time in the life cycle of the pathogen. The kinds of active microorganisms are probably less important than the microbial biomass that competes with the pathogen for carbon, nitrogen, and energy or causes inhibition through more direct forms of antagonism. *Specific suppression* operates against a background of general suppression but involves more specific effects of individual or select groups of microorganisms antagonistic to the pathogen during some stage of its life cycle.

The initial objective of biological control is to maximize soil suppressiveness through the manipulation of *resident antagonists*. These microorganisms probably resided in equilibrium with pathogens and plants before the intervention of agriculture. If the indigenous biological control antagonists are no longer present or are inadequate, introductions of antagonists (one-time, occasional, or repeated) may be helpful. Approaches to biological control of plant disease are summarized in Box 19–2 and are discussed further in the remainder of this chapter.

Box 19–2

Approaches to Biological Control of Plant Diseases with Microbial Antagonists in Soil.

Biological control with resident antagonists:

- pathogen-suppressive soils

- general suppression

- specific suppression

Biological control with introduced antagonists:

- rhizosphere competence

- root colonization process

- introduction with other soil treatments to favor the antagonist

- seed treatments

- composites of antagonists

Biological Control with Resident Antagonists

As a result of certain crop or soil management practices, some field or greenhouse soils are naturally suppressive, even though a susceptible host and soil populations of the pathogen are present. Although the development of biological control systems is often poorly understood, pathogen suppressive soils have been associated consistently with these crop or soil management practices:

- crop rotations with nonsusceptible plant hosts or fallowing (no crop), which provide time for resident antagonists to displace soilborne pathogens or for pathogen propagules to die,

- soil tillage, which accelerates displacement of certain pathogens in crop residues by resident antagonists,

- incorporation of composts and other organic materials, which stimulate the activities of resident antagonists or suppress the activities of pathogens,

- selective soil treatments with steam, biocides, or solarization, which suppress or eliminate the pathogen but not resident soil saprophytes, and

- management of certain cultivars of a host, which select for specific rhizosphere antagonists.

In each case, suppression of the activities of the pathogen by resident antagonists occurs prior to infection of the host root (Fig. 19–1).

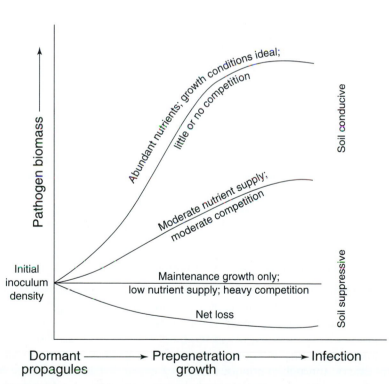

Figure 19–1 Possible changes in pathogen biomass near a host root. As dormant propagules go through prepenetration growth to infection, the pathogen biomass may increase or decrease depending on whether the soil is conducive or suppressive to disease development. *From Cook and Baker (1983). Used with permission.*

The following examples of suppressive soils demonstrate how some of these management practices contribute to the establishment of biological control by general and specific suppression of the pathogen.

General Suppression: *Phytophthora*-Suppressive Soils

Suppressive soils in an avocado grove in Queensland, Australia, demonstrate the role of soil amendments in suppression of the fungal pathogen *Phytophthora cinnamomi*. In contrast to the presence of the pathogen and severe disease in neighboring groves that are conducive to disease, only minor damage from root rot develops in the grove with suppressive soil. This grove is managed with continuous legume-maize cover crops, applications of poultry manure, and dolomitic limestone to maintain the soil pH above 6. The suppressive soils are red clays with abundant soil organic matter, calcium, and organic nitrogen typical of tropical rain forests in the region. Treatment of suppressive soil with aerated steam at 60°C for 30 minutes does not destroy suppressiveness but 100°C for 30 minutes does. This temperature effect suggests that spore-forming bacteria or actinomycetes are the biological antagonists. Broadbent and Baker (1974) demonstrated that suppressiveness of these soils originates in the organic fraction. No single isolate from the soil produces a suppressive effect, although an exceptionally diverse microbial population abounds. The principal infective propagules of *P. cinnamomi* are zoospores released from zoosporangia, but few zoosporangia are formed in suppressive soils. In addition, antagonistic bacteria attach to and lyse zoosporangia. The control mechanism is referred to as *general suppression,* because it is related to the total amount of microbiological activity that occurs at a time and place critical to the pathogen.

Several mechanisms may explain suppressiveness in these soils. High organic matter and calcium improve soil structure and drainage. The combination of organic matter, calcium, and ammonium nitrogen supports a large, active lytic microbial population (Fig. 19–2). The soil pH of 6 to 7 is favorable to bacteria. High calcium concentration may also enhance resistance of the avocado roots to the pathogen, while nitrogen supplied in ammonium form inhibits growth and sporangium formation by *P. cinnamomi*. Suppressiveness may break down when soil is waterlogged because the biological equilibrium among antagonists, the pathogen, and the background microflora is altered.

General and Specific Suppression: Take–All Decline

Take-all of wheat, a disease caused by the fungus *Gaeumannomyces graminis* var. *tritici,* is controlled in many areas of the world by not growing wheat or barley in the same field more often than every other year (Cook and Baker, 1983). The pathogen depletes its nutrient supply in fragments of root and basal stem or is displaced by competing microorganisms in most soils. An alternative approach is to manage the disease through continuous crops of wheat or barley, resulting in take-all decline. This biological control is thought to include the combined effects of competition from fungi and bacteria in root fragments, which reduces the numbers of the pathogen (general suppression), and antagonistic microorganisms in the rhizoplane and in young lesions, which limit infection and secondary spread of the pathogen (specific suppression). The agents responsible for take-all decline are:

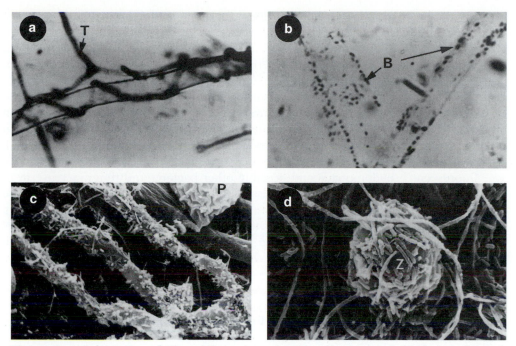

Figure 19–2 Colonization of *Phytophthora cinnamomi* by soil antagonists: (a) hypha colonized by an antagonistic *Trichoderma* sp. (T); (b) lysed hyphae with bacteria (B) colonizing the hyphal surface; (c) scanning electron micrograph (SEM) of hyphae incubated in soil suppressive to *Eucalyptus* root rot showing intense colonization by bacteria and presence of predatory protozoa (P); (d) SEM of *P. cinnamomi* zoospore (Z) intensely colonized by bacteria and actinomycetes in the rhizosphere of a *Eucalyptus* root. *From Malajczuk (1983). Used with permission.*

- transferable by soil from field to field,

- eliminated by aerated steam at 60°C or greater for 30 minutes, and

- sensitive to the biocides chloropicrin and methyl bromide (common soil fumigants).

General suppression of the pathogen in root-fragment inoculum is partly due to nonpathogenic, saprophytic *Fusarium* spp. The fungus *Fusarium oxysporum* reduces root infection by the take-all pathogen when both organisms are added to pasteurized soil. The competitive effects of microorganisms on *G. graminis* var. *tritici* in the root residue is promoted by tillage combined with delayed seeding date. Tillage temporarily increases microbial activity and accelerates residue breakdown by exposing fresh surfaces of the residue for microbial colonization while promoting soil aeration. Some tillage is required to promote decline of take-all in consecutive crops of wheat. Delayed seeding is also beneficial because it allows more time for displacement of the pathogen by the antagonists.

Specific suppression of take-all by antagonistic fluorescent pseudomonads is a phenomenon that results from growing consecutive wheat crops fertilized with ammonium rather than nitrate nitrogen. Plant uptake of ammonium lowers the pH of the rhizosphere but alone does not account for disease suppression in that the effect does not occur in soil fumigated with methyl bromide. Rather, the increased acidity results in a greater frequency of fluorescent pseudomonads inhibitory to *G. graminis* var. *tritici.* Populations of fluorescent pseudomonads inhibitory to *G. graminis* var. *tritici* increase specifically in the presence of the take-all fungus (Box 19–3). Antagonistic bacteria selectively proliferate in root lesions to a level five to ten times greater in take-all decline soil than in soil conducive for the disease. The bacteria are then positioned to inhibit further spread of the pathogen on roots and to carry over as cohabitants with the pathogens in host residues. As the incidence and severity of take-all increases, populations of fluorescent pseudomonads and possibly other bacteria and fungi increase in response to the amount of infected tissue. An equilibrium may then develop between microbial populations that limits future outbreaks of severe disease and the production of enough lesions to sustain populations of the antagonists. Thus, although the pathogen can be isolated from wheat roots—and is virulent in conducive soil—after take-all decline, it cannot cause severe disease in the suppressive field.

Box 19–3

Role of Antibiotic Production. In the laboratory, strain 2-79 of the bacterium *Pseudomonas fluorescens* produces a phenazine-type antibiotic inhibitory to *G. graminis* var. *tritici* at 1 μg ml^{-1} of medium. Antibiotic-negative mutants, produced by inactivating single genes by transposon mutagenesis, are less suppressive to take-all than the parent wild-type strain when introduced with the seed. Restoration of the phenazine-producing ability by complementation with a DNA fragment from the wild type restores ability of the mutant to suppress take-all. The importance of antibiotic production for biocontrol justifies seeking this trait in natural or genetically engineered strains. Manipulation of the rhizosphere environment to influence production or activity of the antibiotic may increase control of the pathogen (Thomashow and Weller, 1988).

Specific Suppression: Nematode Suppressive Soil—Decline of *Heterodera avenae*

Decline of cereal cyst nematode, *Heterodera avenae,* on susceptible cereals grown in intensive rotation throughout Europe is the best documented example of naturally occurring specific suppression of a plant parasitic nematode. Nematode populations peak after two crops and then fall rapidly. Once natural suppressiveness is established, populations of the cereal cyst nematodes are permanently maintained at levels below the economic threshold.

Kerry and Crump (1980) demonstrated that spore populations of two species of nematode-parasitic fungi are much greater in soils where *H. avenae* fails to multiply. These two parasites are widely distributed in cereal-growing soils in England and northern Europe. *Nematophthora gynophila* invades cysts of nematode females as they emerge from roots and destroys them within a week. *Verticillium chlamydosporium* invades virgin females, which produce empty cysts; infected egg-producing females produce fewer eggs than normal and many of the eggs are parasitized. Thus, the effect of fungal parasitism is to reduce the number of females forming cysts and the number of viable eggs that cysts contain. Application of the biocide, formaldehyde, to soil reduces spore populations of the antagonistic fungi and increases nematode multiplication. The biocide effect is observed in soils where *N. gynophila* and *V. chlamydosporium* are present, but it does not occur where they are absent.

Environmental factors influence the level of parasitism. *Nematophthora gynophila* is more active in wet seasons because soil moisture is critical for the dispersal of its zoospores along the root surface. Although limited by lack of moisture in dry seasons and in some free-draining soils, parasitism by *N. gynophila* and *V. chlamydosporium* may still occur in these soils and thereby provides economic benefits.

Biological Control with Introduced Antagonists

Diseases of belowground plant parts, such as seed rots, damping-off of seedlings, foot rots, root rots, and wilts are difficult to control by chemical approaches because large quantities of pesticides and repeated applications are required to treat the soil volume occupied by plant roots. An advantage of introduced antagonists over chemical protectants is that microbial populations can grow from a small quantity of inoculum applied to seeds or cuttings, or as an in-furrow application.

Although the first attempts at biological control were with introduced antagonists, they were generally ineffective unless the soil had been steamed or large quantities of organic matter had been added to support the introduced antagonist. Thus, the concept developed that the soil environment must be modified to permit establishment of an alien microorganism. There are few examples of successful commercial use of introduced organisms. However, recent advances in the technology of introducing biocontrol agents by encapsulation and for improving them by selection or genetic manipulation reflect a renewed interest in the use of introduced antagonists.

Rhizosphere Competence

To achieve protection of subterranean plant parts, the biocontrol organism needs to be **rhizosphere competent,** that is capable of colonizing the expanding root surface (Ahmad and Baker, 1987). Rhizosphere-competent organisms can proliferate along the root as it develops. This ability differs from that of colonization or infection at specific points as the root encounters propagules in bulk soil. Rhizosphere-competent organisms may not only effectively protect roots against pathogens but also have other benefits, such as production of plant-growth promoting substances or enhancement of uptake of nutrients and water by plant roots.

Specific traits that contribute to rhizosphere competence are mostly unknown but probably involve attachment, distribution, growth, and survival of microorganisms. The intrinsic ability of the microorganism to colonize root surfaces should be considered separately from the movement of propagules by mass flow of water in soil or along root surfaces.

Until recently, rhizosphere-competent strains of fungi had not been described despite the well-documented ability of several groups of bacteria to readily colonize plant root systems (Weller, 1988). Certain mutants of the fungus *Trichoderma* that produce more cellulase than their wild-type parent strains are also more rhizosphere competent. Thus, rhizosphere competence of strains of *Trichoderma* spp. may be related to the ability to degrade cellulosic materials on the root surfaces. Genetic changes also have been achieved by protoplast fusion of parental strains of *Trichoderma* spp. to produce progeny with enhanced rhizosphere competence. Although all of the gene products examined are from one parental strain or the other, remarkable variation exists in morphology, growth rate, nutritional requirements, and biocontrol ability. The basis for the variation is not yet known.

Bacterial traits that may contribute to rhizosphere competence are polysaccharide production, fimbriae, flagella, chemotaxis, osmotolerance, and ability to utilize complex carbohydrates. Extracellular polysaccharide (EPS) surrounds and binds bacterial cells together to form microcolonies. The EPS protects cells against desiccation, antibacterial agents, and predators; it also aids the cells by concentrating nutrients and ions. This protection may allow an introduced bacterium to avoid displacement by indigenous microorganisms. Fimbriae are proteinaceous, filamentous appendages that enhance root attachment by, for example, N_2-fixing *Klebsiella* and *Enterobacter* and the take-all suppressive strain of *Pseudomonas fluorescens*. The importance of flagella for bacterial movement along roots depends on bacterial strain, plant species, or physical conditions of the soil. Nonflagellated mutants of *Pseudomonas* spp. colonize roots to the same extent as their respective wild types. Chemotaxis at a soil water potential suitable for motility may enable bacteria to colonize roots, especially when bacteria are added with the seed in the furrow and are not initially in contact with the plant root. Chemotaxis may guide rhizobacteria to infection sites in plant roots up to several centimeters away. Tolerance to dry soil and low osmotic potential may aid the survival of some bacteria in the rhizosphere. A relationship between osmotolerance and population size of root-colonizing strains of *Pseudomonas* species has been observed. Ability to degrade complex carbohydrates might also provide a competitive advantage. While introduced and indigenous bacteria readily use most root exudates, fewer organisms can degrade the root-tip mucilages composed of such complex carbohydrates as cellulose, hemicellulose, and pectin.

The Root Colonization Process

The effectiveness of introduced antagonists for biological control depends on survival, establishment, or proliferation in the infection site on the root. If a biocontrol agent introduced on seed multiplies around the site of application but not along the root surfaces away from the cotyledon attachment, it may suppress pathogens causing seed rot and seedling diseases. However, root diseases are not typically suppressed by seed treatments.

Colonization of roots by introduced organisms occurs in two phases (Weller, 1988):

- Phase I—The bacteria attach to and are transported on the elongating root tip.

- Phase II—The bacteria spread locally and proliferate to the limits of the niche in competition with indigenous organisms and survive.

In Phase I, introduced bacteria on seeds, seed pieces, or tubers come in contact with and attach to emerging roots and gain access to root exudates. As root elongation occurs, some bacteria are carried along with the root tip, while others are left behind as a source of inoculum on older portions of the root. Transport of bacteria continues as long as bacterial proliferation occurs at the root tip; otherwise, without multiplication, transport occurs only until the initial inoculum at the tip is diluted out. Transport of bacteria by the root tip may be inefficient because tips do not always become or remain colonized. Loss of bacteria may occur by physical removal as the tip displaces soil particles, bacteria adsorb to soil particles, or indigenous microorganisms provide competition. However, the primary limitation may be an inability of bacteria to multiply quickly enough to keep pace with the root tip, which extends rapidly through the soil (2 to 9 cm day^{-1}) by expansion of root cells in the zone of elongation.

In Phase II, rhizosphere-competent bacteria multiply and survive on the root, whereas incompetent bacteria are rapidly displaced. The fate of introduced bacteria is ultimately governed by the ability to compete with indigenous microorganisms. Nutrients, rather than space, are thought to be the limiting factor in competition during rhizosphere colonization. Direct observations of roots from soil show that most of the root surface remains noncolonized. Bacteria tend to congregate at cell junctions where nutrients may be most abundant (Fig. 19–3). Since the carrying capacity of the rhizosphere is limited, an introduced strain must preempt indigenous bacteria to become established.

The distribution of bacteria along the root during Phase I and the propagation and survival during Phase II are profoundly affected by abiotic and biotic factors. Soil water potentials optimal for bacterial cell growth (−30 to −70 kPa) result in the greatest population development on roots, but establishment on roots may occur at water potentials as low as −400 kPa. Transport of bacteria along elongating roots does not require water percolation, but such water movement aids in bacterial dispersal. Movement through soil by water percolation is affected by bacterial

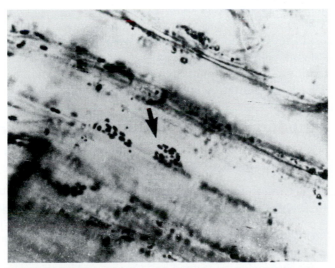

Figure 19–3 Bacteria (arrow) proliferating at cell junctions of *Eucalyptus* roots where root exudates may be most abundant. *Photo courtesy of N. Malajczuk. Used with permission.*

Table 19–1 **Reduction in Fusarium wilt when watermelon cultivars of varying susceptibility were grown in a disease suppressive soil obtained from a field where cultivar Crimson Sweet had been grown previously as a monoculture.**

Soil source*	Wilt per cultivar (%)			
	Florida Giant, susceptible	Charleston Gray, moderately resistant	Crimson Sweet, moderately resistant	California Gray, resistant
Nonfumigated fallow soil	82	72	92	56
Fumigated fallow soil	87	79	100	44
Florida Giant plot	79	80	36	44
Crimson Sweet plot	33	35	32	18

From Hopkins et al. (1987). Used with permission.

*Fallow soil had not been planted previously in watermelon. All four soils infested with Fusarium oxysporum f. sp. niveum at 1.5×10^3 conidia g^{-1} soil.

characteristics, such as cell shape, size, motility, and electrostatic charge. Soil factors that influence movement of bacteria include soil type, pore-size distribution, water content, and pH of the soil.

Optimum temperature for growth of bacteria *in vitro* may not reflect that for maximum root colonization. Optimal growth of *Pseudomonas fluorescens* and *P. putida* occurs at 25 to 30°C, but root colonization is generally greatest below 20°C. Microbial activity in soil increases as soil temperature increases; thus better colonization at lower temperature probably reflects less competition from indigenous microflora.

Plant genotype influences the quantity and composition of the rhizosphere microflora through qualitative and quantitative differences in root exudates. For example, the watermelon cultivar Crimson Sweet exhibited unusually high field resistance against Fusarium wilt during seven years of monoculture, although it is only moderately resistant to the disease in greenhouse tests. Repeated culture of Crimson Sweet results in populations of soil microorganisms suppressive to Fusarium wilt of watermelon, not only on this cultivar but also on other cultivars of watermelon when planted in the same soil (Table 19–1). In contrast, monoculture of other cultivars does not result in disease-suppressive populations of microorganisms. Thus, manipulation of host genotype may offer some opportunity to increase the efficiency or consistency of root colonization by introduced antagonists.

Introduction of Antagonists in Conjunction with Other Soil Treatments

The microbiological balance is altered by various environmental factors such as heat treatments, pesticides, and other chemical treatments that eliminate groups of microorganisms. Such treatment favors introduced strains of fungal species of *Trichoderma* or *Gliocladium* antagonistic to plant pathogens. Introduced *Trichoderma* spp. have a greater tolerance of broad-spectrum biocides than many other soil microorganisms and colonize treated soils more rapidly than other soil competitors. *Trichoderma* spp. predominate in soils after fumigation with dilute formaldehyde or carbon disulfide, or after application of sodium azide to soil.

Biocides can weaken pathogen propagules and render them more susceptible to attack by antagonists. *Armillaria* spp., the cause of mushroom root rot of many woody crops, are susceptible to biocontrol with *Trichoderma* spp. when sublethal doses of soil fumigants interfere with defense mechanisms of the pathogen resting structures.

Trichoderma spp. that are resistant to a chemical when applied in combination with an amount of the chemical that is sublethal to the pathogen, can achieve the same effect as a lethal dose of the chemical applied alone. *Trichoderma* spp. are normally sensitive to benomyl, but a benomyl-tolerant mutant of *T. harzianum* was found to be more rhizosphere competent when benomyl was added to the soil. This mutant has a significant advantage for application with benomyl or to fields previously treated with benomyl. Strains of *T. harzianum* are also significantly more tolerant than the root pathogen *Rhizoctonia solani* to methyl bromide. Treatment of soil with sublethal doses of the fumigant can kill *R. solani* with no effect on naturally occurring *Trichoderma* spp. This allows for rapid colonization of the soil with introduced *Trichoderma* spp. as well.

Trichoderma spp. can also be introduced into soils treated by solarization to control soilborne diseases. **Solarization** is accomplished by covering moistened soil in hot climates with transparent polyethylene plastic sheets, thereby trapping incoming radiation. This increases soil temperatures by 7 to 14°C, depending on soil depth, killing propagules of pathogens and weed seeds. The combination of *T. harzianum* and solarization reduces inoculum density of pathogens such as *R. solani* and *Sclerotium rolfsii* more than either treatment alone.

Seed Treatment with Biocontrol Agents

Biocontrol agents generally provide less effective control of seed and seedling root diseases and are more variable in their performance than are chemical pesticides. The antagonist must establish and grow well at the infection site after introduction on seed. Biocontrol agents frequently have been grown and then applied with systems designed for chemical pesticides.

Improving the efficacy of biocontrol requires alteration of the environment to make it more conducive for the biological control agent. The antagonist is placed on the seed surface, and the seed acts as a carrier or delivery system for the agent. After sowing, the antagonist is present at the infection site between the seed and the soil environment. Biocontrol agents are most often applied to seeds in a slurry. Though the slurry is applied uniformly, the agent may not adhere to the seed surface. Binders, such as methyl cellulose and polyvinylacetate, improve adhesion.

Controlled hydration or dehydration treatments, known as seed *priming*, improve seed performance, but osmotic treatments involved in the process may adversely affect microorganisms. Solid-matrix priming with vermiculite and an osmoregulant provides a suitable medium that is nontoxic, has a high water-holding capacity, and has a favorable chemical and physical environment for the biocontrol agent. Application of *T. harzianum*, combined with solid-matrix priming, provides an effective and reliable biological seed treatment of a number of crops for protection against several pathogens (Table 19–2). The combination of superior strains of fungal antagonists and solid-matrix priming has resulted in better plant stands than with fungicide-treated seeds alone.

Table 19–2 Biological seed treatment combining *Trichoderma harzianum* and solid-matrix priming to control pathogens of a wide range of crops.

Pathogen	Crops
Pythium ultimum	Cotton, Cucumber, Pea, Snap bean, Sweet corn, Wheat
Fusarium graminearum	Wheat
Sclerotium rolfsii	Snap beans
Rhizoctonia solani	Radish, Cucumber

From Harman et al. (1989). Used with permission.

Biological Control of Crown Gall

Crown gall is caused by the soilborne bacterium, *Agrobacterium tumefaciens*. The disease results from disorganized and uncontrolled cell division in the host plant and is usually visible on roots or around the crown. A wide variety of dicotyledonous plants is susceptible to crown gall, especially stone fruits, roses, and grapes.

The majority of the bacterial genes needed to induce crown gall formation are plasmid-borne. A portion of the tumor-inducing (Ti) plasmid, the T-DNA, is transferred to the plant chromosomal DNA, where it is expressed and stably maintained. The T-DNA genes encode for enzymes involved in metabolism of plant growth substances, which lead to production of a tumor on the plant. Other T-DNA genes code for enzymes that produce compounds called opines, which are low-molecular-weight metabolites that act as nutrients for *A. tumefaciens*. Ti plasmids are classified according to the type of opines present in the tumors they induce. Two of the Ti plasmid types are nopaline/agrocinopine A and octopine/agropine.

A highly successful biological control for crown gall has been developed in Australia (Kerr, 1980). The method involves inoculation of planting stock with nonpathogenic *Agrobacterium radiobacter* strain K84. The ability to produce the **bacteriocin** agrocin 84 is an important factor in biological control by strain K84. A bacteriocin is an agent produced by isolates of certain bacteria that inhibits the same or closely related species. The control method involves dipping the plant material in a tap water suspension of strain K84 at 10^7 to 10^8 cells ml^{-1} immediately before sowing or planting. Strain K84 has performed better than commercial chemical treatments in preventing crown gall, often providing 100% control in naturally infested soil. Strain K84 has been sold commercially in Australia since 1973 and is used in many countries.

Agrocin 84 is an adenine nucleotide thought to inhibit DNA synthesis. It is taken up by nopaline/agrocinopine A strains via agrocinopine permease. Part of the agrocin molecule mimics the structure of agrocinopine A, the usual substrate for agrocinopine permease. Bacterial mutant strains that do not take up agrocinopine are not sensitive to agrocin 84 (Box 19–4). Production of agrocin 84 is the major, but not the only, mechanism in the control of crown gall by strain K84. Survival of the bacteria and colonization of the root surface also play important roles.

The chromosomal background for the plasmid coding for agrocin 84 influences the strain's ability to control disease. When agrocin production is transferred from strain K84 to other strains of *A. radiobacter*, the new strains do not control disease as effectively as K84. The new strains cannot match the superior ability of K84 to colonize roots and survive in the soil and rhizosphere of the plant.

Box 19–4

Minimizing the Breakdown of Biological Control. Breakdown in biological control of crown gall by *A. tumefaciens* strain K84 appears to be related to agrocin 84. Pathogenic strains of *A. tumefaciens* acquire resistance to agrocin 84 and produce agrocin because the agrocin plasmid (pAg K84) is transferred from strain K84 to the pathogenic agrobacteria. This phenomenon led to the development of a genetically engineered, transfer-deficient (tra⁻) mutant of strain K84 (K1026), which guards against the breakdown in biocontrol because part of the transfer region in the pAg K84 plasmid has been deleted. Strain K1026 has the same ability to produce agrocin 84 and control crown gall without the ability to transfer the agrocin plasmid to other agrobacteria. The use of strain K1026 in place of K84 minimizes the potential for breakdown of biological control (Jones and Kerr, 1989).

Rhizobacteria and Biocontrol

The term **rhizobacteria** describes bacteria that aggressively colonize roots. The term *plant growth-promoting rhizobacteria* (PGPR) describes bacteria that aggressively colonize roots and cause plant growth promotion, regardless of the mechanisms involved (Kloepper et al., 1980). Root-colonizing bacteria and fungi that are deleterious to plant growth but not parasitic are called *deleterious rhizosphere microorganisms* (DRMO) (Schippers et al., 1987). Deleterious activities include alterations in the supply of water, nutrient ions, and plant growth substances which change root functions or limit growth. The PGPR are thought to improve plant growth by colonizing the root system and preempting the establishment of or suppressing DRMO on the roots. The PGPR may also promote the availability and uptake of mineral nutrients and the supply of plant growth substances. Recently, PGPR have been shown to induce biocontrol through a mechanism known as *systemic resistance*. For example, precolonization of roots by a *Pseudomonas* sp. induced systemic resistance in the stem against Fusarium wilt (Liu et al., 1995).

Pasteuria penetrans—Root-Knot Nematode

The bacterium *Pasteuria penetrans* has several characteristics that make it a potentially useful biocontrol agent against the root-knot nematode *Meloidogyne* spp. as well as other types of nematodes (Oostendorp et al., 1991). Bacterial populations parasitic on root-knot nematode not only prevent reproduction of the pathogen but also reduce infectivity of juveniles. The spores of *P. penetrans* have morphological and biochemical features of endospores; consequently, they can tolerate environmental extremes. Spores can be stored air-dried in soil and roots without loss of infectivity. Spores are also tolerant of temperature extremes. The spores are relatively small and can be distributed by percolating water. Thus they can be applied at or near the surface of sandy soils and will move down through the soil profile where nematodes are present.

However, the obligately parasitic nature of *P. penetrans* may limit its biocontrol potential. The soil eventually reaches a spore density that results in reduced infectivity of spore-encumbered juveniles, decreased production of infected females, and a corresponding decline in spore density of *P. penetrans*. Still, the longevity of *P. penetrans* may override the need for high host densities in some soils.

| Box 19–5 |

Inoculum and Strategies for Using *Pasteuria penetrans*. A dried and finely ground root preparation containing large numbers of females of *Meloidogyne* spp. infected with *Pasteuria penetrans* produces a light, easily handled powder for soil infestation. However, difficulties in mass culturing by other means have limited attempts to test the effectiveness of biocontrol. As a biological nematicide, spore concentrations as low as 0.5 to 5 × 10^5 spores g^{-1} soil produce control. However, it may be more feasible to introduce lesser quantities of inoculum and rely on the nematode's growth and reproduction to build up spore populations. One solution might be to grow crops with some tolerance to the nematode and rely on *P. penetrans* spores to increase to levels that maintain nematode populations below the damage threshold. Nevertheless, lack of mass production remains the major limitation preventing commercialization of *P. penetrans* (Oostendorp et al., 1991).

Biological Control with Composites of Antagonists

Composites of antagonists may be selected for their capacity to inhibit the pathogen through microbial succession and used to reduce disease. This approach requires that the biocontrol agents be compatible. Composites of antagonists should:

- operate at different sites, such as in the rhizosphere or in organic residues,

- effect biocontrol by different mechanisms, such as by competition or antibiosis,

- require different substrates for activity, such as mucigel metabolism by rhizosphere-competent fungi as compared to use of root exudates by pseudomonads, and

- be compatible with the soil environment and enhanced by cultural manipulations.

Composites of microorganisms can control several plant diseases. For example, a composite of fungal antagonists, selected for their ability to establish high populations in the rhizosphere of tomato, can reduce incidence of Fusarium crown rot of tomato. Two other examples are combinations of fluorescent pseudomonads and *Trichoderma koningii* that were more effective in control of take-all of wheat than the pseudomonads alone (Duffy et al., 1996) and the composites of fluorescent pseudomonads and nonpathogenic *Fusarium* spp. that were more effective for control of Fusarium wilt of cucumber than either antagonist alone (Park et al., 1988). In each case, the diversity of strains applied may better simulate the natural microflora responsible for disease suppressiveness than single species.

Commercialization of Biocontrol Agents

Emphasis on Singly Introduced Antagonists

Because of the potential for environmental pollution, the loss of many currently registered fungicides, nematicides, and soil fumigants for control of soilborne diseases from the market is inevitable. As an example, methyl bromide, the most widely used soil fumigant, may be eliminated by the turn of the century. This has intensified interest in biocontrol agents because they have few, if any, negative environmental or toxicological impacts. The translation of recent research findings into products for biological control of plant pathogens has been rapid. The most highly successful and long-standing example is biological control of crown gall. The record for other introduced biocontrol treatments is far less consistent. Nevertheless, the number of private companies with programs in biological control has increased. Continued efforts in the private sector will depend on the economics of the market, future progress in research, and governmental regulation of pesticide use.

The emphasis on biocontrol agents in the marketplace has been almost exclusively on singly-introduced organisms, even though mixtures of microorganisms may be more efficacious. Several reasons account for this focus. First, although single organisms have a relatively narrow spectrum of activity, they are more easily patented and marketed. Second, the time-limited efficacy of single applications ensures the need for repeated applications and subsequent enhanced sales. Third, the development of fermentation and formulation technology and delivery systems for individual biocontrol agents is similar to existing technology for mass production of microorganisms.

Interest in Microorganisms as Gene Vectors

With the advent of molecular techniques for genotypic manipulation of biocontrol agents, there is considerable commercial interest in the incorporation of foreign genes to enhance or alter the efficacy of microbes. In this way, microorganisms act as gene-vectors that express the production of specific antibiotics in the rhizosphere or in the plant. For example, the biopesticide *Clavibacter xyli* subsp. *cynodontis* is an endophytic pathogen of bermudagrass capable of systemically colonizing corn. The bacterium was transformed with the gene from *Bacillus thuringiensis* subsp. *kurstaki* coding for the production of endotoxin (Tomasino and Beach, 1991). The transformed strain of *C. xyli* subsp. *cynodontis* can produce and deliver the toxin to European corn borer in systemically colonized corn. Similar transformations are being evaluated for the management of plant diseases.

Limitations to Commercialization

A critical obstacle to biocontrol by soil amendment or seed treatment is the lack of methods for mass culture and introduction of microorganisms into production systems. Solid media, such as expanded clays, bark pellets, wheat bran, and barley grain, have been used for the experimental production of fungi for soil amendment. However, these solid media are not easily adapted to deep-tank liquid fermentation. Development of improved delivery technology, such as the production of alginates which encapsulates propagules in aqueous solutions of compounds that form gels, offers promise (Fravel et al., 1985). Then it is possible to use large-scale industrial production of propagules in dried forms to mix with gelling agents to prepare pelletized formulations. A further refinement of this approach is the incorporation of nutrient carriers together with the gelling agents to provide a food base in intimate contact with the antagonist.

Summary

Three broad categories of effort in biological control of soilborne pathogens and nematodes are identified in this chapter. First, resident biological controls can be manipulated with cultural practices that influence the antagonist-pathogen relationship. These practices include crops with durable resistance to the pathogen, crop rotations, and other cultural practices that augment and stabilize the biological equilibrium that results in suppressiveness. Second, when indigenous biological antagonists are inadequate, biocontrol agents or genes that tend to be self-maintaining are introduced one time or occasionally. Rhizosphere-competent organisms can be genetically engineered, formulated, and applied to subterranean plant parts in environments most conducive for establishment, growth, and dispersal of the introduced agent. Third, if the first two approaches are ineffective, plant-associated microorganisms, such as seed inoculants, are applied repeatedly with every crop.

Although the emphasis is on the patentability, production, and marketing of singly introduced organisms, mixtures of microorganisms may be more efficacious because they simulate the natural microflora responsible for disease suppressiveness. Expanded use of biocontrol agents will depend on the economics of their production, stricter regulation of chemical pesticides, and joint research ventures between the public and private sectors.

Cited References

Ahmad, J.S., and R. Baker. 1987. Rhizosphere competence of *Trichoderma harzianum*. Phytopathology 77:182–189.

Broadbent, P., and K.F. Baker. 1974. Behaviour of *Phytophthora cinnamomi* in soils suppressive and conducive to root rot. Austral. J. Agric. Res. 25:121–137.

Cook, R.J., and K.F. Baker. 1983. The nature and practice of biological control of plant pathogens. American Phytopathological Society Press, St. Paul, Minn.

Curl, E.A., R. Lartey, and C.M. Peterson. 1988. Interactions between root pathogens and soil microarthropods. Agric. Ecosys. Environ. 24:249–261.

Duffy, B.K., A. Simon, and D.M. Weller. 1996. Combinations of *Trichoderma koningii* with fluorescent pseudomonads for control of take-all on wheat. Phytopathology 86:188–194.

Fravel, D.R., J.J. Marois, R.D. Lumsden, and W.J. Connick. 1985. Encapsulation of potential biocontrol agents in an alginate-clay matrix. Phytopathology 75:774–777.

Harman, G.E., A.G. Taylor, and T.E. Stasz. 1989. Combining effective strains of *Trichoderma harzianum* and solid-matrix priming to improve biological seed treatments. Plant Dis. 73:631–637.

Hopkins, D.L., R.P. Larkin, and G.W. Elstrom. 1987. Cultivar-specific induction of soil suppressiveness to Fusarium wilt of watermelon. Phytopathology 77:607–611.

Jones, D.A., and A. Kerr. 1989. *Agrobacterium radiobacter* strain K1026, a genetically engineered derivative of strain K84 for biological control of crown gall. Plant Dis. 73:15–18.

Kerr, A. 1980. Biological control of crown gall through production of agrocin 84. Plant Dis. 64:25–30.

Kerry, B.R. and D.H. Crump. 1980. Two fungi parasitic on females of cyst nematodes (*Heterodera* spp.). Trans. Brit. Mycol. Soc. 74:119–125.

Kloepper, J.W., M.N. Schroth, and T.D. Miller. 1980. Effects of rhizosphere colonization by plant growth-promoting rhizobacteria on potato plant development and yield. Phytopathology 70:1078–1082.

Liu, L., J.W. Kloepper, and S. Tuzun. 1995. Induction of systemic resistance in cucumber against Fusarium wilt by plant growth promoting rhizobacteria. Phytopathology 85:843–847.

Malajczuk, N. 1983. Microbial antagonism to *Phytophthora*. pp. 197–218. *In* D.C. Erwin, S. Barnicki-Garcia, and P.H. Tsao (eds.), *Phytophthora:* Its biology, taxonomy, ecology and pathology. American Phytopathological Society Press, St. Paul, Minn.

Oostendorp, M., D.W. Dickson, and D.J. Mitchell. 1991. Population development of *Pasteuria penetrans* on *Meloidogyne arenaria*. J. Nematol. 23:58–64.

Park, C.-S., T.C. Paulitz, and R. Baker. 1988. Biocontrol of Fusarium wilt of cucumber resulting from interactions between *Pseudomonas putida* and nonpathogenic isolates of *Fusarium oxysporum*. Phytopathology 78:190–194.

Schippers, B., A.W. Bakker, and P.A.H.M. Bakker. 1987. Interactions of deleterious and beneficial rhizosphere microorganisms and the effect of cropping practice. Annu. Rev. Phytopathol. 25:339–358.

Thomashow, L.S., and D.M. Weller. 1988. Role of a phenazine antibiotic from *Pseudomonas fluorescens* in biological control of *Gaeumannomyces graminis* var. *tritici*. J. Bacteriol. 170:3497–3508.

Tomasino, S.F., and R.M. Beach. 1991. Field activity of a *Clavibacter xyli* subsp. *cynodontis/Bacillus thuringiensis* recombinant against European corn borer. Phytopathology 81:704.

Weller, D.M. 1988. Biological control of soilborne plant pathogens in the rhizosphere with bacteria. Annu. Rev. Phytopathol. 26:379–407.

General References

Adams, P.B. 1990. The potential of mycoparasites for biological control of plant diseases. Annu. Rev. Phytopathol. 28:59–72.

Fravel, D.R. 1988. Role of antibiosis in the biocontrol of plant diseases. Annu. Rev. Phytopathol. 26:75–91.

Harman, G.E. 1992. Development and benefits of rhizosphere competent fungi for biological control of plant pathogens. J. Plant Nutr. 15:835–843.

Hoitink, H.A.J., and P.C. Fahy. 1986. Basis for the control of soilborne plant pathogens with composts. Annu. Rev. Phytopathol. 24:93–114.

Hornsby, D. (ed.). 1990. Biological control of soil-borne plant pathogens. CAB International, Wallingford, Oxon, England.

Stirling, G.R. 1991. Biological control of plant parasitic nematodes. CAB International, Wallingford, Oxon, England.

Study Questions

1. What are the roles of plants in the dynamic equilibrium of living organisms in soil?

2. Which groups of organisms are the most important plant pathogens?

3. What are the methods by which microorganisms limit populations of plant pathogens?

4. How may disease of plants be reduced by biological control in situations in which populations of the pathogen are not reduced?

5. Why is it more important to integrate biological controls, as compared to chemical controls, with other plant disease management practices?

6. Which characteristics may be most important for rhizosphere competence of soilborne microorganisms?

7. What are some of the influences that plant genotype may have on the composition of rhizosphere microorganisms involved in biological control of plant diseases?

8. Which methods of delivery of biological control agents may be of the greatest commercial potential?

9. What are the important mechanisms in the control of crown gall by nonpathogenic strains of *Agrobacterium radiobacter*?

10. Why may use of several biological control agents be more effective than single antagonists in the control of plant pathogens?

11. Considering the control of root-knot nematode with *Pasteuria penetrans* as a model system, why may organisms with the greatest potential for biocontrol of a plant pathogen sometimes be the most difficult to manipulate for commercial use?

Chapter 20

♦

Biochemistry and Metabolism of Xenobiotic Chemicals

William J. Hickey

♦

The microbes will have the last word.
Louis Pasteur

Chapters 11 to 16 described the role of microbial communities in cycling natural organic material. Here we continue this discussion, as metabolism of **xenobiotic** chemicals is in many ways an extension of these activities. The term "xenobiotic" is derived from the Greek words *xeno* (stranger) and *bios* (life), and thus a xenobiotic is literally a chemical that is a stranger to the biosphere. It is important to recognize that xenobiotics can be degraded by abiotic as well as biotic reactions. The biotic (microbial) activities are generally more environmentally significant because their enzyme systems greatly accelerate degradation rates and may result in complete destruction, or mineralization, of organic pollutants. Microbial communities are thus important in the natural dissipation of xenobiotics in the environment and in contaminant cleanup operations, the latter by providing the potential for bioremediation (Chapter 21).

The emphasis of this chapter is the biochemical mechanisms underlying microbial transformations of xenobiotics. Our objective is to establish a framework from which the transformations of xenobiotics may be understood and the likely biodegradation pathways predicted for a given set of environmental conditions. A brief discussion of the physiochemical controls on biodegradation is also provided because of the importance of abiotic interactions on xenobiotic metabolism in soil.

Xenobiotics: What Are They, and Why Are They Problems?

Xenobiotics are anthropogenic chemicals, products of the synthetic chemical industry, with "nonnatural" structural attributes. Xenobiotic character may be conferred on a compound by heteroatoms (e.g., oxygen, nitrogen, sulfur) in carbon backbones, **halogen** substituents, and branching or polymeric structures (Fig. 20–1). Those features may have the adverse effect of increasing the compound's resistance to microbial degradation. These structures are not unique to xenobiotics as similar features, even halogens (Gribble, 1994), occur in natural organic compounds. Thus, a xenobiotic structure is not conferred by a particular chemical trait but rather a combination of structural elements created by an anthropogenic process.

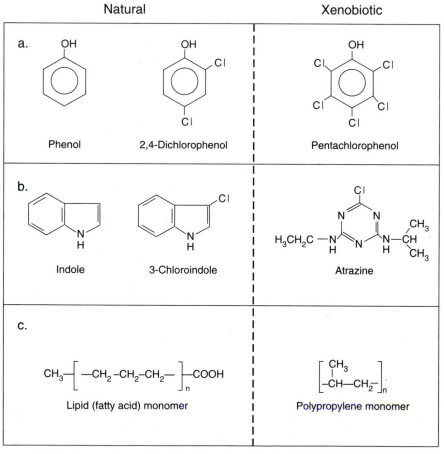

Figure 20–1 Structural comparison of xenobiotic and natural compounds.
(a) Phenol and 2,4-dichlorophenol are naturally occurring compounds, the latter produced by the fungus *Penicillium*. Pentachlorophenol is a xenobiotic biocide.
(b) Contrast between the natural nitrogen heterocycles indole and 3-chloroindole, produced by the marine worm *Ptychorda flava laysanica,* and the xenobiotic herbicide atrazine. (c) Comparison of the polymer subunit for lipids (fatty acids) to that of the widely used plastic, polypropylene.

Furthermore, although resistance to biodegradation, referred to as **recalcitrance,** is commonly associated with xenobiotic substances, this behavior is also exhibited by natural compounds like lignin and humic substances.

The recalcitrance of xenobiotics often reflects incompatibility with microbial metabolic pathways. Here we focus primarily on describing the means by which xenobiotic structures inhibit biodegradation and the biochemical mechanisms developed by microbes to overcome these blockages. Soil physiochemical factors also play a key role in xenobiotic biodegradation by modulating the type and level of microbial activity. Therefore, a brief discussion of these factors provides an environmental context for subsequent discussion of metabolic activities.

Overview of Xenobiotic Degradation

Environmental Influences

Xenobiotic biodegradation involves a potentially complex series of interactions between the xenobiotic microbial populations and the environment. Microbial populations can have the potential to degrade a xenobiotic, but the chemical may persist if environmental conditions are unfavorable for degradative activity. Physiochemical aspects of the environment can influence xenobiotic biodegradation in many ways; for purposes of this discussion, these are categorized as affecting microbial ecophysiology or substrate bioavailability. In the context of xenobiotic degradation, the concept of *ecophysiology* recognizes that microbial populations are reservoirs of catalytic potential from which different types and rates of degradative activity may be expressed, depending on environmental conditions. The interaction between soil water and oxygen levels is a primary ecophysiological variable controlling redox potential. For biodegradation of xenobiotics, this is a critical interaction determining the potential for xenobiotics to be degraded by aerobic or anaerobic mechanisms. Mineral nutrient levels are also important ecophysiology factors, particularly in cases like hydrocarbon spills where xenobiotics create high organic carbon loads in the soil. Finally, in cases where xenobiotics are degraded by cometabolism (discussed later in this chapter), the presence of *cosubstrates* that induce expression of certain degradative enzymes may be an important ecophysiological factor.

Bioavailability is the degree to which a xenobiotic is accessible to microbial cells or extracellular enzymes. Many physiochemical processes can affect bioavailability (Mihelcic et al., 1993). Because microbes interact with the soil environment and access substrates through the aqueous phase, chemical characteristics or physiochemical processes affecting aqueous-phase concentrations are important in determining bioavailability. Thus, nonpolar xenobiotics can have low bioavailability because of low aqueous solubility. Sorption of nonpolar compounds by organic matter and mineral surfaces in soil may also decrease bioavailability.

Polymeric xenobiotics can pose additional bioavailability problems because of the inability of cells to assimilate high-molecular-weight compounds. Effective biodegradation of such materials necessitates the activity of extracellular enzymes like those involved in cellulose and hemicellulose biodegradation. When extracellular enzymes are ineffective against a polymeric xenobiotic, degradation of the chemical is largely dependent on abiotic decomposition. In contrast, some microbial enzymes, notably extracellular peroxidases, may catalyze the opposite reaction and mediate the formation of xenobiotic oligomers or the coupling of xenobiotics to soil humic substances. These transformations have been implicated as mechanisms responsible for the formation of soil-bound residues of xenobiotic compounds and, in some cases, as potential mechanisms for detoxifying contaminated soils. However, the long-term fate of such residues, and thus the efficacy of the approach as a remedial treatment, is unknown.

Pathways and Limiting Factors

Many xenobiotics can support microbial growth by serving as sources of nutrients such as carbon, nitrogen, and phosphorus, as well as energy sources. In these cases, using xenobiotics for growth depends on the microbe's ability to funnel these chemicals into central metabolic pathways (Chapter 10). To do so, microbes must first eliminate blocking groups such as halogens that are commonly associated with xenobiotics and which interfere with central pathway enzymes. Figure 20–2 provides an overview of the general approaches and types of enzymes by which aerobic microbes funnel halogenated **aromatic** xenobiotics into central pathways; details on these mechanisms are discussed later in this chapter. Aerobic haloaromatic metabolism can proceed by several pathways mediated by a diverse collection of enzymes; a basic distinction among these approaches is whether or not the aromatic ring is dehalogenated prior to cleavage. Pathways A1-3 in Figure 20–2 illustrate ring dehalogenation and show that this process can occur through oxidative, hydrolytic, or reductive reactions. In pathway B, the aromatic ring is cleaved directly by oxidative enzymes yielding halogenated **aliphatic** products; dehalogenation of the latter is needed for their entry into the tricarboxylic acid (TCA) cycle and may require additional specialized enzymes.

An attribute common to all of the pathways for aerobic metabolism of aromatics shown in Figure 20–2 is that ring cleavage proceeds from dihydroxylated intermediates, such as catechol and protochatechuic acid. The latter substrates are produced by the activity of oxygenases, which insert two hydroxyl groups in an *ortho* or *para* configuration (Box 20–1). In these reactions, oxygenases mediate the **electrophilic** (Greek, electron loving) addition of O_2 to the benzene nucleus, and ring substituents influence the susceptibility of aromatic compounds to aerobic biodegradation by altering electron distribution (Table 20–1). This substituent effect can help explain and predict the relative biodegradability of xenobiotic compounds.

Cometabolism and Detoxification Reactions

Cometabolism, also called incidental metabolism, fortuitous metabolism, or cooxidation, is a transformation mediated by a single organism leading to dead-end products (Fig. 20–3). At best, the process yields no benefit to the microbe. At worst, it negatively affects the cell by being an unproductive use of reduced cofactors, such as NADH, or by producing toxic metabolites. Cometabolism is usually attributed to the activity of enzymes with "relaxed" substrate specificities. Oxygenases frequently exhibit this trait and are probably the group of enzymes most commonly associated with cometabolic transformations. Other enzymes, such as hydrolases and reductases and even enzyme cofactors like vitamin B_{12}, have also been implicated in cometabolic transformations.

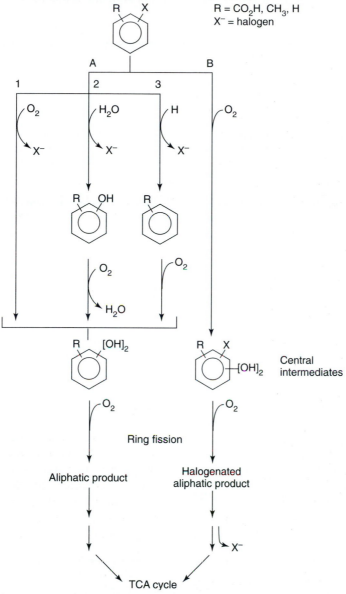

Figure 20–2 Overview of microbial metabolism of halogenated aromatic compounds. The two approaches shown are distinguished by whether dehalogenation occurs before (pathway A) or after (pathway B) ring fission. Three ring dehalogenation mechanisms are shown: oxygenolytic (pathway A1), hydrolytic (pathway A2), and reductive (pathway A3). Common to pathways A and B is the formation of a dihydroxylated intermediate from which ring fission proceeds. To make this model pathway as general as possible, R (e.g., a carboxyl, methyl, or hydrogen group) and X (e.g., a halogen) are representative ring substituents at nonspecified positions.

Table 20–1 Substituent effects on aerobic metabolism of aromatic xenobiotics.

| Activate* (*ortho, para*)[†] | Deactivate | |
	(*meta*)	(*ortho, para*)
—NH$_2$	—NO$_2$	—F
—OH	—CN	—Cl
—OCH$_3$	—COOH	—Br
—CH$_3$	—SO$_3$H	—I
—C$_6$H$_5$	—CHO	

*Activate/deactivate the aromatic ring for electrophilic substitution = facilitate/impede biodegradation.
†Positions to which groups are directed in electrophilic substitution reactions.

Box 20–1

Isomers of Benzene. The three possible isomers of a disubstituted benzene are differentiated by the names (substituent position relative to position 1) *ortho* (2 or 6), *meta* (3 or 5), and *para* (4).

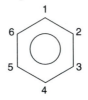

Box 20–2

Using Chemical Rules to Predict the Relative Biodegradability of Substituted Benzenes. How might —OH, —Cl, and —SO$_3$H groups present on a benzene ring be expected to influence the positions at which an oxygenase would attach an oxygen atom to the aromatic ring and the relative ease with which this reaction would occur?

Based on information given in Table 20–1, the substituents indicated above would direct oxygen addition to the positions designated by the arrows. The relative reactivity of the chemicals in this process decreases from left to right.

Cometabolic transformations can facilitate bioremediation by initiating degradation of xenobiotics that would otherwise be slowly degraded by abiotic processes. However, limitations arise when the cometabolic reactions involve enzymes that require cosubstrates, which are needed to induce expression of the biodegradative enzymes and support the microbe's growth and activity. In these cases, the cosubstrates need to be added to soils because their natural levels are generally too low to support rates of biodegradation required in bioremediation. Other limitations for applying cometabolic processes for bioremediation may include competitive inhibition of xenobiotic degradation by the cosubstrate or inactivation of key enzymes by metabolites produced from the xenobiotics.

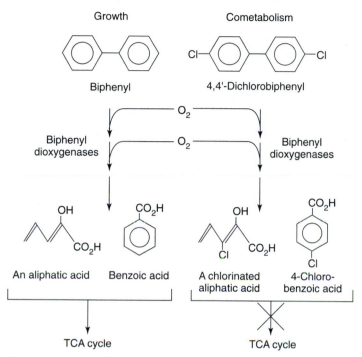

Figure 20–3 Growth-supporting metabolism of biphenyl linked to cometabolic degradation of polychlorinated biphenyls (PCBs). Biphenyl induces formation of oxygenases that oxidize and cause ring cleavage of biphenyl and PCBs. Aliphatic and aromatic acids produced from biphenyl are assimilated through the TCA cycle and support microbial growth. In contrast, PCB metabolism is unproductive in that chlorinated aliphatic and aromatic acids are typically not assimilated by the biphenyl degrader.

Common **detoxification** reactions are methylation and conjugation, which can transform xenobiotics to less toxic or nontoxic metabolites. Detoxification and cometabolic transformations are similar in that they typically do not allow the microbe mediating the process to use the xenobiotic for growth. However, detoxification differs from cometabolism because the organisms presumably benefit from the former by eliminating a toxin from the cell or its environment. The methylation of metals was introduced in Chapter 16. Methylation is also frequently described for phenolic compounds and, although mediated by both bacteria and fungi, is more commonly associated with fungi. There are many kinds of conjugation reactions, but those involving glutathione and glutathione S-transferases (GST) are among the most widespread. Because compounds with electrophilic substituents are common GST substrates, these enzymes can be important in the detoxification of halogenated xenobiotics (Fig. 20–4). Finally, some microbes appear to have capitalized on the dehalogenation capabilities of GST by recruiting these enzymes for use in catabolic pathways (Box 20–3): GST-like enzymes have been identified as mediating dechlorination of xenobiotics such as dichloromethane, 2,4,5-trichlorophenoxyacetic acid (2,4,5-T), and pentachlorophenol (PCP).

Alachlor   Glutathione

Figure 20–4 Glutathione S-transferase mediated dehalogenation of alachlor. The glutathione-alachlor conjugate can be further metabolized resulting in the accumulation of a variety of metabolites. The conjugation reaction and subsequent transformations are considered part of a detoxification mechanism because they typically do not support microbial growth.

| Box 20–3 |

Enzyme Recruitment and the Evolution of Metabolic Pathways. Enzyme recruitment refers to a process in which an enzyme serving a basic cellular function is duplicated, modified, and then used in a different metabolic pathway. This recruitment process is believed to be a mechanism by which organisms evolve metabolic pathways for using xenobiotic chemicals as growth substrates.

Microbial Consortia and Xenobiotic Degradation

Microbial degradative activities can be greatly influenced by interactions with other organisms in the environment. A primary example is the possibility of xenobiotic chemicals being biodegraded by microbial consortia. **Consortia** are functionally defined as two or more populations of organisms with complementary metabolic functions that collectively biodegrade a xenobiotic compound. Some consortia that have been isolated mediate the aerobic degradation of polychlorinated biphenyls and pesticides such as atrazine and parathion. (Fig. 20–5a). In anaerobic environments, which are often characterized by complex microbial food webs, consortia-mediated xenobiotic transformations are perhaps the norm. The most detailed analysis of an anaerobic, xenobiotic-degrading consortium is one that grows on 3-chlorobenzoate (Fig. 20–5b). Given the complexity of most natural microbial communities, consortial activities may be commonplace. However, because traditional laboratory approaches are based on pure-culture studies, the degree to which these processes occur in soil is poorly understood.

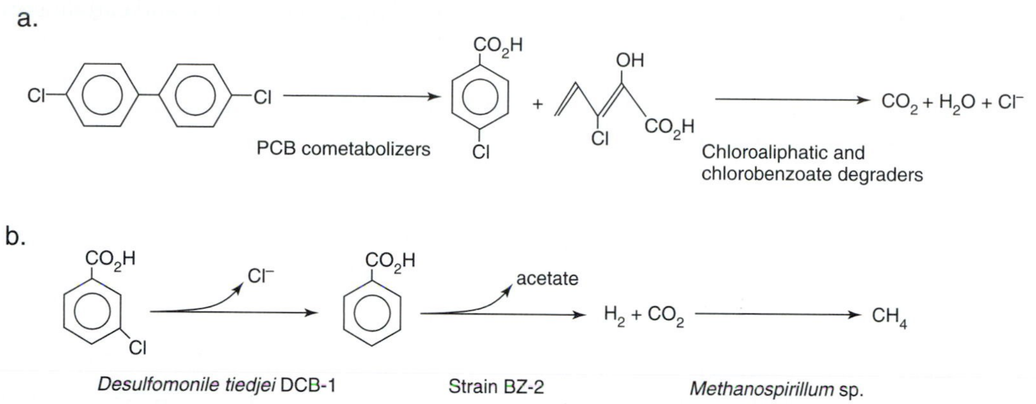

a.

PCB cometabolizers

Chloroaliphatic and
chlorobenzoate degraders

b.

Desulfomonile tiedjei DCB-1   Strain BZ-2   Methanospirillum sp.

Figure 20–5 Consortia in aerobic and anaerobic degradation of chlorinated aromatic compounds. (a) Polychlorinated biphenyl (PCB) cometabolizers degrade PCB (see also Fig. 20–3) and commensal populations of microbes mineralize chlorinated aliphatic and chlorobenzoic acids. The latter groups of microbes typically are unable to initiate PCB degradation. (b) Consortium-mediated anaerobic mineralization of 3-chlorobenzoic acid (3-CBA). *Desulfomonile tiedjei* generates energy from 3-CBA dehalogenation and produces benzoate, which is a growth substrate for strain BZ-2. The H_2 and CO_2 generated by strain BZ-2 supports growth of the methanogen, *Methanospirillum* sp.

Biochemistry of Xenobiotic Metabolism

In this section we discuss the biochemical aspects of reactions important to the degradation of xenobiotics commonly encountered as environmental contaminants. Model reactions are described and illustrated with selected chemicals so as to develop mechanistic frameworks and unifying themes for xenobiotic degradation. Xenobiotics that are not specifically discussed here can be assessed within this framework for probable metabolic pathways and biodegradation potentials under a given set of environmental conditions. The number of xenobiotic chemicals and possible transformations is too great to allow comprehensive treatment here. More detailed information is available from other sources (Alexander, 1994; Barr and Aust, 1994; Hardman, 1991; Mohn and Tiedje, 1992; Rochkind-Dubinsky et al., 1987; Sigel and Sigel, 1992; Walker and Kaplan, 1992).

Hydrolytic Reactions

Hydrolases are commonly used by microbes to biodegrade natural organic polymers such as cellulose or proteins. These enzymes are also important in xenobiotic metabolism because many bonds in anthropogenic chemicals, especially pesticides, are susceptible to hydrolytic cleavage. Also, some *dehalogenases* that funnel xenobiotics into central metabolic pathways use hydrolytic mechanisms. The many types of hydrolytic reactions share the common characteristic that an atom or ligand is replaced with water. In essence, water acts as a **nucleophile** (Greek, nucleus loving) by attacking centers of positive charge. Because O_2 is not a reactant, hydrolytic transformations can occur under anaerobic conditions; the presence of hydrolyzable groups in a xenobiotic can thus be an important factor determining its anaerobic biodegradation potential.

Many hydrolases act on acid derivatives, including esters, amides, and carbamates, which occur commonly in pesticides (Fig. 20–6). As a group, these enzymes have broad substrate ranges and are frequently produced constitutively (always turned on). Thus, xenobiotics may not be the true physiological substrates, and some hydrolytic transformations may be cometabolic processes. Dehalogenases frequently initiate degradation of halogenated aromatics and aliphatics. With the latter compounds, two kinds of halidohydrolases are known: the 2-haloacid dehalogenases, which are active against α-halogenated C_1- and C_2-carboxylic acids (Fig. 20–7a), and the halo-n-alkane dehalogenases (Fig. 20–7b). Halidohydrolases use water to replace halogens with hydroxyl groups, and their ability to do this is affected by the number and type of halogen substituents and by the presence of unsaturated carbon-carbon bonds. For example, halidohydrolase activity decreases as the electronegativity and number of the halogen substituents increases; the former trend reflects increasing carbon-halogen bond strength. Halogenated alkenes are resistant to degradation by halidohydrolases; aerobic biodegradation of these compounds is generally caused by oxygenases in cometabolic reactions or by reductive mechanisms under anaerobic conditions.

As shown in Figure 20–2 (pathway A2), some aerobic bacteria in the initial stages of haloaromatic metabolism may use hydrolytic dehalogenases in the degradation of halogenated benzoates and phenols. For halobenzoates, hydrolytic dehalogenation has been studied most extensively with 4-chlorobenzoate (4-CBA) and proceeds from a 4-CBA-coenzyme A intermediate (Fig. 20–7c). This transformation of 4-CBA is notable for two reasons. First, it represents a degradation mechanism distinct from the dioxygenase-mediated reactions common in aerobic metabolism of other chlorobenzoate isomers. Second, this process is probably another example of enzyme recruitment; the enzymes used in this process are similar to those of the β oxidation pathway (Babbit et al., 1992).

The initial dehalogenation reaction during aerobic metabolism of PCP may be hydrolytic (Fig. 20–7d) or mediated by oxygenases. Using either oxygenases or hydrolases to dehalogenate PCP illustrates the potential for microbes to develop diverse mechanisms for metabolizing such chemicals. It also suggests that the degradative mechanism employed for a xenobiotic compound may be organism-specific.

Molecular Oxygen–Dependent Reactions

The use of molecular oxygen (O_2) as a reactant in xenobiotic metabolism is a hallmark of aerobic microbes and bestows on these organisms a substrate utilization range much greater than that of their anaerobic counterparts. Three types of enzymes that use O_2 as a substrate are oxygenases, oxidases, and peroxidases. Although mechanistically different, oxygenases and peroxidases serve similar degradative functions by generating active oxygen species that oxidize xenobiotics. In contrast, oxidases indirectly interact with xenobiotics by transferring electrons from the substrates to O_2. Here we will focus on the functions of peroxidases and oxygenases in xenobiotic metabolism.

Peroxidases. Peroxidases oxidize xenobiotics by using hydrogen peroxide (H_2O_2) to effect single electron oxidations (Fig. 20–8a). The free radicals generated by this process may react further with O_2 or organic compounds, such as xenobiotic molecules or natural organic matter, to form polymers (Fig. 20–8b). As

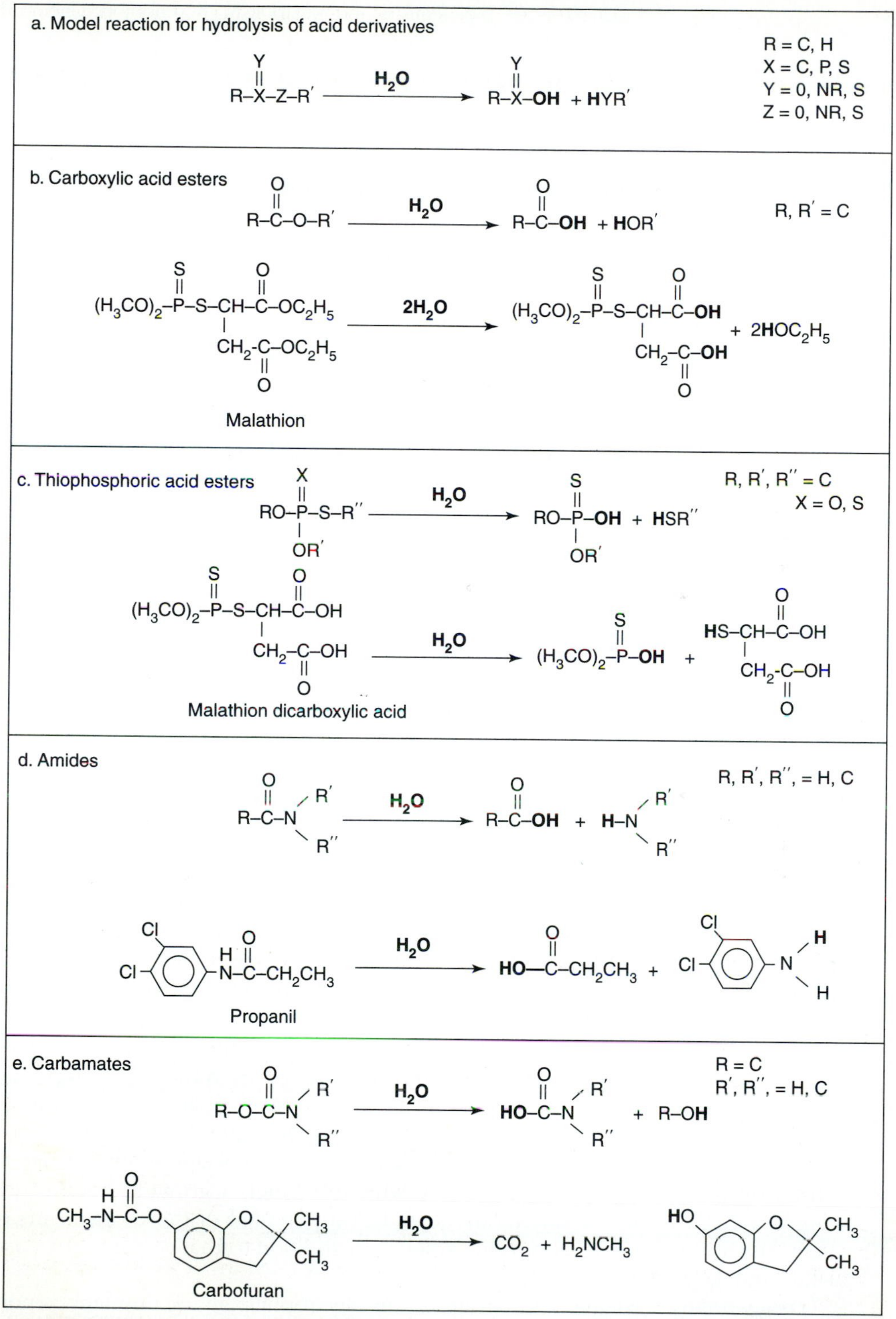

Figure 20–6 Hydrolysis reactions in xenobiotic biodegradation. (a) The model reaction for hydrolysis of acid derivatives. (b–e) Details for some acid derivative structures commonly encountered in xenobiotics; type structures and reactions are shown and illustrated with pesticides. The atoms from the H_2O used in hydrolysis are bolded.

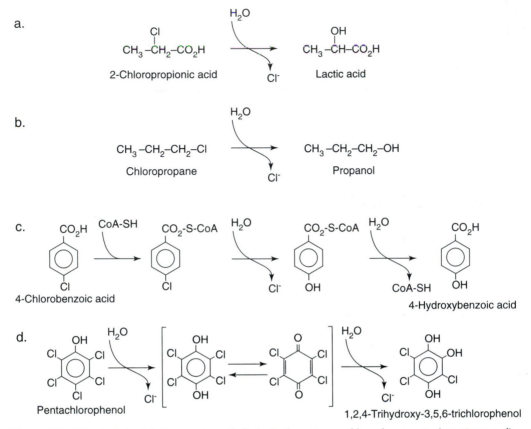

Figure 20–7 Hydrolytic dehalogenation of aliphatic (reactions a–b) and aromatic (reactions c–d) xenobiotics.

described earlier, the latter process might contribute to the formation of soil-bound xenobiotic residues and, because it essentially immobilizes the xenobiotic compound, might be useful to detoxify contaminated soils. Although both bacteria and fungi produce these enzymes, fungal peroxidases have been more widely studied. Lignin and manganese peroxidases produced by the wood-degrading basidiomycetes such as *Phanerochaete chrysosporium* of the **white-rot fungi,** have gained notoriety as nonspecific, extracellular enzymes capable of degrading a wide range of organic chemicals. These reactions are examples of cometabolism since the fungi gain no apparent nutritional benefit. Although attention has focused on the peroxidases produced by white-rot fungi, a growing body of evidence indicates that the xenobiotic degradation potential of these organisms reflects the activity of several enzyme systems in addition to that of the peroxidases (Barr and Aust, 1994).

   Oxygenases. Oxygenases activate O_2 for electrophilic attack on many xenobiotics. These enzymes are unique to aerobic organisms. The "relaxed" specificity of many oxygenases makes possible a variety of cometabolic transformations which extend the range of the hydrocarbon substrates that aerobic organisms can degrade.

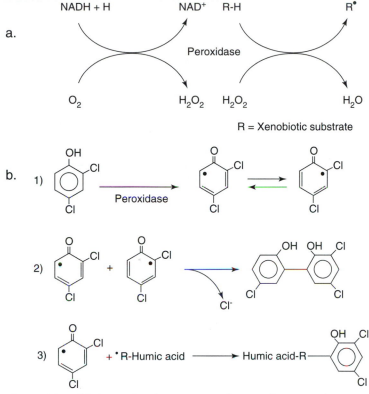

Figure 20–8 (a) Generalized reaction mechanism for peroxidase-mediated radical (R•) generation. (b) Proposed coupling mechanism for xenobiotic radicals (here 2,4-dichlorophenol). Reaction b1 shows radical production and two of four possible resonance structures. 2,4-Dichlorophenol radicals may react with each other forming oligomers (reaction b2) or bind to humic acids in the soil (reaction b3). *Modified From Dec and Bollag (1994).*

These enzymes are categorized as monooxygenases or dioxygenases depending on whether they incorporate one or both O_2 atoms into a substrate. *Monooxygenases* are sometimes called "mixed function oxidases" because they also act as oxidases by reducing one oxygen atom to water (Fig. 20–9). Monooxygenases mediate many transformations, including hydroxylation, epoxidation, and dealkylation.

Hydroxylation reactions are important in the metabolism of aliphatic and aromatic organic compounds. With aliphatic substrates, the alcohols produced by monooxygenases are further oxidized to carboxylic acids, the starting point for β-oxidation (Fig. 20–9a). Hydroxylation is a common reaction for many aromatic compounds and results in the formation of central intermediates, such as catechols, that are funneled into basic energy-generating pathways (Fig. 20–9c). Monooxygenase-mediated dehalogenations may initiate degradation of haloaromatics; for example, *para* dechlorination of PCP by a strain of *Flavobacterium* is known to be monooxygenase-mediated. This *oxygenolytic dehalogenation* rep-

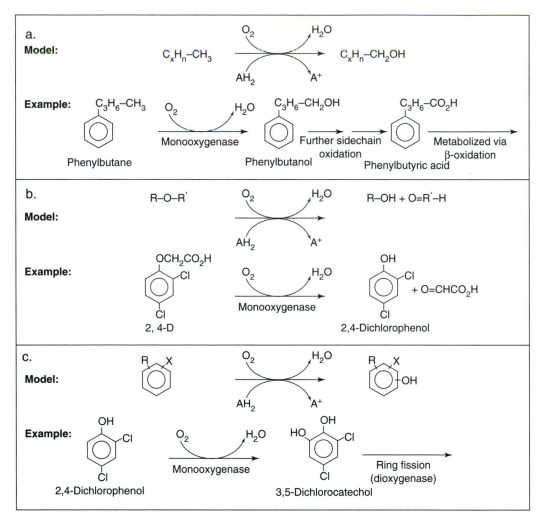

Figure 20–9 Type reactions for monooxygenases showing (a) terminal oxidation of aliphatic compounds, (b) O-dealkylation, and (c) aromatic ring hydroxylation. Common characteristics of these reactions are that a reduced cofactor is required and that one oxygen atom is inserted into the xenobiotic substrate with the other oxygen atom being reduced to water. In (a), C_xH_n is a generalized formula for an aliphatic hydrocarbon chain. In (b), R and R′ indicate alkyl groups. In (c), R (e.g., a carboxyl, methyl, or hydrogen group) and X (e.g., a halogen) are representative ring substituents at nonspecified positions.

resents a second mechanism by which aerobic bacteria eliminate halogens from aromatic xenobiotics. Monooxygenases catalyze removal of alkyl groups bonded to oxygen or nitrogen and thus facilitate biodegradation of many pesticides like 2,4-D (Fig. 20–9b). Epoxidation reactions are also common for these enzymes; an example important in bioremediation is that occurring with trichloroethylene (Fig. 20–10). This cometabolic transformation is mediated by monooxygenases or an aliphatic monooxygenase produced by bacteria growing on compounds

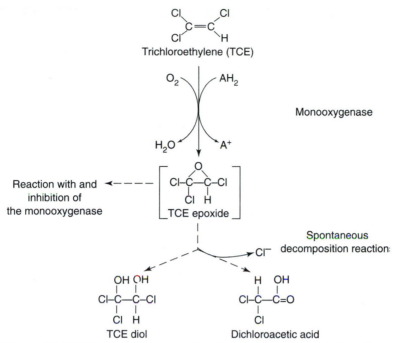

Figure 20–10 Monooxygenase-mediated, cometabolic degradation of TCE. TCE epoxide is an unstable intermediate that can react with and inhibit the oxygenase or spontaneously decompose.

like methane. The TCE epoxide generated by this cometabolic process may spontaneously decompose to innocuous, biodegradable products or react with and inactivate the monooxygenase. When the latter occurs, the epoxide acts as a *suicidal intermediate* (Box 20–4) and can be a factor limiting the effectiveness of TCE biodegradation by aerobic bacteria.

Dioxgyenases catalyze hydroxylation and fission (breakage) of aromatic rings (Fig. 20–11). Dioxygenase hydroxylations of aromatic xenobiotic compounds are functionally similar to those mediated by monooxygenases in serving to create compounds suitable for central metabolic pathways. Dioxygenases usually attach both atoms of O_2 to adjacent ring carbons lacking halogen substituents. Thus, a major limitation for the aerobic metabolism of highly halogenated xenobiotics (e.g., chlorobenzenes and PCBs) is a lack of adjacent, nonsubstituted ring carbons.

In addition to the number of oxygen atoms incorporated into the substrate, dioxygenases and monooxygenases are distinguished by the former's ability to break aromatic rings. The two general mechanisms of ring breakage are *intra-* and *extra-diol cleavage* and give rise to the *ortho-* and *meta*-cleavage pathways, respectively (Fig. 20–11b). The type of ring fission occurring in metabolic pathway is influenced by the occurrence and position of halide atoms in the aromatic substrate and the potential for the formation of suicidal intermediates (Box 20–4).

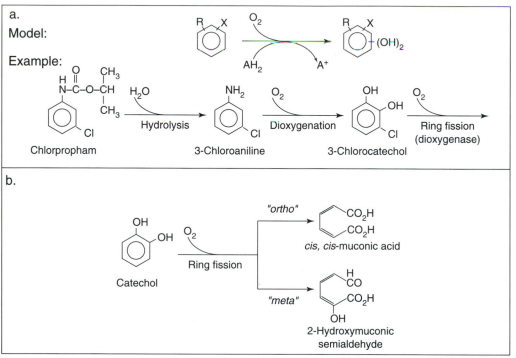

Figure 20–11 Type reactions for dioxygenases including (a) ring dihydroxylation and (b) aromatic ring fission by the *ortho* or *meta* routes. In (a), R (e.g., a carboxyl, methyl, or hydrogen group) and X (e.g., a halogen) are representative ring substituents at nonspecified positions.

Box 20–4

How Suicidal Intermediates Affect Biodegradation of Xenobiotic Compounds.
Suicidal intermediates are compounds produced during the metabolism of a xenobiotic compound that can inhibit the activity of the biodegradative enzymes. Examples are TCE epoxide, described in the preceding text, and *acyl halides,* which are produced by *meta* cleavage of *meta*-substituted halocatechols (reaction a below). Both of these intermediates may inactivate enzymes by bonding with nucleophilic centers, such as thiol groups, at the enzyme's active site. Thus, organisms growing on haloaromatics that form *meta*-halocatechol intermediates generally metabolize these compounds by the *ortho* pathway to avoid the formation of suicidal intermediates (reaction b below).

Reductive Reactions

Reductive reactions are mediated by bacteria under aerobic and anaerobic conditions. Although reductive transformations can yield nutritional or energetic benefits to the organism, some are cometabolic processes. Reductive reactions are usually environmentally beneficial because the xenobiotics are converted to products that are more biodegradable or less toxic than the parent compound. A notable exception to this rule is the reduction of TCE to vinyl chloride, the latter being more toxic than the parent compound. There are many types of reductive reactions, but here we focus on reductive dehalogenation and nitro group reduction because these reactions involve some of the more environmentally important classes of xenobiotic compounds.

Reductive dehalogenation. Two common types of reductive dehalogenation are hydrogenolysis and dihaloelimination. Dihaloelimination occurs when two halogens are lost from adjacent carbon atoms with the concomitant formation of a double bond (Fig. 20–12a). Hydrogenolysis involves the replacement of a halogen with a hydrogen atom and has been documented for aliphatic (alkyl) and aromatic (aryl) compounds (Fig. 20–12b, c). Hydrogenolysis and dihaloelimination can be mediated by both aerobic and anaerobic microbes.

Little is known about the physiological significance of many reductive dehalogenation reactions, particularly in anaerobic environments. These reactions appear to fall into one of two groups: those functioning in growth-supporting reactions and those resembling cometabolic transformations. In aerobes, aryl hydrogenolysis is known to occur during the growth of an *Alcaligenes* strain on 2,4-dichlorobenzoate (Fig. 20–12, reaction c1) and a *Flavobacterium* on PCP (Fig. 20–12, reaction c2). Reductive dehalogenation is thus a third mechanism by which aerobes dehalogenate and grow on aromatic xenobiotics. The *Flavobacterium* dehalogenase is a GST-like enzyme possibly recruited for the catabolic pathway from a detoxification function that GSTs typically serve.

Aryl hydrogenolysis is important in anaerobic transformations of halo-phenols, -benzenes, -benzoates, and -biphenyls (Fig. 20–12, reaction c3). Two lines of evidence have established the potential for this process to be linked to growth-supporting reactions in anaerobic organisms. First, Mohn and Tiedje (1992) documented a pure bacterial culture that could grow with halogenated aromatic compounds as alternate electron acceptors. Specifically, *Desulfomonile tiedjei* strain DCB-1 can generate energy from dehalogenation of 3-CBA to support its growth (see Fig. 20–5b). Second, mixed-culture studies have shown that haloaromatics may enrich for populations of aryl dehalogenators. This enrichment may reflect the ability of some bacteria to derive a competitive advantage from the process, perhaps by using haloaromatics as electron acceptors in a manner similar to that described for strain DCB-1. However, the linkage of aryl dehalogenation to growth-supporting reactions has not been established as a universal occurrence; in some cases, this may be a cometabolic process.

Reductive dehalogenation of haloaliphatic compounds has been reported for many bacterial cultures (Mohn and Tiedje, 1992). In the majority of cases this appears to be a cometabolic process, although some organisms can grow anaerobically using haloaliphatics as electron acceptors (Holliger et al., 1993). Many anaerobic

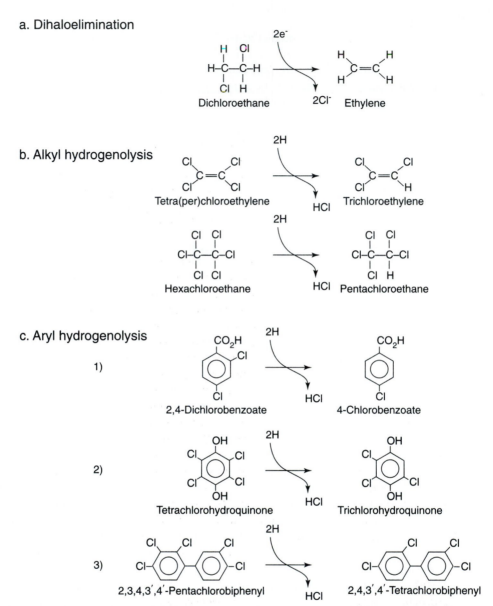

Figure 20–12 Reductive dehalogenation mechanisms. (a) dihaloelimination, (b) alkyl hydrogenolysis, and (c) aryl hydrogenolysis. Alkyl dehalogenation, exemplified in a and b, is thought to be relatively nonspecific and is mediated by aerobic and anaerobic microbes. Aryl hydrogenolysis, in contrast, is a more specific process effected by aerobic (reactions c1, 2) and anaerobic microorganisms (reaction c3).

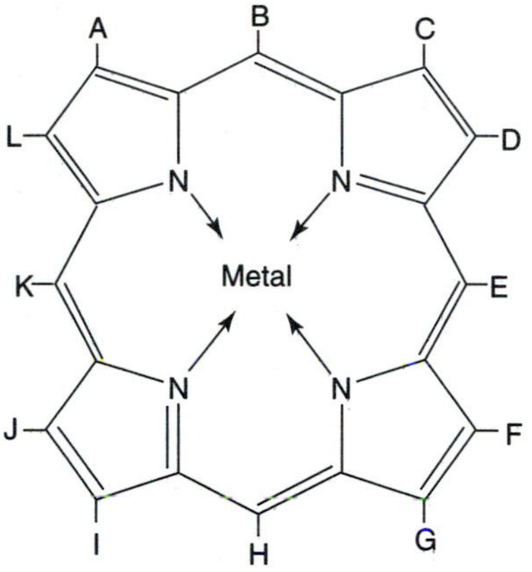

Figure 20–13 The core porphyrin ring structure common to corrinoid cofactors known to mediate reductive dehalogenation. The metals coordinated by the pyrrole nitrogens are (cofactor type): iron (heme), nickel (coenzyme F_{430}), and cobalt (vitamin B_{12}). The cofactors also differ in the substituents at positions A to L.

bacteria exhibit the ability to reductively dehalogenate haloaliphatics, and the process is a common occurrence in anaerobic environments. The mechanistic basis of the latter observation is unclear. However, one theory is that redox-active enzyme cofactors, produced for functions other than xenobiotic metabolism, incidentally dehalogenate haloaliphatics. Microbial cells contain many redox-active molecules but, because of their wide distribution and importance in anaerobic communities, most attention has focused on the metallic corrinoid cofactors vitamin B_{12}, coenzyme F_{430}, and heme (Fig. 20–13). Three lines of evidence support the possibility that these molecules could be environmentally significant in mediating reductive dehalogenation. First, *in vitro* studies show purified cofactors catalyze dihaloelimination, alkyl hydrogenolysis and, to a lesser extent, aryl hydrogenolysis. Second, reductive dehalogenation (mainly alkyl hydrogenolysis) can be maintained in heat-killed cells, a treatment that should inactivate enzymes but not cofactors. Third, the reductive dehalogenation abilities of certain groups of anaerobes are positively correlated with the production of these cofactors.

Nitro group reduction. Nitro groups occur in pesticides (e.g., dinoseb, parathion, and trifluralin) and "energetic compounds," including 2,4,6-trinitrotoluene (TNT). Many pathways are known for aerobic and anaerobic biodegradation of nitrated xenobiotics, most of which involve nitro group reduction. For aerobes, the nitro reduction-deamination process of aromatic compounds is functionally similar to dehalogenation as a reaction yielding an intermediate such as catechol that is funneled into a central pathway. Aerobic biodegradation by this mechanism occurs most readily with aromatic compounds that have a single nitro group and a ring-activating substituent such as CH_3 (see Table 20–1). Highly nitrated compounds such as TNT are more resistant to aerobic biodegradation and are typically only partially transformed to intermediates like 2,6-diamino-4-nitrotoluene (Fig. 20–14). Nitro reduction also occurs under anaerobic conditions. The physiological significance of these reactions is variable; in some cases nitro reduction appears to be cometabolized, while in others the organisms benefit by using the substrate as a nitrogen source or electron acceptor. Nitro reduction could occur with pesticides like trifluralin in flooded soils when denitrifying bacteria become active. Sulfate-reducing bacteria may use nitro groups as electron acceptors in transforming nitrotoluene (Boopathy and Kulpa, 1992). This suggests nitroaromatics could enrich for nitro-reducing activity in anaerobic environments the same way haloaromatics might for reductive dehalogenation.

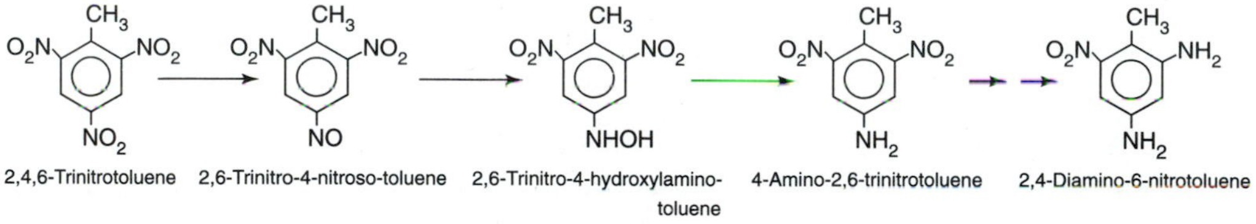

Figure 20–14 Reductive metabolism of TNT that may occur under either aerobic or anaerobic conditions.

Summary

In this chapter we focus on biochemical mechanisms by which microbes metabolize some environmentally important xenobiotics. This overview is necessarily limited in scope, but nevertheless the metabolic versatility and tremendous biodegradation potential of microbial populations should be apparent. Although the emphasis here is on biochemical reaction mechanisms, the importance of environmental factors in modulating microbial biodegradative activities must also be recognized. Further insights into all aspects of the degradation process will enhance the effectiveness with which microbial activities are applied to practical problems such as the remediation of contaminated soils.

Cited References

Alexander, M. 1994. Biodegradation and bioremediation. Academic Press, San Diego, Calif.

Babbit, P.C., G.L. Kenyon, B.M. Martin, H. Charest, and M. Sylvestre. 1992. Ancestry of the 4-chlorobenzoate dehalogenase: Analysis of amino acid sequence identities among families of acyl:adenyl ligases, enoyl-CoA hydratases/isomerases, and acyl-CoA thioesterases. Biochemistry 31:5594–5604.

Barr, D.P., and S.D. Aust. 1994. Mechanisms white rot fungi use to degrade pollutants. Environ. Sci. Technol. 28:79A–87A.

Boopathy, R., and C.F. Kulpa. 1992. Nitroaromatic compounds serve as nitrogen source for *Desulfovibrio* sp. (B strain). Can. J. Microbiol. 39:430–433.

Dec, J., and J.M. Bollag. 1994. Dehalogenation of chlorinated phenols during oxidative coupling. Environ Sci. Technol. 28:484–490.

Gribble, G.W. 1994. The natural production of chlorinated compounds. Environ. Sci. Technol. 28:310A–319A.

Hardman, D.J. 1991. Biotransformation of halogenated compounds. Crit. Rev. Biotechnol. 11:1–40.

Holliger, C., G. Schraa, A.J.M. Stams, and A.J.B. Zehnder. 1993. A highly purified enrichment culture couples reductive dechlorination of tetrachloroethene to growth. Appl. Environ. Microbiol. 59:2991–2997.

Mihelcic, J.R., D.R. Lueking, R.J. Mitzell, and J.M. Stapleton. 1993. Bioavailability of sorbed and separate-phase chemicals. Biodegradation 4:141–153.

Mohn, W.W., and J.M. Tiedje. 1992. Microbial reductive dehalogenation. Microbiol. Rev. 56:482–507.

Rochkind-Dubinsky, M.L., G.S. Sayler, and J.W. Blackburn. 1987. Microbial decomposition of chlorinated aromatic compounds. Marcel Dekker, New York.

Sigel, H., and A. Sigel (eds.). 1992. Metal ions in biological systems. Vol. 28. Degradation of environmental pollutants by microorganisms and their metalloenzymes. Marcel Dekker, New York.

Walker, J.E., and D.L. Kaplan. 1992. Biological degradation of explosives and chemical agents. Biodegradation 3:369–385.

Study Questions

1. Rank each of the chemicals below in terms of its expected aerobic biodegradation potential. For each compound, show a likely initial degradation mechanism and indicate the types of enzymes mediating the postulated reactions and end products. Finally, for each compound, assess the relative biodegradability under anaerobic conditions.

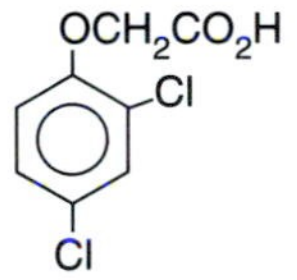

2,4-D

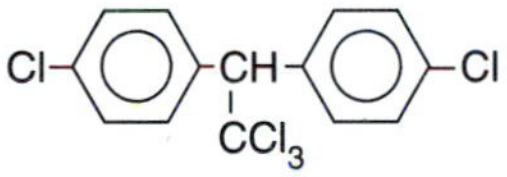

DDT

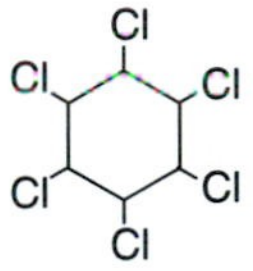

Lindane

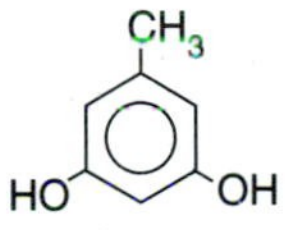

Orcinol

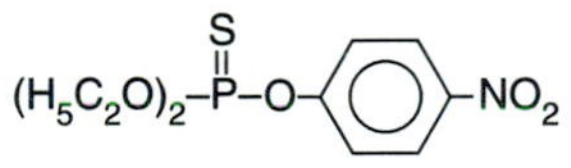

Parathion

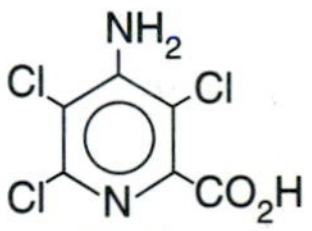

Picloram

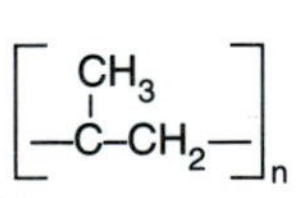

Polypropylene

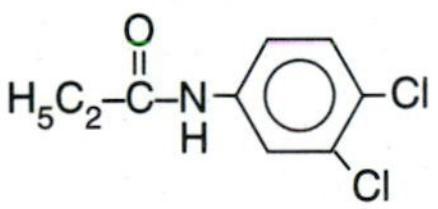

Propanil

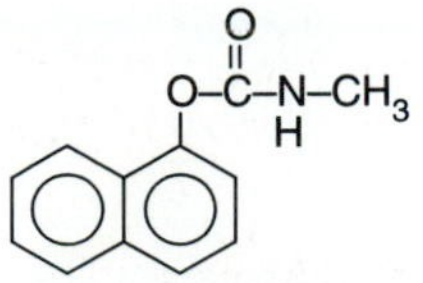

Carbaryl

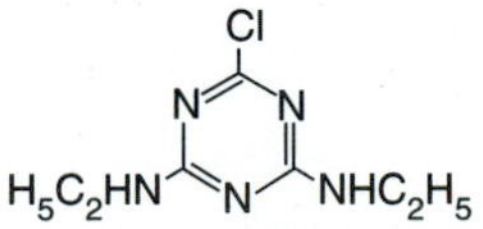

Simazine

2. As an environmental consultant, you are called on to use your knowledge of microbial metabolism to help settle a dispute. At issue is a soil contaminated by polychlorinated biphenyls (PCBs) and chlorobenzenes. Acme Industrial, a PCB manufacturer, is being held responsible for cleanup. Acme's lawyers claim the presence of chlorobenzenes in the soil implicate a neighboring firm that fabricates machine parts (HiTech, Inc.) as another contributor to the contamination and thus equally responsible for the cleanup. Lawyers for HiTech maintain that the chlorobenzenes are metabolites produced during PCB biodegradation, and thus Acme is solely liable for soil remediation. Which party is correct and why? Your answer should include a discussion of the known PCB biodegradation pathways and how environmental conditions likely to be encountered in the soil would affect microbial metabolism of PCB.

3. You are designing a bioremediation system to treat an aquifer contaminated with carbon tetrachloride. Describe the type of microbial metabolism, including types of enzymes and reaction mechanisms, you would attempt to enhance and the effects of the following factors on this process: carbon source (type and amount), concentrations of major electron acceptors (O_2, NO_3^-, SO_4^{-2}, CO_2), and macronutrient levels.

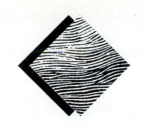

Chapter 21

◆

Bioremediation of Contaminated Soils

Horace D. Skipper

◆

What prevents us from being knee-deep in our own detritus?
Microbes, you reply. Quite so.

J.R. Postgate

In Chapter 20 we considered the biochemistry and metabolism of xenobiotic chemicals. In this chapter we present strategies to restore contaminated soils to a nontoxic, "healthy" state. Healthy soils are essential, not only to sustain production of food and fiber for citizens of the world, but also to provide for a good quality of life. Through the use of case studies in this chapter, we illustrate how restoration strategies are applied in practice.

Diversity and Magnitude of Soil Contaminants

Soil contaminants, also referred to as hazardous wastes and soil pollutants, encompass a large and diverse group of chemicals (Alexander, 1994; Baker and Herson, 1994) including:

- chlorinated solvents, such as trichloroethylene (TCE) and tetrachlorethylene (PCE),

- explosives, such as 2,4,6-trinitrotoluene (TNT),

- metals, such as chromium and lead,

- radionuclides, such as plutonium,

- pesticides, such as atrazine, benlate, and malathion,

- BTEX, an acronym for benzene, toluene, ethylbenzene, xylene,

- PAH, an acronym for polycyclic aromatic hydrocarbons such as creosote mixtures, and

- PCB, an acronym for polychlorinated biphenyl, such as aroclor mixtures.

Contaminants in soil are typically chemicals introduced by human activities that have the potential to cause harm. Contaminants may enter the soil environment purposefully, as in a pesticide application, or accidentally, as spills or storage leaks. As an example of the magnitude and diversity of these contaminants, consider the United States, where more than 100,000 of the estimated 1.4 million buried gasoline and fuel tanks are leaking. According to U.S. Environmental Protection Agency estimates, service station tanks are losing over 40 million liters of gasoline per year (Riser-Roberts, 1992). The United States produces 300 million metric tons of hazardous waste per year—about 1.2 metric tons per person—and about 1,000 new chemicals appear on the market each year (Riser-Roberts, 1992). Because of the diversity and magnitude of these contaminants, there is a critical need for industries to assist in the cleanup and protection of our environment. Estimates put the U.S. commercial market in hazardous waste remediation at about $25 billion per year, and this market is expected to grow at an annual rate of 15% through the year 2005 (Cunningham and Mather, 1995).

Definitions

Bioremediation is a strategy or process that uses microorganisms, plants, or microbial or plant enzymes to detoxify contaminants in the soil and other environments. This concept includes **biodegradation,** which refers to the partial, and sometimes total, transformation or detoxification of contaminants by microorganisms and plants. **Mineralization** is a more restrictive term for the complete conversion of an organic contaminant to its inorganic constituents by a single species or a **consortium** of microorganisms. **Cometabolism** is another more restrictive term referring to the transformation of a contaminant without it providing carbon or energy for the degrading microorganisms. These definitions are all used in the context of converting contaminants to less toxic intermediates or mineralizing them to their inorganic forms (e.g., CO_2, NH_4^+, PO_4^{3-}). Pierzynski et al. (1994), Riser-Roberts (1992) and Skipper and Turco (1995) provide additional reviews on bioremediation.

| Box 21–1 |

Expanding on Proven Success. Bioremediation is hardly new: it has been used for years in wastewater treatment and for cleanup of oily sludges in soils via landfarming. More recently, bioremediation has been applied to other soil contaminants such as pesticides and hazardous wastes.

Criteria for Bioremediation

To seriously consider bioremediation as a practical means of treatment, certain criteria must be met (Alexander, 1994):

- The organisms must have the necessary catabolic activity to degrade the contaminant at a reasonable rate to bring the concentration of the contaminant to a level that meets regulatory standards.

- The target contaminant must be **bioavailable** (e.g., not sorbed as a bound residue).

- The site must have soil conditions conducive to microbial or plant growth or enzymatic activity.

- The cost of the bioremediation must be less than or, at worst, no more expensive than other technologies that can also remove the contaminant.

The failure to meet any one of these criteria may cause rejection of a bioremediation approach. For example, even an abundant source of carbon and energy for microbial growth will not overcome a poor soil abiotic condition like inadequate moisture.

The *Principle of Microbial Infallibility* states that for every naturally occurring organic compound, there is a microbe or enzyme system capable of its degradation (Gale, 1952). In the past a common assumption was that because of their great metabolic flexibility, microbes could degrade all synthetic compounds as well as naturally occurring compounds. This is not true: although the vast majority of synthetic compounds can be degraded, a few defy this principle and cannot be degraded at a reasonable rate. These are often referred to as **recalcitrant** compounds.

| Box 21–2 |

Some Natural Compounds Are Recalcitrant. Even some naturally occurring organic compounds would be poor candidates for bioremediation. For example, if bioremediation of something as chemically heterogeneous as humus were necessary, it is unlikely that it could be degraded at a reasonable rate.

Biological Mechanisms of Transformation

In soil, microorganisms commonly exist as large populations, whether measured as numbers or biomass. Provided with adequate supplies of carbon and energy and environmental conditions conducive to growth, microorganisms have a high level of microbial activity that can be detected by viable count methods, respiration studies, and enzymatic measurements. Thus, great biological potential is generally available to assist in the amelioration of contaminated sites.

Chapter 20 provided an overview of the versatility and biodegradative potential of microbial metabolic processes. These metabolic processes occur under conditions ranging from strictly aerobic to strictly anaerobic. Currently, most efforts and successes in bioremediation have focused on aerobic processes. However, as their utility has gained recognition, anaerobic processes are receiving increasing attention.

Microbial genes code for the degradative enzymes, or biological catalysts, which oxidize, reduce, dehalogenate, dealkylate, deaminate, and hydrolyze hazardous chemicals in the soil environment. Once a contaminant has been enzymatically transformed to a less complex compound, it can often be metabolized further by various pathways, such as via the TCA cycle. Although a single transformation may reduce the toxicity of a contaminant, the complete mineralization of an organic compound to its inorganic constituents typically requires several degradative enzymes produced by multiple genes on plasmids or chromosomes residing in a single species or among different species (Holben et al., 1992). The organism(s) with the degradative enzymes may be either indigenous to the contaminated site or added as an inoculum. The degradative enzyme(s) can be intracellular or extracellular, and each type of enzyme has specific conditions for optimum activity.

The driving force for a metabolic process is to generate energy, carbon, or both for cellular growth and reproduction of microorganisms in the bioremediation process. If cometabolism is involved, then large quantities of available carbon are essential. In some cases, soil organic matter may serve as the carbon source (McCormick and Hiltbold, 1966). For more detailed information on important metabolic processes, see Chakrabarty (1982), Bollag and Liu (1990), and Alexander (1994).

Strategies for Bioremediation

A number of bioremediation strategies can restore soil and environmental quality. For a given contaminant, one or more of the following strategies may be needed to ensure successful bioremediation.

- **Passive** or **intrinsic bioremediation** is the "natural" bioremediation of a contaminated site by indigenous microorganisms. Indeed, unless the environmental conditions are very poor for microbial growth or there is imminent peril from a spilled chemical, many contaminants are degraded by indigenous microbes over time, though the rate of degradation may be too slow for some situations.

- **Biostimulation** is the addition of nutrients, such as nitrogen and phosphorus, to stimulate indigenous microorganisms in soil.

- **Bioventing** is a form of biostimulation where gaseous stimulants, such as oxygen and methane, are added passively or forced into the soil to stimulate microbial activity.

- **Bioaugmentation** is the inoculation of a contaminated site with microorganisms to facilitate biodegradation. The inoculant may contain wild-type or genetically engineered microorganisms either as a single microbial species or a consortium of several species. In any case, organisms are selected for their high potentials to degrade the contaminant of concern.

- **Landfarming** is the application and incorporation of contaminants or wastes into the surface of noncontaminated soil. Typically these are special plots of land underlaid with a thick natural or constructed clay layer to prevent leachates from contaminating the groundwater. The soil is plowed or disked to break it up and to provide uniform mixing, aeration, and moisture. If the concentration of the contaminant is too high for easy biodegradation, plowing or disking also helps reduce its concentration. Finally, if coupled with biostimulation or bioaugmentation, plowing or disking gives a uniform distribution of fertilizer and microbial inoculant, respectively.

- **Composting** is the use of aerobic, thermophilic microorganisms in constructed piles of soils or windrows to degrade contaminants. The piles are physically mixed and moistened periodically to promote microbial activity.

- **Phytoremediation** is the use of plants to remove, contain, or transform contaminants. This can be accomplished directly (e.g., by plants hyperaccumulating heavy metals) or indirectly (e.g., by plants stimulating microorganisms in the rhizosphere).

Case Studies of Bioremediation Strategies

In this section we consider the application of various bioremediation strategies to reduce or remove soil contaminants. It is important to remember that some bioremediation strategies are in their infancy and there is no way to automatically recommend one bioremediation strategy over another. As these case studies demonstrate, using more than one strategy is often useful.

A Case Study Involving Biostimulation

In March 1989, the Exxon Valdez oil tanker hit a reef in Prince William Sound, Alaska, U.S. and released over 40 million liters of crude oil into the sound within five hours. Over 1,500 km of shoreline in the sound and the Gulf of Alaska were contaminated to varying degrees by crude oil. Because oil is inherently high in carbon and low in nitrogen and phosphorus, a portion of the shoreline was selected for biostimulation. After several potential fertilizer candidates were evaluated, a microemulsion, Inipol EAP22™ (henceforth, Inipol), was selected. Inipol, an oleophilic fertilizer, is a stable water-in-oil formulation that yields an N/P/K ratio of 7.3/0.8/0. The nitrogen source is urea, and the phosphorus source is trilaureth (4)-phosphate. At room temperature, Inipol has the consistency and appearance of honey, and it must be heated to 90°C before it can be sprayed on the soil. Inipol was applied as a thin coat to the shore at a rate of approximately 300 ml m^{-2} (Hinton, 1995). As the microemulsion mixed with the weathered crude oil, the crude oil destabilized Inipol to release its urea-nitrogen. In addition, a surface-active organic material, oleic acid, in Inipol served as a readily degradable carbon and energy source to increase the activity and number of indigenoushydrocarbon-degrading bacteria. When the oleic acid was depleted, the increased biomass of hydrocarbon-degrading bacteria supported enhanced biodegradation of the petroleum. Visual observations (Fig. 21–1) and chemical assays showed dramatic

a.

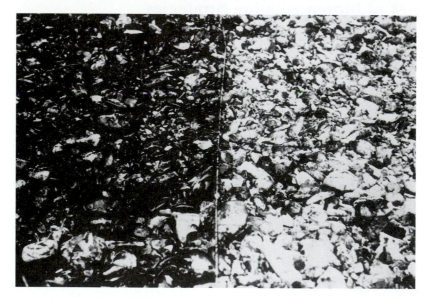

b.

Figure 21–1 Bioremediation of a portion of shoreline contaminated with oil from the Exxon Valdez oil spill in Alaska. (a) Plot (30 x 15 m) treated with the oleophilic fertilizer, Inipol. (b) Close-up of a portion of shoreline not treated (left) or treated with Inipol (right). *Photo courtesy of Steve Hinton, Exxon Corporate Research, Annondale, N.J. Used with permission.*

evidence that biostimulation contributed to the remediation of the site. Although passive bioremediation also occurred in the absence of the fertilizer, the accelerated rate of biodegradation observed with Inipol was critical to a successful bioremediation effort.

Box 21–3

Notable Success. The Exxon Valdez oil spill was a historic event because of the magnitude of the spill, the vastness and isolation of the area to be treated, and the large number of personnel and vehicles ultimately involved (Hinton, 1995). The success of bioremediation, particularly in a climate as cold as Alaska's, prompted U.S. regulatory agencies to view bioremediation much more favorably over previous strategies of physical or chemical "entombment" (storage in cement tombs).

Two Case Studies Involving Bioventing and Biostimulation

At the U.S. Department of Energy's Savannah River Site near Aiken, South Carolina, U.S., an *in situ* bioventing and biostimulation program was used to bioremediate soils contaminated with trichloroethylene (TCE) and tetrachloroethylene (PCE). Horizontal wells were drilled and injected with 1% methane in air (Fig. 21–2). The injected methane increased numbers of methanotrophs by several orders of magnitude. Even though the subsurface soil was aerobic, enough anaerobic zones existed, or were created by the increased biomass of aerobic methanotrophs, to promote anaerobic, reductive dechlorination of PCE to TCE, which could then be oxidized by the methanotrophs. It is estimated that air stripping without methane would require an estimated 10 years or more to achieve 95% removal of the contaminants, whereas air stripping with methane would do this in less than 4 years with a savings of $1.5 million. In fact, bioventing with methane may be the only strategy that can lower TCE concentrations to meet drinking water standards of less than 5 ng TCE per liter (Hazen et al., 1994).

In San Bernardino County, California, U.S., an estimated 4,000 liters (approximately 1,000 gallons) of diesel fuel leaked from an aboveground holding tank. During the investigation of this spill, an underground storage tank was also discovered to be leaking diesel fuel. Even though diesel fuel contains a complex mixture of hydrocarbons (generally 45% cyclic alkanes, 30% normal alkanes, and 24% aromatics), it is biodegradable as are related products, such as gasoline and kerosene (Frankenberger et al., 1989). The State Water Board began its investigation by determining the extent of contamination. Over a six-month period, cores of the subsurface plume showed from less than 10 to 1,500 mg hydrocarbons kg^{-1} of soil. Furthermore, groundwater appeared to be moving a contaminated plume (638 mg hydrocarbons kg^{-1} of soil) downhill toward a small creek where the contaminated plume would presumably be released to surface waters (Box 21–4). Concurrent with the site characterization, laboratory studies indicated that subsoil samples contained a population of diesel-fuel-degrading bacteria, fungi, and yeasts. Because degradative microorganisms were already present, a combination of biostimulation and bioventing was used to promote biodegradation by the indigenous populations. Nutrients (nitrogen and phosphorus

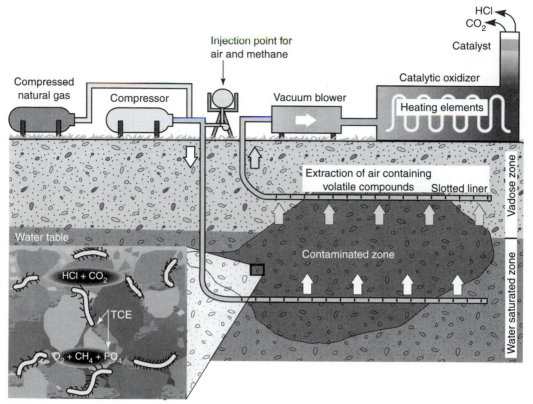

Figure 21–2 Horizontal wells for bioventing soil at a U.S. Department of Energy site in South Carolina. Wells were injected with air and 1% methane to stimulate biodegradation of trichloroethylene (TCE) by methanotrophic bacteria. *Original drawing from Terry Hazen, Westinghouse Savannah River Company.*

in a 5/1 ratio) and hydrogen peroxide (H_2O_2, to provide molecular oxygen) were mixed in water and injected biweekly over a period of four months into seven bore holes. Within six months, no hydrocarbons were detectable (less than 1 mg kg^{-1}). Two years later, a boring within the previously contaminated plume verified no detectable diesel fuel contamination. Thus the bioremediation strategy optimized the soil conditions by supplying nitrogen, phosphorus, molecular oxygen, and moisture and this enhanced biodegradation of diesel fuel. Bioremediation was cost effective by saving the expense of excavation, hauling, and disposal of tons of contaminated soil and subsurface material (Frankenberger et al., 1989).

A Case Study Involving Biostimulation, Bioaugmentation, and Landfarming

At its St. Gabriel, Louisiana, U.S., plant site, Ciba-Geigy Corp. had 19,000 m^3 of atrazine-contaminated soil contained in an 8-ha Biological Cleanup Unit (BCU). The bioremediation strategy was to landfarm the site by plowing or disking the BCU four times a week with weekly soil samplings to measure the disappearance of atrazine (Box 21-5). For biostimulation, 880 kg of fertilizer ha^{-1}(nitrogen-phosphorus-potassium, 13-13-13) were added to the contaminated soil. For bioaugmentation, 2,000 liters of a proprietary, atrazine-degrading consortium of pseudomonads were applied. Initial *s*-triazine con-

┤ **Box 21–4** ├

Hydrological Viewpoint. The San Bernardino County case study is also interesting because it points out the role for hydrology in bioremediation. Historically, hydrogeologists have distinguished between the phreatic zone, which is the saturated zone of soil containing groundwater, and the **vadose zone,** which is the unsaturated zone of soil above the groundwater, extending from the bottom of the capillary fringe (water drawn up into the unsaturated zone from the saturated zone by capillary action) all the way to the soil surface. Interest in the vadose zone has intensified because monitoring this zone will provide an early warning system for groundwater protection (Boulding, 1995). It is also important to note that the subsurface portion of the vadose zone usually contains fewer microorganisms and fewer major nutrients than the surface portion because microorganisms and carbon, nitrogen, and phosphorus concentrations decrease with soil depth. This may affect the bioremediation strategy.

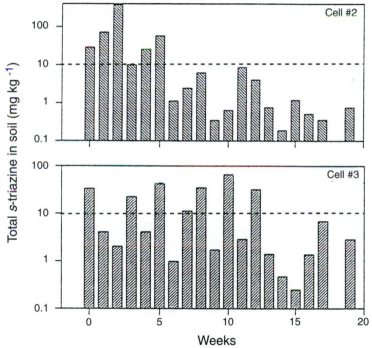

Figure 21–3 Total s-triazine concentrations in two cells of a Biological Cleanup Unit over a 20-week incubation period. The state regulatory cleanup limit is indicated by horizontal dashed lines. *From Finklea and Fontenot, 1995. Used with permission.*

centrations were approximately 100 mg kg^{-1} of soil, but after 20 weeks triazine concentrations were below the target level of 10 mg kg^{-1} of soil (Fig. 21–3). High variability in the BCU Cell #3 was observed from "hot spots" caused by the incomplete mixing and breakdown of clay agglomerates. After some additional time, though, even these hot spots fell below the cleanup standard of 10 mg kg^{-1} (Finklea and Fontenot, 1995). Ciba-Geigy spent approximately $1,050,000 to bioremediate the soil versus $5,300,000 to excavate and to dispose of the atrazine-contaminated soil in an approved hazardous waste disposal facility. Thus bioremediation saved an estimated $4,250,000 at this site.

Box 21–5

Atrazine Degradation. Atrazine (2-chloro-4-ethylamino-6-isopropylamino-*s*-triazine) has been a key herbicide since 1954, especially for managing weeds in corn. To fill the market demand for atrazine and other *s*-triazine herbicides, manufacturing plants have produced millions of kilograms of these products. Despite pollution abatement programs, some contamination of soils at the atrazine manufacturing sites has occurred. The St. Gabriel case demonstrates that all the reactions in atrazine degradation can be biologically mediated. The dechlorination of atrazine to hydroxyatrazine may be catalyzed by certain microorganisms, although it is often described as a chemical process associated with acidic soils. The N-dealkylation processes, which remove the ethyl and isopropyl side-chains, are performed by monooxygenase enzymes in a wide variety of microorganisms. The removal of the amino groups ($-NH_2$) is facilitated by hydrolase enzymes.

A Case Study Involving Phytoremediation

Phytoremediation is the use of plants to remove, contain, or transform contaminants. This area of bioremediation is in its infancy, and good field case studies are still scarce. For indirect plant effects, where root exudates in the rhizosphere stimulate the microbial populations in the root zone to degrade various compounds, most studies have focused on pesticides (Anderson et al., 1995). In one example of phytoremediation of a nonagricultural contaminant, plants facilitated microbial degradation of trichloroethylene (TCE). Microbes in soil with lespedeza, soybean, and lobolly pine mineralized TCE in TCE-contaminated soil twice as fast as in contaminated controls with no plants (Anderson and Walton, 1995). For direct plant effects, most research has focused on *hyperaccumulators,* plants that can accumulate excessive amounts of heavy metals. The plants accumulate the heavy metal (in some instances, up to 3% of their dry weight with no apparent plant damage), and these heavy metals can then be extracted from the plant biomass and recycled (Cunningham and Lee, 1995). It is important to note that for direct plant uptake, small, low-molecular-weight compounds are favored over large, high-molecular-weight compounds because the latter, especially if they are lipophilic, tend to be excluded from the root (Anderson et al., 1995).

Advantages and Disadvantages of Bioremediation

On one hand, the case studies in this chapter demonstrate a number of advantages that bioremediation has over physical and chemical strategies. Bioremediation is often less hazardous to cleanup personnel and less expensive because contaminants can be treated on site. Bioremediation is a natural process and therefore often has lower environmental impact because no wastes are generated.

On the other hand, bioremediation has a number of disadvantages that can be derived from the previously listed criteria for bioremediation. For example, low bioavailability of a contaminant would limit the application of bioremediation. The worst-case scenario is for those sites contaminated with not one compound or even one class of compound, but with a mixture of inorganic and organic compounds. At such sites, one compound might be biodegradable by microorganisms while another might be toxic to microorganisms; one compound might require aerobic conditions for biodegradation while another might require anaerobic conditions; one compound might require neutral to alkaline conditions for biodegradation while another might require acidic conditions. Our knowledge of modeling and manipulating biological, chemical, and physical interactions where sites have complex mixtures of chemicals is still extremely limited.

The Future of Bioremediation

A number of constraints hinder the progress and utilization of bioremediation to clean up environmental contamination. These constraints range from scientific and technological issues to politics, regulatory aspects, and societal inertia. For example, even though we may isolate and identify a microorganism with superior abilities to degrade a contaminant, the desired activity of biodegradation does not occur until the degrader makes contact with the contaminant. Thus, bioremediation may be viewed as analogous to potential energy vs. active energy. In some ways, bioremediation may be constrained by the "needle in the haystack" syndrome where the microorganism, the biological catalyst, must "find" or be brought into contact with the contaminant in order to degrade it.

Even though constraints do exist, bioremediation is poised to be a major, if not the major, strategy in the restoration of many contaminated environments. This is because bioremediation is a natural process that recycles the elements in contaminants rather than simply entombing them as liabilities for future generations. Furthermore, the public views bioremediation favorably, and many agencies worldwide are promoting its use for the restoration of sites damaged by environmental contaminants.

Summary

This chapter discussed bioremediation of contaminated soils and various strategies to restore these soils back to a "healthy" state. Our discussion covered the diversity and magnitude of soil contaminants, the criteria that must be met for successful bioremediation to occur, and the various strategies for bioremediation. These strategies include passive or intrinsic remediation, biostimulation, bioventing, bioaugmentation, landfarming, composting, and phytoremediation. Case histories serve as practical examples of some of these strategies. Despite some success, obstacles remain that limit the use of bioremediation, but its future is promising.

Cited References

Alexander, M. 1994. Biodegradation and bioremediation. Academic Press, San Diego, Calif.

Anderson, T.A., and B.T. Walton. 1995. Comparative fate of [^{14}C] trichloroethylene in the root zone of plants from a former solvent disposal site. Environ. Toxicol. Chem. 14:2041–2047.

Anderson, T.A., E.A. Guthrie, and B.T. Walton. 1995. Bioremediation in the rhizosphere. Environ. Sci. Technol. 27:2630–2336.

Baker, K.H., and D.S. Herson. 1994. Bioremediation. McGraw-Hill, New York.

Bollag, J.-M., and S.-Y. Liu. 1990. Biological transformation processes of pesticides. pp. 169–212. In H.H. Cheng (ed.), Pesticides in the soil environment: Processes, impacts, and modeling. Soil Science Society of America Book Series No. 2. Soil Science Society of America, Madison, Wis.

Boulding, J.R. 1995. Practical handbook of soil, vadose zone, and ground-water contamination: Assessment, prevention, and remediation. Lewis Publishers, Boca Raton, Fla.

Chakrabarty, A.M. 1982. Biodegradation and detoxification of environmental pollutants. CRC Press, Boca Raton, Fla.

Cunningham, S.D., and C.R. Lee. 1995. Phytoremediation: Plant-based remediation of contaminated soils and sediments. pp. 145–156. In H.D. Skipper and R.F. Turco (eds.), Bioremediation: Science and applications. Soil Science Society of America Special Publication No. 43, Madison, Wis.

Cunningham, S.D., and M.S. Mather. 1995. Applying soil science skills in the remediation business. pp. 273–292. *In* H.D. Skipper and R.F. Turco (eds.), Bioremediation: Science and applications. Soil Science Society of America Special Publication No. 43, Madison, Wis.

Finklea, H.C., and M.F. Fontenot, Jr. 1995. Accelerated bioremediation of triazine contaminated soils: A practical case study. pp. 221–236. *In* H.D. Skipper and R.F. Turco (eds.), Bioremediation: Science and applications. Soil Science Society of America Special Publication No. 43, Madison, Wis.

Frankenberger, W.T., Jr., K.D. Emerson, and D.W. Turner. 1989. *In situ* bioremediation of an underground diesel fuel spill: A case history. Environ. Manage. 13:325–332.

Gale, E.F. 1952. The chemical activities of bacteria. Academic Press, New York.

Hazen, T.C., K.H. Lombard, B.B. Looney, M.V. Enzien, J.M. Dougherty, C.B. Fliermans, J. Wear, and C.A. Eddy-Dilek. 1994. Preliminary technology report for in situ bioremediation demonstration (methane biostimulation) of the Savannah River site integrated demonstration project. Published by American Nuclear Society, Inc.

Hinton, S.M. 1995. Bioremediation of Valdez and Prall's Island oil spill. pp. 211–220. *In* H.D. Skipper and R.F. Turco (eds.), Bioremediation: Science and applications. Soil Science Society of America Special Publication No. 43, Madison, Wis.

Holben, W.E., B.M. Schroeter, V.G. Calabrese, R.H. Olsen, J.K. Kukor, V.O. Biederbeck, A.E. Smith, and J.M. Tiedje. 1992. Gene probe analysis of soil microbial populations selected by amendment with 2,4-dichlorophenoxyacetic acid. Appl. Environ. Microbiol. 58:3941–3948.

McCormick, L.L., and A.E. Hiltbold. 1966. Microbiological decomposition of atrazine and diuron in soil. Weeds 14:77–82.

Pierzynski, G.M, J.T. Sims, and G.F. Vance. 1994. Soils and environmental quality. Lewis Publishers, Boca Raton, Fla.

Riser-Roberts, E. 1992. Bioremediation of petroleum contaminated sites. CRC Press, Boca Raton, Fla.

Skipper, H.D., and R.F. Turco. 1995. Bioremediation: Science and applications. Soil Science Society of America Special Publication No. 43, Madison, Wis.

Study Questions

1. Compare and contrast bioremediation, biodegradation, mineralization, and cometabolism.

2. What criteria must be satisfied for bioremediation to be considered a viable option for cleanup of a contaminated site?

3. Compare and contrast biostimulation, bioventing, bioaugmentation, landfarming, and phytoremediation.

4. What are some limitations to the successful application of bioremediation?

5. Why is bioremediation preferred over physical and chemical technologies for remediation of contaminated sites?

Chapter 22

♦

Composting of Organic Wastes

Larry M. Zibilske

♦

. . . Out of its little hill faithfully rise the potato's dark green leaves,
Out of its hill rises the yellow maize-stalk, the lilacs bloom in the dooryards,
The summer growth is innocent and disdainful above all those strata of sour dead. . .
Walt Whitman

Historical records show that ancient Greeks and Romans were familiar with the decay processes that occurred spontaneously in stacks of plant matter or animal wastes. This very early, empirical understanding of composting perhaps led people to view the process as commonplace and unimportant. In retrospect, however, it is clear that composting was one of the first biotechnologies. Today, heightened environmental awareness has sparked renewed interest in composting as a means to prevent or reduce environmental pollution and to manage and recycle wastes produced by society. Recent innovations have brought the process into modern times with computer control and scientific management. Thus at some facilities, composting has become a highly developed enterprise; at the same time, it remains a basically simple process that the backyard gardener can master. At the center of the process are microorganisms and their remarkable decomposition capabilities. Composting demonstrates the powerful metabolic potential of microorganisms and vividly illustrates the vital role of environmental microorganisms in global biogeochemical cycles.

The What and Why of Composting

Haug (1993) describes composting as the biological decomposition and stabilization of organic substrates, under conditions that allow development of thermophilic temperatures as a result of biologically produced heat, to produce a final product that is

482

stable, free of pathogens and plant seeds, and can be beneficially applied to land. The processes are characterized by a period of rapid decomposition and self-heating followed by a cooler, slower decay of remaining organic substrates. Regulating the kinds of organic substrates and controlling the physical and chemical attributes of the decomposition environment in the compost pile facilitate the process. Manipulating moisture content, pH, nutrient concentrations, and oxygen can bring about increased decomposition rates and change the characteristics of the compost. Known to backyard gardeners as a soil-enriching additive, compost promotes plant productivity and soil quality. Today, enhancements in modern engineering have resulted in the growing use of composting for municipal and industrial waste treatment and nutrient recovery. The same microbial processes are used by both large commercial composting industries and by backyard gardeners, although the former may produce thousands of cubic meters of compost while the latter only a few.

Composting to manage organic wastes produces several advantages:

- Compost generally poses a low risk to the environment, assuming it is free of heavy metals or hazardous organic materials.

- Compost improves soil fertility whether in backyard gardens or reclaimed strip-mine soils.

- Composting is cost-competitive with other waste-handling technologies.

The goals of composting range from **sanitization,** eliminating pathogenic or deleterious organisms, insect larvae, intestinal parasites, and weed seeds, to reducing the bioavailable energy content of the waste and reducing waste volume. Composting can even be done in a greenhouse where heat generated by the compost and the large amount of carbon dioxide produced are beneficial to plant production in cold climates. Composting also helps bioremediate wastes containing polycyclic aromatic hydrocarbons (Crawford et al., 1993), pesticides (Michel et al., 1995), and petroleum-contaminated soils (Beaudin et al., 1996). Composting is now accepted as a Process to Significantly Reduce Pathogens (PSRP) in the treatment of wastewater sludges (EPA, 1993).

Substrates Suitable for Composting

Most organic wastes can be successfully composted (Table 22–1). For example, paper and food wastes often comprise 50% of **municipal solid waste.** Both are well suited to composting. **Biosolids** (sewage sludges), animal manures, and agricultural crop and food processing wastes are also excellent candidates for composting. Most of these wastes, however, are presently placed in landfills. In the past, yard wastes (primarily leaves, garden crop residues, grass clippings, and brush) were also deposited in landfills. Several states have recently enacted legislation to limit or prohibit yard-waste disposal at landfills. This has led to a resurgence of interest in community-wide composting facilities, and the number of these facilities is increasing. Many organic waste products of farms, industries, and municipalities and those generated at private residences can be effectively composted, eliminating or greatly reducing a large portion of the bulk delivered to landfills. These materials are suitable substrates for composting because they contain carbohydrates, proteins, and lipids that the microorganisms can readily degrade.

Table 22–1 **Concentrations of selected nutrients in representative compostable substrates.**

Substrate	Nitrogen	Phosphorus	Potassium	Calcium	Magnesium	C/N
			(% of dry mass)			
Chicken manure	4.5	0.8	0.7	1.8	0.4	7
Cattle manure	1.5	0.5	0.6	1.0	0.3	18
Grass clippings	1.2	1.1	2.0	<0.1	0.1	27
Alfalfa	2.4	0.2	1.8	1.4	3.9	15
Corn stover	0.9	0.1	1.2	0.4	0.1	42
Oat straw	0.6	0.1	1.2	0.6	0.1	90
Rice hulls	0.6	0.1	0.7	0.4	0.1	85
Fish processing wastes	9.0	7.0	0.8	1.4	0.1	4
Pine sawdust	0.1	<0.1	0.1	0.1	<0.1	225
Lime-stabilized biosolids	3.6	1.2	0.4	3.6	0.4	14
Starch processing wastes	0.1	<0.1	<0.1	<0.1	<0.1	312
Mixed green weeds	2.3	0.3	1.3	0.1	<0.1	21
Mixed papermill sludge	0.9	0.1	<0.1	6.9	0.3	61

Adapted from Parnes (1986), Seekins (1986), and Haug (1980).

Properties of Compostable Wastes

The chemical and physical attributes of organic wastes affect the composting process and the characteristics of the finished product.

Chemical Properties

The chemical characteristics of the substrates may be considered in terms of nutrient quantity and quality. The range of elemental composition for some wastes that are suitable for composting are shown in Table 22–1. Elemental composition, which is the mass of each element contained in the substrate, describes the quantity of the substrate composition. The relative quantity of carbon, nitrogen, phosphorus, sulfur, and other nutrients is important, but it is also the quality of substrates that determines the rate of decomposition. The quality is a determination or estimation of the kinds of molecules, such as cellulose or lignin, in which the elements are present. For example, cellulose and lignin have similar percentages of carbon, and contain only carbon, hydrogen, and oxygen, but the structures of the molecules that contain the carbon differ radically (Chapter 11). Since lignin undergoes decomposition much more slowly than does cellulose, the quality of lignin as a carbon source is much lower than that of cellulose. Although different compostable substrates often begin with widely divergent nutrient contents, the quantity and quality of the nutrients converge as composting proceeds. This general observation is similar to that seen in the humification processes of plant residues.

The relative quantities and qualities of nutrients in substrates affect the rate of composting and the characteristics of the finished product. Maximum rates of decomposition are observed when substrates contain optimum available carbon, nitrogen, phosphorus, and sulfur ratios. For example, when the C/N ratio of a substrate is between 25 and 35, metabolism of the substrates may proceed rapidly with a high degree of efficiency of nitrogen assimilation into microbial biomass. Narrow C/N ratios, as are often observed in biosolids, can lead to loss of nitrogen from the compost through ammonia volatilization. Wider C/N ratios (more than 40) in substrates such as oat straw promote immobilization of available nitrogen in the compost, resulting in a slowing of the decomposition processes due to limited nitrogen availability. When this occurs, the kinetics of the decomposition process depend greatly on the rate of microbial biomass turnover, which releases more nitrogen. Addition of exogenous sources of nitrogen can overcome such a deficiency and prevent slowing of the composting process.

Substrates vary widely in pH. Initially, many materials such as municipal refuse are acidic. Mixing high pH and lower pH materials moderates the extremes of pH in the materials. Lime-stabilized biosolids have a residual pH of around 8. Additionally, it is important to realize that metabolic processes during composting further affect the pH of the material. Deamination of proteins can rapidly increase pH due to ammonia production, while the production of organic acids during the decomposition of carbohydrates and lipids decreases pH. On average, the pH of inputs is somewhat acidic, while the finished compost may have a near-neutral pH. Finished compost derived from nitrogenous wastes may be acidic due to latent nitrification. Excessive compost acidity may be remedied through the addition of lime to the substrates.

Some types of environmentally harmful chemical compounds may be present in biosolids or industrial sludges. Solvents, pesticides, and other organic chemicals commonly undergo at least some transformation in the composting process but may not be completely degraded. Heavy-metal content of some municipal biosolids can be high, but generally not high enough to affect the microorganisms degrading the wastes. However, metal content is essentially unchanged by composting, and it may become an issue when deciding among alternative uses of the compost.

Physical Properties of Compostable Substrates

The wide range of substrates available for composting suggests an equally wide variety in the physical characteristics of those substrates. The most important physical aspects of substrates, relative to the composting process, are particle size and moisture content.

Particle size is important because it affects the movement of oxygen into the compost pile and microbial and enzymatic access to substrates. The size of individual pieces of substrate ranges from relatively large, such as in wood wastes, to relatively small pieces, such as in biosolids. Large particle sizes promote diffusion of oxygen because of the large mean pore size generated by their presence. However, larger particles also minimize the specific surface area of the substrate which is the ratio of surface area to volume. This means that most of the substrate is not immediately accessible to the microbes or their enzymes. Because efficient composting requires access to oxygen and to nutrients in the particles, it is necessary to balance these two requirements. If the materials being composted do not contain sufficiently large particles to achieve this balance, another material must be added. A common solution is to combine a coarse, relatively insoluble material, called a *bulking agent,* with more soluble substrates. The mixture of particle sizes improves aeration and can be effective in establishing

a.

b.

Figure 22–1 Chipped wood scraps (a) and shredded tires (b) are effective bulking agents for composting.

and maintaining an appropriate moisture content for composting. Common bulking agents are tree bark, sawdust, wood chips (Fig. 22–1a), shredded tires (Fig. 22–1b89), and straw. Shredded tires do not degrade during the process. Sawdust is particularly useful for adjusting the moisture content of very wet substrates such as biosolids.

Moisture is essential for the composting process. Decomposition is carried out in an essentially aquatic environment. Soluble components from the solid substrates and wastes from microbial metabolism diffuse through a film of moisture on the compost solids. Optimum moisture content for composting is generally between 40 and 60% (mass/mass). Too much moisture interferes with oxygen accessibility and slows composting. Too little moisture hinders diffusion of the soluble molecules, likewise slowing the rate of composting. See Chapter 2 for a review of the relationships between particle size, oxygen diffusion, and moisture holding capacity.

Microbial Characteristics of the Composting Process

Most organic substrates carry an indigenous population of microbes from the environment. Representatives of three major groups—bacteria, actinomycetes, and fungi—of soil microorganisms are normally present when the composting process begins. Microbial populations change during the composting process, as shown in Table 22–2, progressing from the *mesophilic stage* (approximately 20 to 40°C) into the *thermophilic stage* (above 40°C) and then through a gradual cooling period, the final *stabilization* or *curing stage*.

Microbes are inefficient in trapping energy released during the oxidation of organic substrates. Energy that is not biochemically captured in the catabolic degradation of organic substrates is dissipated to the environment as heat. When organic matter decomposes over a large area of ground, the heat dissipates into soil, air, and water, and the heat increase is scarcely noticeable. In contrast, compost piles restrict the free dissipation of heat generated during decomposition, resulting in significant temperature increases.

Mesophilic microbiota initiate the decomposition of organic substrates. During the initial stage of decomposition, readily available substrates such as proteins, sugars, and starch, are rapidly oxidized. The physical structure and insulating properties of the organic mass retain much of the heat produced from microbial metabolism, so internal temperatures rise rapidly to high levels, occasionally near 80°C.

Table 22–2 **Microbial population changes during composting.**

Organism	Mesophilic stage	Thermophilic stage	Stabilization/ curing stage	No. species present
	(CFU g^{-1} dry mass)			
Bacteria				
Mesophilic	10^8	10^6	10^{11}	6
Thermophilic	10^4	10^9	10^7	1
Actinomycetes				
Thermophilic	10^4	10^8	10^5	14
Fungi				
Mesophilic	10^6	10^3	10^5	18
Thermophilic	10^3	10^7	10^6	16

Adapted from Poincelot (1982).

| Box 22–1 |

Some Like it Hot. Thermophilic microbes can withstand temperatures that reach 80°C in compost piles. But other microbes can *grow* in hot springs and benthic thermal vents at temperatures greater than 100°C (Chapter 15).

The role of mesophiles is generally considered a preparative one. They bring the compost into a temperature range suitable for thermophiles by rapidly decomposing the readily available substrates present in the compost. Numbers of bacteria, determined by plate counting, can increase by two log units to 10^8 g^{-1} just prior to the thermophilic stage. Mesophilic bacteria and fungi present at the beginning of the composting process give way to thermophiles when the temperature reaches about 40°C. Actinomycetes apparently play little or no role in the preparative processes carried out by other mesophilic bacteria. Their numbers are about 10^4 g^{-1} in the mesophilic stage and in the curing stage, but they rise to 10^8 g^{-1} during the thermophilic stage. Similarly, counts of mesophilic fungi remain stable at around 10^5 to 10^6 g^{-1} in the initial and final stages of the process, both of which are much cooler than the thermophilic stage. Lower populations of mesophilic fungi are found in the thermophilic stage, during which thermophilic fungi flourish. Overall, decomposition rates are most rapid during the thermophilic stage.

During the period of greatest heat production, surprisingly few genera and species of microorganisms are present in the compost. A few species of *Bacillus* (e.g., *subtilis, circulans, stearothermophilus*), thermophilic fungi (e.g., *Aspergillus fumigatus, Mucor pusillus, Chaetomium thermophile, Thermoascus aurantiacus,* and the yeast, *Torula thermophila*), and actinomycetes (e.g., *Streptomyces* spp., *Thermoactinomyces* spp., *Thermomonospora* spp., and *Micropolyspora* spp.) are observed during this stage (Table 22–3). When the temperature exceeds 50 to 60°C, decomposition slows and the process becomes *self-limiting*. It is often difficult to determine whether mesophiles present at the outset merely survive the thermophilic stage on the cooler exterior portions of the compost or if they are reintroduced into the compost from airborne sources. The reintroduction of microbes from the cooler exterior to the warmer central portion can occur during mechanical turning of windrows or piles after the thermophilic stage of the process has been completed.

In addition to the bacteria and fungi, soil animals such as earthworms and arthropods contribute to the decomposition process by reducing the size of organic particles, greatly increasing their surface-area-to-volume ratio and improving microbial access to the substrates. Earthworms can facilitate and expedite the stabilization of compost, a process referred to as *vermicomposting*.

Enzymatic Activities During Composting

Microbes in the compost pile cannot directly metabolize the insoluble particles of organic matter. Rather, they produce hydrolytic extracellular enzymes to depolymerize the larger compounds to smaller fragments that are water soluble. At this point, microbes transport the substrates across the cytoplasmic membrane to complete the degradation process.

Table 22–3 **Microorganisms commonly associated with compost piles.**

	Bacteria	Fungi
Mesophiles		
	Pseudomonas spp.	*Alternaria* spp.
	Achromobacter spp.	*Cladosporium* spp.
	Bacillus spp.	*Aspergillus* spp.
	Flavobacterium spp.	*Mucor* spp.
	Clostridium spp.	*Humicola* spp.
	Streptomyces spp.	*Penicillium* spp.
Thermophiles		
	Bacillus spp.	*Aspergillus fumigatus*
	Streptomyces spp.	*Mucor pusillus*
	Thermoactinomyces spp.	*Chaetomium thermophile*
	Thermus spp.	*Humicola lanuginosa*
	Thermomonospora spp.	*Absidia ramosa*
	Micropolyspora spp.	*Sporotrichum thermophile*
		Torula thermophile (yeast)
		Thermoascus aurantiacus

Adapted from Chang and Hudson (1967) and Strom (1985).

The activities of several enzymes have been monitored during composting (Hankin et al., 1976). For example, cellulase activity decreases the amount of cellulose present by about 25% in about three weeks. Lipase, protease, and amylase activities rise and fall during the successive stages of composting. Activities of all these enzymes decrease sharply during the thermophilic stage, probably because of heat inactivation. The denaturation of enzymes is often correlated with the death of the microbe. This suggests that the apparent recovery of microbial numbers and enzymatic activity in the pile interior after the thermophilic stage is due to reintroduction, during turning, of organisms that survived on the outer, cooler parts of the pile.

So we see that composting is essentially a microbial process, and its rate can be controlled by any factor that affects the microbes involved in the process. Lack of suitable substrates, a moisture content or temperature in the compost outside the optimum range, and problems with oxygen diffusion into the pile are the most common rate-limiting factors in composting.

Progression of the Composting Process

The inputs and outputs (or products) of composting systems are shown in Figure 22–2. As described above, the microorganisms are the driving force or "engine" that substantially transforms the inputs (substrates, water, and oxygen) into products (mainly compost, new microbial cells, carbon dioxide, and heat). Figure 22–3 depicts the two temperature zones often seen in unturned compost piles. High internal temperature is sufficient to eliminate many microbes and other organisms initially found in the substrates. Mesophilic decomposers may be completely eliminated from the compost as heat is generated faster than it can be dissipated from the pile.

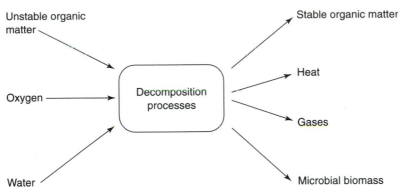

Figure 22–2 Inputs and outputs of the composting process. *Adapted from Diaz et al. (1993).*

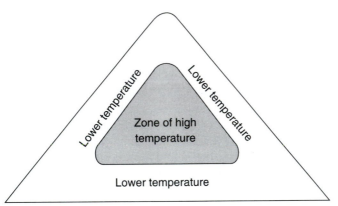

Figure 22–3 Temperature zones in a static compost pile.

This excessive heating may extend the time necessary to achieve optimum decomposition and stabilization of the substrates.

Thermophilic temperatures in the pile are necessary to ensure rapid stabilization and to **pasteurize** the compost, eliminating harmful organisms. United States governmental regulations for biosolids composting require that the internal temperature of the compost pile be 40°C or higher for five days; during that time, the temperature must exceed 55°C for four hours (EPA, 1993). The thermophilic stage may last a number of days, depending on how well oxygen is supplied to the pile and the quality and quantity of substrates. Enteric pathogens, such as *Salmonella* spp., normally survive less than one hour once the compost enters the thermophilic stage. Other noxious organisms, such as weed seeds and parasite eggs, may be more durable, but are often eliminated or greatly reduced during composting. Pasteurization is one reason composting is a popular waste treatment alternative for biosolids and some industrial sludges.

In time, exhaustion of nutrients limits heat production, and the compost pile temperature falls back into the mesophilic range. At this point, composting enters the stabilization or curing phase. Additional decomposition occurs during this phase, further lowering the biochemical energy content of the compost. Organic acids, mercaptans, and ammonia remaining in the compost at this stage often contribute to odor production, but they are normally short-lived. During this stage, reintroduction of mesophilic microorganisms occurs. These organisms grow on remaining polymeric substrates and, more importantly, on the lower-molecular-weight end products that accumulated in previous stages. Stabilization ensures that potentially harmful by-products (e.g., acetate, which acts as a seed-germination inhibitor) and other metabolites with phytoactive effects (e.g., ethylene, ammonia, and amines) are completely vented from the pile or metabolized. This is crucial when the compost will be used for agricultural or horticultural production. Continued ammonification of the substrates may result in the release of ammonia during the latent mesophilic activity in the compost, and nitrification rates may become noticeable.

Figure 22–4 shows the typical progression of substrate transformation into compost. Here, grass clippings represent a relatively intact starting material (Fig. 22–4a) that is not readily water soluble. After 16 days in a compost pile, the physical integrity of the clippings has changed. They are no longer readily recognized as grass clippings, although traces of the original structures remain. When placed in water (Fig. 22–4b), soluble brown-black substances leach from the material, suggesting significant depolymerization and chemical alteration of the molecules originally present in the clippings. After 25 days in the compost pile, the clippings bear no resemblance to the original material, having lost their physical integrity. At this stage, the material contains a large

a.   b.   c.

Figure 22–4 Transformation of grass clippings during composting: (a) original material, (b) appearance after 16 days of composting, and (c) appearance after 25 days of composting.

amount of water-soluble substances (Fig. 22–4c). Some of these compounds would be further decomposed during the curing phase of the composting process.

Compost Systems

Compost systems are categorized on the basis of whether they are batch or continuous processes (Finstein and Morris, 1975). In batch composting, a given quantity of the substrates is allowed to progress through the stages of temperature and microbial succession, finishing with the stabilization stage. This is analogous to growing microorganisms in a flask of nutrients in which no additional nutrients are added or any wastes removed. Eventually growth is limited. In continuous composting, substrates are continually added and the product removed from the reactor. This is analogous to continuous or chemostat culture of microorganisms. Commonly, optimum composting conditions are maintained in the reactor during passage of the substrates, resulting in rapid and thorough decomposition.

Batch Composting Systems

The simplest batch system is the *static pile*, where the materials are not mechanically turned or inverted. Substrates are merely stacked to produce a pile that is roughly triangular or trapezoidal in shape. A completely static pile may eventually lead to oxygen deficiency in the center of the pile as the microorganisms rapidly decompose the substrates. Essentially the center of the pile becomes anaerobic and the rate of substrate decomposition decreases to the rate at which oxygen diffuses into the pile. Exhaustion of oxygen in the pile can greatly extend the time necessary to completely compost the substrates. Inadequate oxygen supply also leads to the production of malodorous end products of anaerobic microbial activity. These are aesthetically unacceptable and are an important concern for operators of compost facilities located near residential communities.

An engineered improvement of the static pile is the *windrow*. A windrow is an elongated pile constructed to handle the larger volumes of compost substrates generated by municipalities and industries. This configuration facilitates turning, which often expedites composting. Turning the windrow periodically ensures that the outer, cooler, and drier parts of the pile are rotated to the center of the pile where the most rapid microbial activity develops. In addition, turning redistributes waste products that accumulate in the central part of the pile. Turning is achieved by simply mixing the contents of the pile with hand tools as in the case of a small backyard pile or heavy equipment for large windrows. Commercial windrow turners are large tractorlike machines that straddle the windrow. In moving from one end of the windrow to the other, rotating "fingers" beneath the center of the machine turn the windrow, bringing about the mixing action required to ensure rapid and complete composting. For smaller-scale composting operations, a tractor-powered turner, such as the one shown in Figure 22–5, is adequate.

Figure 22–5 Yard and kitchen waste compost being turned by a tractor-powered windrow turner at the University of Maine compost facility. Note the steam rising from the heating compost as the pile is turned. The finished compost is used in campus landscaping projects and provided to the public.

Simple, static composting systems often require modification to manage large volumes of organic wastes. Accordingly, many composting designs and strategies have been developed. Some are proprietary, some are not. Some are given the name of the developer or location where they were developed. For instance, the Indore process is named for a region of India in which Sir Albert Howard developed his method in the 1920s. This method involves alternating layers of waste that are periodically turned. The Indore process greatly improved sanitation in rural India and modifications of the process are still in use today.

In other batch systems, more elaborate means of aerating compost piles have been developed. For example, the Beltsville process (developed by the U.S. Department of Agriculture, Agricultural Research Service in Beltsville, Maryland, U.S.) provides aeration to compost piles via a series of pipes underlying the pile (Fig. 22–6). A blower mechanism pulls air through the pile and into pipes near the bottom of the compost pile that run parallel to the length of the windrow. Air may also be pumped from the pipes upwards through the pile. This approach obviates the need for turning the pile to achieve aeration. These kinds of windrows are called *aerated static piles*. Forced aeration, since it allows the compost process to proceed with minimal or no turning, permits piles to be larger. Direction of air flow is sometimes reversed to minimize odors emanating from

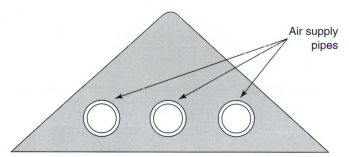

Figure 22–6 Compost pile configured with air distribution pipes to facilitate aeration.

the pile. Air drawn through the pile when odor levels are high can be filtered through activated charcoal or finished compost to absorb the odor-causing compounds. Aerated static piles are popular because they provide a higher level of control of the process, minimize the need to turn the pile, and encourage rapid rates of decomposition. For example, complete composting of biosolids can occur within 21 days.

Continuous Composting Systems

Continuous composting consists of a continuous supply of decomposable substrates kept under strict regulation. This is commonly achieved by enclosing the entire process in a reactor. Substrates are continuously fed into the reactor and move through the reactor as decomposition proceeds. Vertical- or horizontal-flow systems have been designed to aerate and mix the substrates.

Other systems are agitated bed and rotary drum systems. The Dano design is one example of the latter; consisting of a long cylinder (about 3 m in diameter and 30 m in length) that slowly rotates as substrates are fed into one end. The composting substrates move spirally down the slightly inclined cylinder to emerge from the other end as a nearly finished compost. Strict control of the quality and quantity of substrates fed into the reactors and rate of internal movement of the composting wastes allow an equilibrium to develop where thermophilic conditions are continuously maintained.

Grinding of substrates aids in control of the biological process in both static and continuous systems. Maintaining high rates of microbial activity ensures the shortest time to complete the composting process. Rapid composting rates are beneficial for facilities that must process large amounts of wastes in a limited amount of time. Thus large-scale commercial composting facilities often favor reactor-based, continuous process designs. Dozens of modifications of the horizontal or vertical reactor designs are in current use.

Compost Testing and Use

Compost maturity is usually tested by determining oxygen uptake or carbon dioxide evolution. The more mature the compost, the less oxygen is taken up or carbon dioxide evolved. Another simple test of compost maturity is latent heat production. A

Table 22–4 **Selected chemical properties of representative finished composts.***

	pH	Moisture (%)	Bulk density (kg m^{-3})	Nitrogen (%)	Phosphorus (%)	Potassium (%)
Biosolids	6-6.5	20-35	772	2.21	0.97	0.27
Yard wastes	5.8-7.2	34-61	304-400	0.7	0-0.03	0.02-0.36
Food wastes	—	40	352	0.95	0.13-0.46	0.58-0.89
Municipal solid waste	—	45	352-481	—	—	—

Adapted from Diaz et al. (1993) and Satrina (1974).
*For comparison with starting materials refer to Table 22–1.

sample of the compost is placed in an insulated container and the temperature is monitored over time. Raw compost self-heats to 65° or 70°C in two or three days, while a mature compost normally heats to less than 30°C. Biologically based tests of compost maturity often include an examination for the presence of phytotoxins or seed germination inhibitors.

Finished or cured composts are also tested for nutrient content and other characteristics (Table 22–4). Many composts are good sources of nutrients for growing crops, containing useful amounts of nitrogen, phosphorus, and sulfur. They differ from inorganic fertilizers in that compost requires further microbial decomposition to release most of those nutrients. The mineralization of the organic compounds results in a "metering" of the nutrients to the plants. Because composts can fertilize the soil and improve soil physical conditions, testing often includes an estimation of the mineralization rate. This is important for farmers because they must be able to accurately predict the amount of nutrient that will be supplied by the decomposing compost in the field. Supplemental fertilizer often must be added to meet immediate crop demand.

Mineralization of the compost yields plant nutrients to crops over an extended period. On average, 10% of the compost is mineralized during the first year after soil application. In subsequent years, the annual decomposition rates fall to between 1 and 3%, resembling the decomposition rate of soil organic matter. In addition, compost provides the same benefits that are generally ascribed to organic matter. These include increasing soil aggregation and stabilization, soil aeration, water infiltration, water holding capacity, and cation exchange capacity.

The extent of compost use has grown to include fertilization of athletic and recreational turf areas, home lawns and gardens, landfill caps, roadside grade stabilization, and agricultural fields. Some composts have become marketable products for industries and municipalities, further reducing compost operation costs.

Summary

Composting is a microbial process in which organic matter progresses through stages of decomposition and stabilization over a relatively short time. It is characterized by an early period of self-heating in which heat generated by rapid microbial metabolism is produced faster than it can dissipate from the site of decomposition. Temperature increases and substrate changes elicit a succession of microorganisms that reflect the change from mesophilic to thermophilic conditions. During the curing phase, mesophilic conditions return and further decomposition stabilizes the compost.

The purposes of composting wastes range from volume reduction to sanitation of the wastes. Applied to land, finished compost can function as a slow-release fertilizer and soil-improving amendment. Interest in composting has increased greatly and has led to the development of many engineering design modifications. Computerized control systems and large physical facilities designed to enhance the processing of large amounts of municipal and industrial wastes are becoming commonplace.

Cited References

Beaudin, H., R.F. Caron, R. Legos, J. Ransay, L. Lawlor, and B. Ramsay. 1996. Cocomposting of weathered hydrocarbon-contaminated soil. Comp. Sci. Util. 4:37–45.

Chang, Y., and H.J. Hudson. 1967. The fungi of wheat straw compost. I. Ecological studies. Trans. Br. Mycol. Soc. 50:649–666.

Crawford, S.L., G.E. Johnson, and F.E. Goetz. 1993. The potential for bioremediation of soils containing PAHs by composting. Comp. Sci. Util. 1:41–47.

Diaz, L.F., G.M. Savage, L.L. Eggerth, and C.G. Golueke. 1993. Composting and recycling municipal solid waste. Lewis Publishers, Boca Raton, Fla.

Finstein, M.S., and M.L. Morris. 1975. Microbiology of solid waste composting. pp. 113–151. *In* D. Perlman (ed.), Advances in applied microbiology. Vol. 19. Academic Press, New York.

Hankin, L.H., R.P. Poincelot, and S.L. Anagnostakis. 1976. Microorganisms from composting leaves: Ability to produce extracellular degradative enzymes. Microbiol. Ecol. 2:296–308.

Haug, R.T. 1980. Compost engineering: Principles and practice. Ann Arbor Science Publishers, Ann Arbor, Mich.

Haug, R.T. 1993. The practical handbook of compost engineering. Lewis Publishers, Boca Raton, Fla.

Michel, F.C., Jr., C.A. Reddy, and L.J. Forney. 1995. Microbial degradation and humification of the lawn care pesticide 2,4-dichlorophenoxyacetic acid during the composting of yard trimmings. Appl. Environ. Microbiol. 61: 2566–2571.

Parnes, R. 1986. Organic and inorganic fertilizers. Woods End Agricultural Institute, Mt. Vernon, Maine.

Poincelot, R.P. 1982. The biochemistry of composting. pp. 33–39. *In* Composting of municipal residues and sludges, Proc. National Conference on Composting of Municipal and Industrial Sludges, Rockville, Md. Hazardous Materials Control Research Institute, Silver Springs, Md.

Satrina, M.J. 1974. Large scale composting. Noyes Data Corp., Park Ridge, N.J.

Seekins, W. 1986. Usable waste products for the farm. State of Maine, Department of Agriculture, Food, and Rural Resources, Augusta, Maine.

Strom, P.F. 1985. Effect of temperature on bacterial species diversity in thermophilic solid-waste composting. Appl. Environ. Microbiol. 50:899–905.

U.S. Environmental Protection Agency. 1993. Appendix B to Part 503: Pathogen treatment processes. Fed. Reg. 58(32):9404.

U.S. Environmental Protection Agency. 1994. Characterization of municipal solid wastes in the United States, 1994 update. EPA No. 530-S-94-042. Washington, D.C.

General References

Anonymous. 1989. The biocycle guide to composting municipal wastes. JG Press, Emmaus, Pa.

Dickenson, C.H., and G.J.F. Pugh. 1974. Biology of plant litter decomposition. Vol. 1, 2. Academic Press, New York.

Dalzell, H.W., A.J. Biddlestone, K.R. Gray, and K. Thurairajan. 1987. Soil management: Compost production and use in tropical and subtropical environments. FAO Soils Bull. 56. Food and Agriculture Organization of the United Nations, Rome.

Study Questions

1. What is the purpose of composting?

2. How does composting differ from decomposition of organic substrates in soil?

3. What is the role of oxygen in composting?

4. Why is so much heat produced during composting?

5. What chemical and physical characteristics of organic substrates promote rapid composting?

6. What successional changes in microbial populations occur during composting?

7. What is the stabilization or curing phase of composting?

8. What are the differences and similarities between batch and continuous composting systems?

9. Why are engineering modifications of the composting process, such as agitated bed and rotating drum reactors, used in the treatment of large volumes of wastes?

10. What are the advantages and disadvantages of using compost as a fertilizer?

Chapter 23

♦

Global Gases

Joshua Schimel and Elisabeth A. Holland

♦

*Many factors currently limit our ability to project and detect future climate change.
In particular, to reduce uncertainties further work is needed on the following
priority topics: estimation of future emissions and biogeochemical cycling
(including sources and sinks) of greenhouse gases, aerosols and aerosol precursors,
and projections of future concentrations and radiative properties . . .*
Summary for Policymakers: The Science of Climate Change, IPCC Working Group I (1995)

Soil microorganisms play critical roles in the cycling of all the major elements, including carbon, nitrogen, sulfur, phosphorus, and iron. Several of these element cycles have gaseous components such as carbon dioxide, dinitrogen, and sulfur dioxide. Thus soil microorganisms play a role in controlling the composition, chemistry, and physics of the atmosphere. They can affect climate, ozone depletion, smog production, the rate at which organic chemicals are "scrubbed" from the atmosphere, and nutrient inputs to downwind ecosystems. Despite the substantial effects on the atmosphere of gases produced by soil microorganisms, most of these gases are present only in small concentrations in the atmosphere. They are **trace gases.**

Defining Trace Gases

Atmospheric Composition

If it were not for the presence of life on earth, our atmosphere would be largely carbon dioxide (CO_2) with smaller amounts of oxygen (O_2), nitrogen (N_2), and carbon monoxide (CO), traces of nitrous oxide (N_2O), and essentially no methane (CH_4). Instead, our atmosphere is dominated by nitrogen (78%)

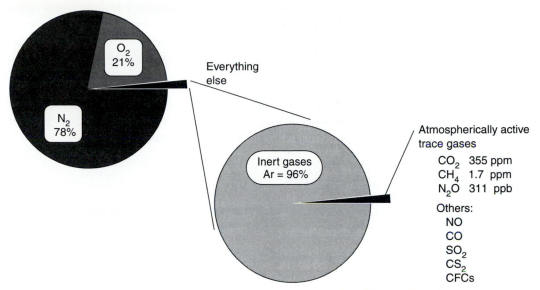

Figure 23–1 Composition of the atmosphere (CFCs = chlorofluorocarbons).

and oxygen (21%). All the rest of the gases put together comprise only about 1% of the atmosphere (Fig. 23–1). However, almost all of this 1% is the inert gas argon (Ar). The rest of the atmosphere (0.04%) comprises a collection of chemically and radiatively active gases that play important roles in atmospheric processes. We call them trace gases because they are present in such small concentrations.

The concentrations of trace gases are variable and are controlled by both natural and anthropogenic factors. Because of anthropogenic emissions, atmospheric concentrations of carbon dioxide, nitrous oxide, and methane have risen at unprecedented rates since the turn of the century (Table 23–1). The rate of increase temporarily slowed to near zero through the early 1990s, though rates of

Table 23–1 **Atmospheric properties of selected trace gases.**

Compound	Concentration (ppb) 1992	Annual increase* (%) 1980s	Annual increase* (%) 1990–1992	Lifetime	Global warming potential relative to CO_2
CO_2	355,000	0.4	0.14	50–200 yr	1
CH_4	1,714	0.8	0.27	11–17 yr	21
N_2O	311	0.25	0.15	120 yr	206
CFC-12	0.5	4.0	0.026	102 yr	15,800
NO_x	0.005–10			1–10 d	
CO	40–200			1–4 mo	

Data from Houghton et al. (1995).
*Note that the reduction in increase during the early 1990s seems to be temporary.

increase appear to be rising again. Possible explanations for the decrease range from changes in atmospheric transport induced by the Mount Pinatubo eruption in the Philippines, which changed the distribution of nitrous oxide and methane in the *troposphere* (region between the earth's surface and below 10 to 20 km into the atmosphere), to a dramatic decline in methane release from industrial sources because of the collapse of the former Soviet economy. The coming years will help sort out these various explanations and any modifying roles played by soil microorganisms. Soil microbes are producers and consumers of each of these gases and contribute to their global budgets.

Importance of Global Gases

If trace gases comprise such a small proportion of the total atmosphere, why do we care about them? Trace gases:

- influence the earth's radiation balance and thereby influence global climate,

- regulate the ability of the atmosphere to cleanse itself of pollutants, and

- are important sources of nutrients to the **biosphere,** particularly nitrogen-containing gases, as represented by the two-way arrows in Figure 23–2.

Long-lived gases (e.g., CO_2, N_2O, CH_4, CH_3Cl, and CH_3Br) play multiple roles in the atmospheric system. The sun's radiation is dominated by the short wavelengths of visible light, but these are absorbed and reradiated at longer wavelengths, largely in the infrared region. The infrared can be absorbed by carbon dioxide, nitrous oxide, and methane in the lower atmosphere, which warms the earth's surface. This is the *greenhouse effect*. The N_2O, CH_4, CH_3Cl, and CH_3Br persist in the atmosphere long enough to be transported to the *stratosphere* (the region 20 to 50 km above the earth's surface) where they are important in ozone (O_3) production and destruction. These gases have been implicated in the Antarctic "ozone hole." The ozone layer present in the stratosphere blocks much of the incident ultraviolet radiation which also influences the earth's radiation balance. Methane is oxidized in the atmosphere to carbon dioxide and water, which is the main source of water vapor in the stratosphere. During the polar winter, the water vapor freezes, forming polar stratospheric clouds. This provides surfaces on which ozone destruction takes place to form the hole in the ozone layer. Nitrous oxide is the main source of reactive nitrogen in the stratosphere, dissociating to form several unusual nitrogen compounds, including NO, NO_2, NO_3, N_2O_5, and HNO_3. These nitrogen species are important in catalyzing ozone destruction and formation. Furthermore, they react with natural and anthropogenic halogens to form halogen nitrates (e.g., $ClONO_2$, $BrONO_2$, and $IONO_2$), which in turn are involved in ozone destruction in the stratosphere, particularly in the presence of polar stratospheric clouds.

Short-lived gases, such as CO and NO_x (NO_2+NO), play an important role in tropospheric chemistry by regulating its oxidizing capacity. In the stratosphere the active nitrogen reactions work primarily to destroy ozone, but in the troposphere NO_x

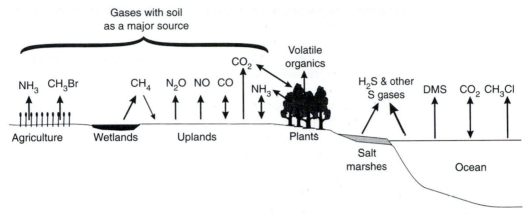

Figure 23–2 Sources of gases contributed to the atmosphere, showing the ecosystems that are dominant sources for each gas (DMS = dimethylsulfide).

can react with methane and other hydrocarbons to form ozone. Tropospheric ozone is an important pollutant, causing crop and ecosystem damage and respiratory illnesses. Production of nitric oxide (NO) in soils and its subsequent release to the atmosphere can strongly influence atmospheric chemistry, particularly in less polluted regions where background concentrations of NO_x are near the threshold (5–10 ppt, parts per trillion) needed for net ozone production. Soils are recognized as secondary sinks for carbon monoxide, consuming 250–640 Tg CO yr^{-1} compared to photochemical destruction of 1,400–2,600 Tg CO yr^{-1} (T = 10^{12}). Carbon monoxide is also an intermediate product in some microbial pathways of methane production. As a result, soils may also be a small source of carbon monoxide. Carbon monoxide is very reactive and is important in atmospheric oxidation chemistry, including ozone production and consumption. Soil microbes regulate atmospheric photochemistry through the production and consumption of these short-lived gases.

Gases Released from Soils

A large number of gases are exchanged between the atmosphere and terrestrial ecosystems. A subset of these, however, is produced in soils (Figure 23–2). These include carbon dioxide, methane, nitrous oxide, nitric oxide, and carbon monoxide. Soil sources of these gases account for roughly 25% of carbon dioxide, 50% of methane, 65% of nitrous oxide, 30% of nitric oxide, and an unknown share of carbon monoxide. Volatile organic compounds are emitted largely by plants, while most of the sulfur emitted to the atmosphere comes from the ocean, though salt marshes have a measurable role as well. Since relatively little is known about the controls on carbon monoxide emissions, our focus in this chapter is on carbon dioxide, methane, nitric oxide, and nitrous oxide.

While trace gas fluxes from terrestrial ecosystems are often large from the perspective of global biogeochemical cycles and atmospheric chemistry, they seem small from the perspective of biogeochemical cycling within an individual ecosystem, with the exception of carbon dioxide and sometimes methane. For example, many ecosystems have net nitrogen mineralization rates as high as 100 kg ha^{-1} yr^{-1}, but nitrous oxide fluxes are rarely greater than 1 kg ha^{-1} yr^{-1}. The production of trace gases can also be extremely variable in space and time; for instance, nitrous oxide from denitrification often increases sharply immediately after a rainfall. These factors can make measuring and modeling trace gas fluxes a challenge.

With the exceptions of methane oxidation and carbon monoxide dynamics, we actually have a good understanding of the basic microbial physiology and processes affecting most trace gas fluxes. The great challenge is to integrate this understanding and then extrapolate it to regional and global scales. To do this, we need to quantify how the specific controls on trace gas dynamics vary in space and time. Then we can estimate large-scale fluxes of trace gases that can be linked to global atmospheric chemistry and climate models while remaining solidly grounded in the basic soil processes. An important component of developing this large-scale perspective is understanding how land use practices and other human activities affect trace gas dynamics. The focus of this chapter is, therefore, to take the more detailed information on specific soil processes, or transformations, presented in Chapters 11 to 16 and discuss how these processes control atmospheric trace gases at a global scale. The essence of this approach is to understand the links between the immediate, physiological controls on a process and the large-scale environmental factors that mediate those controls. For example, oxygen availability controls denitrification rates at the cellular level, but soil moisture and respiration control oxygen diffusion and availability (Figure 23–3). At a larger scale, rainfall and soil texture control soil moisture, while temperature and organic matter quality control respiration. At an even larger scale, Jenny's (1980) fundamental state factors of climate, plant community, and soil parent material control rainfall, soil texture, and organic matter quality. Therefore, we can use a basic understanding of microbial physiology and soil processes to work our way to large-scale, mechanistic models of trace-gas dynamics.

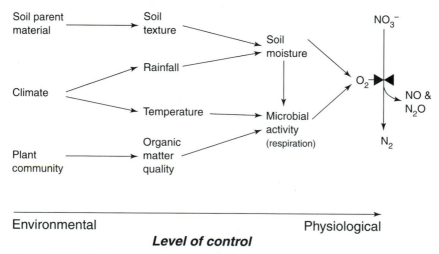

Figure 23–3 Relationship between physiological controls on nitrogen gas production and the large-scale environmental factors that mediate them. This figure is similar to the upper portion of Figure 12-8.

General Controls on Gas Production

Temperature, moisture, redox potential, and substrate availability are probably the four factors that exercise the greatest control on trace gas dynamics across ecosystems.

- Temperature is important because all biological reactions are temperature sensitive. Many reactions have a Q_{10} of approximately 2. However, some processes have Q_{10} values greater than 2. Across seasons and climatic zones, temperature plays a major role in controlling the rates of trace gas production.

- Moisture is important for several reasons. First, all living organisms require water. At low soil moisture contents, microbial activity is limited because the organisms are physiologically stressed from the low water potential and poor substrate diffusion. Second, soil water content controls oxygen diffusion into the soil, thereby affecting soil redox potential and the relative contribution of aerobic and anaerobic processes.

- Redox potential is controlled by soil moisture, which controls oxygen diffusion, and by microbial activity, which consumes the oxygen that diffuses into the soil. Many of the gases discussed in this chapter are produced under anaerobic conditions, and some are consumed under aerobic conditions. Thus the redox state of the soil can act as a master switch on the processes producing and consuming trace gases.

- Substrate availability, particularly carbon quantity and quality, is a key control on trace gas dynamics. Carbon availability directly controls the dynamics of carbon-based gases such as carbon dioxide and methane and the activity of heterotrophic bacteria such as denitrifiers. Similarly, nitrogen availability is a direct control on nitrogen gas dynamics. Carbon availability also controls the production of nitrogen-based gases by mediating inorganic nitrogen availability. The large role of carbon in controlling trace gas dynamics means that plant community composition and ecology are important larger-scale factors to consider in modeling trace gases. The composition of the plant community controls the quantity and quality of organic matter inputs into the soil and thus controls carbon and nitrogen availability. For example, grasslands and some deciduous forests produce relatively nutrient-rich, lignin-poor litter, whereas coniferous forests generally produce nutrient-poor, lignin-rich litter. Nutrient cycling and trace-gas dynamics in these systems differ, in part, because of the differences in plant-litter chemistry.

Carbon-Based Gases

Carbon Dioxide

Carbon dioxide is the most abundant trace gas in the atmosphere (355 ppm, parts per million or μl per liter), and the fluxes of carbon dioxide between the biosphere and the atmosphere are the greatest of any trace gas. Approximately 60,000 Tg CO_2–C yr^{-1} is released from soil organic matter. This is roughly 25% of the total flux to the atmosphere (plant respiration is about 25% and marine sources account for most of the remaining 50%). In context, human inputs from fossil fuel burning are only about 5,000 Tg CO_2–C yr^{-1}.

Carbon dioxide is produced in soil by microbial metabolism during organic matter decomposition. Thus the major controls on its production are essentially those that influence general heterotrophic microbial activity. In the context of understanding carbon dioxide dynamics from a global perspective, climate is the most important factor, with substrate quality second. Decomposition is slow in cool, water-logged environments, leading to the buildup of thick layers of decaying material, or peat. Peat accumulation reduces the amount of carbon dioxide returned to the atmosphere, and peat accounts for roughly 24% of total soil carbon storage worldwide. In warm regions soil-carbon accumulation is often limited. In warm, moist climates, decomposition and carbon dioxide release are usually rapid, but in warm, dry environments, low primary production often limits organic carbon inputs, carbon dioxide production rates, and carbon accumulation. Thus, although changes in carbon storage in peatlands at high latitudes can be important in controlling carbon-cycling at relatively long time scales (in ecological rather than geological terms), the high rate of carbon turnover in tropical systems could dominate short-term changes in global carbon cycling (Townsend et al., 1992).

In the temperate zones, climate has a lesser influence on decomposition rates, and substrate quality becomes relatively more important. Substrate quality is a somewhat ill-defined term. Various measures of litter quality have been suggested, including C/N ratio, lignin content, lignin-to-nitrogen ratio, and the lignin-cellulose index, but different measures work best under different circumstances. The quality of soil organic matter is even harder to measure, but new techniques such as density and isotopic fractionation have shown some promise.

Methane

Budget. The microbial metabolism of methane is more complex than that of carbon dioxide in soils. Soils can both produce and consume methane, even simultaneously, under the proper environmental conditions. Total methane emissions to the atmosphere are approximately 410 Tg CH_4–C yr^{-1}. Direct emissions from wetland soils account for approximately 32% of the total emissions to the atmosphere (Table 23–2). Natural wetlands and rice paddies account for about 86 and 45 Tg CH_4–C yr^{-1}, respectively. When termites, which are predominantly soil fauna, and animal wastes, which are returned to the soil, are included, the total "soil" source increases to about 44%.

In wetlands, methane oxidizers can consume up to 90% of the methane produced in the anaerobic zone before it reaches the atmosphere. Thus methane oxidation in wetlands is one of the most significant factors influencing the global methane cycle. Although any changes have large potential impacts on the net gain of atmospheric methane, it is still often an "invisible" process from the atmospheric perspective.

Unsaturated upland soils, in contrast, actually consume methane from the atmosphere (somewhere between 11 and 34 Tg CH_4–C yr^{-1}) and comprise between 3% and 9% of the total sink for atmospheric methane. Although this value may not sound particularly large, soil consumption of atmospheric methane is sensitive to the human activities of tillage and fertilization, and changes in methane consumption could account for a portion of the increase in atmospheric methane (Ojima et al., 1993).

Table 23–2 **Atmospheric budgets for methane, nitrous oxide, and nitric oxide. All fluxes are in Tg yr^{-1} and are "most likely" estimates.**

Sources		Sinks		Atmospheric increase
Type	Flux	Type	Flux	Flux
Methane				
Soil				
Wetlands	86	Soil	-23	
Rice paddies	45			
Termites	153			
Animal wastes	19			
Others				
Oceans	8	Atmospheric oxidation	-334	
Animals	64			
Biomass burning	30			
Landfills, sewage	49			
Fossil fuels	64			
Total	401		-386	28
Nitrous Oxide				
Natural soil				
Tropical wet forest	3			
Tropical dry savanna	1			
Temperate forests	1			
Temperate grassland	1			
Others				
Cultivated soils	3.5	Atmospheric	-12.3	
Cattle and feedlots	0.4			
Biomass burning	0.5			
Industrial	1.3			
Oceans				
Total	14.7		-12.3	3.9
Nitric Oxide				
Soil	12			
Biomass burning	8			
Fossil fuel combustion	24			
Lightning	5			
NH$_3$ oxidation	3			
Aircraft	0.4			
Total	52			

From Houghton et al. (1995).

Process control: methane production (methanogenesis). Methane is produced exclusively by a group of anaerobic Archaea, known as the **methanogens** (Whitman et al., 1992). Because these bacteria are obligate anaerobes requiring redox potentials less than -100 mV, redox potential acts as an on-off switch for methane production. When redox potential drops low enough, the switch is on. In this case, substrate supply and temperature become major controls on the rate of methane production. In soil systems, the redox potential only drops to values low enough to allow substantial methane production when the soils are flooded, essentially blocking oxygen diffusion into the soil (gases diffuse 10,000 times faster in air than water). Thus, at a large scale, the presence and depth of a water table, given adequate organic matter inputs, controls soil redox potential and the rate of methane production.

Methanogens in soils produce methane by two primary pathways:

$$CO_2 + H_2 \rightarrow CH_4 \qquad\qquad \text{(CO}_2 \text{ reduction)}$$

$$CH_3COOH \rightarrow CH_4 + CO_2 \qquad \text{(acetate fermentation)}$$

Both acetate and hydrogen are by-products of anaerobic fermentation; thus methane is actually produced by a complex food web. Most plant detritus and soil organic matter is made up of polymers, such as cellulose, lignin, and humic compounds. This material must first be broken down into monomeric sugar or phenolic units (usually by extracellular enzymes) before it can be metabolized by soil microbes. The monomers must then be fermented to produce the methanogenic substrates, hydrogen and acetate. As the end product of a complex series of degradation pathways, methane production is sensitive to organic matter supply and quality, even though methanogens themselves use only the simplest of substrates.

Because of its strong dependence on a complex substrate supply, methane production shows a broad array of temperature responses, with Q_{10} values as high as 30 to 40, well out of the normal range for biochemical processes. As a result of this complex temperature response, climate warming would likely increase wetland methane production, but the extent of any increase is difficult to predict. Also, because methane production requires anaerobic conditions, climate changes that alter the extent of flooded land would greatly change methane fluxes. Drying would both slow methane production and increase methane oxidation, thus greatly reducing methane emissions. Increasing the extent of wetlands would increase methane emissions to the atmosphere. Warmer climates could also mean longer growing seasons and greater methane emissions.

Process control: methane oxidation (methanotrophy). Methane is oxidized in nature primarily by a group of bacteria known as **methanotrophs** (Lidstrom, 1992). These aerobic bacteria probably use methane as their sole source of carbon and energy. The overall reaction pathway is:

$$CH_4 + 2\,O_2 \rightarrow CO_2 + 2\,H_2O$$

There is some evidence that ammonium (NH_4^+) oxidizers may oxidize methane in the field. Nitrifiers can oxidize methane because ammonia and methane are very similar in size and shape. Thus, methane can fit into the active site of ammonia monooxygenase, the active enzyme in nitrifiers, and be oxidized by that enzyme. Similarly, ammonia (NH_3) can be oxidized by methane monooxygenase in methanotrophs. Note that ammonia, rather than ammonium, is the active nitrogen form for both monooxygenase enzymes.

Methanotrophs are found in many aerobic environments, including the surface layers of wetland soils and unsaturated upland soils. In flooded environments, primarily natural wetlands and rice paddies, methanotrophs use the methane produced in the flooded, anaerobic zone that underlies the aerobic zone. Methanotrophs at the aerobic-anaerobic interface may be exposed to very high concentrations of methane, 10% or more of the dissolved gases.

In unsaturated soils, mostly forest, grassland, and agricultural soils, there is no belowground source of methane, so methane is instead consumed from the atmosphere. In these environments, methane oxidizers live on atmospheric methane (1.7 ppm) and are strongly substrate limited. Thus, consumption of atmospheric methane is controlled largely by methane diffusion into the soil profile. Vehicular traffic and other activities that promote soil compaction may reduce soil air-filled pore space and therefore reduce methane consumption.

Methane oxidation can also be inhibited by high soil ammonium concentrations, which occur after fertilization or other heavy nitrogen deposition. This reaction has been observed in a wide range of ecosystems, including forests in Alaska, temperate grass and croplands, and tropical rain forests. The mechanism of ammonium inhibition results from the molecular similarity of methane and ammonia. Ammonia can compete with methane for binding at the enzyme's active site, thus inhibiting methane oxidation. Thus, increasing use of nitrogen fertilizers and increased atmospheric nitrogen deposition may decrease global methane consumption, leading to greater atmospheric methane concentrations and greater greenhouse warming.

Process control: overall methane flux from the soil. The net methane flux, movement into or out of the soil, is controlled by the balance of production and consumption. To predict the net flux, we need to understand the individual controlling processes. Vegetation and water table depth are the two factors most critical in controlling the overall flux of methane from a wetland system. The depth of the water table controls the volume of anaerobic soil and thereby the rate of methane production. In a similar manner, the nature of the water table also controls the amount of aerobic soil and the likelihood that methane is oxidized before it reaches the atmosphere.

Vegetation is critical because it affects both the rate of methane production and the fate of any methane that is produced (Figure 23–4). Methane production is sensitive to the supply of easily decomposable (labile) carbon. Often the major source of such carbon is plant roots, either as exudates or from recently dead roots. Because of the importance of plant carbon in driving methane production, strong connections occur between net primary productivity in an ecosystem and the magnitude of methane flux out of that system. However, plants also control methane export from the soil (Schimel, 1995). Many wetland plants have air channels, in specialized tissue called *aerenchyma,* in their roots that serve to supply the roots with oxygen. Aerenchyma allow oxygen to diffuse downward, but also allow methane to diffuse upward through the plant, thereby bypassing the aerobic gauntlet of methane oxidizers. In many wetland systems, 90% or more of the methane that makes it out of the soil is transported through the plants.

Although we understand the basic controls on methane flux, developing and applying models across a range of ecosystems has been challenging. The difficulty may result from the complexity of processes controlling methane flux. First, models

| **Box 23–1** |

What Is the Net Effect of Drying a Peatland on Total Global Warming Potential?
Consider the results of drying a peatland on total greenhouse warming potential. Drying
would reduce methane fluxes, which would reduce the global warming potential. In contrast, drying a peatland usually increases carbon dioxide fluxes, which would increase
global warming potential.

Methane is between 62 and 24.5 times as effective a greenhouse gas as carbon dioxide, depending on whether one integrates over 20 or 50 years; the longer the time, the
less effective methane is relative to carbon dioxide, since it has a shorter lifetime in the
atmosphere (Houghton et al., 1995). If we use the value of 24.5, carbon dioxide fluxes
would have to increase 24.5 times as much as methane fluxes decrease for the changes
in the two gases to balance each other. This is unlikely. Drying a peatland is therefore
likely to reduce total global warming potential. However, changes in carbon dioxide
fluxes, and possibly nitrogen gases, may substantially alter the effect of draining on the
total effect on greenhouse warming and other atmospheric dynamics. Therefore, we
should consider the effects of land use changes on all the important trace gases rather
than just one of particular interest.

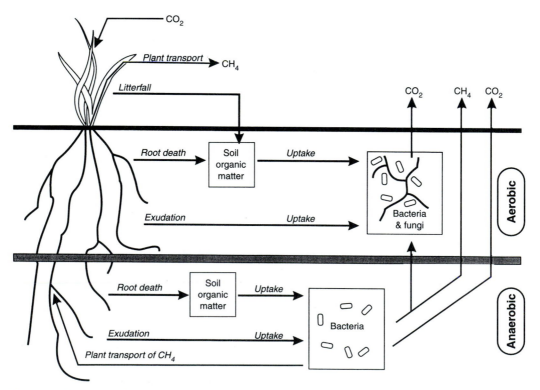

Figure 23–4 Carbon fluxes in a wetland system.

must account for methane production, so they must consider the production and breakdown of organic material and account for its fermentation to methane. Then a model must account for the fate of methane—whether it escapes from the soil or is oxidized by methanotrophs—which requires an understanding of plant transport dynamics. Developing effective integrated models of methane dynamics remains an important area for research, linking soil biology with atmospheric chemistry.

Nitrogen-Based Gases

Nitrous oxide (N_2O) and nitric oxide (NO) are two important nitrogen gases that are produced by soil microorganisms. These gases are produced by both nitrification and denitrification. The physiology of these processes is discussed thoroughly in Chapter 12.

Budgets

Balancing the sources and sinks to develop a global budget of nitrous oxide release and consumption has been a major challenge for researchers. The one thing known for certain is that 3.9 Tg yr^{-1} of N_2O–N was accumulating in the atmosphere at the end of the 1980s. We can quantify the stratospheric sink for nitrous oxide at 12.3 Tg yr^{-1} (Table 23–2). The sum of these provides an estimate of the total sources at between 13 and 20 Tg nitrogen annually. Before 1986, soils and fossil fuel combustion were thought to be the major nitrous oxide sources and the budget was reasonably balanced. However, in 1986, the discovery of an experimental error in previous smokestack measurements virtually eliminated the fossil fuel combustion source. At the same time, estimates of tropical emissions dropped dramatically; early estimates had been based on only a few measurements from active sites and extrapolated to the entire tropics. These two factors left the budget badly unbalanced. Since then, estimates of nitrous oxide emissions from tropical soils have again risen with the completion of more measurements. Compilations of nitrous oxide emissions following fertilizer application have also increased estimates of this source and account for much of the remainder (Table 23–2). After 10 years of struggling with sources and sinks, the budget is reasonably balanced once more. The struggles with the nitrous oxide budget illustrate that science is not always a steady process of enlightenment. Sometimes major advances merely illustrate that our previous understanding was wrong, leaving us in the dark once again.

Nitric oxide is a very short-lived compound in the atmosphere, and its budget is even harder to estimate than for nitrous oxide. Early estimates of soil emissions were from 4 to 16 Tg NO–N yr^{-1}, a large range. Current estimates for global soil nitric oxide production are approximately 12 Tg NO–N yr^{-1} (Table 23–2).

Controls on Nitrogen Gas Production

In an ecological context, the overall controls on the production of nitric oxide and nitrous oxide have best been considered within the "leaky-pipe" model of

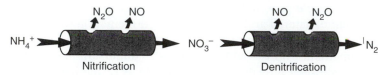

Figure 23–5 "Leaky pipe" model of the control of nitric oxide and nitrous oxide production from nitrification and denitrification. *After Firestone and Davidson (1989). Used with permission.*

Firestone and Davidson (1989; Figure 23–5). This conceptual model links overall nitrogen dynamics with nitric oxide and nitrous oxide production. The basis of this conceptual model is that nitric oxide and nitrous oxide are side products of nitrification and denitrification. Thus the production of these gases is dependent on both the total rate of the processes (flow through the pipe) and the proportion of the nitrogen that is released as either nitric oxide or nitrous oxide (the size of the holes in the pipe). By considering these factors separately, we can build effective models for how soil processes control the emissions of these globally important gases.

Process rates: nitrification. As is true for many soil microbes, the main limitation on nitrifier activity is substrate availability, ammonium in this case. Nitrification rates often correlate well with mineralization rates and ammonium availability. If ammonium availability is great enough, it can even allow nitrifiers to overcome other factors that generally limit nitrification, such as low pH. This explains why the very acid forest soils of New England after clear-cutting exhibit rapid nitrification, which is almost unmeasurable in the native forest: by eliminating plant uptake, ammonium is made available for nitrifiers.

One of the major controls on ammonium availability is plant activity (Figure 23–6). Plant uptake can limit ammonium supply, but plants also release large amounts of organic matter into the soil as leachates, root exudates, or litter. If this material is nitrogen rich, it can stimulate mineralization and nitrification. If the material is nitrogen poor, it stimulates microbial immobilization, reducing the availability of ammonium to nitrifiers.

Water is a major secondary control on nitrification. First, nitrification is an aerobic process, and wet soils slow oxygen diffusion into the soil. Increasing soil moisture, therefore, limits the oxygen supply to nitrifiers and reduces nitrification rates. Second, in dry soils, water films around soil particles become thinner and the water film may become discontinuous. This makes substrate diffusion more difficult and reduces substrate supply (Stark and Firestone, 1995).

Process rates: denitrification. Some of the controls on denitrification are the same as those on nitrification; others are complementary. Denitrification is an anaerobic, heterotrophic process in which nitrate (NO_3^-) is the electron acceptor. Because denitrification is an anaerobic process, redox potential is an important control on the rate. However, denitrification can occur at much higher redox potentials than methane production and can actually occur in anaerobic mi-

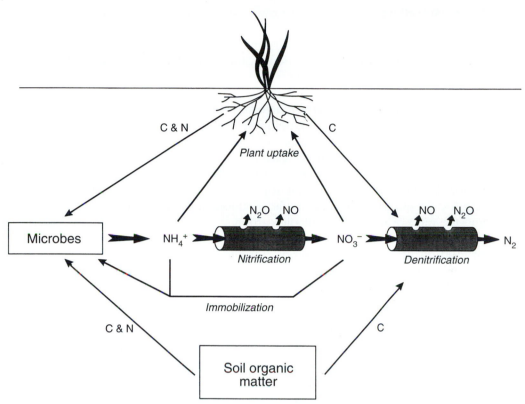

Figure 23–6 Interactions of carbon and nitrogen cycling in controlling production of nitric oxide and nitrous oxide in soils.

crosites in an otherwise aerated soil. Producing these microsites requires that the pores within aggregates are water filled. Thus, soil moisture is a critical control on denitrification rates. When anaerobic microsites are present, nitrate availability becomes a major control on denitrification, so nitrification rates can be important in controlling denitrification rates. Carbon availability can control denitrification rates but is usually secondary to water and nitrate. Thus, at a large scale, water availability and plant activity are major controls on denitrification. Water controls oxygen diffusion, while plants control both nitrogen availability and carbon supply to the denitrifiers.

Holes in the pipe: nitrification. Nitric and nitrous oxide are side products of nitrification, rarely comprising more than about 1% of the total ammonium nitrified. However, the fraction of nitrified nitrogen released as nitric oxide and nitrous oxide is variable and controlled by a number of factors. Oxygen is a major control of the proportion of those gases produced. Although the overall rate of nitrification decreases with decreasing oxygen concentrations, the proportion of the nitrified nitrogen as nitric oxide and nitrous oxide increases at lower oxygen levels. According to one study, nitrous oxide production was maximal at about 1% oxygen; above this only a very small fraction of the product was nitrous oxide, whereas below 1% oxygen, the nitrification rate was too low to produce much nitrous oxide. This study demonstrated the balance between the "size of the pipe" and the "size of the holes in the pipe" in controlling total nitrous oxide production.

Holes in the pipe: denitrification. The factors controlling the proportion of nitrous oxide evolved during denitrification are fairly well established, though those controlling the production of nitric oxide are less well understood. Because nitrous oxide is an obligate intermediate in denitrification, it can vary from being none to all of the final product. Nitric oxide, in contrast, is usually a minor product. Three factors that appear to be particularly important in controlling the proportion of nitrous oxide produced in denitrification are oxygen, pH, and the ratio of nitrate to available carbon.

Although denitrification rates generally decline as oxygen increases, the proportion of nitrogen evolved as nitrous oxide increases as oxygen concentrations rise. Thus, as with nitrification, a balance point exists between maximizing the denitrification rate and maximizing the proportion of product as nitric and nitrous oxide. Also similar to nitrification, the oxygen concentration that maximizes nitrous oxide production from denitrification appears to be about 1%.

Low pH generally inhibits the reduction of nitrous oxide to dinitrogen. Thus, at low pH, nitrous oxide can be the dominant product of denitrification. However, highly acidic soils often have low nitrogen availability and, therefore, low nitrification and denitrification rates. As a result, the highest rates of nitrous oxide production from denitrification generally occur in moist soils that cycle nitrogen rapidly, regardless of pH.

When nitrate, an electron acceptor, is present in excess relative to organic carbon, an electron donor, denitrifiers are "wasteful" of nitrate and generally produce nitrous oxide as a major product. When nitrate is present in limited amounts, however, denitrifiers are more likely to use it to its full potential as an electron acceptor and reduce it all the way to dinitrogen.

Nitrification versus denitrification as sources of nitrogen gases. Because nitric oxide can be consumed by both biotic and abiotic reactions, it is generally only released from the soil in large quantities when the soil is relatively dry. Thus, in most cases, the major source of nitric oxide is nitrification. Nitrous oxide, however, is less reactive in soil and can escape from relatively wet soil. Nitrous oxide can, therefore, be produced in large quantities by either process, though the largest nitrous oxide fluxes are probably from denitrification because the "holes-in-the-pipe" for denitrification are "larger" than for nitrification.

Under extremely dry conditions, microbial activity and gas emissions are limited (Figure 23–7). As water-filled pore space approaches 30% and higher, nitrification increases and nitric oxide is produced. As the soils become still wetter, nitrification continues while denitrification begins and both processes produce

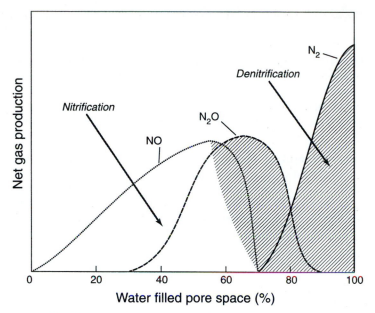

Figure 23–7 Relationship between soil water-filled pore space in soil and the relative fluxes of nitrogen gases from nitrification (unshaded) and dentrification (shaded). *After Davidson (1991). Used with permission.*

nitric and nitrous oxide. As water-filled pore space increases past 60%, nitrification slows, nitric oxide is consumed before it can diffuse out of the soil, and nitrous oxide from denitrification is the main nitrogen gas produced. As soils become completely waterlogged, denitrification may continue rapidly until available nitrate is consumed. Under these conditions, dinitrogen is likely to be the main product and relatively little nitric or nitrous oxide will be produced (Davidson, 1991).

We can combine our understanding of plant-soil carbon and nitrogen dynamics to predict which ecosystems are likely to produce large quantities of nitric oxide and nitrous oxide and which processes are likely to be responsible. For example, the Great Plains of the western United States are semiarid and dominated by grasslands, where decomposition and nitrogen cycling are fairly rapid. Thus, nitrification is relatively rapid, plants use nitrate for a substantial proportion of their nitrogen uptake, and anaerobic microsites are few. Nitrification generally dominates nitrogen gas emissions and the ratio of nitric oxide to nitrous oxide is fairly high, except shortly after a rainstorm. When a rainstorm temporarily saturates the pore spaces within soil aggregates, anaerobic sites become common, and a pulse of denitrification occurs, with nitrous oxide dominating nitric oxide as a product.

Box 23–2

Effects of Converting Primary Tropical Forest to Pasture. Tropical forests are being rapidly cleared for conversion to pastures and farming and for fuel wood. In addition to the tremendous amount of carbon dioxide, carbon monoxide, nitrous oxide, and other gases released during burning, the subsequent state of the land determines trace gas fluxes. Common results include decreasing rates of methane consumption or increases in methane production, and increases in nitric oxide and nitrous oxide production. What factors contribute to this shift in gas fluxes?

- **Changes in nitrogen cycling.** Initially, ash produced during burning and left on the soil surface contains a lot of nitrite which is converted to nitric oxide chemically. This chemodenitrification can increase nitric oxide fluxes by more than tenfold for a short period. Later conversion to pasture can accelerate nitrogen turnover and thereby enhance nitric and nitrous oxide emissions. Both nitrous oxide and nitric oxide emissions are enhanced for years following conversion to pasture. The increased ammonium availability from this accelerated nitrogen cycling can also inhibit methane oxidation and thereby reduce methane consumption.

- **Altered soil microclimate.** Clearing the forest and litter layer removes their buffering effect on soil temperature and moisture. This is especially true after burning, because the black ash layer is particularly good at absorbing sunlight. This can cause desiccation in the surface soil, making it a less favorable zone for microbial activity. Desiccation can be a particular problem for methane-oxidizing bacteria because it may reduce their numbers in the upper soil layers, effectively lengthening the diffusion path of methane, and thereby decreasing methane availability and consumption rates. As a new plant canopy develops, some of these effects can be reduced.

- **Altered soil structure.** Forest clearing and conversion can cause soil compaction. This results from hauling timber during clearing and trampling by animals in pasture. The compaction reduces gas diffusion into the soil and stimulates anaerobic processes when soils are moist, enhancing denitrification and methane production, while inhibiting nitrification and methane oxidation. Compaction also lowers oxygen diffusion into the soil thereby increasing the proportion of products from nitrification and denitrification evolved as nitric oxide and nitrous oxide.

Summary

Soil microbial processes are key players in producing and consuming atmospheric trace gases. Questions remain about the atmospheric budgets of some trace gases (e.g., nitrous oxide, nitric oxide, and carbon monoxide) and the physiology of some key processes (e.g., methane oxidation). However, even for those gases for which the atmospheric budgets and microbial physiologies are fairly well understood (e.g., carbon dioxide, methane production), the challenges of linking the divergent scales from process and site to global dynamics remain. Models of the whole earth system require understanding trace gas fluxes and how they will change under altered climate and land use practices. This in turn requires a firm mechanistic understanding of the processes controlling trace gas production and consumption and how these processes operate at different scales from the microbial to the global.

Cited References

Davidson, E.A. 1991. Fluxes of nitrous oxide and nitric oxide from terrestrial ecosystems. pp. 219–235. *In* J.E. Rogers and W.B. Whitman (eds.), Microbial production and consumption of greenhouse gases: Methane, nitrogen oxides, and halomethanes. ASM Press, American Society for Microbiology, Washington, D.C.

Firestone, M.K., and E.A. Davidson. 1989. Microbiological basis of NO and N_2O production and consumption in soil. pp. 7–21. *In* M.O. Andreae and D.S. Schimel (eds.), Exchange of trace gases between terrestrial ecosystems and the atmosphere. John Wiley and Sons, New York.

Houghton, J.T., L.G. Meira Filho, J. Bruce, H. Lee, B.A. Callander, E. Haites, N. Harris, and K. Maskell. 1995. Climate change 1994. Cambridge University Press, Cambridge, England.

Jenny, H. 1980. The soil resource. Springer-Verlag, New York.

Lidstrom, M.E. 1992. The aerobic methylotrophic bacteria. pp. 431–445. *In* A. Balows et al. (eds.), The prokaryotes. 2nd ed. Springer-Verlag, New York.

Ojima, D.S., D.W. Valentine, A.R. Mosier, W.J. Parton, and D.S. Schimel. 1993. Effect of land use change on methane oxidation in temperate forest and grassland soils. Chemosphere 26:6575–6585.

Schimel, J.P. 1995. Plant transport and methane production as controls on methane flux from arctic wet meadow tundra. Biogeochemistry 28:183–200.

Stark, J.M., and M.K. Firestone. 1995. Mechanisms for soil moisture effects on activity of nitrifying bacteria. Appl. Environ. Microbiol. 61:218–221.

Townsend, A.R., P.M. Vitousek, and E.A. Holland. 1992. Tropical soils could dominate the short-term carbon cycle feedbacks to increased global temperatures. Clim. Change 22:293–303.

Whitman, W.B., T.L. Bowen, and D.R. Boone. 1992. The methanogenic bacteria. pp. 719–767. *In* A. Balows et al. (eds.), The prokaryotes. 2nd ed. Springer-Verlag, New York.

General References

Conrad, R. 1996. Soil microorganisms as controllers of atmospheric trace gases (H_2, CO, CH_4, OCS, N_2O and NO). Microbiol. Rev. 60:609–640.

Schlesinger, W.H. 1991. Biogeochemistry: An analysis of global change. Academic Press, San Diego, Calif.

Study Questions

1. Construct a "level of control" diagram similar to Figure 23–3 for another process involving trace-gas dynamics, such as nitrification, methanogenesis, or methane oxidation.

2. What is the difference between methane production and methane flux? Under what conditions would you use one term or the other?

3. Many atmospheric models assign each ecosystem a specific, fixed flux for each trace gas. What are some of the problems with this approach? How would you improve it?

4. Given that there will be relatively large uncertainties in any global gas budget, what are the scientific reasons for constructing them?

5. Reforestation is one of the proposed solutions for slowing the accumulation of carbon dioxide in the atmosphere. What are the possible implications of a global reforestation program for the other trace gases?

6. Atmospheric deposition of nitrogen, associated with acid rain, has increased in areas of New England. How might this affect gas fluxes from the region? What are the mechanisms involved?

7. Which trace gas fluxes would increase with increasing precipitation? With increasing temperatures? Why?

Glossary* /Index

♦

*Derived from a web-based glossary compiled by David Sylvia,
http://www.ifas.ufl.edu/~dmsa/glossary.htm
*Page number in italics indicates the term is found only in a box, figure
or table; index words automatically include plurals when they are not
given.*

A

absorption Movement of ions and water into an organism as a result of metabolic processes, frequently against an electrochemical potential gradient (active) or as a result of diffusion along an activity gradient (passive). 17, *74*, 83, 396, 409, 416, *422*

Acaulospora, 413

Acetobacter, 308, *314*

acetogenic bacterium Prokaryotic organism that uses carbonate as a terminal electron acceptor and produces acetic acid as a waste product. 61, *62*

acetylene-block assay Estimates denitrification by determining release of nitrous oxide (N_2O) from acetylene-treated soil. *157*, 285

acetylene reduction assay Estimates nitrogenase activity by measuring the rate of acetylene reduced to ethylene. 302, 316, *329*

***N*-Acetylglucosamine** and ***N*-Acetylmuramic acid** Sugar derivatives in the peptidoglycan layer of bacterial cell walls. 52, *264*

acid mine drainage, 106, *349*

acid precipitation, *349*

acid soil Soil with a pH value < 6.6. 29, 84, 106, *334*

acidophile Organism that grows best under acid conditions (down to a pH of 1). 61, *64*

actinomycete Nontaxonomic term applied to a group of high G+C base composition Gram-positive bacteria that have a superficial resemblance to fungi. Includes many but not all organisms belonging to the order Actinomycetales. *7,10*, *27*, 29, 46-49, 65-66, 144, 156, 340, 398, 487-488

actinorhizal, 340-341

activation energy Amount of energy required to bring all molecules in one mole of a substance to their reactive state at a given temperature. 198

active

absorption, 50, 64

dispersal, 88, 400

nutrient pool, 352-*353*

active site Region of an enzyme where substrates bind. 199-201, *231*

adaptive phase. *See* **lag phase**

adenosine triphosphate (ATP) Common energy-donating molecule in biochemical reactions. Also an important compound in transfer of phosphate groups. 190, 203-214, 370, 375

adsorption Process by which atoms, molecules, or ions are taken up and retained on the surfaces of solids by chemical or physical binding. 32, 36, 143

aerated static pile, 493

aerenchyma, 507

aerobic (i) Having molecular oxygen as a part of the environment. (ii) Growing only in the presence of molecular oxygen, as in aerobic organisms. (iii) Occurring only in the presence of molecular oxygen as in certain chemical or biochemical processes such as aerobic respiration. 8, 38-39, 61-63, 207-211, 255

aeroponic culture, *423*

aerotolerant anaerobes Microbes that grow under both aerobic and anaerobic conditions, but do not shift from one mode of metabolism to another as conditions change. They obtain energy exclusively by fermentation. 62

affinity constant (Ks), 155

agar Complex polysaccharide derived from certain marine algae that is a gelling agent for solid or semisolid microbiological media. Agar can be melted at temperatures above 100°C; gelling temperature is 40–50°C. 68, 75, 141, *142*

Agaricus, *80*, 83, 88

agarose Nonsulfated linear polymer consisting of alternating residues of D-galactose and 3,6-anhydro-L-galactose. Agarose gels are often used as the resolving medium in electrophoresis. 172-174

aggregate. *See* **soil aggregate**, *27*, *28*, 29, *49*, 63, 248, *287*

Agrobacterium, 134, 324, *404*, 440

akinete Thick-walled resting cell of cyanobacteria and algae. 105

Alcaligenes, *174*, *183*, 275, 282, *283*, *404*, 463

alga (plural, algae) Phototrophic eukaryotic microorganism. Algae can be unicellular or multicellular. Blue-green algae are not true algae; they belong to a group of bacteria called cyanobacteria.

classification, 94-99

diatoms, 98-99

ecology, 103-111

estimating population, 7, *104*

eutrophication, 377

metabolism, *194*

green, 95

microbiotic crust or mat, 108-109, *110*

physiology, 100-103

pigments, *95*

red, 95, 100

storage products, *95*

structure, 5, 99-100

yellow-green, 95, 98-100

alginate, *443*

aliphatic An organic compound in which the main carbon structure is a straight chain. *45*, 450

alkaline soil Soil with pH value > 7.3. 29, 64, 143

alkalophile Organism that grows best under alkaline conditions (up to a pH of 10.5). 64

alkane Straight chain or branched organic structure that lacks double bonds. 456, 475

alkene Straight chain or branched organic structure that contains at least one double bond. 456

allelopathy, *390*

allosteric site Site on the enzyme other than the active site to which a nonsubstrate compound binds. This may result in a conformational change at the active site so that the normal substrate cannot bind to it. 200

aluminosilicate, 30

aluminum (plural, alumina), 15, 30, 31, 255, 353

amensalism (antagonism) Production of a substance by one organism that is inhibitory to one or more other organisms. The terms antibiosis and allelopathy also describe cases of chemical inhibition. 107, *157*, 160

amino acid, 59, 228, *230, 263, 351, 396*

peptide bond, *229*

amino group An —NH_2 group attached to a carbon skeleton as in the amines and amino acids. 251, *263, 478*

amino sugar, *263, 265*

ammonia, 203, 260, 262, 265, 268, 485, 491, 506

ammonia oxidation, 273

ammonia-oxidizing bacteria, 60, *61*

ammonification Liberation of ammonium (ammonia) from organic nitrogenous compounds by the actions of microorganisms. 83, 260, 262, 265, 266, 271, 491

ammonium, 14, 211, 259, *261, 266, 269,* 271, *395*

amoeba (plural, amoebae) Protozoa that can alter their cell shape, usually by the extrusion of one or more pseudopodia. 115, 117, *118,* 123, 125-127, 129

amphibolic pathway, 215

amylase, 235, 489

amylopectin, 235

amylose, 235

Anabaena, *95-96,* 102-103, *227,* 304, *314, 323, 341, 342*

anabolism Metabolic processes involved in the synthesis of cell constituents from simpler molecules. An anabolic process usually requires energy. 192, *215*

anaerobe, 61-63, 68, 102, 255, 282, 297, *304, 360,* 465

anaerobic (i) Absence of molecular oxygen. (ii) Growing in the absence of molecular oxygen, such as anaerobic bacteria. (iii) Occurring in the absence of molecular oxygen, as a biochemical process. *8, 38, 39,* 63

anaerobic respiration Metabolic process whereby electrons are transferred from an organic or in some cases, inorganic compounds to an inorganic acceptor molecule other than oxygen. The most common acceptors are nitrate, sulfate, and carbonate. 61-63, 207, 209, 282

anamorph Asexual stage of fungal reproduction in which cells are formed by the process of mitosis. 77, 79

anastomosis, 82, 86

anion exchange capacity Sum total of exchangeable anions that a soil can adsorb. Expressed as centimoles of negative charge per kilogram of soil. 30, *33*

anoxic Literally "without oxygen." An adjective describing a microbial habitat devoid of oxygen. 61, *314, 359, 383*

anoxygenic photosynthesis Type of photosynthesis in green and purple bacteria in which oxygen is not produced. 213

anoxygenic phototrophic bacteria, 47

antagonism. *See* **amensalism**

antagonist Biological agent that reduces the number or disease-producing activities of a pathogen. 90, 428, 435

antagonistic, 55, 64

antheridium Male gametangium found in the phylum Oomycota (Kingdom Stramenopila) and phylum Ascomycota (Kingdom Fungi). 81

anthropogenic Derived from human activities. 364, 448, 499

antibiosis Inhibition or lysis of an organism mediated by metabolic products of the antagonist; these products include lytic agents, enzymes, volatile compounds, and other toxic substances. *10, 57,* 64, 66, *390,* 428

antibiotic Organic substance produced by one species of organism that in low concentrations will kill or inhibit growth of certain other organisms. 15, 91, 160

antibiotic resistance, *57, 140*

antibody Protein that is produced by animals in response to the presence of an antigen and that can combine specifically with that antigen. *46,* 142

antigen Substance that can incite the production of a specific antibody and that can combine with that antibody. 142

AP-PCR. *See* **polymerase chain reaction**

apatite, *370*

apoenzyme, 199

apothecium Open ascoma of fungi in the phylum Ascomycota. 82

aquatic, 39, 47, 79, 81, 88, 115

arbuscular mycorrhiza (AM) Mycorrhizal type that forms highly branched arbuscules within root cortical cells. 335, 412-414

arbuscule Special "tree-shaped" structure formed within root cortical cells by arbuscular mycorrhizal fungi. 410, 412-413

arbutoid mycorrhiza, 415

Archaea Evolutionarily distinct group (domain) of prokaryotes consisting of the methanogens, most extreme halophiles and hyperthermophiles, and *Thermoplasma*. 4-6, 44-45, 211

archaebacteria Older term for the Archaea. 45

argon, 499

arid, 29, 64, 109

Armillaria, 77, 78, 88

aromatic Organic compound which contains a benzene ring or a ring with similar chemical characteristics. 225

arsenic, *370*

Arthrobacter, 47, 144, *227*

arthropod Invertebrate with jointed body and limbs (includes insects, arachnids, and crustaceans). 47, 114, 151, 402, 488

ascoma (plural, ascomata) Fungal fruiting body that contains ascospores; also called an ascocarp. 82

ascomycete, 75, 76, *80,* 82-83, 85

ascospore Spores resulting from karyogamy and meiosis that are formed within an ascus. Sexual spore of Ascomycota. *75,* 82

ascus (plural, asci) Saclike cell of the sexual state formed by fungi in the phylum Ascomycota containing ascospores. 82, *89*

aseptic technique Manipulating sterile instruments or culture media in such a way as to maintain sterility. 10

asexual phase, 86

Aspergillus, 82, *87,* 91, 225, 230, 488

assimilation, 63, 268, 269

assimilatory nitrate reduction Conversion of nitrate to reduced forms of nitrogen, generally ammonium, for synthesis of amino acids and proteins. 260, 280-281

associative dinitrogen fixation Close interaction between a free-living diazotrophic organism and a higher plant that results in an enhanced rate of dinitrogen fixation. 295, 308

associative symbiosis Close but relatively casual interaction between two dissimilar organisms or biological systems. The association may be mutually beneficial but is not required for accomplishment of a particular function. 295, *296,* 303, *403*

ATP. *See* adenosine triphosphate

attenuated virus, 138, 147

autochthonous flora. *See* **oligotrophs**

autolysis Spontaneous lysis. 123

autoradiography Detecting radioactivity in a sample, such as a cell or gel, by placing it in contact with a photographic film. 170

autotroph Organism that uses carbon dioxide as the sole carbon source. 192, *194,* 211, 214

autotrophic nitrification Oxidation of ammonium to nitrate through the combined action of two chemoautotrophic organisms, one forming nitrite from ammonium and the other oxidizing nitrite to nitrate. *266,* 271

axenic Literally "without strangers". System in which all biological populations are defined, such as a pure culture. 82, 94, *104*

Azolla, 103, 107, 314, *323,* 341-342

Azorhizobium, 325

Azospirillum, 47, 304, 305, 311, 403, 404

Azotobacter, 297, 302, *304, 305, 307,* 308, *312,* 403

azygospore, 413

B

bacillus (plural, bacilli) Bacterium with an elongated, rod shape. 46

Bacillus, 38, 48, 55, 58, 68, 144, 230, 382, 403, *404,* 488

Bacteria All prokaryotes that are not members of the domain Archaea.

   classification, *6,* 44-47

   ecology, 62-65

   estimating populations, *7,* 66-69

   metabolism, 59-62, *194*

   genetics, 57-59

   structure, *5,* 46-57

bacteriochlorophyll Light-absorbing pigment found in green sulfur and purple sulfur bacteria. 324

bacteriocin Agent produced by certain bacteria that inhibits or kills closely related species. 440

bacteriophage Virus that infects bacteria, often with destruction or lysis of the host cell. 58, 134-136, 138, 144-146

   T4, 133, *136*

bacteroid Altered form of cells of certain bacteria. Refers particularly to the swollen, irregular vacuolated cells of rhizobia in nodules of legumes. 328, 329

base composition Proportion of the total bases consisting of guanine plus cytosine or thymine plus adenine base pairs. Usually expressed as a guanine + cytosine (G+C) value. e.g. 60% G+C. *46, 168*

base pair (bp), 169-170, *172*, 176-177

basidioma (plural, basidiomata) Fruiting body that produces basidia; also called a basidiocarp. 83

basidiomycete, *75*, 76, *80*, 83, 85

basidiospore Spore resulting from karyogamy and meiosis that is formed on a basidium. Sexual spore of the Basidiomycota. *75*, 83

basidium (plural, basidia) Clublike cell of the sexual state formed by fungi in the phylum Basidiomycota. 83

batch composting, 492

Bdellovibrio, 47

Beijerinck's rule, 14

Beijerinck, M.W., 10, 12-14, 94, 297, 323

Bergey's Manual, 45, *47*

binary fission (division) Division of one cell into two cells by the formation of a septum. It is the most common form of cell division in bacteria. 4

binomial nomenclature System of having two names, genus and specific epithet, for each organism. 5

bioaccumulation Accumulation of a chemical substance in living tissue. 384

bioaugmentation, 473, 476

bioavailability

   contaminant, 449, 471

   micronutrient, *248, 254*

biocide, 439

biocontrol. *See* **biological control**

biodegradable Substance capable of being decomposed by biological processes. *26*, 152, 461, 463, 475, 479

biodegradation

   definition, 191, 470

   contaminant, *26*, 182, *403*, 476

   xenobiotic, 454, *462*

biofertilizer, 109, 420

biofilm Microbial cells encased in an adhesive, usually a polysaccharide material, and attached to a surface. 56

biogeochemistry Study of microbially mediated chemical transformations of geochemical interest, such as nitrogen or sulfur cycling. 346

biological control, 18, 65, *145*, 160, *390*, 427-443

biological oxygen demand (BOD) Amount of dissolved oxygen consumed in five days by biological processes breaking down organic matter. *146*

biomagnification Increase in the concentration of a chemical substance as it progresses to higher trophic levels of a food chain. 384

biomass. *See* **microbial biomass**

bioremediation Use of microorganisms to remove or detoxify toxic or unwanted chemicals from an environment.

   advantages and disadvantages, 479

   cometabolism, 452

   case studies, 473-479

   criteria for use, 471

   definitions, 470

   mechanisms and strategies, 471-473

biosolid The residues of wastewater treatment. Formerly called sewage sludge. 28, 145, *146*, 227, 483, *484*, 486, 490, 494, *495*

biosphere Zone incorporating all forms of life on earth. The biosphere extends from deep in sediments below the ocean to several thousand meters elevation in high mountains. *349*, 500

biostimulation, 473, 475, 476

biosynthesis Production of needed cellular constituents from other, usually simpler, molecules. 190, 192, 214, *216, 359*

biotechnology Use of living organisms to carry out defined physiochemical processes having industrial or other practical application. 18

biotrophic Nutritional relationship between two organisms in which one or both must associate with the other to obtain nutrients and grow. 84

bioventing, 475, *476*

blue-green alga. *See* **cyanobacterium**

Bodo, 115, *116*

Bradyrbizobium, 65, 144, 159, 232, 324, *325*

broken edge, 31-32

brown rot fungus Fungus that attacks cellulose and hemicellulose in wood, leaving dark-colored lignin and phenolic materials behind. 230

budding Asexual reproduction (usually for yeast) beginning as a protuberance from the parent cell that grows and detaches to form a smaller, daughter cell. 138

bulk density, soil Mass of dry soil per unit bulk volume (combined volume of soil solids and pore space). 24-26, *248*

bulking agent, 485, *486*

C

calcium, 59, 143, 180, 333

Calvin cycle Biochemical route of carbon dioxide fixation in many autotrophic organisms. 100, 214, *215,* 348

capsid Protein coat of a virus. 134, *135, 136*

capsule Compact layer of polysaccharide exterior to the cell wall in some bacteria. 56

carbon, 22, 59-60, 192-194, 214, 218-246, *254*

carbon cycle Sequence where carbon dioxide is converted to organic forms by photosynthesis or chemosynthesis, recycled through the biosphere with partial incorporation into sediments, and ultimately returned to its original state through respiration or combustion. 83, 218-221, 498, 503-509

carbon dioxide, 100, 108, 207-211, 214, 219, *237,* 240-246

carbon fixation Conversion of carbon dioxide or other single-carbon compounds to organic forms (e.g., carbohydrate). 100, 192, 212, 214

carbon-nitrogen (C/N) ratio Ratio of the mass of organic carbon to the mass of total nitrogen in soil or organic material. 90, 235, 270, *271,* 484

carbon nutrition, *214*

carboxyl group A —COOH group attached to a carbon skeleton as in the carboxylic acids and fatty acids. 32

carcinogen Substance which causes the initiation of tumor formation. Frequently a mutagen. 280

carotenoid, *95,* 99-101

carrier, 338, *339*

catabolism Biochemical processes involved in the breakdown of organic compounds, usually leading to the production of energy. 191, *210, 215*

catabolite repression Transcription-level inhibition of a variety of inducible enzymes by glucose or other readily used carbon sources. 200

catalyst Substance that promotes a chemical reaction by lowering the activation energy without itself being changed in the end. Enzymes are a type of catalyst. 198

cation exchange capacity (CEC) Sum of exchangeable cations that a soil can adsorb at a specific pH. Expressed as centimoles of positive charge per kilogram of soil. 30, *33,* 143, 247, *248,* 252

cell Fundamental unit of living matter.

cell membrane. *See* **cytoplasmic membrane**

cell wall Layer or structure that lies outside the cytoplasmic membrane; it supports and protects the membrane and gives the cell shape. *5, 45, 51,* 52-53, 74, 136, 227

cellobiose, 229-230

cellulase, 203, 229-230, *401,* 435, 489

cellulose Glucose polysaccharide (with $\beta1,4$-linkage) that is the main component of plant cell walls. Most abundant polysaccharide on earth. 74, 79, 84, 99, 203, 222, 226, 227, 229-230, 484, 489

Centrosema, 331

Cepbalosporium, 91

CFU. *See* **colony-forming unit**

Chaetomium, 90, 488

chelate Organic chemical that forms a ring compound in which a metal is held between two or more atoms strongly enough to diminish the rate at which it becomes fixed by soil, thereby making it more available for plant and microbial uptake. *248, 254-255, 376, 377, 403*

chemoautotroph Organism that obtains energy from the oxidation of chemical, generally inorganic, compounds and carbon from carbon dioxide. 12, 60, 193, *194,* 211, 347, 378

chemodenitrification, 280, *281,* 514

chemoheterotroph Organism that obtains energy and carbon from the oxidation of organic compounds. 59-60, 192, *193, 194*

chemolithotroph Organism that obtains energy from the oxidation of chemical, generally inorganic, compounds and uses inorganic compounds as electron donors. 47, *194,* 211-212, 214

chemoorganotroph Organism that obtains energy and electrons (reducing power) from the oxidation of organic compounds. 192, *193,* 208-209

chemostat Continuous culture device usually controlled by the concentration of limiting nutrient and dilution rate. *305, 492*

chemosynthesis, *348*

chemotaxis Oriented movement of a motile organism with reference to a chemical agent. May be positive (toward) or negative (away) with respect to the chemical gradient. 57, 85, 436

chemotroph, 192, *193, 194*

chitin, 74, 79, 227, *264*

chlamydospore Thick-walled resting structure that forms from the cell wall of a fungal hypha; usually formed under conditions where the hypha is no longer able to function optimally. *75,* 79, 83, 87, 413

Chlorella, 7, 95

chloroform fumigation, *237, 374*

chlorophyll Green pigment required for photosynthesis. *95,* 97-100, *101, 104,* 115, 212-213

chloroplast Chlorophyll-containing organelle of photosynthetic eukaryotes. *5,* 54, 97-98, 100

chromatography Any technique that separates different species of molecules (or ions) by subjecting them to two different carrier phases, mobile and stationary. 182

chromosome Genetic element carrying information essential to cellular metabolism. Prokaryotes have a single chromosome, consisting of a circular DNA molecule. Eukaryotes contain more than one chromosome, each containing a linear DNA molecule complexed with specific proteins. 53-54, 58, 85, 133, 177

chrysolaminarin, *95,* 99

chytrid Fungal organism in the phylum Chytridiomycota that consists of a spherical cell from which short thin filamentous branches (rhizoids) grow that resemble fine roots. *74, 75, 80,* 81, 88

ciliate Protozoan that moves by means of cilia on the surface of the cell. *8,* 115, 117-118, 125, *126*

cilium (plural, cilia) Short, threadlike appendages that extend from the surface of some protozoa and beat rhythmically to propel them. 117

citric acid cycle. *See* **tricarboxylic acid cycle**

clamp connection Small branch of a fungal hypha that connects two compartments separated by a septum and helps to maintain a dikaryon in each hyphal compartment; characteristic of fungi in the phylum Basidiomycota. 76, 77

classification (i) Arrangement of organisms into groups based on mutual similarity or evolutionary relatedness. (ii) Systematic arrangement of soils into groups or categories on the basis of their characteristics.

algae, *95*

bacteria, 44-45

enzymes, *202*

fungi, *75*

nutritional groups, 60

microorganisms, 4-5

nematodes, 119

protozoa, 114-115

soil, 23-24

viruses, 134-138

clay Soil particle < 0.002 mm in diameter. *8,* 22, *24-27,* 30-34, 38, 127, 143

cleistothecium Closed ascocarp of fungi in the phylum Ascomycota. 82

climate, 16, 21, 29, 127, 498, 504, 514

climax Most advanced successional community of plants capable of development under, and in dynamic equilibrium with, the prevailing environment. 90, 161, 278

clone (i) Population of cells all descended from a single cell. (ii) Number of copies of a DNA fragment to be replicated by a phage or plasmid. *104*

Clostridium, *47*, 55, *134*, 139, 383

cobalt, 59, *335*

coccus Spherical bacterial cell. 46-48

coenocytic Fungal hypha without crosswalls (septa), so that the nuclei present in the cytoplasm are free-floating and mobile. 79, 85

coenzyme Low-molecular-weight chemical which participates in an enzymatic reaction by accepting and donating electrons or functional groups. 199

coliform Gram-negative, nonspore-forming facultative rod that ferments lactose with gas formation within 48 hours at 35°C. Often used as an indicator organism for fecal contamination of water supplies. *Escherichia coli* and *Enterobacter* are important members. 181-182

Collembola, 90

colloid fraction Organic and inorganic matter with very small particle size and a correspondingly large surface area per unit of mass. 24, 108, 180, 370, *376*, 409

colonization Establishment of a community of microorganisms at a specific site or ecosystem, 104, 123, 184, 400-401, 420

colony forming unit (CFU), 17

colony hybridization, 171, *172*

colony lifts. *See* **colony hybridization**

colony Clone of bacterial cells on a solid medium that is visible to the naked eye. *49*, 66, 68, 85, 171

Colpoda, *118*

cometabolism Transformation of a substrate by a microorganism without deriving energy, carbon, or nutrients from the substrate. The organism can transform the substrate into intermediate degradation products but fails to multiply at its expense. 273, 450-453, 456, 470

commensalism Interaction between organisms where one organism benefits from the association while the second organism remains unaffected. 157

community All organisms that occupy a common habitat and interact with one another. 17, 64, 66, 123, 125, 181

competent In a genetic sense, the ability to take up DNA. 58

competition Rivalry between two or more species for a limiting factor in the environment that usually results in reduced growth of participating organisms. 3, 64, 155, *157*, 160, 428

complementary In reference to base pairing, the ability of two polynucleotide sequences to form a double-stranded helix by hydrogen bonding between bases in the two sequences. 58, 133, 168-170, 174, 176-177

compost Organic residues which have been mixed, piled, and moistened, with or without addition of fertilizer and lime, and generally allowed to undergo thermophilic decomposition until the original organic materials are substantially altered or decomposed. 482

composting

  batch systems, 492-493

  Beltsville process, 493

  bioremediation, 473

  continuous systems, 494

  Dano process, 494

  Indore process, 493

  maturity, 494

  stages, 487

  substrates, 483

  static pile, 492-493

  stabilization stage, 487, 491

conformational protection, *307*

conidium (plural, conidia) Nonmotile, asexual spore resulting from mitotic nuclear division and formed from the ends or sides of a hypha; produced in abundant numbers by the asexual phase of soil fungi in the phyla Ascomycota and Basidiomycota. 65, 75, 79, 83

conjugation In prokaryotes, transfer of genetic information from a donor cell to a recipient cell by cell-to-cell contact. 58, 59

conjugative plasmid Self-transmissible plasmid; a plasmid that encodes all the functions needed for its own intercellular transmission by conjugation. 58

consortium (plural, consortia) Two or more members of a natural assemblage in which each organism benefits from the other. The group may collectively carry out some process that no single member can accomplish on its own. 158, 454, *455*

constitutive enzyme Enzyme always synthesized by the cell regardless of environmental conditions. 54, 200

copiotroph, 156, 238

copper, 59, *370*

cosubstrate, 449, 452

Corynebacterium, 47, *382*

covalent Nonionic chemical bond formed by a sharing of electrons between two atoms. 199, 254, *347*

crop rotation, 9, 87, *248*, 421, 431

cross-feeding (i) Specific type of syntrophy where two populations cooperate to metabolize a compound. (ii) One organism consuming products excreted by another organism. 158

cross-inoculation group, 323, *331*

crown gall, 440

culture medium, 3, 10, 12, 66-69, 142, 180

culture Population of microorganisms cultivated in an artificial growth medium. A pure culture is grown from a single cell; a mixed culture consists of two or more microbial species or strains growing together. 12, 45, 57, 68, 129, 180-182

curing, 487, 491

cyanobacterium Prokaryotic, oxygenic phototrophic bacterium containing chlorophyll *a* and phycobilins; formerly the "blue-green algae." 47, 59-60, 94-109, 162, *194, 304, 314, 323, 403*

cyclic photophosphorylation Formation of ATP when light energy is used to move electrons cyclically through an electron transport chain during photosynthesis. 213

cyst Resting stage formed by some bacteria, nematodes, and protozoa in which the whole cell is surrounded by a protective layer; not the same as endospore. 115, *116*, 127

cytochrome Iron-containing porphyrin rings (e.g., heme) complexed with proteins which act as electron carriers in an electron-transport chain. 100

cytoplasm Cellular contents inside the cell membrane, excluding the nucleus. 4, 7, 54-55, 73, 97, 100, 119, *121*

cytoplasmic membrane Selectively permeable membrane surrounding the cell's cytoplasm. 4-5, 50-54, *73*

D

deamination, 268, 485

decomposer Heterotrophic organism that breaks down organic compounds. 47, 62, 129

decomposition Chemical breakdown of a compound into simpler compounds, often accomplished by microbial metabolism.

  cellulose, 69, 230

  composting, 489-491

  definition, 219

  kinetics, 245

  measurement, *243*, 244

  microbial cells, *227*

  nitrogen requirement, 242

  protozoa effects, 123, 127

  rate, 123, 125, 127, 222, *231*

  rate constant, 226, 242, 245

  residue, 63, 222, 223, *238, 241*, 245, *250*

  soil moisture effects, 246

  soil organic matter, 24, 255

  stages, 22, 128, 239

  temperature effects, 226

degradation Process whereby a compound is usually transformed into simpler compounds. *47*, 125, 180-184, 472, 482-495

dehalogenation

  glutathione S-transferase-mediated, *454*

  hydrolytic, *451, 456, 458*

  oxygenolytic, *451, 459*

  reductive, *451*, 463-465

dehydrogenase, *202*, 268

deleterious rhizosphere microorganism (DRMO), 441

denaturation Process where double-stranded DNA unwinds and dissociates into two single strands. The reverse of DNA-DNA hybridization. 169-170, 174, 176

denitrification Reduction of nitrate or nitrite to molecular nitrogen or nitrogen oxides by microbial activity (dissimilatory nitrate reduction) or by chemical reactions involving nitrite (chemical denitrification). 15, 157, 209, 260, *281,* 282-290, 510

denitrifying bacteria, *62, 283*

deoxyribonucleic acid (DNA) Polymer of nucleotides connected via a phosphate-deoxyribose sugar backbone; the genetic material of the cell. 53-54, 162, 168, 370

derepressible enzyme Enzyme that is produced in the absence of a specific inhibitory compound acting at the transcriptional level. 200

Desmodium, 331

Desulfotomaculum, *359*

Desulfovibrio, *47, 359*

detoxification, *454*

deuteromycete, *79*

diatom Alga with siliceous cell wall that persist as a skeleton after death. Any of the microscopic unicellular or colonial alga constituting the class Bacillariophyceae. 95, 98-100, 105-106, *109*

diatomaceous earth Geologic deposit of fine, grayish siliceous material composed chiefly or wholly of the remains of diatoms. It may occur as a powder or as a porous, rigid material. *109*

diazotroph Organism that can use dinitrogen as its sole source, i.e., capable of N_2 fixation. 295-*296,* 305, *309, 314*

differential medium Culture medium with an indicator, such as a dye, which allows various chemical reactions to be distinguished during growth. 69

diffusion (nutrient) Movement of nutrients in soil that results from a concentration gradient. 38, 39, 40, 50

dikaryon Two nuclei present in the same hyphal compartment; they constitute a homokaryon when both nuclei are genetically the same or a heterokaryon when each nucleus is genetically different from the other. 76

dilution plate count method Method for estimating the viable numbers of microorganisms

in a sample. The sample is diluted serially and then transferred to agar plates to permit growth and quantification of colony-forming units. 66-68

dinitrogen, 16, *261, 266*

dinitrogenase, 300

dinitrogenase reductase, 300

dinitrogen fixation Conversion of molecular dinitrogen (N_2) to ammonia and subsequently to organic combinations or to forms useful in biological processes.

　　actinorhizal, 340-341

　　associative, 308-314

　　Azolla/*Anabaena,* 341-342

　　by cyanobacteria, 101-103, 109

　　discovery, 12, 16, 297

　　effect of environment, 305-306, 332-336

　　free-living, 303-307

　　mechanism, 298-303

　　methods for measuring, 315-317

　　in nitrogen cycle, *260*

　　rates, *312*

　　symbiotic, 159, 322

dinitrogen-fixing bacteria, 63, *304, 325, 341*

diploid In eukaryotes, an organism or cell with two chromosome complements, one derived from each haploid gamete. 79, 81, 82, 83

direct count Method of estimating the total number of microorganisms in a given mass of soil by direct microscopic examination. 68

disease, 9-10, 12, *46,* 59, 72, 119, 123, 127, 134-138, 145, *146, 390, 427, 428*

disease-suppressive soil. *See* **pathogen-suppressive soil**

dispersal, 88, 128

dissimilatory nitrate reduction, 209, 262, 280-281

dissimilatory nitrate reduction. *See* **denitrification.**

dissimilatory nitrate reduction to ammonium (DNRA) Use of nitrate by organisms as an alternate electron acceptor in the absence of oxygen, resulting in the reduction of nitrate to ammonium. 262, 281, *282*

disulfide bond, 346, *347*

divalent cation bridging, 34

diversity

biodiversity, 14, 17

functional, 9, 164

genetic, 7, 58, 161

physiological (metabolic), 14, 62, 164, 189, 303

DNA denaturation. *See* **denaturation**

DNA fingerprinting Molecular genetic techniques to assess possible differences among DNA samples. *178,* 182

DNA polymerase, 133, 174

DNA. *See* **deoxyribonucleic acid**

DNRA bacteria, 282

dolipore septum Specialized crosswall separating compartments of a hypha of fungi in the phylum Basidiomycota; consisting of a central pore covered with perforated membranes on both sides (called a parenthosome). 76, 83, 85

domain Highest level of biological classification, superseding kingdoms. The three domains of biological organisms are the Bacteria, the Archaea, and the Eukarya. 5-6, 44-45, 134

doubling time Time needed for a population to double in number or biomass. 152, *154*

E

earthworm, 7, 114

ecology Science which studies the interrelations among organisms and between organisms and their environment. 14, 62, 90, 103, 149

ecophysiology, 449

ecosystem Community of organisms and the environment in which they live. 44, 62, 72, 123, *124, 125,* 129, 161

ectendomycorrhiza, *410,* 415

ectomycorrhiza (EM) Mycorrhizal type in which the fungal mycelia extend inward, between root cortical cells, to form a network (Hartig net) and outward into the surrounding soil. Usually the fungal hyphae also form a mantle on the surface of the root. 17, 83, 84, 85, 410

ectorhizosphere, *393*

edaphic (i) Of or pertaining to the soil. (ii) Resulting from or influenced by factors inherent in the soil or other substrate, rather than by climatic factors. 95

effectiveness, 328, *331,* 414, 421

E_h Potential generated between an oxidation or reduction half-reaction and the H electrode in the standard state.

electron acceptor Substance that accepts electrons during an oxidation-reduction reaction. An electron acceptor is an oxidant. 60-63, *162,* 203, 206-211, 219, 265, 359-360, 363, 378, 463, 510, 512

electron donor Substance that donates electrons in an oxidation-reduction reaction. An electron donor is a reductant. 60, *194,* 206, *219,* 252, 359-360, 363, 512

electron-transport chain Final sequence of reactions in biological oxidations composed of a series of oxidizing agents arranged in order of increasing strength and terminating in oxygen. 100, 203, *204,* 207, *210*

electron-transport chain phosphorylation. *See* **oxidative phosphorylation**

electrophilic compounds Chemicals that attack or are drawn to regions in other chemicals in which electrons are readily available; oxidizing agents act as electrophilic compounds. 450

electrophoresis Separation of charged molecules, such as nucleic acids, in an electrical field. *172*

ELISA. *See* **enzyme-linked immunosorbent assay**

eluviation Removal of soil material from a layer of soil as a suspension. 23

Embden-Meyerhof-Parnas pathway (Embden-Meyerhof pathway; EMP pathway) Biochemical pathway that converts glucose to pyruvate; the six-carbon stage converts glucose to fructose-1,6-bisphosphate, and the three-carbon stage produces ATP while changing glyceraldehyde-3-phosphate to pyruvate. 205

encyst, 81, 127

endergonic reaction Chemical reaction that proceeds with the consumption of energy. 189, 192, 197

endoenzyme Enzyme that operates along the internal portions of a polymer. *202, 231*

endomycorrhiza Mycorrhizal association with intracellular penetration of the host root cortical cells by the fungus as well as outward extension into the surrounding soil. 17, 81, 409, 412

endonuclease Endoenzyme that cleaves phosphodiester bonds within a nucleic acid molecule. 267

endophyte Organism growing within a plant. The association may be symbiotic or parasitic. 107, 393

endorhizosphere, *393*

endoplasmic reticulum, 5, 54, 73

endospore Differentiated cell formed within the cells of certain Gram-positive bacteria and extremely resistant to heat and other harmful agents. 10, 12, *47,* 55, 56, 68

end product inhibition. *See* **feedback inhibition**

enrichment culture Technique in which environmental (including nutritional) conditions are controlled to favor the development of a specific organism or group of organisms. 69, *104*

enteric bacteria General term for a group of bacteria that inhabit the intestinal tract of humans and other animals. Among this group are pathogenic bacteria such as *Salmonella* and *Shigella.* 177, 490

Enterobacter, *47, 282, 304,* 436

enterovirus, 138, 182

Entrobacter, 436

Entrophospora, 413

enzyme Protein within or derived from a living organism that functions as a catalyst to promote specific reactions or groups of reactions.

   activity during composting, 488-489

   active site, 199-201

   allosteric control, 200

   allosteric inhibitors, 201

   cellobiase, 229

   constitutive enzyme, 54, 200

   endoglucanase, 229, 235

   exoglucanase, 229, 235

   extracellular, 50, *202,* 229, 267, *231*

   function, 198-203

   glucosidase, 229-230, *231*

   hydrolase, 455, 456, *478*

   inducer, 54, 200

   kinetics, 153-154

   oxygenases, 450, 458, 461, *462*

   pectinase, 232

   prosthetic group, 100, 199

   recruitment, 454

   regulation, 200

   sulfatase, 203, *357*

enzyme-linked immunosorbent assay (ELISA) An immunoassay that uses specific antibodies to detect antigens or antibodies. The antibody-containing complexes are visualized through an enzyme coupled to the antibody. Addition of substrate to the enzyme-antibody-antigen complex results in a colored product. 142

episome Plasmid that replicates by inserting itself into the bacterial chromosome. 53

ergosterol, 74

ericaceous mycorrhiza Type of mycorrhiza found on plants in the Ericales. The hyphae in the root can penetrate cortical cells (endomycorrhizal habit); however, no arbuscules are formed. Major forms are ericoid, arbutoid, and monotropoid. 84, 85, 409, *410,* 414

ERIC-PCR. *See* **polymerase chain reaction**

erosion, 40, 108, 247, 372, *422*

Escherichia coli, *136,* 140, 170-171, 176

ethylene, 302, 316, 491

Eubacteria Old term for the Bacteria.

Eucarya Phylogenetic domain containing all eukaryotic organisms. 6

Euglena, *74, 95, 96,* 106

eukaryote Organism having a unit membrane-bound nucleus and usually other organelles. 4-5, 54, 64, 72, 94, 114, 119, 181

eutrophic Having high concentrations of nutrients optimal, or nearly so, for plant or animal growth. Can be applied to nutrient or soil solutions and bodies of water. *228,* 278, 377

exergonic reaction Chemical reaction that proceeds with the liberation of energy. 189, 190, 197, *198*

exobiology Branch of biology concerned with the effects of extraterrestrial environments on living organisms. *103*

exoenzyme Enzyme that acts at the end of a polymer cleaving off monomers and dimers and sometimes larger chain fragments. *202, 231*

exonuclease, 267

exponential growth Period of sustained growth of a microorganism in which the cell number consistently doubles within a fixed time period. 152-153

exponential phase Period during the growth cycle of a population in which growth increases at an exponential rate. Also referred to as the logarithmic phase. 152

extracellular Outside the cell.

DNA, 58

enzyme, 50, *202*, 229, 267, *231*

polysaccharide, 29, 108

protease, 143

viral activity, 133

exudate Low-molecular-weight metabolites that leak from plant roots into soil. 65, 129, 184, *309*, 395-396, 438

Exxon Valdez, 473-475

F

facultative organism Organism that can carry out both options of a mutually exclusive process (e.g., aerobic and anaerobic metabolism).

anaerobe, 38, *41*, 62-63, 209, 281-282, *307*

chemoautotrophs, 358

biotroph, 85

lithotrophs, 61

symbiont. *See* **facultative biotroph**

FAD/FADH₂. *See* **flavin adenine dinucleotide**

fairy ring, *88*

fatty acid. *See* **lipid and signature fatty acid**

feedback inhibition Inhibition by an end product of the biosynthetic pathway involved in its synthesis. 200, *201*

fermentation Metabolic process in which organic compounds serve as both electron donors and electron acceptors. 61-62, 206-207, 506

ferredoxin, *301*

fertilizer Any organic or inorganic material of natural or synthetic origin (other than liming

materials) added to a soil to supply one or more elements essential to plant growth. 15, 23, 36, 109

field capacity Content of water, on a mass or volume basis, remaining in a soil after being saturated with water and after free drainage is negligible. 37

filament, 46, 47, 55, 56

filamentous In the form of very long rods, many times longer than wide (for bacteria), in the form of long branching strands (for fungi). 64, 73, 115, 117

fimbria (plural, fimbriae) Short filamentous structure on a bacterial cell; although flagella-like in structure, generally present in many copies and not involved in motility. Plays a role in adherence to surfaces and in the formation of pellicles. *51*, 55, 56

fingerprinting. *See* **DNA fingerprinting**

first-order reaction, *243*, 245, 246

fission Type of cell division in which overall cell growth is followed by formation of a cross wall which typically divides the fully grown cell into two similar or identical cells. 75

flagellate Protozoan that moves by means of one to several flagella. *8*, 115-117, *124*, 129

flagellum (plural, flagella) Whiplike tubular structure attached to a microbial cell responsible for motility. 51, 56, 79, 81, 115-116

flavin adenine dinucleotide (FAD/FADH₂), 191, 200, 203-204, 207, 210-211

Flavobacterium, 383, 459

flavodoxin, *301*

flavonoid, 327

fluorescence, 142. *See* **immunofluorescence**

fluorescent Able to emit light of a certain wavelength when activated by light of a shorter wavelength. 142, 182, 382, 396, 433

fluorescent antibody Antiserum conjugated with a fluorescent dye, such as fluorescein or rhodamine. 142

flux Rate of emission, sorption, or deposition of a material from one pool to another. For example, the exchange of methane between the land and the atmosphere is a flux, while the production of methane within the soil is not. 218-219, 276, 352, 501, 503, *508*, 512

food chain Movement of nutrients from one life form to another as a result of the different feeding habits and dietary requirements of organisms in an ecosystem. 106, 150

food web Diagram of the interconnections of nutrient flow through a food chain. 129, 151

forest, 22, 23, 77, 90, 103, 125, 150, 233, *279*, 411, 503

fortuitous metabolism. *See* **cometabolism**

Frank, A.B., 12, 17, 408

Frankia, 47, 65, 159, *296, 323,* 340-341

free energy Intrinsic energy contained in a given substance that is available to do work, particularly with respect to chemical transformations; designated ΔG. 36, 197-198, 212

fruiting body Macroscopic reproductive structure produced by some fungi, such as mushrooms, and some bacteria, including myxobacteria. Fruiting bodies are distinctive in size, shape, and color for each species. 115

frustule Siliceous wall of a diatom. 98, *109*

fucoxanthin, 95, 99, 101

fulvic acid Yellow organic material that remains in solution after removal of humic acid by acidification. *247, 252, 254*

fungistasis Suppression of germination of fungal spores or other resting structures in natural soils as a result of competition for available nutrients, presence of inhibitory compounds, or both. 87

fungus Microorganism that contains a rigid cell wall.

   Ascomycota, 82

   Basidiomycota, 83

   Chytridiomycota, 81

   classification, *75,* 79-83

   dispersal, 88-89

   ecology, 90-91, 230

   genetics, 85-87

   nutrition, 83-85, *194*

   Oomycota, 79

   reproduction, 75-79

   structure, 5, *8,* 72-75

   Zygomycota, 81

furrow slice, 23

Fusarium, 38, 82, 87, 90-91, 230, 419, 433

fusiform Spindle-shaped; tapered at both ends. 119

G

Gaeumannomyces, 432

β-**galactosidase,** 178, 180

gametangium A fungal structure that contains one or more gametes, i.e. haploid cells which result from meiosis. 78, 81

gas chromatography Chromatographic technique in which the stationary phase is a solid or an immobile liquid and the mobile phase is gaseous. The gaseous samples are separated based on their differential adsorption to the stationary phase. 316

gene Unit of heredity; a segment of DNA specifying a particular protein or polypeptide chain, a tRNA or an mRNA. 57, 169

gene probe Strand of nucleic acid that can be labeled and hybridized to a complementary molecule from a mixture of other nucleic acids. 169-173, *179,* 180, 182

general suppression. *See* **pathogen-suppressive soil**

generation time Time needed for a population to double in number or biomass. 152

genetic code Information for the synthesis of proteins contained in the nucleotide sequence of a DNA molecule (or in certain viruses, of an RNA molecule). 168-169

genetic engineering *In vitro* techniques for the isolation, manipulation, recombination, and expression of DNA. 109

genome Complete set of genes present in an organism. 133

genotype Precise genetic constitution of an organism. 438

genus (plural, genera) First name of the scientific name (binomial); the taxon between family and species. 5, 48

Geobacter, 380

germ theory, 12

Gibb's free energy. *See* **free energy**

Gibberella, 80, 82, 87

Gigaspora, 414

Gleocapsa, 96

global gas. *See* **greenhouse gas**

global warming, 18, 259, *499, 508*

Glomales, 81, 410

Glomus, 232, 413

glucan, *74*

glycogen, 55, 74

glycolysis Reactions of the Embden-Meyerhof (glycolytic) pathway in which glucose is oxidized to pyruvate. 205, 207, 209, *210,* 214, 215

Golgi body, *5,* 73

Gram stain Differential stain that divides bacteria into two groups, Gram-positive and Gram-negative, based on the ability to retain crystal violet when decolorized with an organic solvent such as ethanol. The cell wall of a Gram-positive bacterium consists chiefly of peptidoglycan and lacks the outer membrane of Gram-negative cells. *53*

grassland, 109, *124-125,* 128, *165,* 312-313, 410, 503

gravitational potential Portion of total soil water potential due to differences in elevation. 36

gravitational water, 36-37

grazing. *See* **predation**

green bacteria, *47,* 60, *162*

greenhouse gas or greenhouse effect, 15, 259, 500

groundwater That portion of the water below the surface of the ground at a pressure equal to or greater than atmospheric. 18, 145, *228, 477*

growth In microbiology, an increase in both cell number and cellular constituents.

characteristics, 151-156

calculating, 154

curve, *152*

efficiency, *243, 244*

exponential, *152,* 153

lag, 152

limited, *152,* 153

mathematical concepts, 152

rate, maximum, *155*

stationary, *152*

growth factor Organic compound necessary for growth because it is an essential cell component or precursor of such components and cannot be synthesized by the organism itself. Usually required in trace amounts. 59, 129

growth rate Rate at which growth occurs, usually expressed as the generation time. 123, 127

growth rate constant (μ) Slope of $\log_{10}$ of the number of cells per unit volume plotted against time. 153

growth yield coefficient (Y) Quantity of biomass carbon formed per unit of substrate carbon consumed. 154, 270

H

Haber-Bosch process, 298

habitat Place where an organism lives. 3, *8,* 21, 62-63, 125

halogen Any of the five elements fluorine, chlorine, bromine, iodine, and astatine that form part of group VII A of the periodic table. 448

halogen substituents, 448

halophile Organism requiring or tolerating a saline environment. 45, 64, *283*

haploid In eukaryotes, an organism or cell containing one chromosome complement and the same number of chromosomes as the gametes. 76, 79-83, 85-86

Hartig net, 410

heavy metal Those metals which have densities of > 5.0 Mg m^{-3}. These include the metallic elements copper, iron, manganese, molybdenum, cobalt, mercury, nickel, and lead. Aluminum and selenium have densities > 5.0 Mg m^{-3} but are also considered heavy metals. 141, 448, 485

helper virus, 136

heme, 100, 199-200, 465

hemicellulose, 84, 226, 230, 232

hepatitis, 138

herbicide. *See* **pesticide**

heterocyst Differentiated cyanobacterial cell that carries out dinitrogen fixation. 97, 102, 107, 341, 342

Heterodera, 434

heterokaryon Hypha that contains at least two genetically dissimilar nuclei. 82-83, 86

heterothallic Hyphae that are incompatible with each other, each requiring contact with another hypha of compatible mating type which, upon fusion, forms a dikaryon or a diploid. 86

heterotroph Organism capable of deriving carbon for growth and cell synthesis from organic compounds; generally also obtain energy and reducing equivalents from organic compounds. 83, 106, 192, *194*, 208, 214

heterotrophic nitrification Biochemical oxidation of ammonium to nitrite and nitrate by heterotrophic microorganisms. 275

high-performance liquid chromatography (HPLC), 182

Hiltner, L., 389, *390*

histosol, 255

holomorph Whole fungus consisting of all sexual and asexual stages in its life cycle. 79

homothallic Hyphae that are self-compatible in that sexual reproduction occurs in the same organism by meiosis and genetic recombination; fusion of hypha results in a dikaryon or diploid. 86

horizon. *See* **soil horizon**

hormone, 65, 129

host Organism capable of supporting the growth of a virus or other parasite. 17, 58, 59, 132, *133*, 138, 331, 334

host specificity, 330, *331*, 341, 414

Howard, A., 493

HPLC. *See* **high-performance liquid chromatography**

humic acid Dark-colored organic material extracted from soil by various reagents (e.g., dilute alkali) and that is precipitated by acid (pH 1 to 2). *247, 252, 254*

humic substances Series of relatively high-molecular-weight, brown-to-black substances formed by secondary synthesis reactions. The term is generic in a sense that it describes the colored material or its fractions obtained on the basis of solubility characteristics, such as humic acid or fulvic acid. 59, 180, 240, 246-*250*, 449

humification Process whereby the carbon of organic residues is transformed and converted to humic substances through biochemical and chemical processes. 225, 244

humin, *247*, 252

humus Total of the organic compounds in soil exclusive of undecayed plant and animal tissues, their "partial decomposition" products, and the soil biomass. The term is often used synonymously with soil organic matter. 15, 63, 90, *247, 251,* 352, 372

hybridization Natural formation or artificial construction of a duplex nucleic acid molecule by complementary base pairing between two nucleic acid strands derived from different sources. 169-172

hydrogen bond Chemical bond between a hydrogen atom of one molecule and two unshared electrons of another molecule. 168, 177, *347*

hydrogen sulfide (H_2S), 61-63, 211, *348*, 364

hydrogen-oxidizing bacterium Facultative lithotrophs that, in the absence of an oxidizable organic source, oxidize H_2 for energy and synthesize carbohydrates with carbon dioxide as their source of carbon. *60*-61, 211-212

hygroscopic water Water adsorbed by a dry soil from an atmosphere of high relative humidity. 37

hymenium Layer of hyphae fertile in producing asci (fungi in the phylum Ascomycota) or basidia (fungi in the phylum Basidiomycota) from the process of meiosis. 83

Hymenoscyphus, 415

hyperparasite Parasite that feeds on another parasite. 428

hypha (plural, hyphae) Long and often branched tubular filament that constitutes the vegetative body of many fungi and funguslike organisms. Bacteria of the order Actinomycetes also produce branched hyphae. *8*, 17, *27, 49*, 73, 81, *116*, 121

hyperaccumulator, 479

I

illuviation Deposition of soil material removed from one horizon to another in the soil. 23

immobilization Conversion of an element from the inorganic to the organic form in microbial or plant biomass. 63, 242, 260, 265, 266, 268, 354, 369, 374

immunity Ability of a human or animal body to resist infection by microorganisms or their harmful products such as toxins. 139

immunofluorescence Technique to visualize specific antibodies and any attached homologous antigens by means of conjugating the antibodies to a fluorescent dye. 142

inducible enzyme Enzyme synthesized (induced) in response to the presence of an external substance (the inducer). 54, 200

infection Growth of an organism within another living organism. 56, 65, 123, 134, 136, 138, 140

infectiveness, *331,* 414, 421

infection thread A cellulosic tube in a root hair through which rhizobia can travel to reach and infect root cells. 325-327, *326, 330-331*

inoculate To treat with microorganisms for the purpose of creating a favorable response. For example, the treatment of legume seeds with rhizobia to stimulate N_2 fixation. 12, *57,* 66, 142, 338-339, 403, 421-422, 435

inoculation, *311,* 336-339, *422*

inoculum Material used to introduce a microorganism into a suitable situation for growth. 108, 109, 338, 421

inoculum potential, 420

inositol phosphate, 371

insertion Genetic mutation in which one or more contiguous nucleotides are added to DNA. 180, 184

insertion sequence (IS element) Simplest type of transposable element. Has only genes involved in transposition. 57

integration Process by which a DNA molecule becomes incorporated into another genome. *139,* 140

intracellular Inside the cell. 133, *202,* 203

in vitro Literally "in glass" it describes whatever happens in a test tube or other receptacle, as opposed to *in vivo*. When a study or an experiment is done outside the living organism, in a test tube, it is done *in vitro*. 12, 173, 341, 358, 438, 465

ion Atom, group of atoms, or compound, that is electrically charged as a result of the loss of electrons (cation) or the gain of electrons (anion). 29, 34, 56, 64, 200, 211, 255

iron, *39, 47,* 59, 299, 301, 335, 379, 498

iron-oxidizing bacteria, 212, 378-379

irrigation Intentional watering of soil. 88, 109, *274, 289,* 429

isolation Any procedure in which an organism present in a particular sample or environment is obtained in pure culture. *10,* 68, 141

isomorphous substitution Substitution in a crystalline clay sheet of one atom by a similarly sized atom of lower valence. 30-31

isotope Different form of the same element containing the same number of protons and electrons, but differing in the number of neutrons. 265, 317

isotope dilution, *279,* 317

J-M

kaolinite, 30-33, 143

karyogamy Fusion in a cell of haploid (N) nuclei to form a diploid (2N). 78, 81-83, 86

Klebsiella, 303, 304, 307, 322, 436

Koch's Postulates Set of laws formulated by Robert Koch to prove that an organism is the causal agent of disease. *10,* 12

Krebs cycle. *See* **tricarboxylic acid cycle**

K-strategy Ecological strategy where organisms depend on physiological adaptations to environmental resources. K strategists are usually stable and permanent members of the community. 156, 239

***lac* gene,** 178, *180,* 184

lag phase Period after inoculation of fresh growth medium during which population numbers do not increase. 152

lambda (λ) phage, 140

lamella (plural, lamellae) (i) Thin layer, platelike arrangement or membrane.
(ii) Layers of protoplasmic membranes within the chloroplast that contain the photosynthetic pigments. 97, 100

landfarming, 473, 476

Latin binomial. *See* **binomial nomenclature**

leaching Removal of materials in solution from the soil. 23, 27, 63, 278, 352, 473

lectin Plant protein with a high affinity for specific sugar residues. 323

leghemoglobin Iron-containing, red pigment(s) produced in root nodules during the symbiotic association between rhizobia and leguminous plants. The pigment is similar but not identical to mammalian hemoglobin. 328

mesophile Organism whose optimum temperature for growth falls in an intermediate range of approximately 15° to 40°C. *63, 487-489*

mesosome, *51*

messenger RNA (mRNA) RNA molecule transcribed from DNA, which contains the information to direct the synthesis of a particular protein. *54, 133, 169, 183*

metabolism All biochemical reactions in a cell, both anabolic and catabolic. *46,* 59, *189, 190, 194, 471*

methane, 16, 39, 45, 61, 211, 214, 504, 506

methanogenesis Biological production of methane. *211, 505*

methanogenic bacterium (methanogen) Methane-producing prokaryote; member of the Archaea. 61, *62,* 505

methanotroph Organism capable of oxidizing methane. *212, 506, 509*

methyl bromide, *433, 442*

methylation, *369, 453*

Michaelis-Menten equation, 153

microaerophile Organism that requires a low concentration of oxygen for growth. 47, 102, 303, *304, 307*

microaggregate Clustering of clay packets stabilized by organic matter and precipitated inorganic materials. *27, 419*

microbial biomass Total mass of microorganisms alive in a given volume or mass of soil.

    assimilation into, 63, 123

    estimates, 7, 236

    extraction of DNA, 181

    fungal, 72

    measurement, *237, 243-244*

    nematode, *125*

    sulfur cycle, 353

microbial infallibility, 471

microbial population Total number of living microorganisms in a given volume or mass of soil. 7, 182, 236, *238*

microbiology Study of microorganisms. 9, 10

Micrococcus, 134, 382

microcosm A community or other unit that is representative of a larger unity. *162*

microenvironment Immediate physical and chemical surroundings of a microorganism. 60, 287

microfauna Protozoa, nematodes, and arthropods generally < 200 μm long. 7, 114

microflora Bacteria (including actinomycetes), fungi, algae, and viruses. 7, 129, 161

microhabitat Clusters of microaggregates with associated water within which microbes function. May be composed of several microsites (e.g., aerobic and anaerobic). 8

micrometer One-millionth of a meter, or 10^{-6} meter (abbreviated μm), the unit usually used for measuring microorganisms. 124

micronutrient Chemical element necessary for growth found in small amounts, usually < 100 mg kg^{-1} in a plant. These elements consist of boron, chlorine, copper, iron, manganese, molybdenum, and zinc. 59

microorganism Living organism too small to be seen with the naked eye (< 0.1mm); includes bacteria, fungi, protozoa, microscopic algae, and viruses. 7

micropore Relatively small soil pore, generally found within structural aggregates and having a diameter < 30μm, 26, 36, *37, 38,* 61

microscopy, 9, 55, 68, 132, 142

microsite Small volume of soil where biological or chemical processes differ from those of the soil as a whole, such as an anaerobic microsite of a soil aggregate or the surface of decaying organic residues. 27, 40, 61, 127, 128, *287,* 511

microsymbiont, 322

mineralization Conversion of an element from an organic form to an inorganic state as a result of microbial decomposition.

    carbon, 228

    definition, 63, 219, 369, 470

    effect on global gases, 510

    nitrogen, 265, 502

    phosphorus, 374

    rhizosphere effect, *403*

    sulfur, 356

minor elements. *See* **micronutrients**

mite, *8,* 114, 402

mitochondrion (plural, mitochondria)
Eukaryotic organelle responsible for processes of respiration and oxidative phosphorylation. *5, 54*

mitosis Highly ordered process by which the nucleus divides in eukaryotes. 4, 79, 82, 83, 86, 114

mixotroph Organism able to assimilate organic compounds as carbon sources while using inorganic compounds as electron donors. Compare with autotroph and heterotroph. 358, 380

model, 82, *162*, 240, *253, 287*, 405, 503

moist soil, 79, 81, 88, 125

moisture-characteristic curve. *See* **water retention curve**

moisture content. *See* **water content**

moisture-release curve. *See* **water-retention curve**

mold A filamentous fungus. 9, 115

moldboard plow, 40, *41*

molybdenum, 59, 200, 299, *301*, 335

monoclonal antibody Antibody produced from a single clone of cells. This antibody has uniform structure and specificity. 142

Monod equation, 154

monokaryon Fungal hypha in which compartments contain one nucleus. 83

monooxygenase, 273, 459, *460*

monotropoid mycorrhiza, 415

morphotype, 83

most probable number (MPN) Method for estimating microbial numbers in soil based on extinction dilutions. 124, *126*

motility Movement of a cell under its own power. 56

mucigel Gelatinous material at the surface of roots grown in normal nonsterile soil. It includes natural and modified plant mucilages, bacterial cells and their metabolic products (e.g., capsules and slimes), and colloidal mineral and organic matter from the soil. *309, 391, 396*, 398

mucilage Gelatinous secretions and exudates produced by plant roots and many microorganisms. *309, 310, 391, 396*, 398

Mucor, 7, *80*, 82, 86, 90, 228

mulch (i) Any material such as straw, sawdust, leaves, plastic film, and loose soil that is spread upon the surface of the soil to protect the soil and plant roots from the effects of raindrops, soil crusting, freezing, or evaporation. (ii) To apply mulch to the soil surface. 40

multinucleate, *74*

municipal solid waste Combined consumer and commercial waste generated within a defined geographic area. 483, *495*

murein. *See* **peptidoglycan**

mushroom Large, sometimes edible, fruiting body of some fungi. 83, 88

mutant Organism, population, gene, or chromosome that differs from the corresponding wild type by one or more base pairs. *57,* 140, 178, *180*

mutation Heritable change in the base sequence of the DNA of an organism. 57, 87

mutualism Interaction between organisms where both organisms benefit from the association. 17, 107, *157*, 159, 322

mycelial strand, 76-77

mycelium (plural, mycelia) Mass of hyphae that form the vegetative body of many fungal organisms. *49*, 75, 76, 81, 416

mycrophage, 135

mycophage. *See* **mycovirus**

mycophagous Organisms that consume fungi, such as mycophagous nematodes. 90, 402

mycorrhiza Literally "fungus root." The symbiotic association between specific fungi with the fine roots of higher plants.

   arbuscular mycorrhiza (AM), 412-414

   arbutoid mycorrhiza, 415

   dependency, 420, 422

   ectendomycorrhiza, *410*, 415

   ectomycorrhiza, 17, 83, 84, 85, 410

   endomycorrhiza, 17, 81, 409, 412

   ericoid mycorrhiza, 84-85, 409, *410*, 414

   function, 17, 159, 376, 402, 415-418

   inoculum production, 421-423

   management, 420

   orchidaceous mycorrhiza, *410*, 415

   structure, 8, *411*

vesicular-arbuscular mycorrhiza, 413

mycorrhization helper bacteria, 418

mycorrhizosphere Unique microbial community that forms around a mycorrhiza. 418

mycovirus Virus that infects fungi. 135

N-Q

^{15}N, 317

NAD+/NADH. *See* **nicotinamide adenine dinucleotide**

NADP+/NADPH. *See* **nicotinamide adenine dinucleotide phosphate**

nanopore Soil pore having dimensions measured in nanometers. Materials encased in nanopores are beyond the reach of microorganisms and enzymes. *26*

necrosis Damage of living tissues because of infection or injury. 133

necrotrophic Nutritional mechanism by which an organism produces a battery of hydrolytic enzymes to kill and break down host cells and then absorb nutritional compounds from the dead organic matter. 84

nematode Multicellular eukaryote defined as an unsegmented, usually microscopic roundworm. Various species feed on plants, animals, fungi, and bacteria. *7, 8,* 119-123, 125, 127-129, 135, 402, 434

nematode-trapping fungus, *116, 123*

Nematophthora, 434

niche Functional role of a given organism within its habitat. 62, 72, 86, 88, 156

nickel, *335*

nicotinamide adenine dinucleotide (NAD$^+$) Important coenzyme, functioning as a hydrogen and electron carrier in a wide range of redox reactions; the oxidized form of the coenzyme is written NAD$^+$, the reduced form as NADH. 61, 190, *191,* 203, *204,* 206, 207, *210,* 211

nicotinamide adenine dinucleotide phosphate (NADP$^+$) Important coenzyme functioning as a hydrogen and electron carrier in a wide range of redox reactions; the oxidized form of the coenzyme is written NADP$^+$, the reduced form as NADPH. *191,* 213, 214

nif **gene,** 342

nitrate, 14, 60, *61,* 209, 211, *228,* 259, *261, 266,* 510

nitrate reduction (biological) Process whereby nitrate is reduced by plants and microorganisms to ammonium for cell synthesis (nitrate assimilation, assimilatory nitrate reduction) or to various lower oxidation states (N_2, N_2O, NO) by bacteria using nitrate as the terminal electron acceptor in anaerobic respiration. 280-281

nitrate respiration, 209

nitric oxide, *266,* 498, 509

nitrification Biological oxidation of ammonium to nitrite and nitrate, or a biologically induced increase in the oxidation state of nitrogen. 14, 38, 182, 260, 262, 275, 491, 510

nitrification inhibitors, *274*

nitrifying bacteria Chemolithotrophs capable of carrying out the transformations from ammonia to nitrite or nitrite to nitrate. 14, 60-61, 211-212, 272, 276, 376

nitrite, 14, 60, *61,* 209, 211, *266*

Nitrobacter, 5, 14, *47,* 60, *272*

nitrogen, 12, 59, *254,* 259-265, 498, 509

nitrogen cycle Sequence of biochemical changes wherein nitrogen is used by a living organism, transformed upon the death and decomposition of the organism, and converted ultimately to its original state of oxidation. 259, 260, *262*

nitrogen fixation. *See* **dinitrogen fixation**

nitrogenase Specific enzyme system required for biological N_2 fixation. 203, 298-303, *419*

Nitrosomonas, 14, *47,* 60, *272*

nitrous oxide, 15, 157, *266,* 499, 509

nod gene, *328*

nod factor, *320*

nodulation, *331, 340*

nodule bacteria. *See* **rhizobia**

nodule. *See* **root nodule**

nodulin Unique protein produced in root hairs or nodules in response to rhizobial infection. 330

nomenclature System of naming organisms. 5

nonhumic substance, *247,* 248

nonpathogenic, 138, 440, 442

nonpolar Possessing hydrophobic (water-repelling) characteristics and not easily dissolved in water. 34, 449

nonrespiratory denitrification, *328*

nonspecific amplification, 176

Nostoc, *95,* 96, 102, 107

nuclear polyhedrosis virus, 145

nucleic acid Polymer of nucleotides. 168, 203, *263*

nucleoid Aggregated mass of DNA that makes up the chromosome of prokaryotic cells. *51,* 53, 54

nucleophilic compound Chemical that attracts or is drawn to electron-deficient regions in other chemicals; reducing agents act as nucleophilic compounds. 455, 462

nucleoside Nucleotide without the phosphate group. 268

nucleotide Monomeric unit of nucleic acid, consisting of a sugar (pentose), a phosphate, and a nitrogenous base. *46,* 133, 170, 176

nucleus Membrane-enclosed structure containing the genetic material (DNA) organized in chromosomes. 4, 50, 53, 72, 98, 116

nutrient Substance taken by a cell from its environment and used in catabolic or anabolic reactions. 15, 59, 63, 98, 127, 155

nutrient depletion zone, 416

obligate (i) Adjective referring to an environmental factor (for example, oxygen) that is always required for growth. (ii) Organism that can grow and reproduce only by obtaining carbon and other nutrients from a living host, such as obligate symbiont.

    aerobe, 60, 61, 68, 84, 359

    anaerobe, 38, 61, *62,* 63, 68, 211, *282,* 303

    parasite, 133

    symbiont, 85, 418

oligonucleotide Short nucleic acid chain, either obtained from an organism or synthesized chemically. 173

oligotrophs Microorganism specifically adapted to grow under low nutrient supply. Thought to subsist on the more resistant soil organic matter and be little affected by the addition of fresh organic materials. Sometimes a synonym for autochthonous. 156, 239

oligotrophic, 180

Olpidium, 81

oogonium Specialized sexual structure formed as a female gametangium by funguslike organisms in the phylum Oomycota. 81

oomycete, 75, 79-81, 84, 85, 88

oospores Thick-walled spore formed within an oogonium by funguslike organisms in the phylum Oomycota. 75, 79, 81

operon Cluster of genes whose expression is controlled by a single operator. Typical in prokaryotic cells. 178

orchidaceous mycorrhiza, *410,* 415

organelle Membrane-enclosed body specialized for carrying out certain functions; found only in eukaryotic cells. 50

organic matter. *See* **soil organic matter**

organic soil Soil that contains a high percentage (>200 g kg^{-1} or >120–180 g kg^{-1} if saturated with water) of organic carbon. 22-24, 255

organotroph Organism that obtains reducing equivalents (stored electrons) from organic substrates. 192

orthophosphate. *See* **phosphate**

osmotic potential Portion of total soil water potential due to the presence of solutes in soil water. *See* **soil water potential**

ostiole, 83, *89*

oxalate, 417

oxidation Process by which a compound gives up electrons, acting as an electron donor, and becomes oxidized. 38, 194, 197, *250,* 378

oxidation-reduction (redox) reaction Coupled pair of reactions, in which one compound becomes oxidized, while another becomes reduced and takes up the electrons released in the oxidation reaction. 38-39, 194, 196-197, 369, 503

oxidation state Number of electrons to be added (or subtracted) from an atom in a combined state to convert it to the elemental form. 195-196

oxidative phosphorylation Synthesis of ATP involving a membrane-associated electron-transport chain and the creation of a proton-motive force. Also called electron-transport chain phosphorylation. 50, 106, 203, *204,* 207, *210,* 211

Oxisol, 379

oxygen, 38, 57, 59, 61, 62, 146, *157*, 203, 209, 213

oxygenic photosynthesis Use of light energy to synthesize ATP and NADPH by noncyclic photophosphorylation with the production of oxygen from water. 47

ozone, 209, 259, *266*, 500

^{32}**P-,** 170, *171*, *179*

paramylon, 95

parasexual cycle Nuclear cycle in which genes of haploid nuclei recombine without meiosis. 86

parasite, 114, 122, 133, 134

parasitism Feeding by one organism on the cells of a second organism, which is usually larger than the first. The parasite is, to some extent, dependent on the host at whose expense it is maintained. 108

Parasponia, 324

parent material, 21, 23, 29, 63, 125, 128

parenthosome, 76

particle density Density of the soil particles, the dry mass of the particles being divided by the solid (not bulk) volume of the particles, in contrast with bulk density. 24, 25, 26

particle size Effective diameter of a particle measured by sedimentation, sieving, or micrometric methods. 24

passive

bioremediation, 472

dispersal, 88, 400

nutrient absorption, 50

nutrient pool, 352-353

Pasteur, L., 10-11

Pasteuria, 441

pasteurization Process using mild heat to reduce microbial numbers in heat-sensitive materials. 490

pathogen Organism able to inflict damage on a host it infects. 59, 127, *403*, 427

pathogenic Ability of a parasite to inflict damage on the host. *46*, 59, 84, 129, 138, 397, 400, *441*, 483

pathogen-suppressive soil Soil where a pathogen does not establish or persist, a pathogen establishes but causes little or no damage, or a pathogen causes disease for a while, but the disease becomes less important even though the pathogen persists in soil. 429-435

peat Unconsolidated soil material consisting largely of undecomposed, or only slightly decomposed, organic matter accumulated under conditions of excessive moisture. 24, 255, 504, 508

peatland. *See* **peat**

pectin Important component of the plant cell walls containing chains of galacturonic acid that is often esterified with a methyl group. 84, 232

Penicillium, 38, 82, *87*, 91, *227*, 230, *448*

peptidase, *202*, 229, 267

peptidoglycan Rigid layer of cell walls of bacteria, a thin sheet composed of N-acetylglucosamine, N-acetylmuramic acid, and a few amino acids. Also called murein. 45, 52, 138, 227, *264*

peribacteroid membrane Plant-derived membrane surrounding one to several rhizobia within host cells of legume nodules. 327

periplasmic space Area between the cell membrane and the cell wall in Gram-negative bacteria, containing certain enzymes involved in nutrition. 53

perithecium Flask-shaped ascocarp open at the tip; containing asci of fungi in the phylum Ascomycota. 82, 83, *89*

permanent wilting point Greatest water content of a soil at which indicator plants growing in that soil wilt and fail to recover when placed in a humid chamber. Often estimated by the water content at −1.5 MPa soil matric potential. 37

peroxidase, 456, 458, *459*

pesticide,

adsorption by SOM, 255

degradation, 84, 456, 470, 477-479

effect on microorganisms, 107

PFU. *See* **plaque-forming unit**

pH Negative logarithm of the hydrogen ion activity. The degree of acidity (or alkalinity) of a soil as determined by means of a glass or other suitable electrode or indicator at a specified moisture content or soil-water ratio and expressed in terms of the pH scale.

definition, 29

effect on decomposition, 485

effect on exchange capacity, 31, 32, 34

effect on microorganisms, 64, 69, 106, 143

effect on N_2 fixation, 102, 332-333

effect on nitrification, 277

effect of nitrogen source, *395*

phage conversion, 138

phage. *See* **bacteriophage**

phagotrophic Form of feeding where animals, such as protozoans, engulf particulate nutrients, such as bacterial cells or detritus. 119

phenotype Observable properties of an organism. 5, *324*

phenotypic, 5

phenylpropene, 233

pho **gene,** 178

Phoma, 87

phosphatase, 178, 203, 375, 417

phosphate, 63, 164, 168, 190, *228*, 370, *376*, 380, 416

phosphobacterium Bacterium that is especially good at solubilizing the insoluble inorganic phosphate in soil. *377*

phosphodiester bond Type of covalent bond linking nucleotides together in a polynucleotide. 375

phospholipid Lipids containing a substituted phosphate group and two fatty acid chains on a glycerol backbone. 50, 164, 370

phosphorus, 55, 59, 106, *228*, *248*, *254*, 335, 370-371, 498

phosphorus cycle Sequence of transformations undergone by phosphorus where it is transformed between soluable and insoluble, and organic and inorganic forms. 372, *373*

photic zone Uppermost layer of a body of water or soil that receives enough sunlight to permit the occurrence of photosynthesis. 105-106

photoautotroph Organism able to use light as its sole source of energy and carbon dioxide as sole carbon source. 59, *60*, 94, 106

photoheterotroph Organism able to use light as a source of energy and organic materials as carbon source. 106

photolithotroph, 194

photophosphorylation Synthesis of high-energy phosphate bonds, as ATP, using light energy.

anoxygenic, 214

cyclic, 213, 214

microbial activity, 52, 100

noncyclic, 213, 214

oxygenic, 213

photosynthesis Process of using light energy to synthesize carbohydrates from carbon dioxide.

dark reactions, 212, 214

microbial activity, 59, 100

light reactions, 212, *213*

photosystem I, 213

photosystem II, 102, 213

phototroph An organism that uses light as the energy source to drive the electron flow from the electron donors, such as water or hydrogen sulfide. *47*, 89, 192, *194*, 212, 214

phycobilin Water-soluble pigment that occurs in cyanobacteria and functions as the light-harvesting pigments for Photosystem II. 100

phycocyanin, 95, 97

phycoerythrobilin, *101*

phylogenetic, 5, 6, 45, 168, 169, *324*

phylogeny Ordering of species into higher taxa and the construction of evolutionary trees based on evolutionary (genetic) relationships. 5, 6, 45, 164, 168, 273, 274

phylum (plural, phyla), 74, *115*, 119

Phytophthora, 79, 432

phytoremediation, 473, 479

phytotoxin, 495

phytotoxicity, *390*

pilus (plural, pili) Fimbria-like structure that is present on fertile cells involved in DNA transfer during conjugation. Sometimes called sex pilus. 55, *51*

Pisolithus, 412

plant growth-promoting rhizobacteria (PGPR) Broad group of soil bacteria that exert beneficial effects on plant growth usually as root colonizers. Many members of the genus *Pseudomonas*. 403, 441

plant growth promotion, 65, 248, 402-403, 441

plaque Localized area of lysis or cell inhibition caused by virus infection on a lawn of cells. 142

plaque forming unit (PFU), 142

plasma membrane. *See* **cytoplasmic membrane**

plasmid Covalently closed, circular piece of DNA which as an extrachromosomal genetic element is not essential for growth. 54, 58, 171, 172, *173*, *174*, 177

plasmid profile, *173*, 177

plasmogamy Fusion of the contents of two cells, including cytoplasm and nuclei. 78, 81, 86

plastid Specialized cell organelles containing pigments or protein materials. 115

plate count Number of colonies formed on a solid culture medium when uniformly inoculated with a known amount of soil generally as a dilute soil suspension. The technique estimates the number of certain organisms present in the soil sample. 66-69

polar Possessing hydrophilic characteristics and generally water soluble. *272*

pollutant, 84, 107, 365, 469, 475, 476, 500

pollution, 109, 322, 365, 377, 442, 482

poly-beta-hydroxybutyrate (PHB) Common storage material of prokaryotic cells consisting of beta-hydroxybutyrate or other beta-alkanoic acids. 55

polyclonal antiserum Mixture of antibodies to a variety of antigens or to a variety of determinants on a single antigen. 142

polymer Large molecule formed by polymerization of monomeric units.

    carbohydrate, 55, 223

    cell wall, 52, 229-235

    polypropylene, *448*

    synthesis, 249, *250*

    uptake, 50, *231*

    xenobiotic, 448

polymerase chain reaction (PCR) Method for amplifying DNA *in vitro,* involving the use of oligonucleotide primers complementary to nucleotide sequences in target genes and the copying of the target sequences by the action of DNA polymerase. 17, 64, 143, 164, 169, 173-*175*

polysaccharide Long chain of monosaccharides (sugars) linked by glycosidic bonds. *27*, 52, 56, 180, 229, 235

polysome Strings of ribosomes attached by strands of mRNA. *51*, 54

pore space Portion of soil bulk volume occupied by soil pores. 24-26, *25*, 29, 40, 41, 117, *248*

porin Protein channel in the lipopolysaccharide layer of Gram-negative bacteria. *52*, 53

porosity Volume of pores in a soil sample (nonsolid volume) divided by the bulk volume of the sample. 38

porphyrin, *465*

pot culture, 421

potassium, 59

pour plate Method for performing a plate count of microorganisms. A known amount of a serial dilution is placed in a sterile Petri dish, and then a melted agar medium is added and the inoculum mixed well by gently swirling. After growth the number of colony-forming units is counted. 67

poxvirus, 132, 133

predation Relationship between two organisms whereby one organism (predator) engulfs or captures and digests the second organism (prey). 56, 107-108, 123, 143, *157*, 160

predator paradox, 128

primary producer Organism that adds biomass to the ecosystem by synthesizing organic molecules from carbon dioxide and simple inorganic nutrients. 62, 94

primary production, *348*

primary treatment. *See* **sewage treatment**

primer Molecule (usually a polynucleotide) to which DNA polymerase can attach the first nucleotide during DNA replication.

    annealing, 174, 176

    application, 182-183

    definition, 173-174

    dimer, 177

reducing equivalent (power) Electrons stored in reduced electron carriers such as NADH, NADPH, and FADH$_2$. 191, 192, 194, 211

reducing power. *See* **reducing equivalent**

reduction Process by which a compound accepts electrons. 38, 39, 194, 197, 378

reduction (redox) potential Inherent tendency of a compound to act as an electron donor or an electron acceptor. Measured in millivolts. 38, 196-197, 369, 503

reductive dehalogenation. *See* **dehalogenation, reductive**

reductive dechlorination Removal of Cl as Cl$^-$ from an organic compound by reducing the carbon atom from C–Cl to C–H. 459

remediation. *See* **bioremediation**

replication Conversion of one double-stranded DNA molecule into two identical double-stranded DNA molecules. 54, 57, *126, 133, 139*

REP-PCR. *See* **polymerase chain reaction**

repression Process by which the synthesis of an enzyme is inhibited by the presence of an external substance (the repressor). 200

reporter gene Gene that signals the activity of an associated gene of interest. For example, microbes containing the *lux* genes emit light when the operon containing the genes is switched on. 184, 277-278, *280*

respiration Catabolic reactions producing ATP in which either organic or inorganic compounds are primary electron donors and exogenous compounds are the ultimate electron acceptors. 61-63, 203, 207, 209, 211, 218

respiratory protection, *307*

resting structure, 75, 79, 87-88, 99, 115, 439

restricted nodulation, *340*

restriction endonuclease (restriction enzyme) Enzyme that recognizes and cleaves specific DNA sequence, generating either blunt or single-stranded (sticky) ends. 177, *178*

restriction fragment length polymorphism (RFLP) Means to identify differences between similar genes from different organisms. Digestion of genes with restriction endonucleases followed by separation of the resulting fragments by gel electrophoresis yields banding patterns that are characteristic of the individual gene. 177-179, *324*

reverse transcription Process of copying information found in RNA into DNA. 182

RFLP. *See* **restriction fragment length polymorphism**

rhizobacteria Bacteria that aggressively colonize roots. *311*, 441

rhizobia Bacteria capable of living symbiotically in roots of leguminous plants, from which they receive energy and often fix dinitrogen. Collective common name for *Rhizobium* and closely related genera. 16, 145, *172, 173, 296,* 322-343, 419

Rhizobium, 12, 50, 65, 142, 144, 159, 160, 172, 232, 322, 325

rhizoid Rootlike structure that helps to hold an organism to a substrate. 75, 81

rhizomorph Mass of fungal hyphae organized into long, thick strands usually with a darkly pigmented outer rind and containing specialized tissues for absorption and water transport. 76, 77, *78*, 411

rhizomycelium, 81

Rhizophydium, *80*, 81

rhizoplane Plant root surfaces and usually strongly adhering soil particles. *309, 390*, 391, 393, *399*

Rhizopus, 90, 228

rhizosphere Zone of soil immediately adjacent to plant roots in which the kinds, numbers, or activities of microorganisms differ from that of the bulk soil.

    colonization, 400-402

    effect on biological control, 428

    effect on bioremediation, 479

    effect on microorganisms, 150, 398-400, 402

    exudates, 228, 395

    modeling function, 405

    N$_2$ fixation, 308, *309*

    pH, 394, *395*

    terminology, 389, *391*, 392-393, *396*

rhizosphere competence Ability of an organism to colonize the rhizosphere. 398, 435

ribonucleic acid (RNA) Polymer of nucleotides connected via a phosphate-ribose backbone, involved in protein synthesis. 133, 168

slime mold Nonphototrophic eukaryotic microorganism lacking cell walls, which aggregate to form fruiting structures (cellular slime molds) or simply masses of protoplasm (acellular slime molds). 115

smectite, 30, 33, *34*, 143

sodium, 59, 180

soil (i) Unconsolidated mineral or material on the immediate surface of the earth that serves as a natural medium for the growth of land plants. (ii) Unconsolidated mineral or organic matter on the surface of the earth that has been subjected to and influenced by genetic and environmental factors of: parent material, climate (including water and temperature effects), macroorganisms and microorganisms, and topography, all acting over a period of time and producing a product—soil—that differs from the material from which it is derived in many physical, chemical, biological, and morphological properties and characteristics.

   aeration, 38-39, 276, 285

   chemistry, 29, 35, 127

   color, 39

   density, 24, 26

   depth, 39, *40, 41*

   disturbance, *421*

   health. *See* soil quality

   moisture, 35, 104

   name, 23

   profile, 23, 36

   temperature. *See* temperature

soil aggregate Unit of soil structure generally < 10 mm in diameter and formed by natural forces and substances derived from root exudates and microbial products which cement smaller particles into larger units. 27-29, *49*, 72, 76, 105, 108, 128, *157*

soil atmosphere Gases occupying the pore space in soil. Generally characterized as having a greater percentage of carbon dioxide and a lesser percentage of oxygen than the overlying air. 36, 38, 39

soil classification. *See* classification

soil extract Solution separated from a soil suspension or from a soil by filtration, centrifugation, suction, or pressure. *204, 261, 374*

soil formation, 29, 63, 64, 104, 108

soil habitat, 3, *8*

soil horizon Layer of soil or soil material approximately parallel to the land surface and differing from adjacent genetically related layers in physical, chemical, and biological properties or characteristics such as color, structure, texture, consistency, kinds and number of organisms present, and degree of acidity or alkalinity. 22, 23

soil microbiology Branch of soil science concerned with soil-inhabiting microorganisms and their functions and activities. 3, 9

soil organic matter (SOM) Organic fraction of the soil exclusive of undecayed plant and animal residues. Often synonymous with humus.

   aggregate formation, *27, 29*, 247

   cation exchange capacity, 32-33, 252

   chemical effects, 247

   clay complex, *252*

   decomposition, 482-496

   fractionation, 247, 252

   plant residues, 63, 221, *222*

   properties, 15, 218, 247, 248, 252, *254*, 262, 372, 495

   synthesis, 63, 108, 218, *239*, 246, 249, *250-251, 403*

   water-soluble compounds, 228

   soil pH, 29, 277, 332

soil population (i) All the organisms living in the soil, including plants and animals. (ii) Members of the same taxa. 7, 17, 64, 66, *67*, 68, 182

soil pore That part of the bulk volume of soil not occupied by soil particles. Soil pores have also been referred to as interstices or voids. 26, 36, 37

soil quality Continued capacity of soil to function as a vital living system to sustain biological productivity, maintain the quality of the environment, and promote plant, animal, and human health. 108, 109, 125, 221

soil salinity Amount of soluble salts in a soil. The conventional measure of soil salinity is the electrical conductivity of a saturation extract. 17, 36, 64, 336

soil science Science dealing with soils as a natural resource on the surface of the earth, including soil formation, classification, and mapping and physical, chemical, biological, and fertility properties of soils and these properties in relation to their use and management. 3

soil series Lowest category of U.S. system of soil taxonomy; a conceptualized class of soil bodies (polypedons) that have limits and ranges more restrictive than all higher taxa. The soil series serve as a major vehicle to transfer soil information and research knowledge from one soil area to another. 23

soil solution Aqueous liquid phase of the soil and its solutes. 50, 53, 56, 63, 117, 122, 143

soil structure Combination or arrangement of primary soil particles into secondary particles, units, or peds. The secondary units are characterized and classified on the basis of size, shape, and degree of distinctness into classes, types, and grades, respectively. 17, 29, 63, 108, *248*

soil texture Relative proportions of the various soil separates in a soil. The major textural classes are sand, silt, and clay. 24-25, 32, 36-38, *401*

soil water potential The amount of energy that must be expended to extract water from soil. The total potential of soil water consists of the following: gravitational potential, matric potential, and osmotic potential. 35-38, 117, *395*

solarization Method to control pathogens and weeds where moistened soil in hot climates is covered with transparent polyethylene plastic sheets, thereby trapping incoming radiation. 439

solubilization, 369, 375

somatic cell. *See* **vegetative cell**

somatic embryo, 422

Sordaria, 82, 89

Southern blot Hybridization of single-stranded nucleic acid (DNA or RNA) to DNA fragments immobilized on a filter. 171, 173

spatial variability Variation in soil properties (i) laterally across the landscape, at a given depth, or with a given horizon, or (ii) vertically downward through the soil. 104, 288-290, *289*

species In microbiology, a collection of closely related strains sufficiently different from all other strains to be recognized as a distinct unit. 5, 45

specific activity Amount of enzyme activity units per mass of protein. Often expressed as micromoles of product formed per unit time per milligram of protein. Also used in radiochemistry to express the radioactivity per mass of material (radioactive and nonradioactive). 177

specific epithet Designation of a particular organism in the binomial nomenclature system. For example, *coli* is the specific epithet of *Escherichia coli. See* **binomial nomenclature**

specific suppression. *See* **pathogen-suppressive soil**

specificity, 134, 330, 414

spermosphere Area of increased microbial activity around a germinating seed. 389, *391*

spirillum (plural, spirilli) (i) Bacterium with a spiral shape which is relatively rigid. (ii) Bacterium in the genus *Spirillum.* 46

spontaneous generation, 10, *11*

sporangiospore Spore formed within a sporangium by fungi in the phylum Zygomycota. 82

sporangium Fungal structure which converts its cytoplasm into a variable number of sporangiospores; formed by fungi in the phylum Zygomycota. *47, 75,* 79, 81-82, *115*

spore Specialized reproductive cell. Asexual spores germinate without uniting with other cells, whereas sexual spores of opposite mating types unite to form a zygote before germination occurs.

  algal, 99

  asexual, *87*

  bacterial, 55

  discharge, *89*

  fungal, 73

spread plate Method for performing a plate count of microorganisms. A known amount of a serial dilution is spread over the surface of an agar plate. After growth the number of colony-forming units is counted. 68

springtail. *See* ***Collembola***

stationary phase Period during the growth cycle of a population in which growth rate equals the death rate. 487, 491

stabilization stage, 487, 491

staphylococcus (plural, staphylococci), 48

starch, 63, 95, 235-236

Starkey, R., 14

systemic resistance Not localized in a particular place of the body; an infection disseminated widely through the body is said to be systemic. 441

T-Z

take-all decline, *160,* 432

Taq **polymerase,** 64, 176

tardigrade, 114, 122

taxon (plural, taxa) Group into which related organisms are classified. 45, 103, 164

taxonomy Study of scientific classification and nomenclature. 5, 10, 73, 95, 413

TCA cycle. *See* **tricarboxylic acid cycle**

teichoic acids All wall, membrane, or capsular polymers containing glycerophosphate or ribitol phosphate residues. 52

teleomorph Sexual stage in reproduction in which cells are formed by the process of meiosis and genetic recombination. 77-79

temperate virus Virus which upon infection of a host does not necessarily cause lysis but whose genome may replicate in synchrony with that of the host. 138

temperature

    effect on decomposition, 226, 487-490

    effect on microbial growth, 39, 63-64, 106, 125, 143

    effect on N_2 fixation, 315, 333

    regime for polymerase chain reaction, 176

    soil temperature, *40*

template denaturation. *See* **denaturation**

terminal-electron acceptor External oxidant (often oxygen) that accepts the electrons as they exit from the electron-transport chain. 61-62, 203, *204,* 207, 209, 211

tertiary treatment. *See* **sewage treatment**

test Hard external covering or shell. 116

texture. *See* **soil texture**

tfd **gene,** *174,* 178, 182

thallus Vegetative body that is not differentiated into tissue systems or organs. 74

thermal cycler, 176

Thermoactinomyces, 488

Thermomonospora, 488

thermophile Organism whose optimum temperature for growth lies between 45 and 85°C. 64, 90, 487-490, 489

Thermus aquaticus, 64, 176

Thiobacillus, 5, 47, 60, *61,* 359, *360,* 362, 379

Ti plasmid Conjugative tumor-inducing plasmid present in the bacterium *Agrobacterium tumefaciens* which can transfer genes into plants. 440

tillage, 39-40, *41,* 106, 221, *248,* 255, 420

tobacco mosaic virus, 133, 143

toluene, 183

topsoil (i) Layer of soil moved in cultivation. (ii) The A horizon. (iii) Presumably fertile soil material used to topdress roadbanks, gardens, and lawns. 218, 420

toxin Microbial substance able to induce host damage. 65, 108, 122, 138, 139, 141

trace gas Gas other than nitrogen and oxygen in the atmosphere, particularly those gases that are active in the chemistry or radiation balance of the atmosphere. 498

transcription Synthesis of an RNA molecule complementary to one of the two strands of a DNA double-stranded molecule. 169, 182

transduction Transfer of host genetic information via a virus or bacteriophage particle. 58, 140, *141,* 145

transfer RNA (tRNA) Type of RNA that carries amino acids to the ribosome during translation. 45, 169

transformation Transfer of genetic information into living cells as DNA. 58, 378

translation Synthesis of proteins using the genetic information in mRNA as a template. 133, 183

transposable element Genetic element that can move (transpose) from one site on a chromosome to another. 57

transposition Movement of a piece of DNA around the chromosome, usually through the function of a transposable element. 57

transposon Transposable element of which, in addition to genes involved in transposition, carries other genes; often confers selectable phenotypes such as antibiotic resistance. 57, 178, 180

transposon mutagenesis Insertion of a transposon into a gene; this inactivates the host gene, leading to a mutant phenotype and also confers the phenotype associated with the transposon gene. *434*

trehalose, 79

tricarboxylic acid cycle (TCA cycle, citric acid cycle, Krebs cycle) Series of metabolic reactions by which pyruvate is oxidized completely to carbon dioxide, also forming NADH, which allows ATP production. 207-211, 215

Trichoderma, 90, 230, *404, 433,* 435, 442

trichome Row of cells which have remained attached to one another following successive cell divisions. Trichomes are formed by many cyanobacteria and by species of *Beggiatoa*. *97,* 102, 360

trophic level Describes the residence of nutrients in various organisms along a food chain ranging from the primary nutrient assimilating autotrophs to the predatory carnivorous animals. 62, 129, 150

tropics, 101, 103, 111, 125, 255, 365, 379, 420, 504

troposphere, 500

truffle, 82

***uid* gene,** 178

unlimited growth phase. *See* **exponential growth**

urea, 267

urease, 267, *272*

uronic acid Class of acidic compounds of the general formula HOOC(CHOH)$_n$CHO that contain both carboxylic and aldehydic groups, are oxidation products of sugars, and occur in many polysaccharides, especially in the hemicelluloses. 232

vacuole, *5,* 73

vadose zone Unsaturated zone of soil above the groundwater, extending from the bottom of the capillary fringe all the way to the soil surface. 146, *477*

vanadium, 299, *301*

vector (i) Plasmid or virus used in genetic engineering to insert genes into a cell. (ii) Agent, usually an insect or other animal, able to carry pathogens from one host to another. 135

vegetative cell Growing or feeding form of a microbial cell, as opposed to a resting form such as a spore. 86

vermicomposting, 488

Verticillium, 434

vesicle Spherical structure, formed intracellularly, by some arbuscular mycorrhizal fungi. *116, 117, 118*

vesicular-arbuscular mycorrhiza. *See* **arbuscular mycorrhiza**

viable Alive; able to reproduce. 17, 68

viable but nonculturable Organisms that are alive but cannot be cultured on laboratory media. 17, 180

viable count Measurement of the concentration of live cells in a microbial population. 66

vibrio (i) Curved, rod-shaped bacterial cell. (ii) Bacterium of the genus *Vibrio*. 47, 382

virion Virus particle; the virus nucleic acid surrounded by protein coat and in some cases other material. 132, 139

viroid, 136

virulent, 138, 145, 429

virus Any of a large group of submicroscopic infective agents that typically contain a protein coat surrounding a nucleic acid core and are capable of growth only in a living cell. 7, 44, 58, 108, 132

vitamin, 59, 200

volatilization, *266,* 271, 349, 352, 364, 382, 485

Waksman, S., 10, 13-15

wastewater, 145, 146

water activity, 37, *38*

water-holding capacity, 26, 37

water content Water contained in a material expressed as the mass of water per unit mass of oven-dry material. 35, 36, *37,* 248, 315, 487

water potential. *See* **soil water potential**

water-retention curve Graph showing soil-water content as a function of increasingly negative soil water potential. 36-37

wax, 63

weathering All physical and chemical changes produced in rock by atmospheric agents. 63, *248*

wet soil. *See* **moist soil**

wetland, *162,* 255, *314,* 364, 377, 382, 504-508

white rot fungus Fungus that attacks lignin, along with cellulose and hemicellulose, leading to a marked lightening of the infected wood. 230, 458

Wilconina, 415

wild type Strain of microorganism isolated from nature. The usual or native form of a gene or organism. *180*

wilting point. *See* **permanent wilting point**

windrow, 492, 493

Winogradsky column Glass column with an anaerobic lower zone and an aerobic upper zone, which allows growth of microorganisms under conditions similar to those found in nutrient-rich water and sediments. *162-163*

Winogradsky, S., 5, 10, 13-14, 155, 297

Woronin body Spherical structure associated with the simple pore in the septa separating hyphal compartments of fungi in the phylum Ascomycota. 76, 77

xenobiotic Compound foreign to biological systems. Often refers to human-made compounds that are resistant or recalcitrant to biodegradation and decomposition.

   aromatic, 461

   aromatic degradation, *451,* 461

   atrazine, *448,* 470, 477, *478*

   biodegradation, 449, 455, *460*

   cometabolism of poloychlorinated biphenyl (PCB), *455*

   comparison with natural compounds, 228, *448*

   definition, 447

   degradation by monooxygenases, *461*

   detoxification, 453

   halogenated, 450

   hydrolytic degradation, 457

   molecular tools, 183

   nitro group reduction, 465, *466*

   polychiorinated biphenyl (PCB), *453,* 470

   polycyclic aromatic hydrocarbons (PAH), 470

   polymeric, 449

   substituent effects for aromatic compounds, *452*

   trichloroethylene (TCE), 475, *476,* 479

xerophile Organism adapted to growth at low water potentials, i.e., very dry habitats. 64

xylanase, 84

yard wastes, *493, 495*

yeast Fungus whose thallus consists of single cells that multiply by budding or fission. 75, 83, 119, 122, 125

yield coefficient. *See* **growth yield coefficient**

zinc, 59, *370*

zoospore Asexual spore formed by some fungi that usually can move in an aqueous environment via one or more flagella. *75,* 79, 81, 82, 85

zygomycete, *75,* 76, 81

zygospore Thick-walled resting spore resulting from fusion of two gametangia of fungi in the phylum Zygomycota. 81

zymogenous flora Refers to microorganisms, often transient or alien, that respond rapidly by enzyme production and growth when simple organic substrates become available. Also called copiotrophs. 156, 180, 228

Conversion factors for SI and non-SI units.

To convert column 1 into column 2, multiply by	Column 1 (SI unit)	Column 2 (non-SI unit)	To convert column 2 into column 1, multiply by
Length			
0.621	kilometer, km (10^3 m)	mile, mi	1.609
1.094	meter, m	yard, yd	0.914
3.28	meter, m	foot, ft	0.304
1.0	micrometer, μm (10^{-6} m)	micron, μ	1.0
3.94×10^{-2}	millimeter, mm (10^{-3} m)	inch, in	25.4
10	nanometer, nm (10^{-9} m)	Angstrom, Å	0.1
Area			
2.47	hectare, ha	acre	0.405
247	square kilometer, km^2 (10^3 m)^2	acre	4.05×10^{-3}
0.386	square kilometer, km^2 (10^3 m)^2	square mile, mi^2	2.590
2.47×10^{-4}	square meter, m^2	acre	4.05×10^3
10.76	square meter, m^2	square foot, ft^2	9.29×10^{-2}
1.55×10^{-3}	square millimeter, mm^2 (10^{-3} m)^2	square inch, in^2	645
Volume			
9.73×10^{-3}	cubic meter, m^3	acre-inch	102.8
35.3	cubic meter, m^3	cubic foot, ft^3	2.83×10^{-2}
6.10×10^4	cubic meter, m^3	cubic inch, in^3	1.64×10^{-5}
2.84×10^{-2}	liter, L (10^{-3} m^3)	bushel, bu	35.24
1.057	liter, L (10^{-3} m^3)	quart (liquid), qt	0.946
3.53×10^{-2}	liter, L (10^{-3} m^3)	cubic foot, ft^3	28.3
0.265	liter, L (10^{-3} m^3)	gallon	3.78
33.78	liter, L (10^{-3} m^3)	ounce (fluid), oz	2.96×10^{-2}
2.11	liter, L (10^{-3} m^3)	pint (fluid), pt	0.473
Mass			
2.20×10^{-3}	gram, g (10^{-3} kg)	pound, lb	454
3.52×10^{-2}	gram, g (10^{-3} kg)	ounce (avdp), oz	28.4
2.205	kilogram, kg	pound, lb	0.454
0.01	kilogram, kg	quintal (metric), q	100
1.10×10^{-3}	kilogram, kg	ton (2000 lb), ton	907
1.102	megagram, Mg (tonne)	ton (U.S.), ton	0.907
1.102	tonne, t	ton (U.S.), ton	0.907
Yield and rate			
0.893	kilogram per hectare, kg ha^{-1}	pound per acre, lb acre^{-1}	1.12
7.77×10^{-2}	kilogram per cubic meter, kg m^{-3}	pound per bushel, bu^{-1}	12.87
1.49×10^{-2}	kilogram per hectare, kg ha^{-1}	bushel per acre, 60 lb	67.19
1.59×10^{-2}	kilogram per hectare, kg ha^{-1}	bushel per acre, 56 lb	62.71
1.86×10^{-2}	kilogram per hectare, kg ha^{-1}	bushel per acre, 48 lb	53.75
0.107	liter per hectare, L ha^{-1}	gallon per acre	9.35
893	tonnes per hectare, t ha^{-1}	pound per acre, lb acre^{-1}	1.12×10^{-3}
893	megagram per hectare, Mg ha^{-1}	pound per acre, lb acre^{-1}	1.12×10^{-3}
0.446	megagram per hectare, Mg ha^{-1}	ton (2000 lb) per acre, ton acre^{-1}	2.24
2.24	meter per second, m s^{-1}	mile per hour	0.447
Specific surface			
10	square meter per kilogram	square centimeter per gram	0.1
1,000	square meter per kilogram	square millimeter per gram	0.001
Pressure			
9.90	megapascal, MPa (10^6 Pa)	atmosphere	0.101
10	megapascal, MPa (10^6 Pa)	bar	0.1
1.00	megagram per cubic meter, Mg m^{-3}	gram per cubic centimeter, g cm^{-3}	1.00
2.09×10^{-2}	pascal, Pa	pound per square foot, lb ft^{-2}	47.9
1.45×10^{-4}	pascal, Pa	pound per square inch, lb in^{-2}	6.90×10^3